THE DIFFUSE INTERSTELLAR BANDS

ASTROPHYSICS AND SPACE SCIENCE LIBRARY

VOLUME 202

THE DIFFUSE INTERSTELLAR BANDS

Edited by

A. G. G. M. TIELENS

NASA Ames Research Center,
Moffett Field, CA, U.S.A.

and

T. P. SNOW

Center for Astrophysics and Space Astronomy,
University of Colorado,
Boulder, CO, U.S.A.

KLUWER ACADEMIC PUBLISHERS

DORDRECHT / BOSTON / LONDON

Library of Congress Cataloging-in-Publication Data

The diffuse interstellar bands / edited by A.G.G.M. Tielens and T.P. Snow.
p. cm. -- (Astrophysics and space science library ; v. 202)
Includes index.
ISBN 0-7923-3629-1 (alk. paper)
1. Stars--Spectra--Congresses. 2. Astronomical spectroscopy--Congresses. 3. Astrophysics--Congresses. I. Tielens, A. G. G. M. II. Snow, Theodore P. (Theodore Peck) III. Series.
QB870.D54 1995
523.1'125--dc20 95-30317

ISBN 0-7923-3629-1

Published by Kluwer Academic Publishers,
P.O. Box 17, 3300 AA Dordrecht, The Netherlands.

Kluwer Academic Publishers incorporates
the publishing programmes of
D. Reidel, Martinus Nijhoff, Dr W. Junk and MTP Press.

Sold and distributed in the U.S.A. and Canada
by Kluwer Academic Publishers,
101 Philip Drive, Norwell, MA 02061, U.S.A.

In all other countries, sold and distributed
by Kluwer Academic Publishers Group,
P.O. Box 322, 3300 AH Dordrecht, The Netherlands.

Coverphoto: Composite Diffuse Interstellar Band spectrum
courtesy of Peter Jenniskens and F.-X. Désert.

Printed on acid-free paper

Printed in the Netherlands

CONTENTS

Preface

A.G.G.M. Tielens and T.P. Snow ix

CHAPTER I: Observational Characteristics of the Diffuse Bands

The Diffuse Bands: A Review with Speculations 3
Wm. Smith

Diffuse Interstellar Bands in Individual Sightlines 13
J. Krełowski and C. Sneden

Ultra-High-Resolution Spectroscopy of Diffuse Bands 25
R. Hibbins, J. Miles, P. Sarre, S. Fossey, and W. Somerville

The Diffuse Interstellar Bands Towards Southern Stars 31
S. Fossey and W. Somerville

The Diffuse Interstellar Band Spectrum 39
P. Jenniskens and F.-X. Désert

Revisiting Diffuse Interstellar Bands in the Satellite Ultraviolet with the Goddard High Resolution Spectrograph 53
J. Cardelli

Diffuse Interstellar Bands in the Near Infrared - A Dedicated Search for Polycyclic Aromatic Hydrocarbon and Fullerene Cations 65
B. Foing and P. Ehrenfreund

A Search for Polarization in Diffuse Interstellar Bands Towards HD197770 73
A. Adamson and D. Whittet

CHAPTER II: Environmental Behavior of the Diffuse Interstellar Bands

Collective Behaviour of the Diffuse Interstellar Bands 83
W. Somerville

DIBS and UV Extinction 97
F.-X. Désert and P. Jenniskens

The Environment Dependence of Diffuse Interstellar Bands 105
P. Ehrenfreund and P. Jenniskens

Diffuse Interstellar Bands in the Young Galactic Clusters M16 and M17 113
W. Pfau and Th. Henning

The Interstellar Space Mosaic: A User Interface for Accessing ISM Data World-Wide Distributed 121
I. Porceddu, S. Corda, F. Pasian, and R. Smareglia

Diffuse Circumstellar Bands 129
C. Seab

Red Rectangle Emission Bands 143
J. Miles, P. Sarre, and S. Scarrott

HAC and the ERE 149
A. Witt and D. Furton

Visible Luminescence From PAH Cations and the Diffuse Bands in Emission 157
C. Joblin, F. Salama, and L. Allamandola

Carrier of the Emission λ5797 DIB in the Red Rectangle: Suprathermally Rotating PAH? 165
D. Rouan and A. Léger

CHAPTER III: Laboratory Studies of Proposed Carriers

Laboratory Studies of DIB Carriers 175
L. Allamandola

Laboratory Studies of Highly Unsaturated Carbon Chains and the Diffuse Interstellar Bands 199
J. Maier, P. Freivogel, J. Fulara, and D. Lessen

Absorption Spectroscopy of Neutral and Ionized PAHs. Implications for the Diffuse Interstellar Bands 207
F. Salama, C. Joblin, and L. Allamandola

A Critical Assessment of the PAH Hypothesis: Revisited 219
B. Donn

Photofragmentation of Isolated PAH Cations 221
P. Boissel

Experimental Study of Naphthalene and Anthracene Reactions 231
A. Canosa, D. Pasquerault, D. Parent, C. Rebrion, J. Gomet, and B. Rowe

CHAPTER IV: Diffuse Bands and Interstellar Dust

The Diffuse Interstellar Bands and Interstellar Dust 241
J. Mathis

Dust Grain Carriers and the Diffuse Band Profiles 253
P. Shapiro

Interstellar Polarization 263
P. Martin

Polarization of the 2175 Å Feature 271
P. Martin, W. Somerville, D. McNally, D. Whittet, R. Allen, J. Walsh, and M. Wolff

CHAPTER V: Theoretical Studies of Molecular Carriers

Spectroscopy and Photophysics of Possible Large Molecule Carriers of the Diffuse Interstellar Bands 281
S. Leach

Large Molecules in the Interstellar Medium 307
E. Herbst

Charge Variation of Polycyclic Aromatic Hydrocarbons in the Diffuse Interstellar Medium 315
E. Bakes and A. Tielens

CHAPTER VI: Proposed Carriers

Observational Constraints on the Carriers of the Diffuse Bands 325
T. Snow

Porphyrins Within Interstellar Grains 341
F. Johnson

Fulleranes and the Diffuse Interstellar Bands 349
A. Webster

DIB Carriers on Grain Surfaces 359
W. Duley

PAH Cations as Candidates to the DIB Carriers 363
A. Léger

Carbon Chains and the Diffuse Interstellar Bands 369
P. Thaddeus

A Critique of Suggested Diffuse Band Carriers 379
T. Snow

Summary 395

A. Tielens

Addresses of Participants 407
Index 411
Index of Specific Diffuse Bands 415

PREFACE

The mystery of the diffuse interstellar bands has been variously a curiosity, a conundrum, and a nuisance for astronomers in the seven decades since the features were first noticed, but recently they have become a forefront issue in astrophysics. Ever since Paul Merrill, in a series of papers starting in 1934, pointed out the interstellar and unidentified nature of the bands, a Who's Who of twentieth century astronomers have tried their hands at solving the problem of identifying the carriers. Henry Norris Russell, Pol Swings, Otto Struve, Paul Ledoux, W. W. Morgan, Walter Adams, Jesse Greenstein, Lawrence Aller, and Gerhard Herzberg all briefly entered the stage, only to move on quickly to other problems where the chances for progress appeared more realistic. In more recent times a number of equally prominent scientists have pursued the bands, but generally only as a sideline to their real astronomical research.

But in the past decade, and particularly in the past three years, the view of the search for the diffuse band absorbers as an interesting but perhaps quixotic quest has changed. Today there are several astronomers, as well as laboratory chemists, who are devoting substantial research time and resources to the problem and, as perhaps the most reliable indicator of the newly elevated status of research in this field, some research grants have now been awarded for the study of the bands. The mystery of the diffuse bands is reaching unprecedented levels of awareness in the general scientific world; it has risen above the recognition thresholds of a growing number of astronomers and chemists.

On May 16 through 19, 1994, a conference on The Diffuse Interstellar Bands was held at the University of Colorado, in Boulder. The major goal of this conference was to draw a coherent picture of the observational characteristics of the diffuse bands and the physical and chemical properties of their proposed carriers. The timing of the meeting was especially propitious due to recent advances in observational technology, the enticement of several prominent physical chemists into the study of the diffuse bands, and the new climate of acceptance of the formerly heretical notion that large molecules might play a very important role in the diffuse interstellar medium. Following in the footsteps of the late A. E. Douglas, astronomers are developing a growing sense that the carriers of the diffuse bands may be complex molecules. Moreover, thanks in large part to revolutionary advances in detector technology, our observational knowledge of the diffuse bands has expanded at a rapid pace in recent years. At the same time, several groups have begun dedicated laboratory studies of proposed molecular carriers. It is clear that we are at an exciting juncture where spectroscopy from the UV to the millimeter-wave regime, in combination with detailed laboratory and theoretical studies, can be expected to lead to significant advances in our understanding of the diffuse bands and their carriers.

Surprisingly, while the problem of the diffuse bands has baffled scientists for most of this century, no major conference has ever been devoted to this topic. In view of the rapid progress being made, the Scientific Organizing Committee believed it very timely to organize a conference on *The Diffuse Interstellar Bands* in order to bring scientists working in these separate areas together to focus on the mystery features and to catalogue our current knowledge and chart future research. The study of the diffuse bands is a highly interdisciplinary field where astronomy, physics and chemistry meet. This is reflected in the varied background of the roughly 70 participants from 10 different countries. Besides astronomers involved in observations and/or analysis of the bands, chemists and physicists - studying proposed carriers spectroscopically - were active participants. We hope that this encounter in Boulder will foster an ongoing, probing dialogue among these scientists with diverse backgrounds but a common interest in the diffuse bands. In this way, fundamental progress towards the solution of this problem can be anticipated.

A major aim of this meeting was to produce conference proceedings which reflect the current situation regarding the diffuse bands. This book contains the invited papers presented at the meeting. The contributed papers have been published separately as *NASA Conference Publication 10144*.

This conference was made possible by a generous grant from the Astronomy and Relativity Branch of the National Aeronautics and Space Agency. We are also grateful to the support and services rendered by the dedicated staff at NASA Ames Research Center (California) and at the Center for Astrophysics and Space Astronomy (Boulder Colorado) without whom this conference would not have been possible. In particular, we would like to single out Sue Barnes at the Center for Astrophysics and Space Astronomy who has borne the brunt of the difficult tasks associated with final preparation of the manuscript, and has serenely and cheerfully tolerated the meddling of the editors in the process. We are also grateful for the assistance provided by the Office of Conference Services of the University of Colorado, especially Julia Gidney, who coordinated the local arrangements. Invaluable help was provided during the conference by Maribeth Heitzmann of the Center for Astrophysics and Space Astronomy, who, along with Sue Barnes, helped participants solve day-to-day problems such as transportation, preparing visual aids for presentations, and general logistics.

Xander Tielens
Ted Snow

April 14, 1995

Scientific Organizing Committee

H.J. Habing (The Netherlands)
J. Krelowski (Poland)
P. Martin (Canada)
W.B. Somerville (UK)
A.G.G.M. Tielens (US; Chair)
Th. Henning (Germany)
S. Leach (France)
T. Onaka (Japan)
T.P. Snow (US)

CHAPTER I

Observational Characteristics of the Diffuse Bands

THE DIFFUSE BANDS: A REVIEW WITH SPECULATIONS

WM. H. SMITH
McDonnell Center for Space Sciences, Washington Unviversity, St. Louis, MO 63130

ABSTRACT. I review briefly the principal observational properties of the diffuse interstellar bands (DIBs), with particular emphasis on those that appear to provide useful constraints on the physics of the carriers. The latter portion of the paper provides discussion and speculation on the nature of possible molecular carriers.

1. THE OBSERVATIONAL DATA FOR DIBS

Extensive observations of the diffuse interstellar bands (DIBs), along with a variety of laboratory measurements, have led to the present theories concerning the DIB carriers, the mechanisms of their formation, and reasons for their correlations with other observables. The discovery that DIBs are broad spectral features from unknown carriers that are closely reproducible from direction to direction in the interstellar medium raises interesting questions as to what species could be responsible, how they might be formed, and how they survive in sufficient abundance in interstellar space to produce the numerous DIB spectral features.

Observations have yielded an abundance of DIB information. Lines of sight towards stellar types as late as G4 have shown DIBs, although the detection of DIBs becomes progressively more difficult for later type stars. Observers (Jenniskens and Désert, this volume) have now found as many as 160 DIBs. Of these, a carrier for two has almost certainly been identified; the DIBs at 9577Å and 9632Å closely agree with expected spectral features due to the C_{60} fullerene ion (Foing and Ehrenfreund, 1994). This detection was personally satisfying to me, since my discussion of buckyballs, their ions, and inclusions related to DIBs was rejected due to the lack of a plausible ISM formation mechanism (Smith, 1985). How they form is still an important question, but is no longer an objection, if the Foing and Ehrenfreund result is confirmed.

From the wavelengths of the narrower DIBs, it has been ascertained that the DIBs are interstellar and not stellar features, and that the wavelengths of the narrow features vary only very slightly from direction to direction. Krełowski and Walker (1987) and others have sorted the DIBs into a few families that appear related from direction to direction. Other classifications depend upon profile types, e.g. narrow and/or broad features that appear to associate with each other. These associations are taken to mean that each "family" of DIBs arises from one or a few particular species that form or are stable under similar conditions. In a few cases, there is

A. G. G. M. Tielens and T. P. Snow (eds.), The Diffuse Interstellar Bands, 3–12.

evidence, discussed at this meeting by Miles and Sarre, that circumstellar emission found in the Red Rectangle, in particular, arises from a family of DIBs. *These data imply an important spectroscopic result; that DIB transitions from vibrationally and rotationally excited states are being observed.* Analysis of these transitions eventually may be a key to the identification of further DIB carriers.

Examination of the profusion of DIBs and the regional characteristics where they are found has revealed correlations that suggest the nature of the response of the DIB carrier to ionizing radiation, to shielding by interstellar grains, and possibly to chemistry. Herbig (1993) has updated and analyzed many correlations in detail, showing that modern photoelectric data remove the historical zero reddening offset for the strength of DIBs (see also Fossey and Sommerville, this volume). Herbig's plots show the direct correlation of increased DIB strength with increased E(B-V), H I, Na I, K I, and C I column densities, and a lack of correlation with H_2 column density. Along with the lack of a polarization signature (see Adamson and Whittet, this volume) and non-correlation with UV extinction, these results may be taken to imply that DIBs arise from a gas-phase species and not from a species embedded on grain surfaces or in the grains themselves, particularly not the small grains. This does not preclude a critical role for grains in the formation of the DIB carrier. At the very least, dust reduces the rate of the DIB carriers' (precursors) photodestruction.

The nature of the DIB carrier might be constrained by the abundances of elements in the lines of sight where they are detected. From the total equivalent widths of the DIBs and assuming the DIB absorptions fall onto the linear portion of the curve of growth, the abundance of the DIB carriers falls between about 10^{12} and 10^{15}cm^{-2} in a ζ Oph-like line of sight for oscillator strengths between 1 and 10^{-3}. These column abundances pose no serious constraints on DIB carriers, given the interstellar abundances of carbon, oxygen, and nitrogen, unless the mean number of (carbon) atoms in the DIB carriers exceeds about ten.

The extant knowledge of the interstellar extinction curve has grown quickly as UV spectra from space platforms have become available. The foremost of diffuse spectral features is the 2175Å bump, although this feature is not usually thought of as a DIB (see remarks by Mathis, this volume). Otherwise, extensive searches have located only a feature at 1369.1Å that has been reliably proposed as a DIB (see Cardelli, as reported in this volume). More results from *HST* can be anticipated, but new, major ultraviolet DIBs appear to be excluded by the existing data.

Concerning IR DIBs, the interstellar 3.3 μm - 3.4 μm features' connection to DIBs is uncertain. The (corresponding?) cometary 3.4 μm feature is thought to arise from UV-pumped fluorescence of unspecified organic species. The detection of DIB carrier features at infrared wavelengths is of great potential value since in that wavelength region vibration-rotation features dominate. The identification of the DIB carriers' infrared vibrational features may be simpler than the interpretation of DIB features arising from low-lying electronic transitions in the visible-near IR; however, the smaller oscillator strengths of vibrational-rotational features makes their telescopic detection and laboratory study more difficult.

The spectroscopic identification of DIB carriers can be done, in principal, from resolved DIB fine structure. Fine-structure searches in DIBs were done by Smith *et al.* (1981), by Herbig and Soderblom (1982), and, more recently, by Hibben *et al.* (this volume). Smith *et al.* found that HD187982 yielded a very narrow 6197Å DIB (0.45Å). Herbig and Soderblom found that HD187982 has a structured profile for the

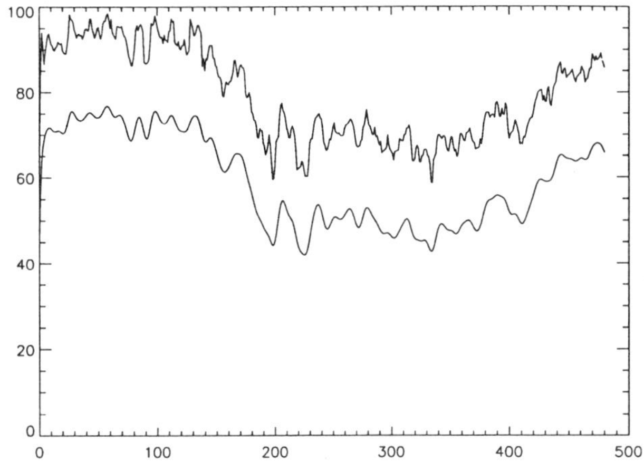

Fig. 1. Spectrum of the 6613Å DIB towards HD187982 normalized to 100, plotted blue to red. The bottom plot is smoothed by a 5 pixel wide Gaussian and shifted for clarity.

6613Å DIB. In Figure 1 above, a preliminary plot (Smith and Schempp, 1984) for the 6613Å DIB towards HD187982, at a spectral resolution of 700,000 (0.35 km s^{-1}), has structure qualitatively similar to that seen by Herbig and Soderblom, but possibly better resolved due to a five-fold increase in velocity resolution (and very similar to that reported by Hibben *et al.*). Our profile does not vary smoothly, but has small scale structure that is largely, but not all, noise. Our profile for the 6197Å DIB for HD187982, Figure 2, shows little structure above the noise. The S/N of the data is too low to reveal detailed fine structure of the 6613Å and 6197Å DIBs.

A serious barrier to the search for fine structure in DIBs is illustrated in Figure 3. The velocity structure along the line of sight towards HD 187982 is quite complicated. Lacking an accurate assessment of which or how many of the velocity components contribute to the DIB profile, we are not certain that high velocity resolution can resolve the DIB into its fine structure (if any) in the presence of the blending in the line of sight. To remove this complication, we observed less-reddened lines of sight in a search for a simple velocity structure. All observed lines of sight showed complex profiles. The narrowest features had a FWHM near 1 km s^{-1}. This includes the widely-studied ζ Ophiuchi, as illustrated in Figure 4 for K I and a Phillips C_2 band feature.

2. DIFFUSENESS IN SPECTRA

Diffuseness in the spectroscopic sense means that homogeneous broadening is occurring for the transition observed. This can arise in a number of ways. In the present context, diffuseness may arise due to a truncated upper-state lifetime for the states reached from lower (ground) states by absorption of a photon. This broadening can be very large when a non-radiative process limits the excited state lifetime so that a transition into a dissociative state naturally results in a broadened spectral feature. The greater the broadening, the less the structure in the resulting spectra. The auto-ionization continua of the rare gases are well-known examples of this effect. Attempts to identify the DIBs with broadened atomic transitions have, however, been

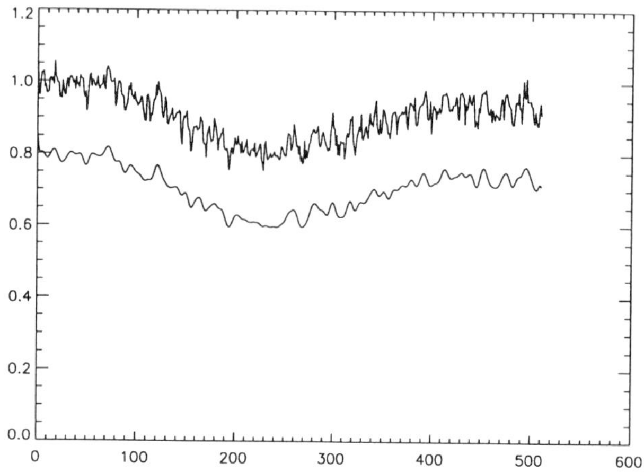

Fig. 2. Spectrum of the 6197Å DIB towards HD187982, normalized to unity and plotted blue to red. The lower plot is smoothed by a 5-pixel wide Gaussian and shifted for clarity.

unsuccessful to date. Similar, but more complicated processes occur in molecules due to dissociative states reached through the excitation of electronic, vibrational, and rotational degrees of freedom. Diffuseness due to dissociative processes is equated to the rapid destruction of the species (molecule) undergoing the observed transition. Consequently, such diffuse transitions exhibit weak or no emission.

Spectra can appear diffuse as a result of spectral complexity. Complex spectra may be resolved at sufficiently high spectral resolution (and sufficiently low temperature). Then, the inherently heterogeneously broadened nature of the spectra is revealed. This consists of a large number of resolvable spectral transitions that are broadened only by radiative processes and by their velocity distribution (at sufficiently low pressures). Thus, in the ISM, the practical limit to fine structure resolution may be the line-of-sight velocity structure.

It is worth mentioning a mechanism that might prevent dissociative processes from destroying a species undergoing a diffuse spectral transition. The discovery of a fullerene in the ISM prompts this discussion. A "cage" effect, discussed in Smith (1985), is illustrated in Figure 5 below. The suggestion that the DIB carriers may be trapped within solids is not new. The detection of the interstellar fullerenes, however, refocuses the idea into a concrete entity. The carrier of a DIB might be trapped inside a fullerene and yet have only weak (e.g. van der Waals or dispersion) interactions with the walls. Such a carrier would tend to exhibit spectra very similar to its own gas-phase spectra. Even transitions that were dissociative would not necessarily destroy the species then, since its fragments would have ample opportunity to recombine while the walls acted as a third body to carry away excess energy. It would be difficult, even with the new data described above, to exclude this option completely, so there may yet be a place for the notion that DIBs arise directly from the grains, if one describes a species like a "fullerene" as a grain.

The DIB band shapes are relatively stable, like their wavelengths but, as the S/N of the spectra has improved, small differences have emerged. The implication of small shape changes, spectroscopically, may be: (1) temperature changes, resulting in a redistribution of oscillator strength; (2) more than one species; (3) more than one

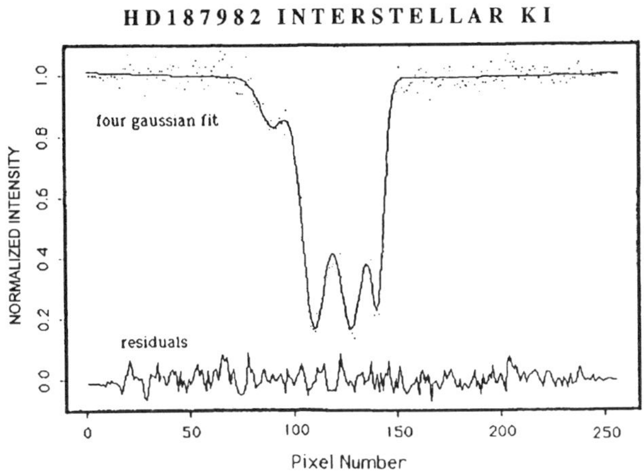

Fig. 3. Velocity profile for HD 187982 K I 7799Å transition, showing complex line-of-sight structure.

environment (cage) is present, resulting in a variable profile; or (4) all of the above could apply.

Molecular structures that lead to stability against photodestruction have been the subject of extensive laboratory and theoretical study (see Herbst, this meeting). Given the harsh ISM radiation field, molecules are not proof against photodestruction. The question devolves to the rates of photodestruction versus formation and other processes that define the competition that governs the abundances among these species. There are three main classes of molecules under consideration (Figure 6). First, linear (cumulated) chain molecules; second, the ring molecules; and finally, the aromatic molecules, all of which have relevant properties. The fullerenes have introduced a hybrid into the mix. Most early laboratory studies have naturally centered on neutral species, while ISM circumstances dictate that the most likely gas phase species will be singly ionized molecules that are stabilized by resonance structures and electron delocalization.

Laboratory methods described in detail in this meeting have begun to crack the difficult problems of (1) forming, (2) trapping, and (3) cooling an ion of a 4) specified large molecule and then (5) studying its spectrum in absorption or emission with great sensitivitiy at high spectral resolution. Note that this work requires a series of steps, each of which is very difficult and all of which must be successful to provide the required spectroscopic search information.

DIBs have been observed in many locations outside the solar system. The only solar system object that is readily observable and that may contain primitive material or at least material reminiscent of the proto-solar nebula is a comet. For some time, strong arguments were made for a grain mantle site as the location of the DIB carrier. Under the assumption that the DIBs might originate in a species contained in a grain mangle, a search in the cometary (primitive) material for the DIB carrier is sensible. We have carried out a search for the 6196Å, 6203Å, the 6284Å, and the 6613Å DIBs towards the nucleus of Comet Halley. In Figure 7, a spectrum at 0.33Å resolution of the 6613Å DIB region is plotted. The 6613Å DIB, which would be located near pixel 45, is not obviously present. The detailed removal of the reflected solar spectrum and

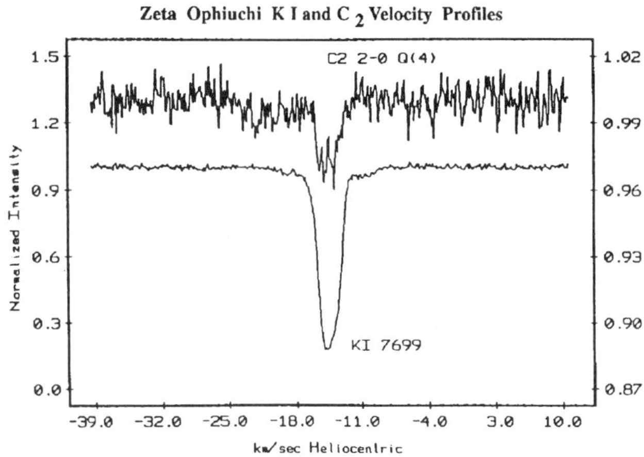

Fig. 4. ζ Ophiuchi line-of-sight velocity profile in K I and C_2 with a FWHM over 1 km/sec. The asymmetry implies the presence of more than one component.

the cometary features which vary across the spatial region of the comet being observed makes their complete cancellation difficult to achieve. We are left with the conclusion that the DIB carrier is weak or absent in Comet Halley at the level we could detect. A brighter, first-pass comet may make a much better choice for such observations in the future. The lack of prominent DIBs in the optically thick, dusty material between us and the comet nucleus at least implies that either spontaneous decay or other processing has destroyed the DIB carriers in cometary grains. A hypothesis, then, is that the lack of DIBs in P/Hally, if not unique to P/Hally, implies that the DIB carriers require refreshment in some fashion to maintain a typical abundance. This condition implies further that DIBs would be weakened under any condition, such as dark regions or in the interior of large grains, where refreshment is inhibited, as is generally observed.

3. CONCLUSION

My invitation to present this opening review lecture was based upon my experience in the observational study of DIBs and laboratory spectroscopic research experience relevant to dissociative spectra. Also, George Herbig was unable to attend due to a schedule conflict. I have appreciated the opportunity to review recent DIB research, present a few preliminary results from my own earlier work that appear have renewed relevance, and to assess the present status of the effort to understand DIBs from the detachment of over 8 years absence. From that perspective, I believe the major advances in the last decade to be:

(1) The near certain identification of at least one DIB carrier, a first break in the dike.

(2) The highly probable identification of one or more DIB families in emission.

(3) The growing certainty that the DIB carriers are in the gas phase and not directly on or in grains.

(4) The strengthened correlations of DIBs with other species and conditions via

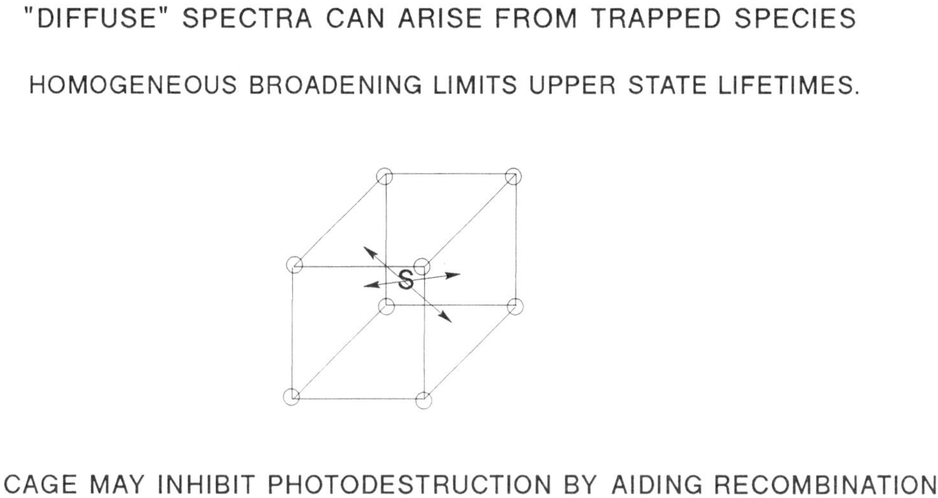

Fig. 5. A diffuse spectrum can arise from trapped species in a cage.

better observations, e.g. the non-correlation of DIBs with $N(H_2)$.

(5) The vastly improved capability in the laboratory to obtain ions and measure their spectra, making possible advances like that listed as No. 1.

3.1. SUGGESTIONS AND SPECULATIONS

The most dramatic advances in the our ability to identify DIB carriers unequivocally show that laboratory studies and observations, along with quantum theoretical insights, can advance the understanding of the DIB problem. No one kind of effort has been able to solve the complex DIB questions alone. The rapidly expanding capability of laboratory instruments to acquire and trap for extended observation an increasingly broad and known mass selection of ions while conduction exquisitely precise spectroscopic observations presents an unprecedented opportunity to select and measure the spectra for relevant species (predicted, guessed, or serendipitous, like the fullerenes).

From an observational perspective, the full confirmation of the discovery that a DIB family exhibits spectra in emission will open an entirely new regime of observational research. Herbig (1975) stated that no DIBs had been observed in emission. Subsequently, Smith *et al.* (1977) and, independently, Douglas (1977) proposed internal conversion as a means to suppress emission while preserving the DIB carrier. Both the observational facts and the physics required to explain them may now have fundamentally changed. The detection of a DIB family in emission implies that radiative relaxation competes effectively with all dissociative or other non-radiative relaxation process for that particular (family of) DIB carrier(s). Such a result has far-reaching consequences for the nature of the DIB carrier. One consequence is that, unless the DIB carrier is trapped, as in a fullerene, the DIB spectra can, in principle, be resolved spectroscopically, either in emission towards the Red Rectangle or in absorp-

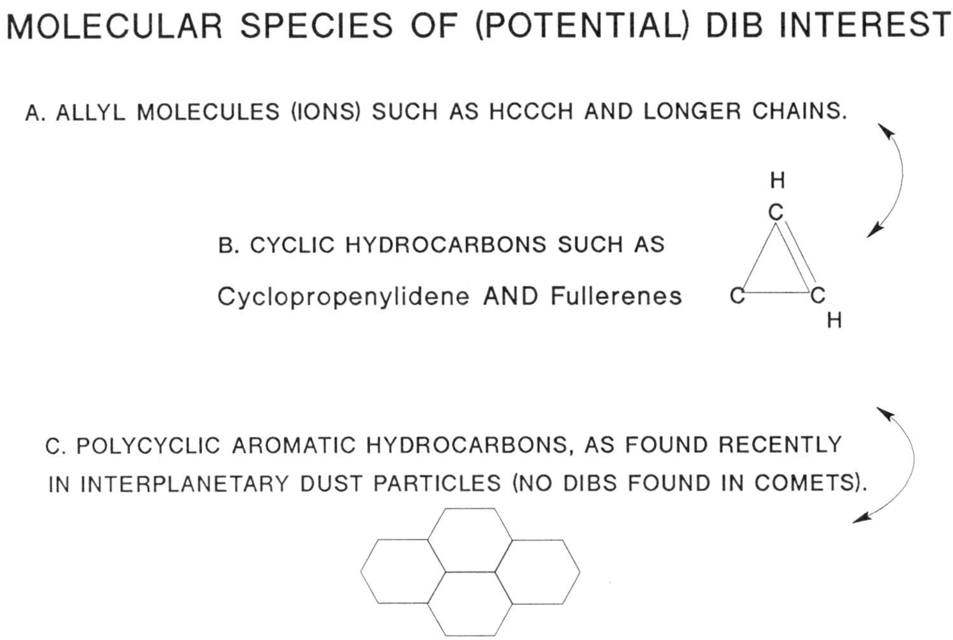

Fig. 6. Examples of species of a type observed or predicted to exist in the ISM.

tion towards lines of sight carefully selected for their low complexity. Since the DIB emission may originate in a single cloud (velocity component), rather than reflect the complicated velocity structure typically seen in absorption, a spectrum more easily resolvable into fine structure may be observed. Indeed, the interpreted temperature gradient of the Red Rectangle emission features adds an additional degree of freedom to the analysis.

Spectroscopically, it is plausible that vibrational sequences from upper and lower (maybe, the ground) DIB carrier states are being observed and resolved. Better spectra followed by analysis can confirm this possibility. Then, an attempt to resolve rotational structure for the emission features (or even rotational envelopes) can provide the rotational constants for the involved moments of inertia. Together with quantum theoretical methods, the range of possible DIB carriers may be narrowed drastically so that powerful, new laboratory methods can be brought to bear on a tractable set of candidates that, in turn, will guide observations. The synergism here is especially compelling.

The means to conduct the implied observations is in hand or close to realization. Very large spectroscopic telescopes are coming into use. If equipped with a high spectral resolution Fourier Transform Spectrograph (FTS), the full light-gathering power of these telescopes can provide the required fluxes to achieve high signal-to-noise very high spectral resolution spectra for the emission features. In absorption, our 1984 data shown here can readily be improved upon with the modern 1 e^- read noise, a/r coated, small-pixel CCDs that reach 80% quantum efficiency in the red. The remarkable Tull echelle at the McDonald 2.7 meter, with which we acquired the

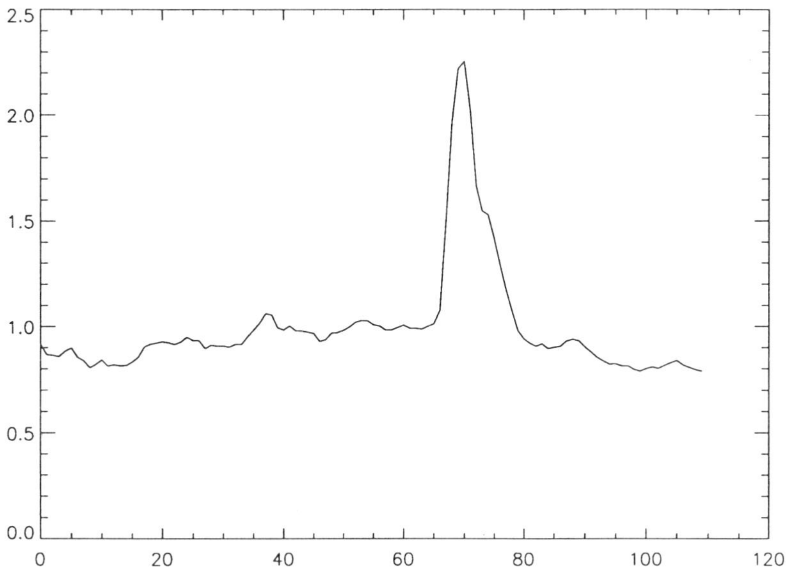

Fig. 7. The NH_2 A-X (0,7,0) $7_{07} - 7_{17}$ features dominate the spectrum towards the nucleus of p-Halley. The 6613 Å DIB would be located near pixel 45 at our spectral resolution.

spectra shown here, can reach a resolving power well over 10^6. That spectrograph is yet to be much improved upon.

Finally, the observation that the strongest DIB lies at the shortest wavelength is a puzzle that has always been intriguing. No light has been shed on this subject over the past decade. In a typical spectral series, the longest wavelength feature has the greatest oscillator strength after which the f-values decline in strength towards the continuum level. A vibrational series or a series of rotational levels, generally of lower excitation, can rise to a Boltzmann maximum and then decline, although at a high-enough temperature an increase could occur right up to a dissociative level. So the fact that the strongest DIB feature is found at the highest energy does not immedially suggest a solution.

For the sake of argument, let us assume that the DIBs are carried by gas-phase species and that the discovery of the interstellar fullerene ion confirms that, in general, the DIB carriers are very stable molecules with significant numbers of atoms, possibly heterocyclics, if the correlation with CN is taken to imply CN (photo)chemistry. The DIB carriers, then, are not a single species, but families, as is implied by the associations already described. These families are degraded incrementally, but with bottlenecks at certain stable structures. That is, a few species end up with the lion's share of the abundance; these are carriers of the prominent DIBs, until finally a single, last, most-favored species is formed that resists rapid destruction. This last species would then be the smallest and most abundant of the DIB carriers, and the carrier of the 4428Å transition. Then, the strength of the 4428Å feature can be due to the abundance of the carrier as well as its intrinsic strength. The wavelength of the 4428Å DIB, in the simplest model, may be coupled to a small size, as in a particle-in-a-box, that is also consistent with the concept of a conjugated molecule with delocalized electronic density. Can it be discovered what exact molecule that might be? Clearly, such a jigsaw puzzle is likely to yield its secrets slowly within the intervention of serendipity.

REFERENCES

Douglas, A.E. 1977, *Nature*, **218**, 130.

Foing, B.H. and Ehrenfreund, P. 1994, *Nature*, **369**, 296.

Herbig, G.H. 1975, *Ap. J.*, **196**, 129, 1993, *Ap. J.*, **407**, 142, and preprint, 1994.

Herbig, G.H. and Leka, K.D. 1991, *Ap. J.*, **382**, 193.

Herbig, G.H. and Soderblom, D.R. 1982, *Ap. J.*, **252**, 610.

Krelowski, J. and Walker, G.A.H. 1987, *Ap. J.*, **312**, 860.

Smith, W.H., 1985, McDonnell Center for Space Science Reprint.

Smith, W.H., Snow, T.P., Cochran, W.D., and Jura, M. 1981, *Ap. J.*, **248**, 128.

Smith, W.H., Snow, T.P., and York, D. 1977, *Ap. J.*, **248**, 124.

Smith, W.H. and Schempp, W.V. 1984, unpublished observations of DIBs and other interstellar absorption features at 0.35 km/sec resolution using the Tull Coudé Echelle spectrograph.

Witt, A.N., Stecher, T.P., Boronson, T.A., and Bohlin, R.C. 1989, **336**, L21.

DIFFUSE INTERSTELLAR BANDS IN INDIVIDUAL SIGHTLINES

J. KREŁOWSKI
Institute of Astronomy, Nicholas Copernicus University, Chopina 12/18, Pl-87-100 Toruń, Poland
C. SNEDEN
Department of Astronomy and McDonald Observatory, University of Texas, Austin TX 78712

ABSTRACT. This paper reviews briefly the history of observations of diffuse interstellar bands (DIBs) — features covering the spectral range from the far UV to near IR — in individual sightlines, and describes their relations to other spectroscopic features originating in the same clouds. Four different kinds of inter– and circumstellar clouds are briefly characterized. DIB intensity ratios are not constant between ISM clouds, thus several carriers are responsible for the whole DIB spectrum. Several recent surveys have indicated the existence of hundreds of very weak interstellar features – the patterns of which are clearly related to the well–known, strong DIBs. These patterns may facilitate the identification of the DIB carriers, which most likely are big molecules or very small dust grains. In single cloud cases, the DIB intensity ratios clearly split into 3-4 groups, most probably characterizing different cloud environments. The ISMs of single OB stellar associations probably are quite homogeneous – the DIB spectra of all members of such aggregates are similar. The DIB intensity ratios apparently correlate with the shapes of vacuum UV extinction curves, as well as with absorption line strengths of simple interstellar molecules when individual clouds are considered. The patterns of the recently discovered weak DIBs as well as the strong DIB ratios suggest that more than one cloud and of different optical parameters exist along most of individual sightlines. The Be stars, characterized by the emission in the *HeI* D_3 line, are apparently immersed in circumstellar shells that do not contain the DIB carriers. The carriers are apparently absent or extremely underabundant also in high velocity clouds.

1. INTRODUCTION

Diffuse interstellar bands (DIBs) were originally discovered by Heger (1922). The interstellar natures of several strong DIBs were established by *e.g.* Merrill (1934,1936), Beals and Blanchet (1937), York (1971) and Herbig (1975). At present no reliable identification of the carrier(s) of DIBs has been found. Many potential carriers have been proposed: dust grains, whole varieties of free molecules, even the hydrogen negative ion. It is commonly believed now that DIBs originate in "something in between" the identified molecules and dust grains. Such compounds can share some quantum

A. G. G. M. Tielens and T. P. Snow (eds.), The Diffuse Interstellar Bands, 13–24.

as well as classical spectral properties, i. e. they can be responsible both for DIBs and some segments of the interstellar extinction curve. Their internal structures can be very complicated and the possible presence of many species makes the task of identification very difficult.

The most recently discovered DIBs (usually in the near infrared) are typically very weak (Herbig, 1988, Herbig and Leka, 1991). Moreover, the problem of DIB carrier identification is complicated by the possible existence of multiple overlapping features comprising each major DIB (first mentioned by Herbig, 1975)). In particular, the prominent 5780 DIB (henceforth, major bands will be labeled by their approximate central wavelengths in Å) is situated inside a very broad ($\simeq$ 20 Å) but extremely shallow 5778 DIB. Further structure within the general profile of this broad band, especially a weak feature at $\simeq$ 5774.0 Å, is easily seen in Herbig's Fig. 3.

The commonly observed Doppler splitting of the sharp profiles of ISM atomic lines (reflecting the fact that the interstellar medium is not continuous but rather is built of separate clouds) exists also in DIB profiles. Herbig and Soderblom (1982) showed that the extremely narrow 6196 DIB has a complex, Doppler-split profile toward some lines-of-sight. Since that paper it has been clear that the DIBs observed in spectra of heavily reddened, distant stars may be composed of components originating in individual clouds with different radial velocities. In cases of broader DIBs, Doppler splitting is not seen directly, but the resultant profiles are much broader than in single-cloud cases (e. g. Westerlund and Krełowski, 1988a,b). Thus the DIB spectrum toward a heavily reddened line of sight is only an ill-defined average over all the intervening ISM clouds. Until the early 1980's no one had observed any differences in the DIB spectra of individual interstellar clouds. However, a first survey of ISM extinction curves in the vacuum UV (Bless and Savage, 1972) showed large differences between extinction curves derived from spectra of bright, nearby stars – the objects very likely to be obscured by only single, individual clouds. Then Krełowski and Westerlund (1988) and Krełowski and Walker (1987) (see also Josafatsson and Snow, 1987) combined high resolution spectra of the *NaI D*-lines and the major 5780 and 5797 DIBs to show that the DIB ratios were quite different in the spectra of σ Sco, ζ Oph, ζ Per and HD 40111.

DIBs can now be easily detected in spectra of very lightly reddened stars, whose intervening ISM can be considered as fairly homogeneous media. It is important to analyze the DIB strengths toward these stars together with the much sharper ISM atomic lines, in order to reveal the structure of ISM along the chosen lines of sight and possible relations of DIB spectra to the elemental depletion patterns or their relations to simple molecules. Correlations of DIBs with other ISM parameters are essential to the identification of the DIB carriers. In the next sections we summarize the current state-of-the-art in DIB observations and discuss the correlations of DIBs with other ISM features.

2. BASIC OBSERVATIONAL FACTS

2.1. VARIABLE RATIOS OF DIBS

The striking variability of DIB strength ratios has been demonstrated in many studies (see previous citations). For example, the ratio of 5780 to 5797 varies greatly in different lines-of-sight, as does, but in another fashion, the ratio of 6284 to 6270 (Krełowski and Sneden, 1994)). One supposes that sets of DIBs which appear always with the same intensity ratios may be of the same origin. Any such DIB sets (hereafter called families) may constitute the spectrum of a single species, as first suggested by Krełowski and Walker (1987) and Josafatsson and Snow (1987). More recent attempts at DIB genealogy include the papers of Krełowski (1989), Porceddu, Benvenuti and Krełowski (1991), Westerlund (1989). To first approximation, many sightlines can be divided into those, called σ family members, whose 5780/5797 strength ratios resemble those in the spectrum of σ Sco (HD 147165), and those, called ζ family members, whose spectra resemble that of ζ Per (HD 24398).

Unfortunately, the limited amount of high-quality (adequate resolution and S/N) spectra makes the lists of family members very unstable. Spectra of new sightlines usually imply that well-known, strong DIBs very rarely belong to the same family. This has been already demonstrated for neighboring 5844 and 5850 DIBs by Krełowski *et al.* (1993). The 5844 feature is hardly detectable in certain spectra, irrespective of the strength of the 5850 DIB. Thus the ratio of both features must be variable along different sightlines, implying different carriers for these two DIBs. However although the 5850 DIB may be detected in the absence of 5844, the reverse situation has not yet been observed; no object with strong 5844 but undetectable 5850 has been observed.

Another example of a variable ratio of two neighboring, partially overlapping features is discussed by Porceddu, Benvenuti and Krełowski (1991). The ratio of 6203 and 6205 DIBs changes dramatically in their sample. The recent observations of the Orion nebula by Jenniskens and Désert (1994) confirm this result. The Orion nebula produces a unique diffuse spectrum, seen until now only in this region (see the extensive discussion by Jenniskens, Ehrenfreund, and Foing 1994). The band 6205 is stronger than the neighbor only in the dense core of the Orion Cloud (in the spectra of the Trapezium stars). In the same spectra the DIBs 6196, 5797, 5850 and 6270 are below the level of detection and the major 5780 DIB is very weak. On the other hand the major DIB at 6284 Å is strong, proving that the 5780 and 6284 diffuse bands are not of the same origin as originally suggested by Krełowski and Walker (1987).

2.2. DIFFUSE BANDS VS. OTHER INTERSTELLAR FEATURES

Some attempts to identify DIB carriers have centered on comparisons of DIB strengths with those of various other ISM spectral features (broad-band dust extinction and polarization, absorption features of atomic species and diatomic molecules, broad spectral features in the infrared).

The traditional reddening parameter $E(B-V)$ loosely correlates with all conceivable characteristics of the interstellar medium. However all the correlations suffer much scatter and thus $E(B-V)$ is of little help in classifying the properties of diffuse interstellar clouds (see the figures and discussion of Herbig 1975). However, the detailed dust extinction properties, in the UV and IR regimes, are seemingly related to the intensity ratios of diffuse bands. Krełowski and Westerlund (1988) showed that the DIB strength ratios in σ Sco (HD 147165) and ζ Oph (HD 149757) correlate well with the ultraviolet extinction curve peculiarities (Bless and Savage 1972) in these lines-of-sight. Unfortunately, no significant correlations between ISM polarization curves and the DIB have yet been uncovered.

The atomic and molecular abundances can vary significantly from cloud to cloud (e.g. Bless and Savage 1972; Cardelli 1988; Joseph *et al.*, 1986; Crawford 1989 and 1992; Cowie *et al.*, 1979; Joseph, Snow and Seab 1989; see also the summary by Snow

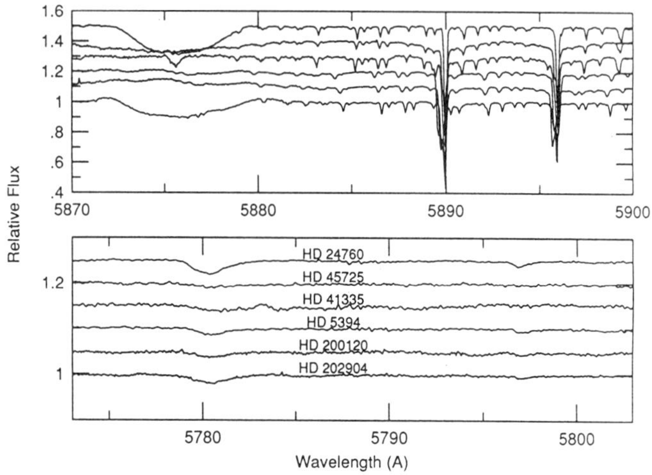

Fig. 1. The spectra of six slightly reddened stars, in the region of the atomic *NaD* lines (upper panel) and in the prominent 5780 and 5797 DIBs (lower panel). Even relatively weak *NaD* lines such as these show clear Doppler splitting, suggesting that these targets suffer obscuration by more than one ISM cloud.

1992). Some correlations among these quantities have been suggested. For example, gas phase depletions, implied by neutral atomic abundances, have been linked to the shape of the UV extinction curve (Joseph, Snow, and Seab 1989; Cardelli, Federman, and Smith 1991). A further link between depletions and molecular abundances, probably indicative of an underlying relationship between depletion and UV extinction curve shape, was uncovered by Joseph *et al.* (1986). Similar conclusions can be inferred from the recent HST observations of ζ Oph – a star obscured by two clouds of slightly different radial velocities (Savage, Cardelli, and Sofia 1992). The clouds apparently differ in the ratios of atomic line strengths as well as in the intensities of the features originating in simple molecules such as CN, CH (Crawford *et al.*, 1994) or CO (Lambert et al. 1994).

These tentative relationships clearly suggest that all of the interstellar cloud properties are somehow linked; it is difficult to change one without also affecting the others. The properties can change very rapidly from place to place in a dense cloud, changing severely the picture of interstellar line ratios along neighbor sightlines (e.g. Snow *et al.*, 1994). Diffuse bands appear to be very sensitive probes of the physical conditions inside interstellar clouds. Some relations between DIBs and spectral features of identified species may be of decisive help in identifying DIB carriers, their formation in certain conditions, etc. For example, Krełowski *et al.* (1992) suggest that the strengths of the CN features around 3875 Å as well as those of CH between 3878 and 3890 Å may be positively correlated with the 5797/5780 DIB intensity ratio. Krełowski *et al.* (1987) showed that the ratio of famous 4430 DIB absorption to $E(B-V)$ was found to be inter–related with that of the prominent 5797 and 5780 DIBs, as well as to a parameter characterizing the far-UV rise of the extinction curve.

Are the diffuse bands related to the well–known, strong sodium D_1 and D_2 features? The correlation published recently by Herbig (1993) may suggest so. However, in most cases, spectra of the observed stars contain very strong, saturated sodium lines and the relation may be difficult to obtain. In Fig. 1 we plot the spectra of

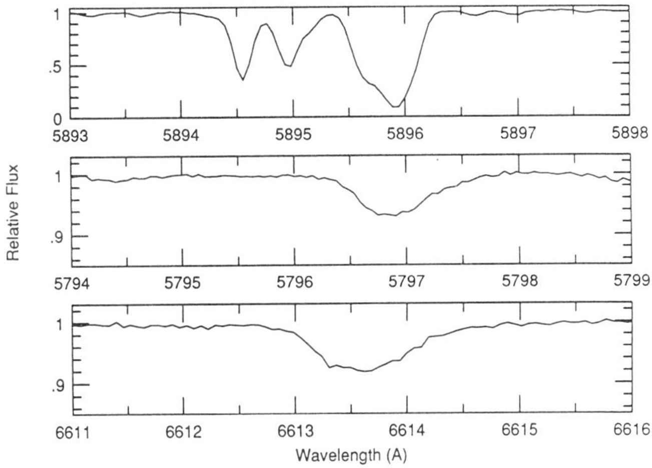

Fig. 2. DIB spectra of HD 8065, showing the existence of high velocity cloud absorptions in the *NaD* lines but not in the DIB features.

slightly reddened stars observed with the echelle spectrograph of the McDonald Observatory 2.1m telescope (for the details of the data acquiring and reducing procedures see Krełowski and Sneden (1993)). Diffuse bands, weak but evident, can be traced in four out of these six spectra. Two of the targets hardly show any detectable DIBs; their strengths do not correlate with either $E(B-V)$ or the observed intensities of *NaD* lines; see, eg. the discussions of Krełowski and Westerlund (1988) in the case of HD 139094 and by Porceddu *et al.* (1992) in the spectrum of HD 63462.

What about the so called high velocity clouds, which sometimes can be observed in spectra of early type stars? In our spectrum of HD 8065 (see Fig. 2), two such clouds are seen. The *NaD* lines contain one strong and broad component and two well separated, weaker components. Inspection of the profiles of two relatively sharp 5797 and 6614 DIBs in the spectra of HD 8065 reveals that high velocity components cannot be detected in the DIB profiles. This may not be significant, however, because the central NaI component is strongly saturated, and therefore may have orders of magnitude greater total column density. But, if the DIBs really are absent, the phenomenon can be interpreted in terms of the hypothesis described by Jenkins (1989). Shock fronts, very probable inside high velocity clouds, can destroy small grains and, probably, also the DIB carriers. However, the mechanism of DIB absence in some slightly reddened stars and in high velocity clouds is probably different. It is not possible to check whether high velocity clouds cause any reddening.

We conclude that the intensity of *NaD* lines (or the *NaI* column density) does not allow precise prediction of the diffuse band strengths – especially in individual clouds. But the presence (along a line-of-sight) of a cloud like any of these obscuring HD 41335 or HD 45725 or a high velocity gas, must make the observed DIB strengths lower in relation to $E(B-V)$ than along other lines of sight. This problem is considered in detail in the subsection below.

2.3. DIFFUSE BANDS IN SPECTRA OF BE STARS

The objects of Fig. 1 in which the DIBs are very weak or completely absent are classified as Be stars. The DIB spectrum of Be stars was considered by Porceddu,

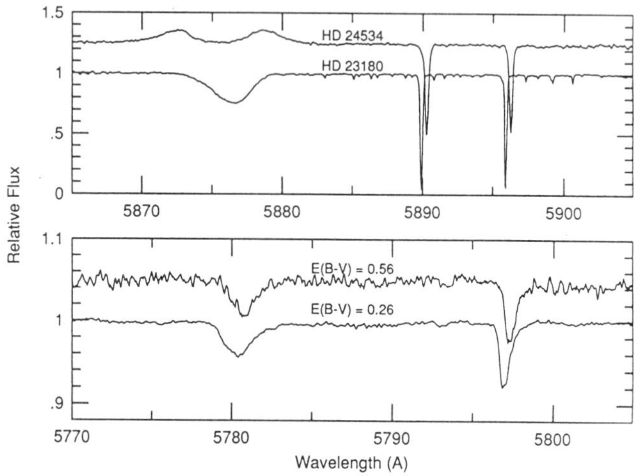

Fig. 3. A comparison of a Be star (HD 24534) and an ordinary B star (HD 23180). Note the near equality of the DIB strengths in the two stars, which have very different $E(B-V)$ values.

Benvenuti and Krełowski (1992), but deserves further comment here. On our spectra (with approximate spectral range 5600–7000 Å) there are only three strong stellar lines which can be seen in all our targets: H_α, and the HeI D_3 5876 Å and 6678 Å lines. The H_α line is in emission in many of the spectra, but the HeI lines are in emission only in a few of them.

It is tempting to label the characteristic DIB spectrum of Be stars as another sightline family (hereafter called the CS family), and probably it may not contain DIB carriers at all. In our sample of Be stars, the DIB spectrum is always weaker in relation to $E(B-V)$ when the helium D_3 line is observed in emission. We illustrate this in Fig. 3 with spectra of two stars from the Per OB2 association. It is obvious that the major diffuse bands 5797 and 5780 are nearly the same strength in the association members HD 23180 and HD 24534. However, the color excess in the latter object is twice as big as in the former one. Other comparisons of Be and normal stars yield quite similar results. The intrinsic flux distributions of Be and B stars are quite similar (eg. see the review of Krełowski and Papaj 1993) so the blame for the differences in observed colors probably lies with differences in the extinction laws governing the circumstellar envelopes. Apparently HD 24534, the Be star, is additionally obscured by some (circumstellar) matter producing the reddening but not containing the DIB carriers.

2.4. *FURTHER REMARKS*

We must reemphasize that the attempted division of DIB spectra into families probably will be helpful in the unraveling of DIB carriers, but the DIB spectra of most sightlines cannot be described adequately with such simple taxonomy. However, the spectra of a great majority of our ~80 targets can be interpreted as compositions of the three types of obscuring clouds discussed above. For example, the star HD 24534 is evidently obscured by one ζ and one CS cloud, both of approximately the same $E(B-V)$. HD 202904 is an example of a σ and CS combination – the DIB ratio is exactly like that in HD 144217 but the intensity of both features much weaker in relation to $E(B-V)$. The sightline towards HD 184915 is seemingly occupied by a mixture of σ and ζ. The examples of DIB spectrum convolution could be continued indefinitely. It is interesting that in our whole sample of about 80 sightlines we have found only one case (HD 207198) which may be considered as a composition of several ζ clouds, almost without any other contamination. In this case the NaD profiles are clearly broadened (multiple components along the line of sight) and the ratio of DIBs 5797/5780 is just slightly lower than in ζ Per and the strengths of both DIBs in relation to $E(B-V)$ is also similar. Also only one case resembling a pure σ composition was found: HD 21389, the star clearly observed through more than one cloud (the NaD profiles are broad) and the ratio of 5797 to 5780 is just a little bit higher than in σ Sco.

The cloud(s) obscuring the Orion Trapezium stars do not resemble any of the above mentioned examples. The unusual ratio of the two neighboring, partially overlapping DIBs 6203 and 6205 as well as the lack of strong 5797 band in this region was reported by Porceddu, Benvenuti and Krełowski (1991); the results have recently been confirmed and extended by Jenniskens *et al.* (1994). This might signal the existence of a fourth kind of cloud, most probably also associated with circumstellar material. However, it will be difficult to prove this case definitively, as our survey contains most of the bright stars – most likely to be obscured by one cloud. The probability of finding more physically different individual interstellar clouds is very small.

3. WEAK INTERSTELLAR FEATURES

3.1. THE DISCOVERY OF WEAK INTERSTELLAR FEATURES

Herbig (1975) and later Chlewicki *et al.* (1987) commented on some very weak features in the vicinity of, and thus possibly related to known, strong DIBs. They both pointed out possible substructures inside the broad profile of the 5778 band, and demonstrated the existence of the shallow 5795 DIB, partially overlapping the strong 5797 feature. The presence of the 5795 absorption makes the measurements of the 5797 equivalent width a difficult deconvolution task. Both Herbig and Chlewicki *et al.* assumed the 5795 feature to be broad and thus 5797 to be completely superimposed onto it. Recently Jenniskens and Desért (1993) examined spectra of lines of sight toward heavily reddened O type stars which differ greatly in rotational velocities (and thus in FWHM of the stellar absorption features). Their spectra show invariant profiles of the alleged 5778 substructures, suggesting strongly that this shallow band is completely interstellar in origin.

Finally, Krełowski and Sneden (1993) have considered the problem of positive identifications of very weak DIBs, and suggest that the following criteria be applied to suggested new DIB features: a) they should appear in spectra of the same targets obtained with different spectrographs, to avoid instrumental effects; b) they should be stationary in heliocentric velocity space, detached from the velocity of the target star, and their velocities should not change toward different lines of sight, or with time in the spectrum of a short-period binary star: c) they should be clearly seen in several reddened stars strongly differing in spectral type and luminosity class; d) they should be invariant in shape irrespective of the target star rotational velocity; e) they should not be seen in any of the unreddened stars with spectral types and luminosities similar to the reddened targets. Future searches for very weak DIB features should apply all of these criteria for certain identification.

Krełowski and Sneden (1993) used these criteria to search very high resolution (R ~ 60,000) spectra for very weak DIBs in the 5760–5830 Å and 6100–6200 Å spectral ranges. They confined their target lines of sight to those in which the *NaD* lines were very narrow and symmetric, thus very likely to be created only in single clouds. The Doppler shifts of the clouds, measured in the sodium lines, have been applied to the weak interstellar features. Many of the features identified by Krełowski and Sneden coincide with those of Jenniskens and Desért (1993). Herbig's (1975) 5795 feature seems to be a composite of two weak DIBs. One of them appears only when the 5797/5780 ratio is high. Thus the broad features upon which the major DIBs are superimposed appear themselves to be complicated structures, contaminated by many barely detectable absorption features.

The recent survey of weak interstellar features made by Jenniskens and Desért (see this volume) indicates that hundreds of such features can be found in high S/N spectra. Some of them are so weak that they are barely seen even in very high quality spectra. And, as shown by Krełowski and Sneden (1993), some additional weak DIBs can be added to the list of Jenniskens and Desért. Surveys of these types are very helpful when trying to identify the DIB carriers; it is thus quite important for multiple groups of investigators to undertake such surveys until a reasonably complete list of very weak DIBs can be agreed upon.

3.2. WEAK INTERSTELLAR FEATURES IN INDIVIDUAL CLOUDS

Are the differences exhibited by strong DIB features (like 5780 and 5797) along different lines of sight repeated in the weaker substructures (as suggested by Chlewicki *et al.*, 1987)? Complete lists of DIB family members, both strong and weak, are necessary in the DIB carrier identification effort. Many possible molecular DIB carriers give rise only to one strong feature, and all other features are very weak (J. Maier, private communication). In such cases a spectroscopic identification is possible only when weak members are assigned to the known families – a single feature does not allow definitive carrier identification.

For this correlation work, we again must appeal to the spectra of bright stars that probably are obscured by one cloud only. In Fig. 4 we show examples, with objects mostly from the Sco OB2 and Per OB2 associations, that have apparently single-component *NaD* lines. The intensity ratios of the strong 5780 and 5797 DIBs vary strongly from one of our samples to another. The sample of Sco OB2 objects is typical σ family – with the deep and strong 5780 DIB. The second sample shows the opposite extreme: the 5797 DIB is here much deeper than 5780; in no other case the 5797/5780 ratio is so high. In Fig. 4 the σ spectra are plotted in the upper panel, whereas ζ ones – in the lower panel.

It is clear that the pattern of weak interstellar features changes between σ and ζ samples. The most characteristic is the behavior of the "triplet" of 5762.80, 5766.08, 5769.09Å weak DIBs. They are all clearly seen in the ζ case, but in the σ objects only the 5766.08 feature can be traced. Thus the new weak interstellar features share the behavior of the strong DIBs and can belong to the same families *i.e.* to together form spectra of single species. Also, the weak features seem not to be just substructures of the broad 5778 DIB: most of them get stronger when the broad feature is very weak (in the "zeta" case). However, in the latter case, the triplet of weak features can easily be included into the equivalent width of 5778, especially while using low resolution spectra.

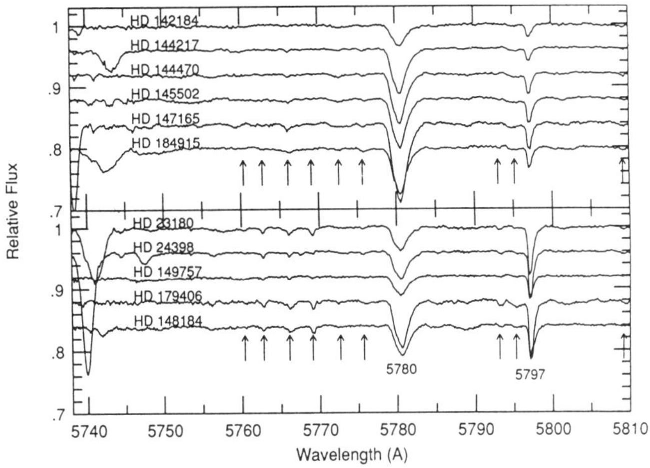

Fig. 4. The very weak DIB spectrum of samples of σ and ζ family sightlines.

If a star is obscured by two clouds, one of them of the σ and one of the ζ type, in the resultant weak DIB spectrum the features that seem to originate in both types of clouds should be noticeably stronger than those that arise in only one type. Some of our stars with multiple NaD components do seem to show this effect in such weak DIBs as that at 5776.08 Å. The same effect probably exists in the spectrum of HD 147889 shown in Herbig's (1993) Fig. 4. More sightlines are needed to clearly delineate this phenomenon, but the first indications are consistent with expectations.

4. DISCUSSION

The rapidly growing number of known unidentified DIBs brings us probably closer and closer to a solution of the problem of their identification – the longest standing unsolved problem in all of spectroscopy. Recent HST observations (Tripp, Cardelli and Savage 1994)) have proven that unidentified features can be observed down to the vacuum ultraviolet. It would be important to compare the spectra of σ and ζ clouds in this spectral range. The observations in the near infrared are also urgent, but this spectral range is difficult to search for weak interstellar features due to severe telluric contamination. The overall goal of production of full sets of DIB attributes (central wavelengths, relative intensities, profile details and behavior in different types of dark clouds) remains clear. We emphasize the necessity of observing lines of sight obscured by only one cloud wherever possible. Current observations suggest the existence of only a few discrete sets of cloud DIB properties, not a broad, continuous "spectrum".

In Fig. 5 we summarize the properties of two major classes of clouds along the sightlines we have already sampled (see also Snow 1992). *σ type clouds:* The spectrum of such a cloud contains a very strong 5780 DIB, much deeper than the neighbor 5797. The extinction curve produced in such a cloud shows a rather low far–UV branch and a narrow 2200 Å bump; the R_V value is quite high – close to 4.5. The features of simple molecules such as CN are usually below the level of detection in spectra of such objects; some of the CH features (like that near 4300 Å) can be detected but they are relatively very weak (Danks *et al.*, 1984). Only a very limited number of weak interstellar features can be observed in σ clouds. The typical examples are the clouds

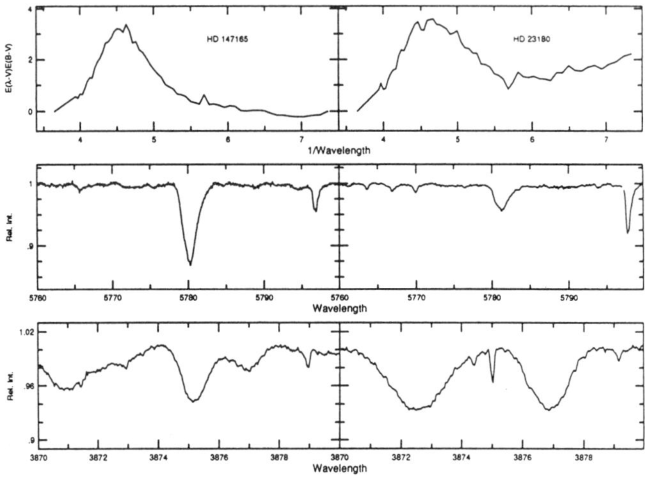

Fig. 5. The UV extinction (upper panel), 5780 Å region DIB spectrum (middle panel), and CN spectrum (lower panel) of HD 147165, a typical σ family member, and HD 23180, a typical ζ family member.

obscuring HD 144217 or HD 147165. *ζ type clouds:* The spectrum is characterized by very deep 5797 (deeper than or comparable to 5780). The far–UV extinction is relatively high, the 2200 Å bump rather broad. The R_V value is close to the most commonly accepted one – 3.1. Spectral features of simple molecules (CN, CH) are strong. The spectrum is full of relatively strong weak interstellar bands, and the whole DIB spectrum is much stronger in relation to $E(B-V)$ than in σ type clouds. The typical examples are the clouds obscuring HD 23180 or HD 24398.

We remind the reader of two other apparent cloud types, whose spectra have not been examined well yet. *CS:* This type of cloud is most easily detected around Be stars, and may not contain DIBs at all – the 5780 and 5797 features are hardly visible. Simple molecular species have not been detected in sightlines containing only circumstellar extinction (examples are HD 45725 or HD 41335); in spectra of circumstellar and interstellar extinction sightlines (such as HD 202904, HD 200120 and HD 10516) the features of CN and CH in the near UV range are also not observed (Szczerba *et al.* – this meeting). The UV extinction curve lacks the prominent 2200 Å feature (Papaj *et al.*, 1991). The far–UV branch of the curve is flat which causes a "gray" extinction in this spectral range, characteristic of rather big grains; the R_V value is hard to obtain – the observations suggest a very high value but this is most probably due to circumstellar emission. In spectra of the observed stars helium lines are observed in emission as well as the prominent H_α feature which suggests the circumstellar origin of the observed absorptions. *Orion:* The cloud(s) obscuring the Orion Trapezium stars do not resemble any of the above mentioned examples (again, see Jenniskens *et al.*, 1994). The 5797 DIB is below the level of detection, whereas 5780 is relatively weak in relation to $E(B-V)$. As shown in Fig. 1 of Krełowski and Sneden (1994) the only DIB which remains strong in Orion Trapezium is λ6284. The UV extinction curves (Fitzpatrick and Massa 1990) show very weak 2200 Å bump and a flat far–UV branches. The S/N values of our Orion spectra are too poor to attempt detection of the very weak features, but the extreme weakness of the major DIBs does not give one hope of the presence of very weak DIBs. Nor have simple molecules been detected yet. Usually very narrow emissions are observed in helium lines, which suggests that the Orion clouds are circumstellar. CS and Orion type clouds both show emissions in HeI lines and thus they are very likely to be circumstellar shells in different evolutionary stages or observed under a different inclinations.

4.1. CONCLUSIONS

The DIB spectra of the great majority of the $\sim$ 80 target sightlines observed in our program to date can be decomposed into some linear combination(s) of the simple σ and ζ families that we have discussed. Only the Orion cloud members seem to have different DIB spectra. In stars surrounded by circumstellar material, the DIBs are anomalously weak in relation to $E(B-V)$ – such clouds might not contain DIB carriers. DIB strengths do not relate in a simple way to any other ISM characteristics (intensities of molecular features, interstellar extinction parameters); thus average relations, even when fairly tight correlations are seen, cannot be blindly applied to any individual case. Such relations cannot be extrapolated – the behavior of the spectral features may change outside of the observed range(s) of their intensities. The presence of only a few kinds of clouds in the ISM, the kinds differing strongly in their optical properties, makes any relation of any two spectroscopic characteristics very difficult for any physical interpretation. Individual sightlines intersect, in a great majority of cases, several different clouds. Neither DIB ratios nor $E(B-V)$'s nor molecular line strengths alone can characterize uniquely any individual sightline. Very sharp, atomic or molecular features, can be sometimes resolved because of the Doppler split, revealing their different patterns along different sightlines. In cases of DIBs this is usually not possible as the intrinsic widths of diffuse features are too large.

More observational data, especially in the still poorly explored near-IR spectral range (where many features of the more complex PAH molecules are predicted to reside) are clearly necessary. Also the weak interstellar features are to be listed as completely as possible in as broad as possible spectral range: spectral features originating in the same molecules can be separated by as much as $\sim$ 2000 Å. More complete lists of very weak features belonging to existing DIB families must be assembled. Very high resolution profiles of atomic and molecular features should allow more accurate assessment of the truly individual cloud cases from the existing lightly to moderately reddened targets. A more complete survey of lightly reddened, early type stars is necessary to search for possible additional types of interstellar clouds.

ACKNOWLEDGEMENTS. We thank D. Hiltgen for assistance at the telescope and for reduction of some of our data. This project has been supported by the Polish National Committee for Scientific Research grant No. 684/2/91 to J.K., and by National Science Foundation grants AST91-15026, AST93-15068, and Texas Advanced Research Program grant 003658-585 to C.S. J.K. thanks the Batory Foundation for the grant allowing his participation in the Colloquium.

REFERENCES

Beals, C. S. and Blanchet, G. H. 1937, *Pub. A. S. P.*, **49**, 224.

Bless, R. C. and Savage, B. D. 1972, *Ap. J.*, **171**, 293.
Cardelli, J. A. 1988, *Ap. J.*, **335**, 177.
Cardelli, J. A., Federman, S. R., and Smith, V. V. 1991, *Ap. J. (Letters)*, **381**, L17.
Chlewicki, G., de Groot, M. S., van der Zwet, G. P., Greenberg, J. M., Alvarez, P. P., and Mampaso, A. 1988, *Astr. Ap.*, **173**, 131.
Cowie, L. L., Songaila, A., and York, D. G. 1979, *Ap. J.*, **230**, 469.
Crawford, I. A. 1989, *M. N. R. A. S.*, **241**, 575.
Crawford, I. A. 1992, *M. N. R. A. S.*, **254**, 264.
Crawford, I. A., Barlow, M. J., Diego, F., and Spyromilio, J. 1994, *M. N. R. A. S.*, **266**, 903.
Danks, A. C., Federman, S. R., and Lambert, D. L. 1984, *Astr. Ap.*, **130**, 62.
Fitzpatrick, E. L. and Massa, D. 1990, *Astr. Ap.*, **72**, 163.
Heger, M. L. 1922, *Lick Obs. Bull.*, **10**, 146.
Herbig, G. H. 1975, *Ap. J.*, **196**, 129.
Herbig, G. H. 1988, *Ap. J.*, **331**, 999.
Herbig, G. H. 1993, *Ap. J.*, **407**, 142.
Herbig, G. H. and Leka, K. D. 1991, *Ap. J.*, **382**, 193.
Herbig, G. H. and Soderblom, D. R. 1982, *Ap. J.*, **252**, 610.
Jenkins, E. B. 1989, in *IAU Symposium 135, Interstellar Dust*, eds L. J. Allamandola and A. G. G. M. Tielens, (Dordrecht: Kluwer), p. 23.
Jenniskens, P. and Désert, F.-X. 1993, *Astr. Ap.*, **274**, 465.
Jenniskens, P. and Désert, F.-X. 1994, *preprint*.
Jenniskens, P., Ehrenfreund, P., and Foing, B. 1994, *Astr. Ap.*, **281**, 517.
Josafatsson, K. and Snow, T. P. 1987, *Ap. J.*, **319**, 436.
Joseph, C. L., Snow, T. P., and Seab, C. G. 1989, *Ap. J.*, **340**, 314.
Joseph, C. L., Snow, T. P., Seab, C. G., and Crutcher, R. M. 1986, *Ap. J.*, **309**, 771.
Krełowski, J. 1989, in *IAU Symposium 135, Interstellar Dust*, eds L. J. Allamandola and A. G. G. M. Tielens, (Dordrecht: Kluwer), p. 67.
Krełowski, J. and Papaj, J. 1993, *Pub. A. S. P.*, **105**, 1209.
Krełowski, J. and Sneden, C. 1993, *Pub. A. S. P.*, **105**, 1141.
Krełowski, J. and Sneden, C. 1994, in *First Symposium on the Infrared Cirrus and Diffuse Interstellar Clouds, ASP Conf. Ser. vol. 58*, eds R. M. Cutri and W. B. Latter, p. 12.
Krełowski, J., Snow, T. P., Papaj, J., Seab, C. G., and Wszołek, B. 1993, *Pub. A. S. P.*, in press.
Krełowski, J., Snow, T. P., Seab, C. G., and Papaj, J. 1992, *M. N. R. A. S.*, **258**, 693.
Krełowski, J. and Walker, G. A. H. 1987, *Ap. J.*, **312**, 860.
Krełowski, J., Walker, G. A. H., Grieve, G. R., and Hill, G. M. 1987, *Ap. J.*, **316**, 449.
Krełowski, J. and Westerlund, B. 1988, *Astr. Ap.*, **190**, 339.
Lambert, D. L., Sheffer, Y., Gilliland, R. L., and Federman, S. R. 1994, *Ap. J.*, **420**, 756.
Merrill, P. W. 1934, *Pub. A. S. P.*, **46**, 206.
Merrill, P. W. 1936, *Pub. A. S. P.*, **48**, 179.
Papaj, J., Wegner, W., and Krełowski, J. 1991, *M. N. R. A. S.*, **252**, 403.
Porceddu, I., Benvenuti, P., and Krełowski, J. 1991, *Astr. Ap.*, **248**, 188.
Porceddu, I., Benvenuti, P., and Krełowski, J. 1992, *Astr. Ap.*, **257**, 745.
Savage, B. D., Cardelli, J. A., and Sofia, U. J. 1992, *Ap. J.*, **401**, 706.
Snow, T. P. 1992, *Australian J. Phys.*, **45**, 543.
Snow, T. P., Hanson, M. M., Seab, C. G., and Saken, J. M. 1994, *Ap. J.*, **420**, 632.
Tripp, T. M., Cardelli, J. A., and Savage, B. D. 1994, *Astr. J.*, **107**, 645.
Westerlund, B. 1989, *Rev. Mexic. Astron. Astrof.*, **19**, 39.
Westerlund, B. and Krełowski, J. 1988a, *Astr. Ap.*, **189**, 221.
Westerlund, B. and Krełowski, J. 1988b, *Astr. Ap.*, **203**, 134.
York, D. G. 1971, *Ap. J.*, **166**, 65.

ULTRA-HIGH-RESOLUTION SPECTROSCOPY OF DIFFUSE BANDS

R.E. HIBBINS, J.R. MILES AND P.J. SARRE
Department of Chemistry, University of Nottingham, University Park, Nottingham NG7 2RD, UK
S.J. FOSSEY AND W.B. SOMERVILLE
Department of Physics and Astronomy, University College London, Gower Street, London WC1E 6BT, UK

ABSTRACT. We have recorded high-resolution spectra of the λ5797 and λ6614 diffuse bands using the Ultra-High-Resolution Facility (UHRF) on the 3.9m Anglo-Australian Telescope. These bands are among the most prominent of the sharper diffuse bands (typical FWHM *ca.* 1 Å) and have been reported to have relatively steep short wavelength profiles; they also belong to the important Family 3 classification of Krełowski and Walker (1986,1987). The data were recorded along lines of sight towards lightly reddened stars at spectroscopic resolutions of 3.6×10^5 and 9×10^5. In the majority of spectra it is found that the profiles are asymmetric with a steep short-wavelength side and in the case of λ6614 there is some evidence of fine structure within the profile.

1. INTRODUCTION

Of the observational characteristics of the diffuse interstellar bands which could provide major clues as to the nature of the carriers, the determination of intrinsic profiles and the search for fine structure must surely rank amongst the most important. Much of the debate concerning a grain or molecular origin has been based on the ability of the two hypotheses to satisfy constraints imposed by the observed profiles. While the lack of any widely accepted assignment means that a grain origin cannot yet be ruled out, at present the evidence for molecular cariers is growing and there is some expectation that, even if not exhibiting resolved rotational lines, high-resolution spectra might show some structure which could provide a test of molecular theories. It was with this aim that observations were undertaken at the AAT using the newly installed Ultra-High-Resolution Facility (UHRF).

The main factors which may inhibit the resolution of fine structure are the velocity dispersion within the interstellar cloud, the presence of multiple clouds along the line of sight, spectral rotational line congestion and any intramolecular broadening process. While the last two are not within the control of the observer, care was taken to choose relatively lightly reddened lines-of-sight to reduce the probability of multiple intervening clouds, and also to select those targets which exhibit a single or few velocity components in the spectra of well-known atomic species such as Na.

A. G. G. M. Tielens and T. P. Snow (eds.), The Diffuse Interstellar Bands, 25–30.

2. BACKGROUND

Two types of study are of particular importance in relation to this work: high-resolution observational work where information on intrinsic 'shapes' or 'profiles' is sought, and those investigations in which recorded spectra are compared with the results of rotational band contour calculations. We review briefly here the main studies which have addressed these aspects.

The work of Danks and Lambert (1976) marked the first detailed attempt to examine the hypothesis that the bands arise from electronic transitions in large molecules through comparison of observed (resolution *ca.* 0.2 Å) and calculated rotational contour profiles for the λ5780 and λ5797 bands. While not advancing particular molecules they were able to demonstrate that the observed profiles could arise from molecules containing about 3 – 20 'heavy' atoms (i. e. cosmically abundant but not including hydrogen), but with this number also depending on the molecular geometry. The band profile of λ5780 was also examined by Savage (1976) at a similar resolution and discussed in terms of molecular preionisation and predissociation, clustering of molecular lines, and grain models; the same band was studied at a higher resolution of 0.05 – 0.09 Å by Snell and Vanden Bout (1981) but no fine structure was observed. The work of Savage (1976) was extended to the λ6379 and λ6614 diffuse bands (Welter and Savage, 1977), and for the two stars with a single dominant Na component, they noted that the short-wavelength side of λ6614 was especially steep. The narrow λ6196 band has also been studied (Smith *et al.*, 1981). Some interesting observations towards HD 29647 have been reported in which the λ5797 and λ5780 bands were found to be anomalously narrow (Snow and Seab, 1991); it was inferred that the observed profiles may be explained as due to unusual rotational excitation in molecules.

Turning to work where attempts have been made to determine the existence of intrinsic structure by taking account of the different velocity components along a given line of sight, the pioneering study was that of Herbig and Soderblom (1982) who recorded spectra of λ6196 and λ6614 as well as the interstellar K I line in a number of lines-of-sight, including HD 187982 and HD 186745 which had been studied previously at lower resolution by Welter and Savage. The most important results were that the λ6614 band showed distinct intrinsic structure including the steepness of the short wavelength side, noted for these lines of sight earlier by Welter and Savage (1977), and "an inflection on the shortward side as if a weaker component were superposed upon the main line, with an offset of about −0.3 Å".

The most wide-ranging set of investigations has been undertaken by Westerlund and Krełowski (1988a,b; Krełowski and Westerlund, 1988), who measured the Na line profiles in a number of lines-of-sight and performed a Gaussian decomposition in order to determine the different velocity components. Under the assumption that the diffuse bands have some similarities to the sodium lines in terms of the distribution of their carriers along the lines-of-sight, it was possible to determine intrinsic profiles for a number of bands including λ5797 and λ5780. The work showed that the λ5797 band has a steep short-wavelength side, similar to that of the λ6614 band as presented by Herbig and Soderblom.

Finally we summarize the calculations of rotational band contours for specific plausible diffuse band carriers. Cossart-Magos and Leach (1990) examined the anticipated profiles for polycyclic aromatic hydrocarbon (PAH) molecules and found that in general there are expected to be multi-peak structures which are not observed.

The first contour relating to a C_{60}-based entity was presented by Scarrott *et al.*(1992) and more detailed calculations have been published by Edwards and Leach (1993).

3. UHRF AND OBSERVATIONAL ASPECTS

The UHRF is an important new development at the AAT which allows unprecedented optical resolution to be achieved. Its main characteristics have been described (Diego *et al.*, 1994) and include operation with a resolving power of up to 10^6 and the incorporation of a new type of image slicer which provides substantial signal enhancement and permits observation of relatively weak objects. While major objectives of the instrument include the study of known atomic and molecular lines in order to investigate velocity dispersion effects and shocks (Crawford *et al.*, 1994; Barlow *et al.*, 1994), its introduction has also offered a unique opportunity to probe an age-old question: Is there fine structure within any of the unidentified "diffuse" interstellar bands? It is worth noting that at least for the narrower bands, there is a possibility that the width may be attributable to rotational line congestion alone, without the presence of any, often invoked and discussed, line-broadening mechanism.

The data were recorded in lines of sight towards lightly reddened stars in 1994 January 1–3 in two of the available resolution modes, 3.6×10^5 and 9×10^5. The observing conditions were of only moderate quality owing to cloud and poor seeing; it is perhaps worth noting that these types of studies are strongly affected by poor seeing, which results in substantial signal reduction.

4. RESULTS

We describe here a subset of the results obtained and concentrate on data for the λ6614 diffuse band recorded toward HD 75149. The spectrum of this B3 Ia star (V = 5.5, $E_{(B-V)} = 0.39$) exhibits a λ6614 band with a central depth of 0.1. The Na (D2) line is dominated by a component which is saturated, but confined in velocity space to a few km/s wide, and there have been two studies of molecular components along this line-of-sight (van Dishoeck and Black, 1989; Danks, Federman and Lambert, 1984). The spectrum recorded at a resolution of 350 000 and smoothed to an effective resolution of 100 000 is shown in Figure 1. The principal characteristics of the profile are the steepness of the short-wavelength side, the relatively gradual slope of the redward wing and, perhaps most interestingly, the appearance of a short-wavelength component separated by ~ 0.7 cm^{-1} (*ca.* 0.3 Å) from the rather broader main peak. The profile has the general appearance of a better resolved version of that obtained for λ6614 by Herbig and Soderblom (1982) towards HD 187982. In order to compare the two profiles we have taken our data for HD 75149 and degraded the resolution to that of the earlier spectrum for HD 187982 (62 500); the result is shown in Figure 2. It appears that the lower resolution spectrum toward HD 187982, importantly a different line-of-sight, can be 'recovered' by degradation of our higher-resolution data for HD 75149. Consequently, though recognising that a wider data set is essential, there are good grounds for considering the structure in the high-resolution spectrum to be "intrinsic", especially as two independent lower-resolution studies prior to this have exhibited similar characteristics (Herbig and Soderblom, 1982, Welter and Savage, 1977).

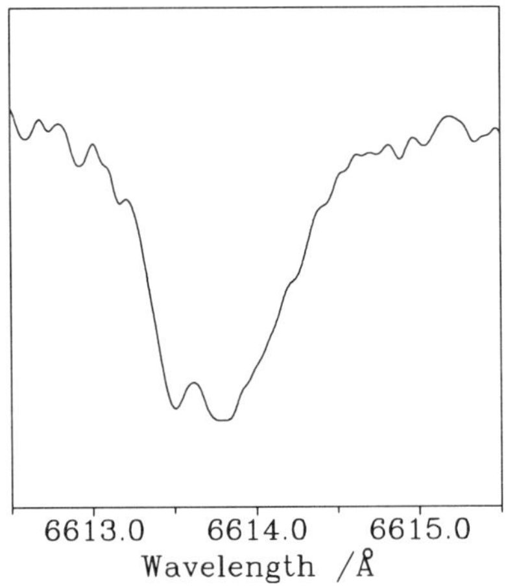

Fig. 1. Spectrum of λ6614 towards HD 75149.

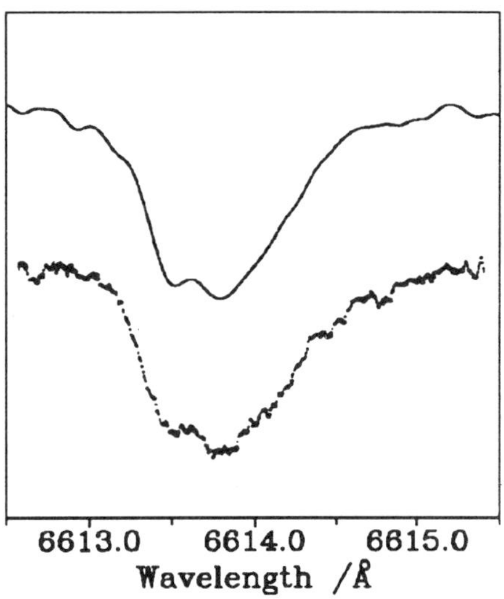

Fig. 2. Comparison of the profile of λ6614 towards HD 75149 (degraded to a resolution of 62 500) with that for the same band towards HD 187982 (Herbig and Soderblom, 1982).

5. DISCUSSION AND CONCLUSION

As outlined earlier, the determination of an interstellar absorption profile for λ6614 allows comparison between the data and calculated profiles. We consider here a small subset of all possible molecular carriers and types of electronic transitions, namely some of the simpler transitions which could reasonably be expected to occur in carbon chain molecules.

At a qualitative level the presence of a steep short-wavelength side is consistent with an origin in a molecular electronic transition with the steepness being attributable to a molecular rotational band head. It is also tempting to consider the sharper peak/shoulder to be the R-head itself, and the second broader peak to be made up of contributions from Q, P and returning R branches. In a very simplified model, we have considered transitions of $^1\Sigma$–$^1\Sigma$ and $^1\Pi$–$^1\Sigma$ symmetry with a range of B values and range of changes in B value on electronic excitation. All effects other than nuclear end-over-end rotation are neglected and a rigid rotor model is assumed.

The range of B'' values considered is between 0.1 and 0.0003 cm^{-1}, with changes (reductions) in B value on excitation between 0.1 and 3%, and equilibrium rotational temperatures between 10 and 500 K. It is found that quite a reasonable fit can be achieved for a $^1\Pi$–$^1\Sigma$ transition as shown in Figure 3, where $B = 0.003\,cm^{-1}$, $\Delta B = -0.5\%$, the rotational temperature is 200 K and a Gaussian broadening with a FWHM of 0.55 cm^{-1} has been applied. The very wide range of parameters and possible transitions which are employed in a search for a match is such that it is not possible to draw a firm conclusion about the importance of this result and we do not claim that these parameters represent a unique solution. We would, however, argue that is of some significance that a profile similar to that of the observed λ6614 diffuse band can be generated within the molecular model.

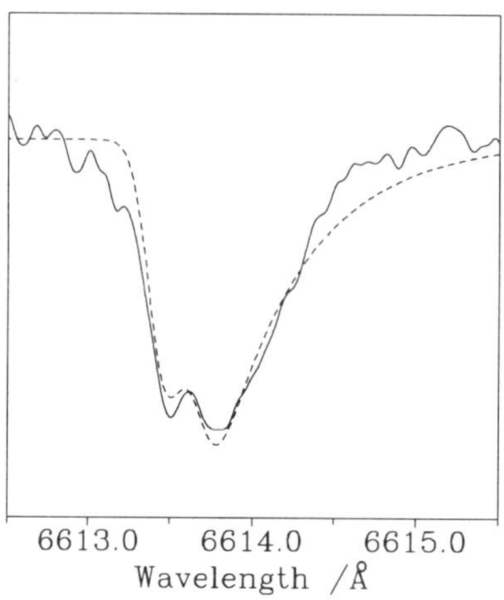

Fig. 3. Observed (solid line) and calculated (dashed line) λ6614 profiles. For details see text.

ACKNOWLEDGEMENTS.

We wish to thank PATT for the award of telescope time, Dr. J. Spyromilio for assistance in making these observations, and Dr. G. Herbig for permission to reproduce the spectrum of HD 187982. JRM and REH wish to thank PPARC and EPSRC for a Research Fellowship and Studentship respectively.

REFERENCES

Barlow, M.J., Crawford, I.A., Diego, F., Dryburgh, M., Fish, A.C., Howarth, I.D., Spyromilio, J., and Walker, D.D. 1994, *M. N. R. A. S.* (in press).
Cossart-Magos, C. and Leach, S. 1990, *Astr. Ap.*, **233**, 559.
Crawford, I.A., Barlow, M.J., Diego, F., and Spyromilio, J. 1994, *M. N. R. A. S.*, **266**. 903.
Danks, A.C., Federman, S.R., and Lambert, D.L. 1984, *Astr. Ap.*, **130**, 62.
Danks, A.C. and Lambert, D.L. 1976, *M. N. R. A. S.*, **174**, 571.
Diego, F., Fish, A.C., Barlow, M.J., Crawford, I.A., Spyromilio, J., Dryburgh, M., Brooks, D., Howarth, I.D., and Walker, D.D. 1994, *M. N. R. A. S.* (in the press).
Herbig, G.H. and Soderblom, D.R. 1982, *Ap. J.*, **252**, 610.
Krełowski, J. and Walker, G.A.H. 1986, . *J. R. A. S. C.*, **80**, 274.
Krełowski, J. and Walker, G.A.H. 1987, *Ap. J.*, **312**, 860.
Krełowski, J. and Westerlund, B.E. 1988, *Astr. Ap.*, **190**, 339.
Savage, B.D. 1976, *Ap. J.*, **205**, 122.
Scarrott, S.M., Watkin, S., Miles, J.R., and Sarre, P.J. 1992, *M. N. R. A. S.*, **255**, 11P.
Snell, R.L. and Vanden Bout, P.A. 1981, *Ap. J.*, **244**, 844.
Snow, T.P. and Seab, C.G. 1991, *Ap. J.*, **382**, 189.
van Dishoeck, E.F. and Black, J.H. 1989, *Ap. J.*, **340**, 273.
Welter, G.L. and Savage, B.D. 1977, *Ap. J.*, **215**, 788.
Westerlund, B.E. and Krełowski, J. 1988a, *Astr. Ap.*, **189**, 221.
Westerlund, B.E. and Krełowski, J. 1988b, *Astr. Ap.*, **203**, 134.

THE DIFFUSE INTERSTELLAR BANDS TOWARDS SOUTHERN STARS

S.J. Fossey and W.B. Somerville
Department of Physics and Astronomy, University College London, Gower Street, London WC1E 6BT, UK

ABSTRACT. Observations of the diffuse bands at 5780, 5797, 5849 and 6614 Å have been obtained towards a number of stars in the southern Milky Way. The sample includes lines of sight passing through predominantly diffuse cloud material, stars lying behind denser clouds, and a few stars at high Galactic latitude. The effect of the interstellar environment on the strengths and relative strengths of the diffuse bands is considered. The relationship of the diffuse bands to reddening varies significantly between the local environments sampled. The $\lambda5797$ and $\lambda6614$ diffuse bands appear to arise from different carriers. It is found that the $\lambda5780$ diffuse band is weakened with respect to reddening and with respect to $\lambda5797$ and $\lambda5849$ along lines of sight passing through the denser clouds sampled here.

1. INTRODUCTION

The dependence of the diffuse band strengths on the interstellar environment provides valuable evidence on the relationship of the carriers of the diffuse bands to other components of the interstellar medium. Recent studies have examined the diffuse bands in Taurus (Adamson *et al.*, 1991) and Orion (Jenniskens *et al.*, 1994), and their relationship to other interstellar atomic and molecular species (Herbig, 1993).

Here, we present a study of the diffuse bands towards a number of stars in the southern Milky Way up to a reddening of $E(B-V) = 0.6$. These include stars in various OB associations, such as CMa OB1, Vela OB1 in the local spiral arm, and a few stars in the Carina-Centaurus arms of the southern Milky Way. We have also observed stars lying behind more local, denser clouds (e.g., Taurus, Southern Coalsack), and a few stars at high Galactic latitude. Broadly, the sample can be divided into two: relatively local stars lying within 500 pc of the Sun; and distant (> 1 kpc) stars lying in the 3rd and 4th Galactic quadrants. We consider here the effect of the interstellar environment on diffuse band strengths.

The observations were made with the coudé spectrograph of the 74-inch telescope at Mount Stromlo Observatory, Australia, in December 1993. The wavelength region principally observed was chosen to include the diffuse bands at 5780, 5797 and 5849 Å. A number of weaker bands in the 5780 Å region were detected also; these have been catalogued recently by Jenniskens and Désert (1993) and Krełowski and Sneden (1993). The diffuse band at 6614 Å was observed also in a few stars. The spectral

A. G. G. M. Tielens and T. P. Snow (eds.), The Diffuse Interstellar Bands, 31–38.

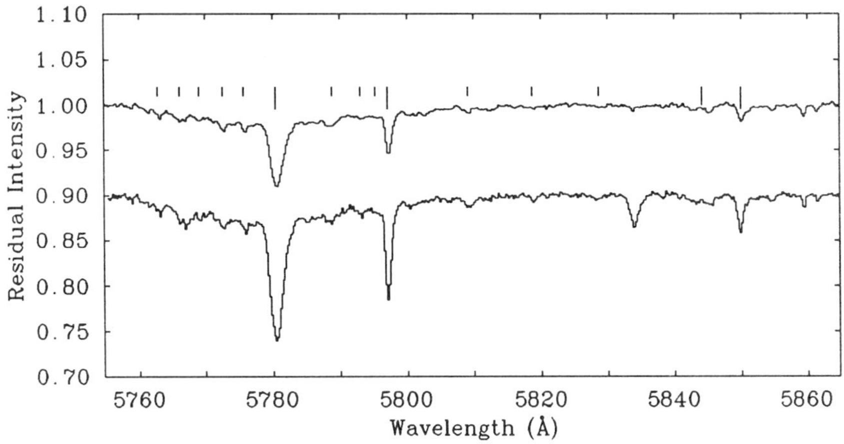

Fig. 1. The diffuse bands in the region of 5780 and 5797 Å. The spectra have been obtained by averaging stars of similar spectral type to reveal the weaker bands more clearly. The upper spectrum is the average of six early B-type dwarf spectra; the lower is the average of five early B-type supergiant spectra and has been shifted down in relative intensity by 0.1. The weak bands listed by Jenniskens and Désert (1993) and Krełowski and Sneden (1993) are indicated by short tick marks.

resolution was 0.30 Å, and the signal-to-noise ratio in the final averaged spectra ranged from about 200 to 500. Figure 1 illustrates the 5780 Å region.

The data for the target stars and the equivalent widths of the diffuse bands will be presented elsewhere (Fossey and Somerville, 1994). In the following sections we examine some of the statistical trends in these data.

2. THE RELATIONSHIP WITH $E(B-V)$

The relationship between the equivalent widths of λ5780 and λ5797 and reddening, $E(B-V)$, shows significant scatter (Figure 2), indicating wide variations in the efficiency of diffuse band production with respect to the column of dust along the lines of sight in this sample. However, as noted above, the sample includes stars seen through very different interstellar environments. Considering these separately, we find the following results for these two bands:

- Towards stars behind local, denser concentrations of interstellar material, λ5780 and λ5797 are weak with respect to $E(B-V)$. This has been noted before (e.g., Snow and Cohen, 1974). Examples are HD27778, which lies behind the Taurus dark cloud (see also Adamson *et al.*, 1991), and HD110432, behind the Coalsack. For λ5780 (but not λ5797), other stars in the Taurus-Auriga region which lie behind concentrations of CO, as observed in CO $J=1-0$ emission (Ungerechts and Thaddeus, 1987), are weak compared to stars seen through holes in the CO emission, which fall closer to the mean "diffuse medium" line defined by HD183143 (Figure 2).

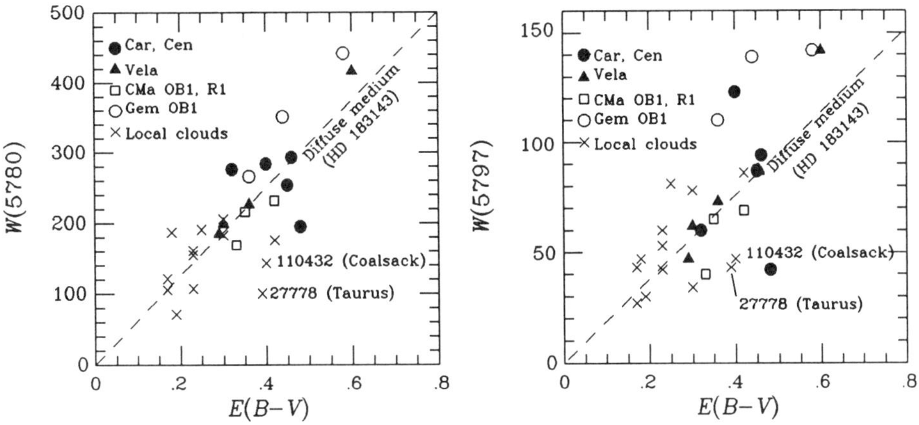

Fig. 2. The relationship between the equivalent widths of λ5780 and λ5797 and interstellar reddening, $E(B-V)$. The symbols differentiate between stars in different associations, as indicated. The trend expected for "diffuse medium" lines of sight is derived from the equivalent widths observed in HD183143 (Herbig, 1993).

- The diffuse band strengths towards the Vela stars are generally consistent with the diffuse medium line of sight towards HD183143. The Vela stars are seen in the opposite direction to HD183143 along the local spiral arm; the diffuse band absorption probably arises from diffuse cloud material associated with the Vela Molecular Ridge which is most prominent in CO emission between 3 and 10 $\mathrm{km\,s^{-1}}$ (LSR) in this direction (May *et al.*, 1988). Indeed, a component of interstellar CH at 5 $\mathrm{km\,s^{-1}}$ (LSR) has been observed in absorption towards HD75149 (Danks *et al.*, 1984), indicating a likely association with this material.
- The diffuse bands towards the stars of the Gem OB1 association are strong with respect to $E(B-V)$, especially λ5797.
- The Carina-Centaurus arm stars are scattered across the plot of equivalent width *vs.* $E(B-V)$. For λ5797, the two most discrepant stars are part of the star-forming regions and molecular cloud complexes (e. g., Grabelsky *et al.*, 1988) of the Car OB1 association, suggesting that the local conditions have strongly altered the relationship between the diffuse band carriers and the dust.

Clearly, the relationship of diffuse band strength with $E(B-V)$ varies significantly from one region to another, a result which has been established by others for different samples of stars (e. g., Herbig, 1975; Wampler, 1966). It will be interesting to investigate these variations further with respect to the measured properties of the interstellar medium as data become available.

3. INTERCORRELATIONS OF THE DIFFUSE BANDS

The existence of families of diffuse bands indicates a common response of the carriers of certain bands to differences in the interstellar environment from one line of sight to another. The 5797, 5849 and 6614 Å diffuse bands have previously been assigned to the same family (Krełowski and Walker, 1987). Here, the data for the 6614 Å band are few; but there is significant scatter in the correlation between the equivalent widths of $\lambda5797$ and $\lambda6614$ (Figure 3a). In fact, $\lambda6614$ appears to follow $\lambda5780$ more closely in these data (Figure 3b). The $\lambda5797$ and $\lambda6614$ diffuse bands respond differently to the various physical conditions sampled in these lines of sight; it therefore seems likely that they arise from different carriers, a point discussed also by Scarrott *et al.* (1992) who considered the emission counterparts of these bands in the Red Rectangle.

The correlation between $\lambda5797$ and $\lambda5849$ is reasonably tight, in contrast to the correlation between $\lambda5849$ and $\lambda5780$, which shows significant scatter (Figure 3d). If we distinguish between lines of sight which sample predominantly either local material (open symbols in Figure 3) or material over extended path lengths (filled), it is found that $\lambda5780$ is relatively weaker with respect to $\lambda5849$ in the local clouds.

We should like to characterize the physical state of the material sampled by these lines of sight in order to investigate these differences between different bands. However, relatively few data related to the interstellar environments are available for this sample of stars. A crude means of characterizing the interstellar clouds is to derive the mean line-of-sight hydrogen density, $\langle n_{\rm H} \rangle$ ($\equiv N_{\rm H}/r$), for each sight-line. Spitzer (1985) has interpreted $\langle n_{\rm H} \rangle$ as an indicator of the predominance of different types of interstellar cloud sampled along the line of sight. Here, the total hydrogen column density, $N_{\rm H}$, has been estimated from $E(B-V)$ using the relationship given by Bohlin *et al.* (1978) and Somerville and Smith (1989), and the distance, r, has been taken from the association distances (Humphreys, 1978) or derived from the adopted photometry and spectral types (Fossey and Somerville, 1994).

We find that the ratios $W(5780)/W(5797)$ and $W(5780)/W(5849)$ of diffuse band equivalent widths are significantly anticorrelated with $\langle n_{\rm H} \rangle$ (Figure 4). These anticorrelations imply that $\lambda5780$ is weaker with respect to $\lambda5797$ and $\lambda5849$ in the densest local clouds than in the more diffuse clouds sampled by the longer lines of sight.

4. OTHER TRENDS

The weakening of $\lambda5780$ with respect to other bands in certain environments has been noted elsewhere. For example, $\lambda5780$ is weak with respect to $\lambda5797$ along certain lines of sight which intersect significant concentrations of diatomic molecules such as CH and CN (Krełowski *et al.*, 1992). With a larger dataset, we are able to examine this trend statistically. In the present sample of data, molecular column densities are available for a few stars only; we have therefore included the equivalent width data for those stars in the sample studied by Herbig (1993) for which molecular column densities are available. The diffuse band ratio $\log W(5780)/W(5797)$ is then found to be significantly anticorrelated with the logarithm of the column density of CH with respect to $E(B-V)$ (Figure 5).

From the intersection of Herbig's (1993) sample of data with other studies of the interstellar medium, we may examine the dependence of the diffuse band strengths on other indicators of local conditions. We have taken the abundance of Mg from Jenkins *et al.* (1986), one of the species for which column densities are among the most reliably determined in their data. We have also taken the rotational temperature of molecular hydrogen from Savage *et al.* (1977), which reflects the kinetic temperature of the clouds for these lines of sight. We have derived a column density, $N(\lambda)$, of

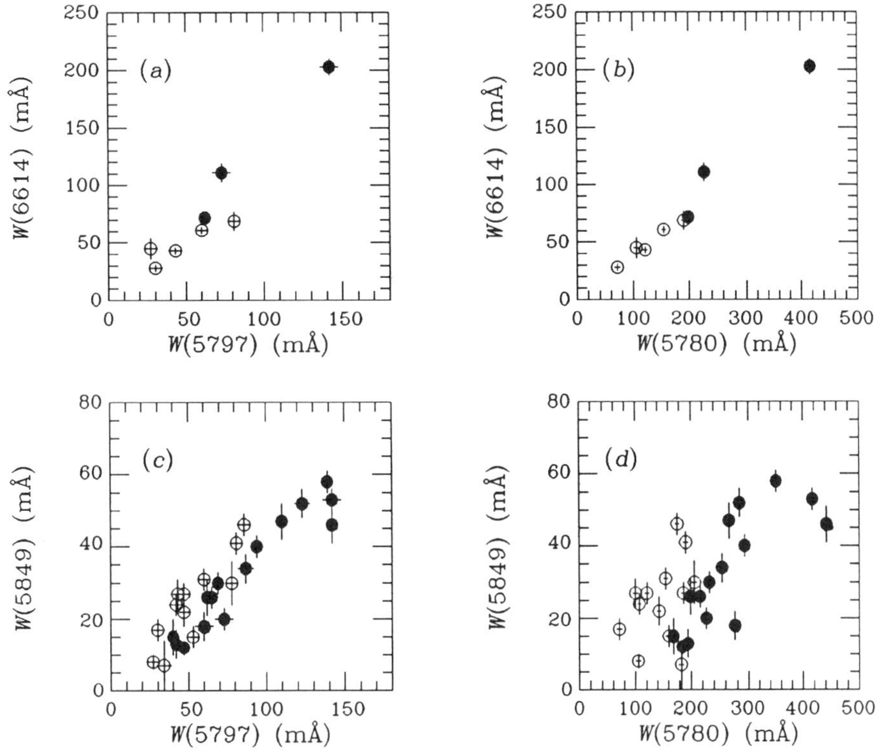

Fig. 3. The intercorrelations between the diffuse band equivalent widths: (*a*) λ6614 *vs.* λ5797; (*b*) λ6614 *vs.* λ5780; (*c*) λ5849 *vs.* λ5797; (*d*) λ5849 *vs.* λ5780. The open symbols denote relatively local stars ($r < 500\,\mathrm{pc}$), and the filled symbols, more distant stars ($r > 1\,\mathrm{kpc}$).

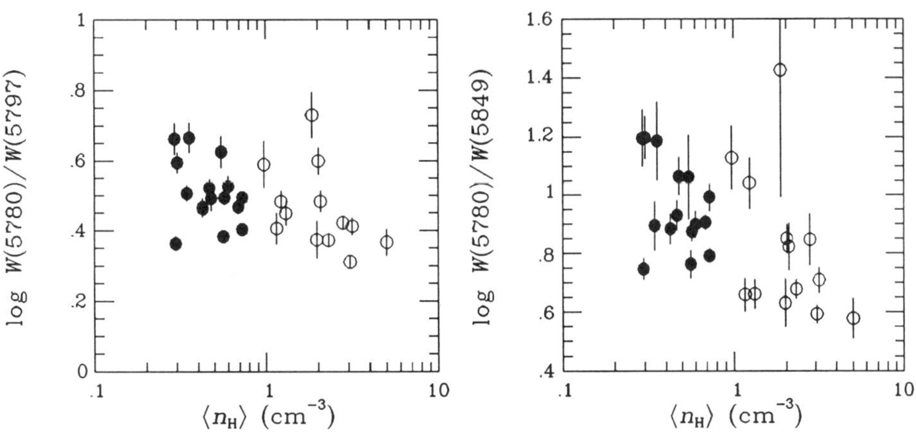

Fig. 4. The correlation of the equivalent width ratios $W(5780)/W(5797)$ and $W(5780)/W(5849)$ with the mean line-of-sight hydrogen density, $\langle n_{\mathrm{H}} \rangle$. Symbols are as in Figure 3.

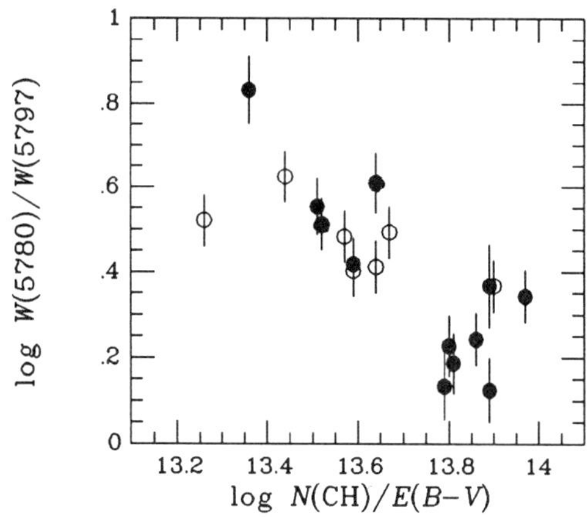

Fig. 5. The anticorrelation between the diffuse band ratio, log $W(5780)/W(5797)$, and the abundance of CH, as expressed by the column density of CH with respect to $E(B-V)$. The open symbols are data from this work; the filled symbols are the data of Herbig (1993).

the diffuse band carriers by assuming unsaturated diffuse bands and an oscillator strength f of unity (for simplicity). Then we determine the logarithmic abundance of the diffuse band carriers, $A(\lambda)$, with respect to the total column density of neutral hydrogen, N_H, given by Jenkins *et al.* (1986).

It is found that the abundance of the carrier of λ5780 is positively correlated with the rotational temperature of H_2 and with the gas-phase abundance of Mg (Figure 6). [The correlations remain statistically significant after correction for correlated errors, where appropriate (e.g., Jenkins *et al.*, 1986).] For λ5797, neither of these correlations is found to be significant. These results are consistent with those of Herbig (1993), who found that λ5780 was better associated with atomic than molecular hydrogen. Taken together, these results imply that the carrier of the λ5780 diffuse band is most abundant in lower density diffuse clouds than in the cooler, denser clouds where the fraction of molecular hydrogen is greater and refractory atomic species such as Mg are depleted from the gas-phase onto grains.

These trends are not reflected for λ5797, which suggests that its carrier persists in the denser material where λ5780 is weakened. However, we have noted above that there are indeed specific lines of sight passing through dense cloud material where the abundance of the carrier of λ5797 is diminished also (e.g., HD27778, HD110432).

5. CONCLUSIONS

The effect of the interstellar environment on the strengths and relative strengths of the diffuse bands has been considered, using observations of stars in the southern Milky Way and stars lying behind relatively local clouds. We summarize the principal results of this paper below:

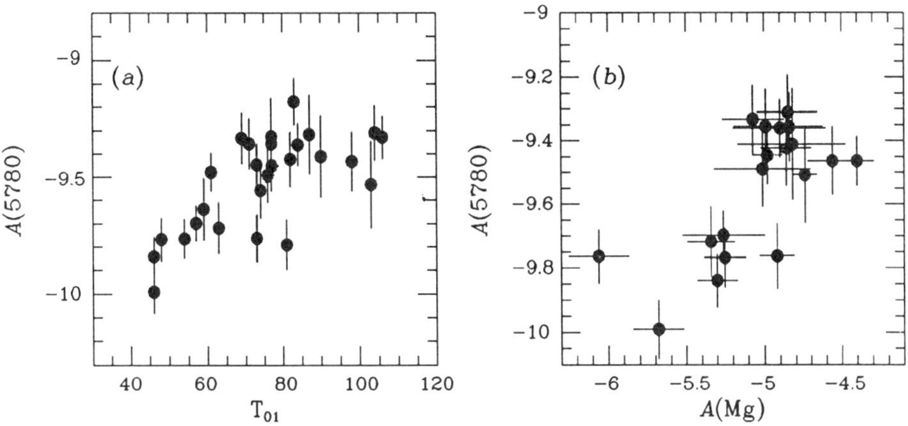

Fig. 6. The correlations of the abundance of the carrier of the λ5780 diffuse band, $A(5780)$, with (a) the rotational temperature of molecular hydrogen, T_{01}, and (b) the abundance of Mg. The diffuse band data are from Herbig (1993).

1. The relationship to reddening is found to vary significantly between the local environments sampled. In particular, we note that the λ5780 diffuse band is weak with respect to $E(B-V)$ along lines of sight lying behind more local, dense cloud material compared to lines of sight passing through predominantly diffuse clouds.

2. Of the diffuse band intercorrelations, the λ5797 and λ6614 bands are not well correlated here in the few stars for which measurements are available; it is likely that they are due to different carriers.

3. The λ5780 band is found to be weakened with respect to the λ5797 and λ5849 bands towards stars lying behind the densest local clouds, which are characterized by a larger mean hydrogen density.

4. When the data obtained here are combined with additional data from a recent study by Herbig (1993), it is found that the λ5780 band is weakened with respect to the λ5797 band where the column density of CH with respect to reddening is increased.

5. Using Herbig's (1993) data, the carrier of the λ5780 band is found to be most abundant in warmer diffuse clouds, where the gas-phase abundances of refractory atomic species such as Mg are enhanced.

The distribution of carriers of different diffuse bands is clearly sensitive to the astronomical environment. Any proposed identification of the carrier of the diffuse bands must be able to explain these results. In particular, models of diffuse band production must account for the enhanced abundance of the carriers of bands such as λ5780 in predominantly diffuse atomic material. It seems that the diffuse band carriers are not synthesized in the reaction networks in diffuse clouds which involve H_2 (e.g., Herbig 1993). Therefore, if the diffuse bands are due to polyatomic molecular species, the question of how these species are formed and how their abundance is sustained in these environments must be addressed.

ACKNOWLEDGEMENTS. We thank the Director and staff of the Mount Stromlo Observatory for the allocation of observing time and technical support, and UK PATT for travel funds. SJF thanks the Royal Society, University College London and the conference organizers for financial support.

REFERENCES

Adamson, A.J., Whittet, D.C.B., and Duley, W.W. 1991, *M. N. R. A. S.*, **252**, 234.
Bohlin, R.C., Savage, B.D., and Drake, J.F. 1978, *Ap. J.*, **224**, 132.
Danks, A.C., Federman, S.R., and Lambert, D.L. 1984, *Astr. Ap.*, **130**, 62.
Fossey, S.J. and Somerville, W.B. 1994, in preparation.
Grabelsky, D.A., Cohen, R.S., Bronfman, L., and Thaddeus, P. 1988, *Ap. J.*, **331**, 181.
Herbig, G.H. 1975, *Ap. J.*, **196**, 129.
Herbig, G.H. 1993, *Ap. J.*, **407**, 142.
Humphreys, R.M. 1978, *Ap. J. Suppl.*, **38**, 309.
Jenkins, E.B., Savage, B.D., and Spitzer, L. 1986, *Ap. J.*, **301**, 355.
Jenniskens, P. and Désert, F.-X. 1993, *Astr. Ap.*, **274**, 465.
Jenniskens, P., Ehrenfreund, P., and Foing, B. 1994, *Astr. Ap.*, **281**, 517.
Krełowski, J. and Sneden, C. 1993, *Pub. A. S. P.*, **105**, 1141.
Krełowski, J., Snow, T.P., Seab, C.G., and Papaj, J. 1992, *M. N. R. A. S.*, **258**, 693.
Krełowski, J. and Walker, G.A.H. 1987, *Ap. J.*, **312**, 860.
May, J., Murphy, D.C., and Thaddeus, P. 1988, *Astr. Ap. Suppl. Ser.*, **73**, 51.
Savage, B.D., Bohlin, R.C., Drake, J.F., and Budich, W. 1977, *Ap. J.*, **216**, 291.
Scarrott, S.M., Watkin, S., Miles, J.R., and Sarre, P.J. 1992, *M. N. R. A. S.*, **263**, L27.
Snow, T.P. and Cohen, J.G. 1974, *Ap. J.*, **194**, 313.
Somerville, W.B. and Smith, C.A. 1989, *M. N. R. A. S.*, **238**, 559.
Spitzer, L. 1985, *Ap. J.*, **290**, L21.
Ungerechts, H. and Thaddeus, P. 1987, *Ap. J. Suppl.*, **63**, 645.
Wampler, E.J. 1966, *Ap. J.*, **144**, 921.

THE DIFFUSE INTERSTELLAR BAND SPECTRUM

P. JENNISKENS[1], F.-X. DÉSERT[2]
[1] *NASA/Ames Research Center, Mail Stop 239-4, Moffett Field, CA 94035*
[2] *Inst. d'Astroph. Spatiale, Bat 121, Université Paris XI, F-91405 Orsay*

ABSTRACT. The location of diffuse interstellar bands (DIBs) in the electromagnetic spectrum, their shape and relative position, as well as their relative intensities contain information about the nature of the DIB carrier. What follows is a brief summary of such features of the DIB spectrum as they emerge from a recent systematic survey of DIBs and from related studies in the literature.

1. INTRODUCTION

Diffuse interstellar bands were first identified as interstellar features in the spectra of reddened stars that were recorded on photographic plates most sensitive in the near-UV and visible (Merrill 1934; 1936). Shortly before solid state detectors were introduced in the 1970's, already some 39 DIBs were known (Herbig 1975) but signal-to-noise constraints forced many studies of DIB characteristics to concentrate on a few strong DIBs, notably those at 4430, 5780, 5797, and 6284 Å(e.g. Wu 1972, Snow 1977, Josafatsson & Snow 1987, Benvenuti & Porceddu 1989). Note that many attempts to identify the DIB carrier in the laboratory, too, have focused on these strong DIBs (e.g. McKellar 1955, Stoeckly & Dressler 1964, Johnson 1967, Manning 1970, and recently Wdowiak 1980, Miles & Sarre 1993, Wdowiak *et al.* 1993). These strong DIBs may not be the most important in solving the DIB mystery. The collection of positions, shapes, and relative intensities of all DIBs together carry important clues to the nature of the DIB carrier.

The introduction of solid state detectors has allowed the detection of new DIBs at longer wavelength. Sanner *et al.* (1978) were the first to open the near-infrared, but notably Herbig (1988) and Herbig & Leka (1991) were very succesful in finding new DIBs there. At the same time there was a shift in focus from impurities in solid grains to large organic molecules as the most likely DIB carrier (Herbig 1975, Danks & Lambert 1976, Smith *et al.* 1977, Van der Zwet 1987, Herbig 1993). From 1975 to 1983 mid-infrared spectral signatures of PAHs or related macromolecules were observed in various astrophysical environments. Ions of PAHs have strong absorptions in the visual and some members of this vast expanse of possible DIB carriers have subsequently been studied through matrix isolation spectroscopy (e.g. Salama & Allamandola 1992, Ehrenfreund *et al.* 1992, Fulara *et al.* 1993, Lee & Wdowiak 1994). A need was born for a systematic survey of DIBs that would allow a direct comparison of laboratory spectra to the complete astronomical DIB spectrum.

For this purpose, we did a systematic DIB survey out to 8680 Å at the Observatoire d'Haute Provence in France (Jenniskens & Désert 1992; 1994). The study has resulted

A. G. G. M. Tielens and T. P. Snow (eds.), The Diffuse Interstellar Bands, 39–51.

in many new detections and offers an unprecedented look at the diffuse interstellar band spectrum. This paper is devoted to the statistical aspects that emerge.

2. THE DIB SURVEY

The DIB survey covers a wide spectral range (3800-8680 Å) at a high spectral resolution ($\Delta\lambda = 0.3$Å), but consequently considers only a small sample of four reddened stars and two comparison stars for telluric and stellar line identifications. The four program stars have been selected in order to fulfill a number of conflicting constraints: allow an easy detection of DIBs; allow a certain discrimination against other sources of absorption; and (with less weight) have a representative DIB spectrum for average diffuse medium conditions. Suitable background stars for this type of study are of spectral type earlier than A3 (I-V), where stars of type B3-5 V have least disturbing stellar photospheric absorptions. By choosing a sequence of progressively reddened stars of different stellar type, with radial velocities different from that of the interstellar matter, it is possible to identify the DIBs as those features that increase systematically with reddening and have a radial velocity coinciding with that of the interstellar matter. In order to have a systematic increase of DIB strength with reddening in a small sample of stars, suitable lines of sight are those that pass through the diffuse medium, because significant variations of DIB strength per unit extinction exist towards high-density molecular clouds (e.g. Snow & Cohen 1974, Adamson *et al.* 1991), photodissociation regions, and lines of sight with very low reddening (Jenniskens *et al.* 1994).

The spectral resolution of the DIB survey was chosen such that the narrowest DIBs known (about 0.4 Å wide) are resolved and only slightly broadened. The exposure times were chosen such that the sensitivity limit is fairly constant over the spectral range. The weakest DIBs detected have an equivalent width (W) per unit reddening (E_{B-V}) of about $W/E_{B-V} = 0.008$ for narrow 1 Å wide DIBs. Broader DIBs need higher equivalent width to be detected, with a limit approximately proportional to $\Delta W \sim \sqrt{FWHM}$ (Full Width at Half Maximum). The detection limit of the broadest DIBs is mainly limited by the accuracy of placing the baseline and is proportional to the depth of the band, or $\Delta W \sim FWHM$. Individual spectra cover a range of 200 Å. Second and third order baseline variations limit the detection of the widest DIBs to those narrower than FWHM = 50 Å . These detection limits are approximately constant over the complete spectral range except for a number of wavelength regions where stellar and telluric lines lower the sensitivity, e.g., between 5900 and 6000 Å .

3. STATISTICAL ASPECTS OF THE SURVEY

Solid state detectors have not only opened the near-infrared part of the spectrum, but have resulted also in a higher signal to noise at given exposure time. This again has allowed a quick recording of calibration sources that enables an accurate removal of telluric lines. The result is a hyper-inflation of new detections (Figure 1). The works of Herbig (1988) and Herbig & Leka (1991) brought the total number of known DIBs to about 100. The DIB survey has added to that another 64 certain DIBs and 52 probable DIBs. In addition, three DIBs are known outside our survey limits, up to 13175 Å (Joblin *et al.* 1990, Gredel & Münch 1986). The total number of known

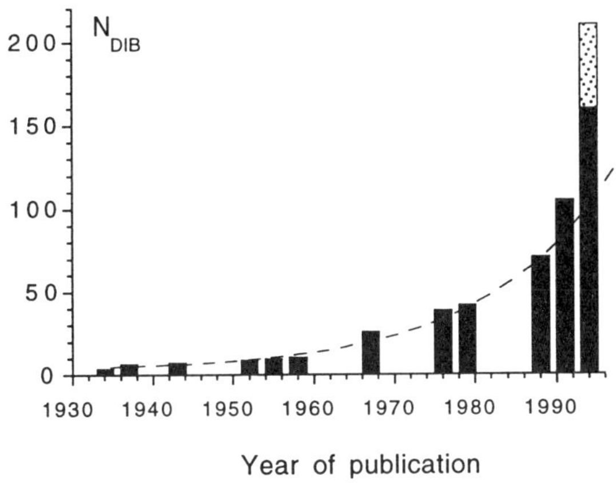

Fig. 1. The number of known DIBs versus year of publication. Recent surveys have caused a hyper-inflation of new discoveries.

DIBs is now about 160 and rising. Figure 2 is a composite DIB spectrum of all certain and probable identifications.

3.1. *LOCATION IN THE SPECTRUM*

Figure 3 shows the distribution of DIBs in a diagram of FWHM versus band position (λ), where the energy in electron volt E is equal to: $E(eV) = 1.2396/\lambda(\mu m)$. Vertical bars are proportional to the equivalent width (W) of the bands. Figure 3 is a pseudo energy diagram (Herbig 1975, Herbig & Leka 1991) if it is assumed, as is likely, that all observed electronic transitions occur from ground states that are within a few cm^{-1} of zero. The wavenumber of a DIB is then proportional to the energy level of an excited state. The distribution of transitions, shown in the right section of Fig. 3, could represent the distribution of states in a sample of large interstellar molecules, ions, or radicals, for example.

The DIBs are not homogeneously spread throughout the spectrum, but are found mainly between 1.56 and 2.25 eV (5500-7940Å). The presence of residual telluric lines does not fully account for the lack of DIBs above 7940 Å . The windows below 5500 Å are remarkably devoid of narrow DIBs (Herbig 1975), a conclusion that is unaffected by the presence of steller lines. No new narrow DIBs were identified with certainty in this part of the spectrum in a star like HD30614, which has few and very broad photospheric lines. While several very broad DIBs are found below 5500 Å, there is no concentration of wide DIBs at short wavelengths as expected if some neutral molecules contribute significantly to the DIB spectrum. There is a relative absence of DIBs below 4430 Å (Herbig 1967). The first (broad and weak) DIB in this region has now been found. It is at 4176 Å. Four more were announced at this conference (Hibbins *et al.* 1994).

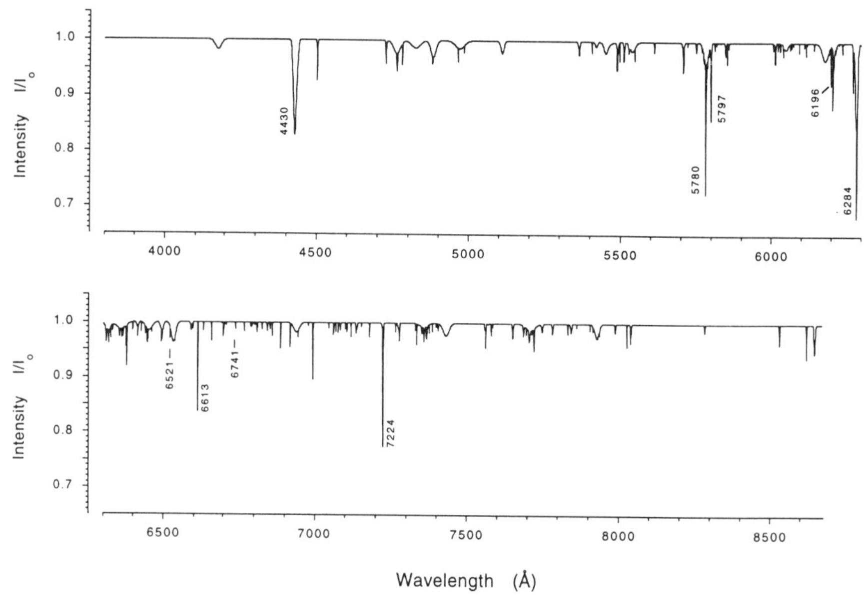

Fig. 2. The Diffuse Interstellar Band spectrum. The graph is a composite of gaussian profiles fitted to all certain and probable DIBs that are found in the DIB survey and forms what could be regarded as a photometric tracing of the DIB spectrum normalised to unity. This spectrum is thought to be characteristic of diffuse medium matter.

3.2. THE BAND WIDTH

Diffuse bands have a characteristic full width at half maximum (FWHM), which varies little from one line of sight to the other. Reported variations of FWHM, after allowing for the Doppler structure of interstellar clouds in the line of sight, are small, perhaps not significant, and could be related to different rotational band populations (Snell & Vanden Bout 1981, Smith *et al.* 1981).

The distribution of FWHM (Figure 3) has two, or probably three, components. For the purpose of this paper, we introduce three broadness classes based on this diagram: DIBs that are between 1-3 cm^{-1}, between 3-15 cm^{-1}, and between 15 and 200 cm^{-1}, and refer to these as narrow, broad, and very broad.

Few DIBs are as very broad as what could be detected in our 200 Å windows. The distribution falls off above FWHM = 20 Å. Five DIBs have FWHM in the range 30-40 Å. Note that the very broadband structure in the extinction curve (Hayes *et al.* 1973) does not behave like DIBs to environmental changes and, therefore, is probably not due to a collection of very broad DIBs (Jenniskens 1994).

Few DIBs are as narrow as the spectral resolution of the survey. No more than 25 DIBs have FWHM < 0.70 Å and most of these are very weak, suggesting that the FWHM is uncertain and may be underestimated. λ6196 is the narrowest of the strong DIBS with an intrinsic width of only 0.4 Å (Smith *et al.* 1981).

There is a marked difference in the relative strength of narrow and very broad

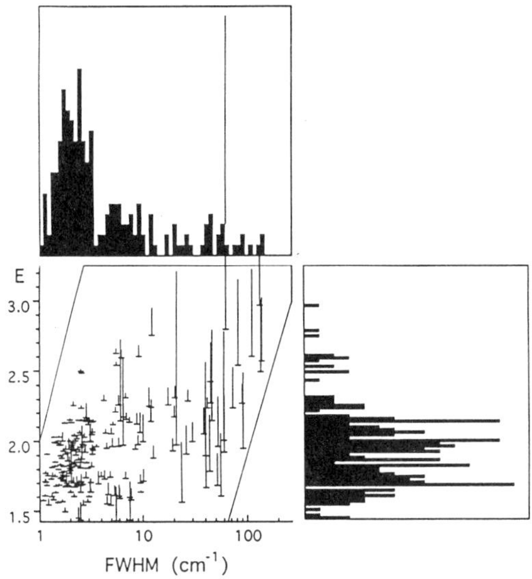

Fig. 3. Pseudo energy diagram, showing the location of bands (in units of eV) versus their full width at half maximum (FWHM) in the parameter space covered by the DIB survey. Histograms of the energy and FWHM distributions are given. The FWHM distribution has two or perhaps three peaks, while the energy distribution has a broad maximum between 1.56 and 2.25 eV.

DIBs. Many of the broadest DIBs are much more intense than most of the narrow DIBs. This is shown more clearly in Figure 4, which plots the equivalent width (W) versus FWHM. There are no narrow DIBs that are as strong as the typical very broad DIB.

The summation of all equivalent width results in a total of $W_{DIB}/E_{B-V} = 22$ Å, compared to 13.5 Å for the DIBs listed in Herbig (1975).

3.3. SEQUENCES

No significant variations in the position of DIBs have been reported after allowing for cloud radial motion as evident in the KI or NaD lines.

Several attempts have been made to find patterns in the distribution of DIBs in the spectrum. Herbig (1975) and Herbig & Leka (1991) found no sequences based on wavenumber regularities, allowing for anharmonicity, for gaps due to missed lines, and for positional uncertainties, if continuity in width or strength along the sequence is demanded. Such sequences are expected if a series of simply spaced vibronic levels are excited from a single ground state.

We attempted to find sequences of the form $E(n) = a + b/n$ (n=i,i+1,i+2,i+3 and i an integer between 1 and 6), where a and b are free variables. Such a relationship holds, approximately, for polyynes where each additional acetylene group added to the chain (n-1 $\rightarrow$ n) increases the absorbed energy to $E = 0.415 + 8.9/n$ eV (Braga *et al.* 1991). No such series was found, allowing for uncertainties in wavelength positions.

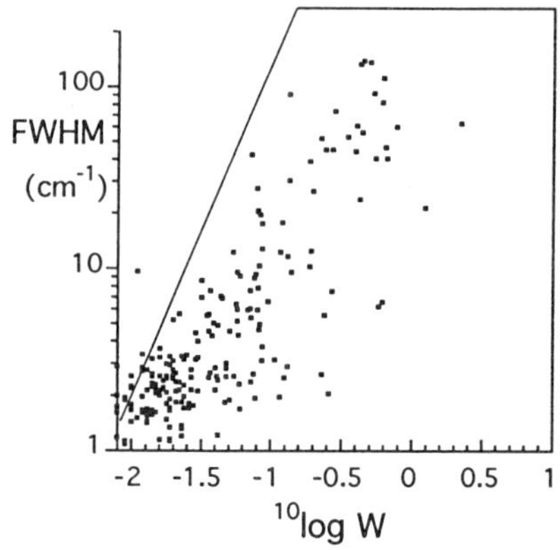

Fig. 4. The distribution of band width (FWHM) versus band strength in terms of equivalent width (W). Approximate detection limits are indicated. There are very few narrow DIBs as strong as the typical very broad DIB.

Another approach is to look at the distribution of energy differences, assuming that transitions are possible between all levels in the pseudo energy diagram. Figure 5 shows this distribution of all possible energy differences for the three broadness classes of DIBs. Herbig & Leka (1991) reported that the number of differences is high between 1 and 35 cm^{-1} (>280 micron) and the distribution has a broad peak between 900-1100 cm^{-1} (9-11 micron) and attributed these effects correctly to artifacts of the concentration of many DIBs in the 7400, 6800, and 6300 Å regions in combination with gaps caused by the difficulty of detecting DIBs in some regions with heavy telluric absorptions. Our more complete DIB survey does not avoid the problems with telluric lines completely but the nearly complete coverage removes the spurious peak in the frequency distribution in the range 1-35 cm^{-1}. The broad peak at 2-7 micron (or 9-11 micron in Herbig & Leka) merely reflects the wavelength range in which DIBs have been found and the different position of the peak for different broadness classes is due to the different ranges in which these bands are found. There are some similarities in the 5780 and 6284 windows with spacings of 1400 cm^{-1} (Cohen & Jones 1987, Jenniskens & Désert 1993), but the DIB spectrum has no significant concentrations of energy differences near the UIB band positions at 3.4, 6.2, 7.7, and 11.3 micron. Note that Figure 5 does show a spike for the interval 3.43-3.48 micron for the broadness class of $3 < \mathrm{FWHM} < 15$ cm^{-1}, but none of these pairs is convincing when demanding a constant FWHM for both bands and decreasing intensity to higher wavelengths.

Although the DIB spectrum is not dominated by vibrational sequences, some particular cases may still be relevant. Notably, Herbig (1988) found a series of narrow DIBs in the window 6770-6862 Å. However, these features are not exactly equally spaced (resp. 34.43, 34.59, 35.48, and 38.07 cm^{-1}) and, there is no extension to other bands at higher and lower frequency. Jenniskens & Désert (1993) reported a similar series of nearly equally spaced DIBs superposed on the broad λ5778, which was confirmed by Krelowski & Sneden (1993). However, that also provides no evidence for a vibrational progression of low frequency or for excited rotation-vibrational structure within the lowest electronic state.

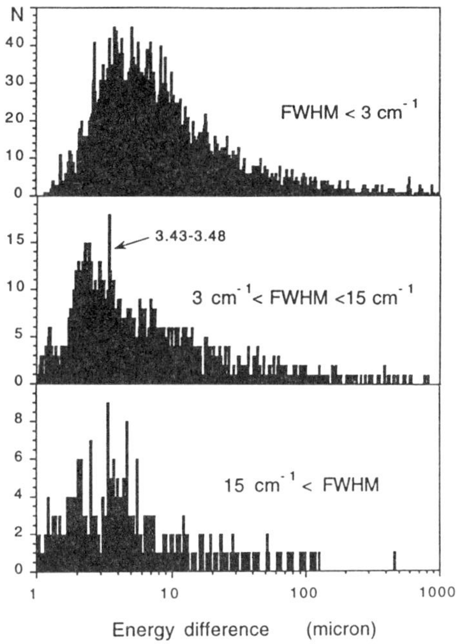

Fig. 5. The infrared spectrum that would arise if transitions were possible between all levels of the pseudo energy diagram of Figure 3. The figure is a histogram of the number of differences between wavenumber positions for the three broadness classes with FWHM as indicated.

3.4. SHAPES OF BANDS AND SUBSTRUCTURE

It is possible to identify a number of different band shapes. Narrow DIBs show a range in asymmetries with a shading towards the red. The shading is particularly strong in λ5797 and λ6614 but weaker in λ6196, for example. No asymmetries are found with a shading towards the blue.

The same is true for very broad DIBs, although the more symmetric shapes seem to dominate. A group of highly asymmetric, red shaded, very broad DIBs has nearly constant width in wavelength, not in frequency, and is grouped parallel to the right border of the box of Figure 3.

Some broad DIBs have a narrower component superposed on a broader feature (Figure 6) or a triangular instead of a gaussian shape. These DIBS often have a slight asymmetry towards the red also, but asymmetries towards the blue are found too. Two bands in this class, λ5780 and λ6284, show a surprising substructure (Jenniskens & Désert 1993). Figure 6 shows both band contours in the spectrum of HD21389. To guide the eye, stellar lines and DIBs that show deviating relative band intensities in different lines of sight have been removed. The full spectrum is shown in Jenniskens & Désert (1993).

Among band shapes there is a conspicuous absence of resolved rotational-vibrational (rovib) bands. Our DIB survey does not contain clear cases of Type A rovib bands (with a strong Q branch) and only a few possible cases of Type B (no Q branch), but these resemble two nearly overlapping DIBs. Herbig (1975) and Herbig & Leka

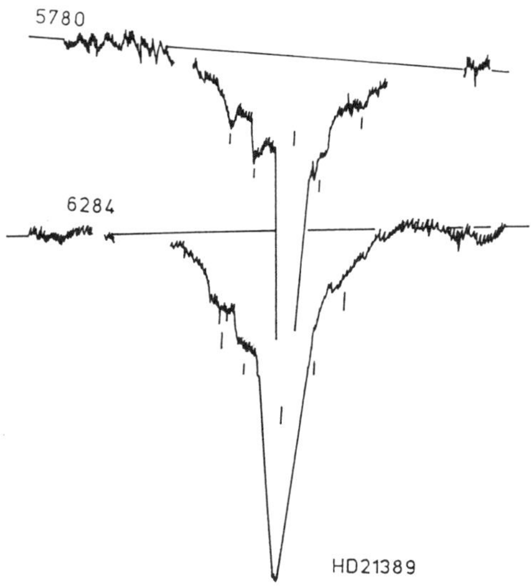

Fig. 6. The substructure in the bands of λ5780 and λ6284. The band shapes are unusual and show a number of features that are shared by both bands (marked). Stellar lines and features that do not correlate with the strength of either band have been removed (from: Jenniskens & Désert 1993).

(1991) found fifteen bands occurring in pairs with a narrow DIB located on the long-wavelength edge of the broader band (e.g. λ5778/λ5797 and λ6177/λ6203). As a rule, the narrow and broad components do not correlate in a large enough sample of lines of sight, and it is not clear whether these bands together form a single rovib band.

3.5. FAMILIES

There are significant variations of band strength from one line of sight to another (Snow 1973, Herbig 1975), where most of the well known DIBs can be grouped in three families with representative members λ5797, λ5780, and λ4430 respectively (Chlewicki *et al.* 1986, Krelowski & Walker 1987, Benvenutti & Porceddu 1989, Josafatsson & Snow 1987, Porceddu *et al.* 1991). DIBs in a family correlate better with members of their own group than with members of another group. These families of bands correlate with DIB band behavior in IRAS sources with strong Unidentified Infrared Band emission (Cohen & Jones 1987, Cohen *et al.* 1989) and with the bands identified in emission in the Red Rectangle (Sarre 1991, Fossey 1991). Notably, the 5797 family is the only group of DIBs that can be related to the yellow and red emission bands (Miles & Sarre 1993).

It is often mistakenly assumed that the family members are necessarily due to a single carrier because, at first sight, they behave as a single spectrum. In fact, most DIBs within a family show significant relative intensity variations from one line of sight to another. Figure 7 summarizes the information available for those band pairs that have been studied. All combinations that do not tightly correlate are highlighted

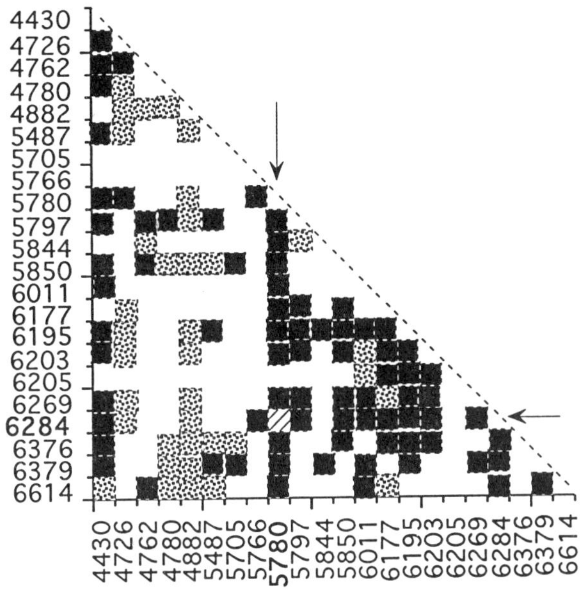

Fig. 7. Family behaviour is lost if one insists on a perfect correlation of band intensities. Black entries mark the band pairs that show significant band strength variations in the lines of sight studied. Dotted black entries possibly also do not correlate. Only one entry (marked by arrows and shaded) may represent two bands that perfectly correlate and perhaps are part of the same (or closely related) molecule(s).

as black (or dotted when less certain). Only one case exists where two DIBs do show exactly the same ratio in most lines of sight (Jenniskens & Désert 1993). These are λ5780 and λ6284 (marked by arrows). However, there is some evidence that the line ratio of these two bands varies significantly in the extreme line of sight to HD37022. In conclusion, true cases of perfect correlations may exist but have not yet been reported. The message is clear: adding several bands into a complex spectrum to account for a single carrier is bound to be incorrect and should be done with care if the band ratios are thought to be carrier specific.

The broadness classes, identified earlier in this paper, seem to be identical to the DIB families, where very broad, broad, and narrow DIBs correspond to the λ4430, λ5780, and λ5797 family respectively. The most obvious exception, at first sight, is that the very broad λ5778 (Herbig 1975) correlates better with the broad λ5780 than with the very broad λ4430. However, it was recently shown that λ5778 is part of the λ5780 profile and not an independent DIB (Jenniskens & Désert 1993). A second deviating case is the strong and narrow DIB λ6195, which is not seen in emission in the Red Rectangle and does not belong to the "narrow" λ5797 family. However, λ6196 does not correlate well either with the two other families and may form a separate family. This DIB is also exceptionally narrow and thus could similarly constitute a fourth broadness class.

4. DISCUSSION

How do all these observations relate to the macromolecule hypothesis of DIBs?

The absence of DIBs narrower than 0.4 Å may indicate that the intrinsic width of the narrowest DIBs is due to rotational broadening (Danks & Lambert 1976). The strong red-degrading in some of these band countours suggests that there is significant geometry change during excitation and an increase in moment of inertia. However, this change should be small enough not to excite a vibrational squence. The broad and very broad DIBs are probably broadened mainly by lifetime broadening. The profile shapes and widths are consistent with non-radiative coupling rates in macromolecules (Leach 1989; 1994). Structure in the density of states of the electronic state to which the energy is leaked can perhaps account for the structure observed in the λ5780 and λ6284 profiles.

The DIB spectrum seems to be dominated by vibrationless transitions, because there is a remarkable absence of vibrational sequences. Another argument for the preponderance of vibrationless transitions is the one-on-one correspondence between emission and absorption for the small number of DIBs seen in emission. Such vibrationless transitions are not unexpected. Note that laboratory results of cold large molecular ions similarly do not show a preponderance of vibrational sequences and often there are only one or two strong absorptions for any given molecule (e.g. Shida & Iwata 1972, Fullara *et al.* 1993). Reason for this is that the bond strengths do not change much in an electronic excitation and Franck-Condon factors are small.

The absence of DIBs in the near-UV excludes many large neutral molecules as the carrier, but perhaps not those that have large conjugated systems or electron donors and acceptors as in organic-metal complexes. Radicals and ions with an open shell are possible candidates, amongst which are the PAH ions, inferred to be ubiquitous in space from mid-infrared observations (Van der Zwet & Allamandola 1985, Crawford et al 1985, Leger & d'Hendecourt 1985). We recently did find some evidence for an ion-like behavior of two DIBs, those at 6196 and 6284 Å, where λ6196 seems to be ionized and double ionized by photons of higher energy than λ6284 (Jenniskens *et al.* 1994). Herbig's (1993) argument for neutral molecules does not hold if all matter in the line of sight is considered and not only HI.

The DIB transitions of highest energy (3.0 eV) are at the intriguing value of about half the energy associated with absorptions that give rise to the bump at 2180 Å, i.e. 5.69 eV. These absorptions are thought to be due to $\pi \rightarrow \pi^*$ transitions in large aromatic molecules (or small carbonaceous grains). Perhaps, the DIBs are due to the ionized version of the bump carrier.

The existence of broadness classes, and the observation that the very broad DIBs are found at higher frequency than the narrow DIBs, may bear some relation to a feature of PAH ions pointed out by Salama (1994). Their absorption spectra are dominated by either one of two types of transitions: $\pi \rightarrow \pi$ (as for the naphthalene cation) or $\pi \rightarrow \pi^*$ (as for the pyrene cation). Salama found that the $\pi \rightarrow \pi^*$ type of transition is typically at shorter wavelength and has a much larger oscillator strength. We can add to this that, because $\pi \rightarrow \pi^*$ is a transition to higher energy, it is prone to stronger non-radiative coupling, resulting in broader bands. This distinction in transition type may account for the different strength and width of narrow DIBs on one side and broad and very broad DIBs on the other.

TABLE 1

Three broadness classes of DIBs

Class	FWHM cm^{-1}	N	Σ W	$<W>$	f	% C	$<\%C/N>$
narrow	1-3	130	9.8	0.037	$\sim$ 0.0043	8.5	0.06
broad	3-15	65	15.8	0.124	$\sim$ 0.014	4.2	0.06
very broad	15-200	30	41.7	0.852	$\sim$ 0.1	1.6	0.05

The summation of all equivalent width for each broadness class is listed in Table 1. Assuming oscillator strengths that scale with the median value of W and cover the typical range of PAH ions (f = 10^{-4} to 0.6; Parisel *et al.* 1992), we find that the carbon locked up in DIB carriers may well be spread over all three broadness classes, a conclusion similar to the conclusion by Salama & Allamandola (1992) that the naphthalene ion, if it accounts for λ6741 and λ6521, has a similar abundance as the pyrene cation if that accounts for the strong λ4430. In conclusion, weak and strong DIBs may well be equally important in the scheme of Nature.

5. CONCLUSION

Our current knowledge of the DIB spectrum in absorption in the interstellar medium can be summarized as follows:

- Total number of known DIBs is some 160 and rising.
- Reported variations in position and width of the absorption bands have not yet been established.
- DIBs concentrate between 1.56 and 2.25 eV; few narrow DIBs above 2.25 eV; and few very broad DIBs above 3.0 eV.
- Very few DIBs are narrower than 0.4 Å and few DIBs are broader than 20 Å .
- The FWHM distribution shows three maxima suggesting that there are three broadness classes of DIBs: very broad (15-200 cm^{-1}), broad (3-15 cm^{-1}), and narrow DIBs (1-3 cm^{-1}). These broadness classes bear some relation to the DIB families.
- Many DIBs, if not all, show some form of red shading. Only rarely is there an asymmetry towards the blue.
- No sign of resolved rovib bands, yet.
- There are no obvious vibrational sequences in the DIB spectrum nor sequences that suggest a progressive addition of conjugating groups.
- There is no preference in the pair distribution of energy differences and no unidentified infrared bands in the pseudo infrared spectrum.
- No DIBs are yet known that correlate perfectly in intensity from one line of sight to another.

- Up to 10% of cosmic carbon may be locked up in the DIB carriers with no preference for any of the broadness classes.

The identification of DIBs is still an open question. The observations are consistent with a collection of vibrationless electronic transitions from at least a hundred large molecular ions or radicals. Broad and very broad DIBs are broadened by a form of lifetime broadening, while the narrow DIBs may include a signature of an unresolved rotational-vibrational profile.

ACKNOWLEDGEMENTS. We thank D. Gillet and coworkers of O.H.P., H. Habing of Leiden Observatory, and M. Dennefeld of I.A.P., as well as P. Ehrenfreund and B. Foing, for their support of this project. Part of this work was performed while PJ held a National Research Council - ARC Research Associateship.

REFERENCES

Adamson A.J., Whittet D.C.B., Duley W.W., 1991, *MNRAS*, **252**, 234.
Benvenuti P., Porceddu I., 1989, *Astr. Ap.*, **223**, 329.
Braga M., Larsson S., Rosen A., Volosov A., 1991, *Astr. Ap.*, **245**, 232.
Chlewicki G., Van der Zwet P., Van IJzendoorn L.J., Greenberg J.M., Alvarez P.P., 1986, *Ap. J.*, **305**, 455.
Chlewicki G., de Groot M.S., van der Zwet G.P., Greenberg J.M., Alvarez P.P., Mampaso A., 1987, *Astr. Ap.*, **173**, 131.
Cohen M., Jones B.F. 1987 *Ap. J.*, **321**, L151.
Cohen M., Jones B.F., Walker H.J. 1989, *Ap. J.*, **341**, 908.
Crawford M.K., Tielens A.G.G.M., Allamandola L.J. 1985, *Ap. J.*, **293**, L45.
Danks A.C., Lambert D.L., 1976, *MNRAS*, **174**, 571.
Duley W.W. 1982, *Astroph. & Space Sci.*, **88**, 501.
Ehrenfreund P., d'Hendecourt L., Verstraete L., Léger A., Schmidt W., Defourneau D., 1992 *Astr. Ap.*, **259**, 257.
Fossey S.J., 1991, *Nature*, **353**, 393.
Fulara J., Lessen D., Freivogel P., Maier J.P. 1993, *Nature*, **366**, 439.
Gredel R., Münch G., 1986, *Astr. Ap.*, **154**, 336.
Hayes, D.S., Marko, G.E., Radick, R.R., and Rex, K.H. 1973, in *Interstellar Dust and Related Topics*, eds. J.M. Greenberg and H.C. Van de Hulst (Dordrecht:Reidel), p. 83.
Herbig G.H. 1967, in *Radio Astronomy and the Galactic System*, IAU Symp. 31, ed. H. van Woerden, p. 85.
Herbig G.H. 1975, *Ap. J.*, **196**, 129.
Herbig G.H. 1988, *Ap. J.*, **331**, 999.
Herbig G.H., Leka K.D. 1991, *Ap. J.*, **382**, 193.
Herbig G.H., 1993, *Ap. J.*, **407**, 142.
Hibbins R.E., Miles J.R., Sarre P.J., Herbig G.H., 1994, in *The Diffuse Interstellar Bands: Contributed Papers*, NASA Conf. Publ. 10144, p. 31.
Hollas J.M., 1982, High Resolution Spectroscopy, Butterworths.
Jenniskens P., Désert F.-X. 1992, in *Organic matter in interstellar extinction*, PhD Thesis, Leiden University, p. 95.
Jenniskens P., Désert F.-X. 1993, *Astr. Ap.*, **274**, 465.
Jenniskens P., Ehrenfreund E., Foing B. 1994, *Astr. Ap.*, **281**, 517.
Jenniskens P. 1994, *Astr. Ap.*, **284**, 227.
Jenniskens P., Désert F.-X. 1994, *Astr. Ap. S.*, **106**, 39.
Joblin C., Maillard J.P., d'Hendecourt L., Léger A., 1990, *Nature*, **346**, 729.
Johnson F.M., 1967, *Astron. J.*, **72**, 305.
Josafatsson K., Snow T.P. 1987, *Ap. J.*, **319**, 436.
Krelowski J., Walker G.A.H. 1987, *Ap. J.*, **312**, 680.
Krelowski J., Sneden C., 1993, *PASP*, **105**, 1141.
Leach S., 1989, in *Interstellar Dust*, L.J. Allamandola and A.G.G.M. Tielens (eds.), p. 155.
Leach S., 1994, at this conference.
Lee W., Wdowiak T.J., 1994, *Ap. J.*, (in press).
Léger A., d'Hendecourt L. 1985, *Astr. Ap.*, **146**, 81.

Manning P.G., 1970, *Nature*, **226**, 829.
McKellar A., Welsh H.L., Stephenson F.C., 1955, *Astron. J.*, **60**, 170.
Merrill P.W. 1934, *PASP*, **46**, 206.
Merrill P.W. 1936, *Ap. J.*, **83**, 126.
Miles J.R., Sarre P.J. 1993, *J. Chem. Soc. Faraday Trans*, **89**, 2269.
Parisel O., Berthier G., Ellinger Y. 1992, *Astr. Ap.*, **226**, L1.
Salama F., Allamandola L.J. 1992, *Nature*, **358**, 42.
Salama F., Allamandola L.J., 1992, *Ap. J.*, **395**, 301.
Salama F., 1994, at this conference.
Sanner F., Snell R., Vanden Bout P., 1978, *Ap. J.*, **226**, 460.
Sarre P.J., 1991, *Nature*, **351**, 356.
Shida T., Iwata S., 1972, *J. Am. Chem. Soc.*, **95**, 3473.
Smith W.H., Snow T.P., York D.G., 1977, *Ap. J.*, **218**, 124.
Smith W.H., Snow T.P., Jura M., Cochran W.D., 1981, *Ap. J.*, **248**, 128.
Snell R.L., Vanden Bout P.A., 1981, *Ap. J.*, **244**, 844.
Snow T.P., 1973, *Astron. J.*, **305**, 455.
Snow T.P., Cohen J.G., 1974, *Ap. J.*, **194**, 313.
Snow T.P., York D.G., Welty D.E., 1977, *Astron. J.*, **113**, 113.
Stoeckly R., Dressler K., 1964, *Ap. J.*, **139**, 240.
Van der Zwet G.P., Allamandola L.J., 1985, *Astr. Ap.*, **146**, 76.
Van der Zwet G.P., 1987, in *Polycyclic Aromatic Hydrocarbons and Astrophysics*, A. Léger *et al.* (eds.), p. 351.
Wdowiak T.J., 1980, *Ap. J.*, **241**, L55.
Wdowiak T.J., Beegle L.W., Lee W., Robinson M.S., 1994, in *The Diffuse Interstellar Bands: Contributed Papers*, NASA Conf. Publ. 10144, p. 65.
Wu C.-C., 1972, *Ap. J.*, **178**, 681.
Wu C.-C., York D.G., Snow T.P., 1981, *Astron. J.*, **86**, 755.

REVISITING DIFFUSE INTERSTELLAR BANDS IN THE SATELLITE ULTRAVIOLET WITH THE GODDARD HIGH RESOLUTION SPECTROGRAPH

JASON A. CARDELLI
University of Wisconsin, Madison, WI

ABSTRACT. Given the number and diversity of diffuse interstellar bands (DIBs) in the optical, we might expect many similar features in the satellite UV. However, despite considerable effort over the years, analysis of data from both *Copernicus* and *IUE* (resolutions of 12 and 25 km/s, respectively) have produced very little in the way of convincing results, with of course the notable exception of the 2175 Å feature. However, 3.5 km/s resolution echelle observations with the *Goddard High Resolution Spectrograph (GHRS)* aboard the *Hubble Space Telescope* has led to the identification of a probable weak DIB at 1369.1 Å in the spectrum of ζ Oph (Tripp, Cardelli, and Savage 1994). This conclusion was reached only after careful consideration of other possible origins of the feature like unidentified atomic (neutral and ionized) and molecular features, including a comparison to *GHRS* echelle data of ξ Per in which the feature is not seen.

The observed 1369.1 Å feature is significantly broader than the atomic and molecular features observed toward ζ Oph and is noticeably asymmetric to the red. Although narrower in velocity width than optical DIBs, the width of the feature in energy units, a more relevant basis for comparison than velocity width, is very similar to DIB 5797 Å and is twice as broad DIB 6196 Å observed toward ζ Oph.

If UV DIBs are typically narrow in energy width, this may be a major reason they have been difficult to identify in the past. Such features would be at or below the resolution limit of *Copernicus* and especially *IUE* (a far-UV DIB with an energy width similar to DIB 5780 Å toward ζ Oph and even HD 183143 would be at the resolution limit of *IUE*) and would subsequently be characterized as unidentified lines. Furthermore, if UV DIBs are also typically weak (DIB 1369.1 Å toward ζ Oph has an equivalent width of 5.9 mÅ), this places an additional constraint on their detection in existing data, especially *IUE*.

1. INTRODUCTION

Given the number and diversity of diffuse interstellar bands (DIBs) in the optical, one might expect many similar features in the satellite ultraviolet. However, despite considerable effort over the years, little evidence for an ultraviolet DIB shortward of 4400 Å (energies greater than 2.8 eV) has been found.

A. G. G. M. Tielens and T. P. Snow (eds.), The Diffuse Interstellar Bands, 53–63.

In an analysis of data (R $= \lambda/\Delta\lambda = 25{,}000$; 12 km/s resolution) from the *Copernicus* satellite in several sight lines, including ζ Oph, Snow, York, and Resnick (1977) reported a possible DIB at 1416 Å with a central depth of about 8% of the continuum. However, their analysis excluded the existence of any DIBs in the wavelength range 1114 to 1450 Å with strengths greater than 10% for sight lines with $E(B-V) = 0.3$.

From low dispersion (R $= \lambda/\Delta\lambda = 300$; 900 km/s resolution) data obtained with the *International Ultraviolet Explorer*, Seab and Snow (1985) searched for UV DIBs in an expanded data set of 58 moderately reddened stars (e. g. $0.14 < E(B-V) < 1.13$). Through an analysis that centered on extinction curves derived via the pair method, they found no evidence for small scale structure, placing 2σ limits on central depths of less than 10% for $1150 < \lambda < 2000$ Å and 16% for $2000 < \lambda < 2300$ Å.

Carnochan (1989) identified a broad shallow feature centered near 1700 Å from and analysis of extinction obtained from low dispersion *IUE* data. Unfortunately, this result has never been independently confirmed. The origin of this structure is likely stellar mismatch, which can be a serious problem in this general wavelength region, especially for luminous stars (see Cardelli, Sembach, and Mathis 1992). A famous example of this can be seen in the extinction curve of Savage and Mathis (1979) near 1600 Å.

In this paper, we revisit UV DIBs by examining the issue of their possible nature through comparison to optical analogs and explore reasons why previous attempts may have been unsuccessful. In addition, we present high dispersion (R $= \lambda/\Delta\lambda =$ 90,000; 3.5 km/s resolution) data of a probable DIB obtained with the *Goddard High Resolution Spectrograph (GHRS)* aboard the *Hubble Space Telescope*. This feature, which is comparable (see below) to a number of well known optical DIBs, would not have been identifiable in the previous studies discussed above.

2. ASSESSING THE NATURE OF ULTRAVIOLET DIBS THROUGH OPTICAL ANALOGS

The lack of UV DIBs might result from the fact that the carriers of the optical features have dissociation energies that are less than 3-4 eV. However, this is difficult to understand considering that comparatively strong optical DIBs are seen in sight lines with relatively low reddening/density (Benvenuti and Porceddu 1989) and showing evidence of processed dust (e. g. HD 167756; Cardelli, Sembach, and Savage 1994) where UV radiation is abundant. Assuming they should exist, a more plausible explanation is that UV DIBs are comparatively narrower and weaker than in the optical and that limitations in the capabilities of previous spectrographs (i. e. resolution and/or signal-to-noise) or analysis techniques have prevented their detection or identification.

2.1. ENERGY VERSUS VELOCITY WIDTH

When considering a potential UV DIB characteristic line width, it is important to remember that the analog features in the optical are quite broad. As their name implies, most optical DIB transitions are band-like in that their velocity widths typically exceed the Doppler (thermal plus turbulent) broadening for the ISM atomic absorption lines seen in the same sight line by a substantial amount. Since the widths of

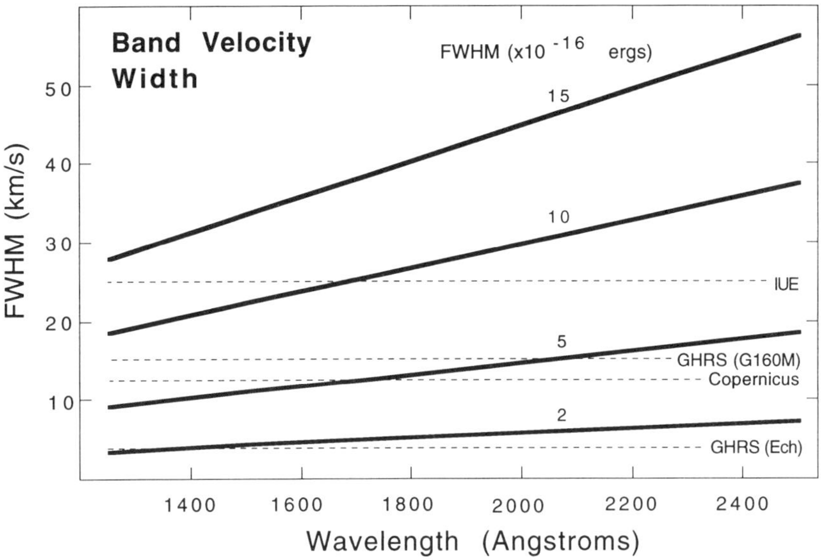

Fig. 1. The velocity FWHM (calculated from Eqn. 2) expected for UV DIB transitions with intrinsic upper energy level widths (x10^{-16} ergs) of 2, 5, 10, and 15, plotted as a function of UV wavelengths from 1300 Å to 2500 Å. The resolution limits of various UV spectrographs are indicated as dashed lines.

most DIBs are dominated by intrinsic (natural) broadening of the upper energy levels, it is incorrect to assume a constant velocity width as a function of wavelength as is done for atomic features (where the natural broadening for all levels is extremely small). Consequently, in considering the potential velocity width of a UV DIB from an optical analog, we must consider the width of the feature in equivalent energy units. The relationship between energy and velocity width is derived from $E = hc/\lambda$ and takes the form

$$FWHM(velocity) = (\lambda/h)FWHM(energy) \quad (1)$$

or

$$FWHM(km/s) = 1.509\text{x}10^{13}\lambda(\text{Å})FWHM(ergs) \quad (2)$$

Consider for example a fictitious DIB at 6000 Å with a velocity width of 50 km/s and therefore an energy width of 5.5x10^{-16} ergs (Eqn. 2). An analog UV DIB with this energy width located at 1400 Å would have a subsequent velocity width of about 12 km/s. Such a DIB would be at the resolution limit of *Copernicus* and a factor 2 below that of the high dispersion mode ($R = \lambda/\Delta\lambda = 12{,}000$; 25 km/s) of *IUE*.

The above is generalized in Figure 1 where we show the velocity widths that UV DIBs would have for a range of energy widths characteristic of a number of optical DIBs including DIB 6196, 5797, 6203, 6270, 5780, and 6284 [energy widths (x10^{-16} ergs) of 2.6, 4.8, 6, 7, 12, and 15, respectively; Benvenuti and Porceddu 1989]. While Figure 1 indicates that some of these optical analogs would be resolvable in high dispersion (25 km/s) *IUE* data, there are many that would only be at or near the resolution limit of *Copernicus*. Therefore, they would not be distinguishable from atomic absorption lines, unlike the case for the many broad, asymmetric optical DIBs. In addition, some would be narrow enough that their widths could be comparable to or dominated by turbulent broadening in the interstellar gas.

2.2. *EQUIVALENT WIDTHS*

In addition to velocity width, the expected strength of a UV DIB will ultimately determine its detectability. For the 6 optical DIBs analyzed by Benvenuti and Porceddu (1989), the typical equivalent widths, W_λ, range between about 50 and 500 mÅ. For an absorption feature, the relationship between column density and equivalent width, *assuming no line saturation*, is

$$N(cm^{-2}) = 1.13\text{x}10^{20} W_\lambda / f\lambda^2 \quad (3)$$

where $f =$ transition oscillator strength (f-value), and W_λ and λ are in units of Å. If we adopt these optical DIBs as analogs and assume that the UV DIB carriers are the same material such that they correspond to the same column density, N, and that the transitions have the same f-values (i. e. N$f =$ constant), then

$$W_\lambda(UV) = W_\lambda(optical)[\lambda(UV)/\lambda(optical)]^2 \quad (4)$$

The application of Eqn. 4 is shown graphically in Figure 2, again assuming a fictitious

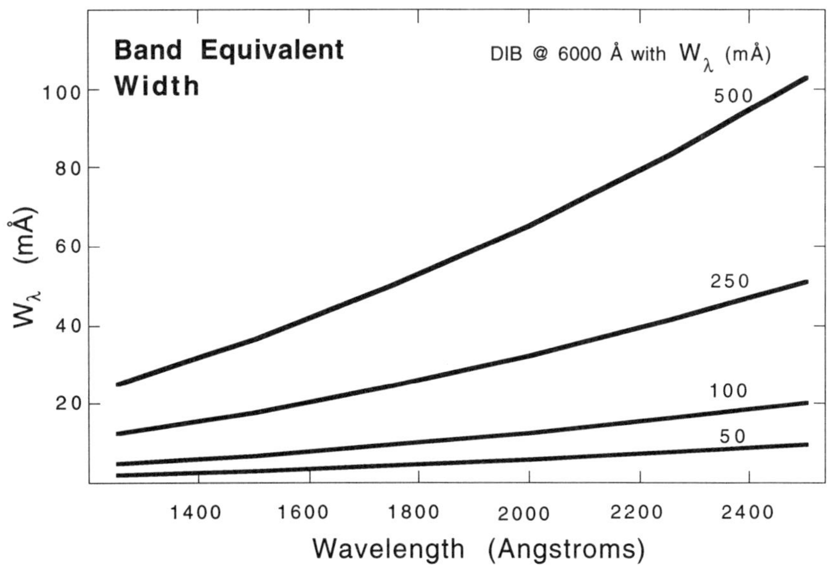

Fig. 2. The expected equivalent widths that UV DIBs would have if they had the same characteristics as some fictitious DIB at 6000 Å (i. e. we assume that the column density and f-value for the UV features are the same as the optical feature). The curves were calculated from Eqn. 4 for four values of W_λ assumed to characterize our fictitious DIB at 6000 Å.

optical DIB at 6000 Å with a range of equivalent widths from 50 to 500 mÅ. As can be seen, the *UV DIBs would be considerably weaker than their optical analogs*, especially at the shorter wavelengths.

2.3. *ON THE LACK OF SUCCESS IN PREVIOUS SEARCHES*

Given the relatively narrow widths and weak absorption strengths that might characterize UV DIBs, especially at the shorter wavelengths, the resolution and signal-to-noise (S/N) characteristics of *IUE* are generally insufficient to detect such features. This is especially true for the study of Seab and Snow (1985) who used the low dispersion mode of *IUE*. While many of these features would be detectable in *Copernicus* data, many would appear as unresolved weak features and likely be cataloged as unidentified (UID) lines. Numerous examples of such UIDs in *Copernicus* data have been noted (see Morton 1975, 1991).

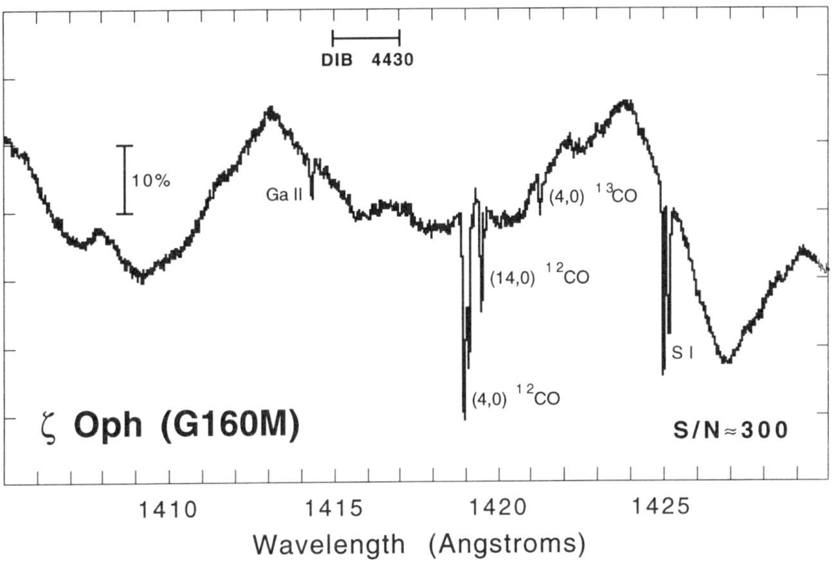

Fig. 3. A high S/N spectrum of ζ Oph obtained with the intermediate dispersion (15 km/s resolution) *GHRS* grating G160M. The data show no evidence of the possible DIB at 1416 Å reported by Snow *et al.* (1977). Aside from the labeled interstellar lines, the structure seen in the continuum can be attributed to stellar absorption. Also shown as a horizontal bar is the width that a DIB comparable to DIB 4430 would have at this wavelength, calculated from the intrinsic energy width of DIB 4430 using Eqn. 2.

Another potential problem in past searches for UV DIBs centers on the analysis methodology itself. For example, in the study of *Copernicus* data by Snow *et al.* (1977), the data were analyzed by deriving extinction curves via the pair method [i. e. comparing a reddened program star to a lightly or unreddened standard star of (nearly) identical spectral type]. While this is a usable technique for extracting information on very broad shallow profiles, it is not a good method for studying narrow weak features. In addition, in order to increase the statistical significance of the data, Snow *et al.* binned the data to an effective resolution of 2 Å, corresponding to velocity widths of 400-500 km/s over the wavelength range they studied. If Figures 1 and 2 are representative of what might be expected for typical UV DIBs, then such a bin size would completely smear out any such features.

Snow *et al.* did note a possible feature at 1416 Å and it is possible that there are UV DIBs that are considerably broader than implied by our scaling of optical analogs. For example, DIB 4430 is considerably larger than the other optical DIBs we consider, with a FWHM = 1350 km/s (20 Å) and $W_\lambda = 3400$ mÅ in the sight line toward HD 183143. From Eqns. 2 and 4, this would correspond to a feature at 1416 Å with a width of 430 km/s and $W_\lambda = 350$ mÅ which would be unresolved, but detectable (e. g. such a feature would have a depth of 18%). However, data obtained by the *GHRS* using grating G160M (15 km/s resolution) in the sight line to ζ Oph, one of the stars analyzed by Snow *et al.*, does not confirm the existence of such a feature. The *GHRS* data, which are shown in Figure 3, show a rather highly structured continuum with numerous interstellar absorption lines. There is some structure at 1416 Å, but it is much weaker than the 8% feature cited by Snow *et al.*. However, this type of structure occurs elsewhere in the spectrum and is likely a weak stellar line. Given the complexity of the ζ Oph spectrum in this region, the apparent feature seen by Snow *et al.* could have resulted from a subtle spectral mismatch with the adopted standard star along with some contribution from ISM CO absorption.

3. THE CASE FOR THE FIRST UV DIB

3.1. GODDARD HIGH RESOLUTION SPECTROGRAPH DATA

If UV DIBs are typically narrow and weak, then they are best studied using the same techniques employed to study atomic and molecular absorption lines (i. e. direct analysis of the spectrum). In this regard, the best strategy is to use data obtained at the highest possible resolution and with a moderate-to-high S/N.

During the *HST* science verification program, high dispersion echelle (R = $\lambda/\Delta\lambda$ = 90,000; 3.5 km/s resolution) observations of ζ Oph were obtained with the *GHRS* for calibration purposes. In one of these observations centered near 1370 Å, several spectral features were seen for which no atomic or molecular identifications were found (Cardelli, Savage, and Ebbets 1991). A portion of this observation is shown in the bottom panel of Figure 4. These data are characterized by a $S/N \simeq 70$.

For 2 of the 3 unidentified lines (UIDs), the widths are similar to those of the atomic lines seen in other *GHRS* data of this sight line (Savage, Cardelli, and Sofia 1992). However, the one feature located at a rest wavelength of 1369.13 Å (derived assuming the feature arises in the principal ζ Oph cloud at a heliocentric velocity of -15 km/s) is considerably broader than any of the other lines. This is shown more clearly in Figure 5 where this feature is compared to the weak absorption line of O I] λ1356 Å. In addition to its width, this feature also exhibits an asymmetry to the red, a property characteristic of many optical DIBs. The behavior exhibited by this feature led Tripp, Cardelli, and Savage (1994) to conclude that this feature must be a DIB (see below).

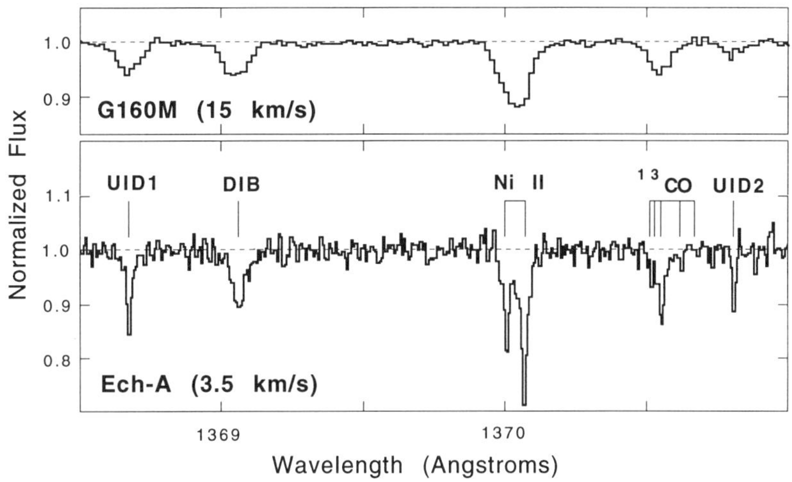

Fig. 4. A portion of the *GHRS* Ech-A (3.5 km/s resolution) spectrum of ζ Oph (bottom panel) in the region of DIB 1369 (Tripp *et al.* 1994). The feature is notable in that it is much broader than the other spectral lines seen in this sight line and is asymmetric to the red (see also Figure 5), a common characteristic among optical DIBs. The top panel shows a spectrum of the same region obtained with the *GHRS* intermediate dispersion (15 km/s resolution) grating G160M. At this resolution, the characteristics that suggest that the feature at 1369 Å is a DIB are not obvious.

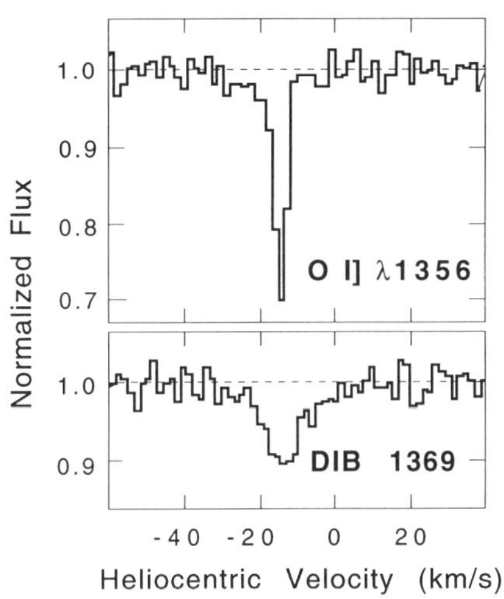

Fig. 5. DIB 1369 from the data in the lower panel in Figure 5 plotted against heliocentric velocity (km/s). The rest wavelength for this feature was calculated assuming that the absorption arises in the ζ Oph principal velocity component at -15 km/s. For comparison, we also show the profile of O I] λ1356 Å also observed with the *GHRS* Ech-A grating (Cardelli *et al.* 1991).

Given the relative narrowness of this feature, this conclusion could have only been reached through data of this resolution. Consider for comparison, the high S/N spectrum in the top panel of Figure 4, obtained with the intermediate dispersion (R $= \lambda/\Delta\lambda = 20{,}000$; 15 km/s resolution) *GHRS* grating G160M. The diagnostics that are important to determining the nature of the feature at 1369 Å (i. e. its width and asymmetry) are not at all readily apparent at this resolution. In fact the feature does not appear substantively different from the other UIDs or the weak ^{13}CO feature also present in the spectrum.

3.2. *COMPARISON TO UV ABSORPTION LINES AND OPTICAL DIBS*

The width and asymmetry exhibited by the feature at 1369 Å strongly suggests that it must arise from a DIB. However, this conclusion was reached by Tripp, Cardelli, and Savage (1994) only after a careful consideration of the feature's observed characteristics in comparison to atomic, molecular, optical DIBs also observed in the ζ Oph sight line.

Table 1

Atoms, molecules, and DIBs observed toward ζ Oph[a]

Species	λ (Å)	FWHM (km/s)[b]	FWHM (10^{-16} ergs)[c]
UID1	1368.74[d]	4.7	2.3
DIB	**1369.13**[d]	**10.7**	**5.2**
UID2	1370.87 [d]	3.1	1.5
Neutrals and Molecules			
Na I	2853.65	5.2	1.2
	2853.85	5.5	1.3
Mg I	1827.94	4.6	1.7
$^{12}CO^{e}$	1367.53	3.8	1.8
$^{13}CO^{e}$	1370.62	4.7	2.3
Dominant Ions			
O I	1355.60	5.5	2.7
Ni II	1370.13	5.2	2.5
Optical DIBs			
	5780.4	94.5	11.0
	5797.0	40.1	4.6
	6195.9	22.6	2.4

[a]All data except the optical DIBs were obtained with GHRS at 3.5 km/s resolution
[b]Line FWHM in velocity units, determined at the features *observed* wavelength
[c]Line FWHM in energy units using Eqn. 2
[d]Rest wavelengths for the UIDs and DIB determined assuming the absorption arises in the principal velocity component located at a heliocentric velocity of -15 km/s
[e]Corresponds to the $A^1\Pi$ - $X^1\Sigma$ R(0) line of the (6-0) band

An example of this broad comparison is shown in Table 1, where we present the observed velocity width as well as the more relevant energy width. In line with our previous discussions, while the observed velocity width of DIB 1369 is quite small, its energy width is actually quite comparable to 2 of the 3 optical DIBs listed in the table.

A similar comparison is also shown in Figure 6, where we plot line equivalent widths as a function of velocity width. The data plotted as filled symbols correspond to the atomic and molecular data from Table 1 for lines observed near 1370 Å. The open symbols correspond to data of 5 optical DIBs scaled to 1370 Å using Eqns. 2 and 4. These data correspond to an average of observations from 26 sight lines obtained by Benvenuti and Porceddu (1989). The error bars correspond to the calculated standard deviations about the average values in both W_λ and energy width. For the sample of 26 stars, $< E(B-V) > \simeq 0.41$ with a standard deviation of 0.2.

As seen in Figure 6, the width of DIB 1369 clearly stands out from the atomic and molecular lines but is quite comparable to 4 of the 5 optical DIBs shown in terms of both width and strength. A direct comparison between the profiles of DIB 1369 and DIB 5797 show them to be nearly identical in shape as well. We conclude that DIB 1369 is a true DIB and is comparable in characteristics to a number of DIBs observed in the optical.

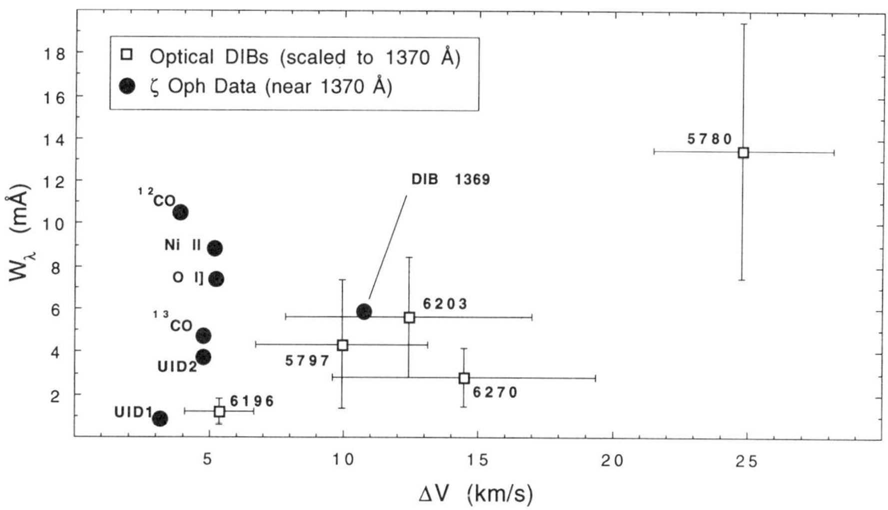

Fig. 6. W_λ plotted against FWHM (km/s) for atomic and molecular features observed with the *GHRS* Ech-A grating in the region near 1370 Å along with the data for the 2 UIDs and DIB 1369 (see Table 1). Also plotted are the values for 5 optical DIBs, scaled to 1370 Å using Eqns. 2 and 4. The optical DIB data represent an average of observations from 26 sight lines (Benvenuti and Porceddu 1989). DIB 1369, which clearly stands out in width in comparison to the other lines observed toward ζ Oph, is quite comparable to 4 of the 5 scaled optical DIBs shown.

4. SUMMARY AND RECOMMENDATIONS

In this paper, we have revisted the issue of diffuse interstellar bands in the satellite ultraviolet. We have concentrated on aspects related to the types of characteristics expected for UV DIBs by employing observed optical DIBs as analogs, and how capable past and present UV spectrographs are for detecting such features. We summarize these discussions as follows:

1) When considered in terms of energy widths, UV analogs of optical DIBs can be expected to be relatively narrow in velocity space and may also be comparatively weak in strength as well.

2) If the intrinsic energy width of the upper level of a UV DIB transition is smaller than optical analogs, in some instances UV DIBs may appear no broader than atomic lines, even at moderate-to-high dispersions.

3) Historical limitations in instrumental resolution and S/N capabilities as well as analysis techniques are most likely responsible for the lack of positive detections of UV DIBs. In some cases, there may be UV DIBs that have been classified as unidentified lines.

4) High resolution (3.5 km/s) data obtained with the *Goddard High Resolution Spectrograph* aboard the *Hubble Space Telescope* in the sight line to ζ Oph has produced the first definitive detection of a UV DIB at 1369 Å.

5) DIB 1369 is relative narrow in width, but is comparable in characteristics (e. g. central depth, energy width, and scaled equivalent width) to optical DIBs such as 5797, 6203, 6270, and 6196.

6) The existence of DIB 1369 indicates that at least some DIB carriers have dissociation energies of at least 9 eV, suggesting that additional features may exist in the interval $1370 < \lambda < 4430$ Å.

7) If UV DIBs are generally comparable in width (i. e. within a factor 2-3) to DIB 1369, then they are *best identified and studied by the same techniques used to study atomic and molecular transitions.*

8) In lieu of traditional random search methods, future searches for UV DIBs should concentrate on spectroscopic features noted as unidentified lines in past studies with *Copernicus* and *IUE*, as well as new unidentified features found with *HST*. In regards to *HST*, archival searches for existing moderate-to-high S/N data would be a good starting point.

ACKNOWLEDGEMENTS. I would like to J. Mathis, T. Tripp, and B. Savage for useful discussions on this topic. I especially thank T. Tripp for his hard work and wonderful insight in putting together the refereed paper on DIB 1369 which helped make this presentation relevant. The author acknowledges support from NASA/LTSARP Grant NAGW-2520.

REFERENCES

Benvenuti, P., and Porceddu, I. 1989, *Astr. Ap.*, **223**, 335.
Cardelli, J. A., Savage, B. D., and Ebbets, D. C. 1991, *Ap. J. (Letters)*, **383**, L23.
Cardelli, J. A., Sembach, K. R., and Mathis, J. S. 1992, *A. J.*, **104**, 1916.
Cardelli, J. A., Sembach, K. R., and Savage, B. D. 1994, submitted to *Ap. J.*.
Carnochan, D. J. in *IAU Symposium 135, Interstellar Dust*, eds. L. J. Allamandola and A. G. G. M. Tielens, NASA Conf. Proc. No. 3036, 11.
Morton, D. C. 1975, *Ap. J.*, **197**, 85.
——. 1991, *Ap. J. Suppl.*, **77**, 119.
Savage, B. D., Cardelli, J. A., and Sofia, U. J. 1992, *Ap. J.*, **401**, 706.
Savage, B. D., and Mathis, J. S. 1979, *Ann. Rev. Astr. Ap.*, **17**, 73.
Seab, C. G., and Snow, T. P. 1985, *Ap. J.*, **295**, 485.
Snow, T. P., York, D. G., and Resnick, M. 1977, *Pub. A. S. P.*, **89**, 758
Tripp, T. M., Cardelli, J. A., and Savage, B. D. 1994, *A. J.*, **107**, 645.

DIFFUSE INTERSTELLAR BANDS IN THE NEAR INFRARED - A DEDICATED SEARCH FOR POLYCYCLIC AROMATIC HYDROCARBON AND FULLERENE CATIONS

B.H. FOING [1] & P. EHRENFREUND[2]

[1] *Solar System Division , ESA/SSD, ESTEC (SO) Postbus 299, 2200 AG Noordwijk, The Netherlands,*

[2] *Leiden Observatory, PO Box 9513, 2300 RA Leiden, The Netherlands*

ABSTRACT. We review observations of diffuse interstellar bands (DIBs) in the near infrared and discuss specific aspects about infrared DIBs, their detectability, criteria for program selection and observing strategy. We report on our recent infrared observations dedicated to the search for polycyclic aromatic hydrocarbon (PAH) and fullerene cations and discuss in particular the detection of two diffuse bands as first evidence for interstellar $C_{60}{}^{+}$.

1. INTRODUCTION

1.1. SPECIFIC ASPECTS ABOUT OBSERVATIONS OF INFRARED DIBS

Astronomical observations have revealed up to the present 150 diffuse interstellar bands (DIBs) at wavelengths ranging from the blue part of the visible spectrum to the near infrared (Herbig, 1975, Herbig & Leka, 1991, Jenniskens & Désert, 1994). Most of the diffuse bands are found between 4000-8000 Å. Their number towards the blue and the red is strongly decreased. According to recent results ionized carbon-containing species are the favorite carrier molecules for the diffuse interstellar bands. The first transitions of large molecules, such as fullerenes and PAHs are expected in the near-infrared (Leach, 1987). In this context, a search for infrared DIBs is of particular interest.

An advantage of near infrared observations is that there is significantly less continuum extinction than in the visible or in the blue part of the spectrum; the extinction at 1 μm is typically one third of the extinction in the visible (thus a star with an $E_{(B-V)}$ of 2 mag, has an A_V of 6 mag and an A_B of 8 mag but an A_J (at 1 micron) of only 2 mag). This means that very extincted stars inaccessible in the blue/visible can be measured in thc near infrared. In the infrared there is also a gain in line equivalent width W, as $\Delta W_\lambda = \lambda^2 N f \pi e^2/mc^2$, for a given oscillator strength f (Spitzer, 1961). Finally, the near infrared domain is largely unexplored for DIB search when compared to the visible domain (only 8 DIBs beyond 8500 Å were reported in 1992).

A. G. G. M. Tielens and T. P. Snow (eds.), The Diffuse Interstellar Bands, 65–72.

The feasibility of such observations is determined by the state of available detectors. Reticons and CCDs have a good efficiency below 9000 Å, and only a low efficiency until 1.1 μm. Very recently, infrared arrays (InSb, or Hg:Cd:Te) became operational for use up to 5 μm, but they had until now only a small size (typically 64x64), and allowed only a limited spectral resolution (R=1500) and wavelength coverage. Moreover, such IR detectors suffer from non-linearities and hystereses which render difficulties in high signal to noise measurements. Fourier Transform Spectrographs (FTS) give a useful/versatile solution for high resolution, high S/N infrared spectroscopy, but lack sensitivity (for large wavelength coverage). In the future, one must look forward to cross dispersed echelle spectroscopy with large NICMOS infrared arrays.

Limitation on infrared observations comes from telluric absorption, especially by water in bands at 0.73, 0.76, 0.95, 1.1, 1.45, 1.9, 2.5-3.3 μm. Skyglow in the near IR (OH and other constituents) must be corrected on faint sources, and thermal background contributes significantly above 3 microns. Another difficulty in the near infrared comes from stellar Paschen lines which are not easy to correct and vary within the same spectral type (due to asymmetries, winds and NLTE effects), making the choice of comparison stars more delicate.

1.2. DETECTABILITY OF DIBS, PROGRAM SELECTION AND STRATEGY

The detectability of DIBs is a function of DIB band width and strength. It depends on instrumental resolution and S/N, as well as errors arising from residuals of telluric /stellar correction, continuum determination and other systematic errors. Signal to noise higher than a few hundred is required for detecting weak DIBs beyond the current limit in the visible (typically 10 mÅ per unit reddening), and also for correct DIB inter-correlations and line profile studies. The minimum number of program stars must be more than 10 for a survey of new DIBs, and above 20 for environmental studies with limited bias on $E_{(B-V)}$ or stellar type.

Comparison stars must be selected within a relevant grid of spectral types (T_{eff}, gravity, abundance) and rotational v sin(i), compared to the program reddened stars. Corrections are required for velocity shifts and v sin(i). Telluric atmospheric correction is particularly delicate in the infrared (especially for strong water variable bands). It requires immediate observations of hot fast rotators at the same airmass. High spectral resolution is needed to limit non-linear averaging and wavelength drifts, to ensure a good correction and high S/N (in the core of atmospheric bands that dominate spectral divisions). For survey work, an optimal choice of comparison stars at the same airmass may allow a direct correction of both telluric and stellar lines.

1.3. DATA ANALYSIS OF DIFFUSE INTERSTELLAR BANDS

Data reduction includes the offset, flat field, background, cosmic ray and other instrumental corrections. Scattered light and a long dark background have to be subtracted (especially to avoid catastrophic divisions in atmospheric bands). Velocity corrections are important to distinguish telluric lines, stellar lines and interstellar absorption at cloud velocities. Weak DIBs can be detected by averaging residual spectra, and confirmed by further tests against DIB criteria (correlation with $E_{(B-V)}$, spectral type, DIB/star velocity).

In order to determine DIB intrinsic profiles it is required to correct for cloud velocities and to deconvolve for cloud distribution using atomic/molecular lines or H I profiles. However, it must be considered that large opacity saturation effects in atomic lines (such as Ca II, K I resonance line) prevent reliable estimates of column densities and velocity distribution in different clouds towards distant objects. Thus single clouds are preferable for such intrinsic profile substructure studies.

1.4. PREVIOUS INFRARED DIB SEARCHES

The observations of infrared DIBs are quite recent, and still largely unexplored. Sanner, Snell, Vanden Bout (1978) surveyed the wavelength range from 6500-8900 Å with a reticon detector, at a low spectral resolution (0.94 Å), but with high S/N. They detected 4 new DIBs (at 7562, 7570, 8572 and 8621 Å) and identified 2 possible DIBs (at 8384 and 8650 Å). Gredel and Münch (1986) measured a DIB at 8763.5 Å with a strength of 5 mÅ. Joblin *et al* (1990), using an FTS, detected 2 new DIBs in the infrared at 11797 Å (FWHM=2.7 Å and W=0.048 Å) and at 13175 Å (FWHM=4.0 Å and W=0.097 Å). Herbig and Leka (1991) measured 20 new DIBs in the red, down to W=10mÅ. Recently, Jenniskens and Désert (1992,1994) found 41 new DIBs and 14 possible DIBs in the range 6944 to 8621 Å. They also remeasured 16 previously proposed DIBs (Herbig & Leka, 1991, Sanner *et al.*, 1978).

2. NEW RESULTS FROM OBSERVATIONS IN THE NEAR INFRARED

2.1. OBSERVATIONS

Our recent observations in the near infrared covered the range from 8600-10200 Å. The data were obtained with the 1.52 m f27.6 coudé telescope at the Observatoire de Haute Provence, France. The spectra were recorded by a 2048 diode linear Reticon array read by two 1024px CCDs. The resolving power of the Aurelie spectrograph was set to R = 36000 ($\Delta\lambda = 0.3$ Å), using a 1200 lines/mm grating and a dispersion of 7.2 Å/mm for high resolution measurements. We also observed at medium resolving power of R=9000 in order to cover a larger wavelength range. We present in this paper high and medium resolution spectra between 9500-10200 Å towards various lines of sight.

2.2. SEARCH FOR SPECIFIC CARRIER MOLECULES

Polycyclic aromatic hydrocarbons (PAHs), fullerenes and their ions have been suggested to be interesting candidates for the carriers of the diffuse interstellar bands (Léger & d'Hendecourt, 1985, van der Zwet & Allamandola, 1985, Léger *et al.*, 1988, Kroto & Jura, 1992). Laboratory measurements, using rare gas matrix isolation techniques, have recently provided a set of spectra of PAHs, PAH ions and fullerenes which allow good comparisons with interstellar data (Salama & Allamandola, 1991, 1992, Ehrenfreund *et al.*, 1992, 1994, Fulara *et al.*, 1993). We have searched for new DIBs in the near infrared in order to identify specific molecules like $C_{60}{}^{+}$, and the coronene and ovalene cations.

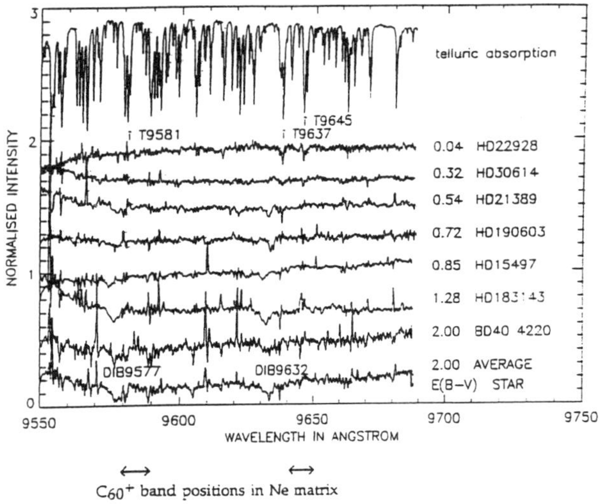

Fig. 1. Sequence of reduced spectra (normalized to unity) of program stars with increasing reddening $E_{(B-V)}$ centred at 9650 Å. Each observed star spectrum has been divided after intrumental corrections by a spectrum of a reference star of similar spectral type. The top trace is a reference spectrum of η Ursae Majoris showing the strength of the telluric bands. Slight residuals from telluric corrections are indicated (T). Two new diffuse bands (DIBs) are detected at 9577 and 9632 Å, increasing with $E_{(B-V)}$ (Foing & Ehrenfreund, 1994). The slight wavelength shifts up to 1.3 Å and different line profiles are partly associated with the velocity distribution of interstellar clouds along the line of sight and possibly reflect intrinsic changes in the line broadening. An average composite spectrum corresponding to $E_{(B-V)}$ =2 is shown at the bottom. We also give the position of the two $C_{60}{}^+$ main bands measured in a neon matrix by Fulara *et al.* (1993).

2.3. FIRST EVIDENCE FOR DIFFUSE ABSORPTION OF INTERSTELLAR $C_{60}{}^+$

C_{60} and its polyhedral geometry were discussed for the first time by Kroto *et al.* (1985). A few years later the synthesis of C_{60} in macroscopic quantities was achieved by Kraetschmer *et al.* (1990). Recently, laboratory spectra of neutral and ionized C_{60} in various matrices have been published (Kato *et al.*, 1991, Gasyna *et al.* 1992, d'Hendecourt *et al.* 1992, Fulara *et al.* 1993), making a search for interstellar C_{60} possible. The inhibition of the fullerene growth mechanism by hydrogen will limit the formation of C_{60} to hydrogen-depleted environments in space (Goeres & Sedlmayr, 1992). The first and second ionization potential of C_{60} are 7.61 eV and 11.39 eV, respectively. C_{60} will therefore exist mostly in ionized form outside dense clouds. Contrary to PAHs, the dication $C_{60}{}^{++}$ may be stable against fragmentation due to the fullerene cage structure and revert to the cation through electronic recombination. Though only small quantities of C_{60} may be formed in the envelopes of R Coronae Borealis stars, C_{60} may still survive long enough to be cycled into the diffuse medium.

The spectrum of $C_{60}{}^+$ in rare matrices shows two dominant peaks at 9663 ± 9 Å and 9724 ± 4 Å in argon and at 9580 ± 4 Å and 9642 ± 3 Å in neon (Gasyna *et al.* 1992, d'Hendecourt *et al.* 1992, Fulara *et al.* 1993). In our observational data we find two diffuse bands in the near infrared at 9577 and 9632 Å that are coincident (within 0.1 %) with laboratory measurements on $C_{60}{}^+$ in a neon matrix (Foing &

Ehrenfreund 1994). Figure 1 shows a sequence of reduced spectra of program stars with increasing reddening $E_{(B-V)}$. The good correlation (correlation coefficients 0.91 and 0.86) between the equivalent widths of these bands and the reddening $E_{(B-V)}$, with slopes of 0.39 and 0.67 Å per unit $E_{(B-V)}$, confirmed the detected bands as new DIBs. The even tighter correlation between the bands themselves (coefficient 0.98) with a ratio of 1.6 ± 0.15 suggest a similar or a single carrier. The intensity ratio of the two new interstellar bands of 1:1.6 is consistent with measurements of $C_{60}{}^+$ in neon by d'Hendecourt *et al.* (1992) and Fulara *et al.* (1993), which represents another criterion for the identification. The observed bands also appear with the same average separation in Ar, Ne and in space of approximately 60 Å and are probably intrinsic to two major ground state geometries favoured by Jahn-Teller types of distortion (Bendale *et al.*, 1992).

Assuming an oscillator strength for $C_{60}{}^+$ of 0.1 according to theoretical estimates (Bendale *et al.*, 1992), we calculate, after the Spitzer formula (1961), $\Delta W_\lambda = \lambda^2 N f \pi e^2/mc^2$, (assuming a ratio of C/H = 3.7 x 10^{-4} and hydrogen column density per unit interstellar absorption, N_H/A_V of 2.10^{21} cm^{-2}, with $A_V = 3\ E_{(B-V)}$), that 0.035 % of the cosmic carbon could be in the form of $C_{60}{}^+$. Taking the lower oscillator strength limit 0.004-0.012 derived from experiments of Kato *et al.* (1992), which is likely the more accurate estimate, this becomes a maximum of 0.3-0.9 % of the cosmic carbon in $C_{60}{}^+$. This indicates that $C_{60}{}^+$ could play an important role in interstellar chemistry.

Two more bands due to vibrational excitation in the upper electronic state should be apparent with 15 % strength of these main bands around 9420 and 9365 Å. It will be difficult to observe these weak bands due to strong telluric and stellar contamination in this range. However, these lines should be searched for from very dry sites.

2.4. SEARCH FOR POLYCYCLIC AROMATIC HYDROCARBONS

Ehrenfreund *et al.* (1992) have measured spectra of coronene and ovalene cations in solid neon. The main laboratory transitions for the coronene cation in a neon matrix fall at 4590 and 9465 Å with respective band widths of 28 and 16 Å and oscillator strengths of 0.012 and 0.0018. We searched for this particular PAH cation in both wavelength domains. The wavelength region around 9465 Å is very difficult to observe due to strong telluric absorption. However, together with observations in the blue, an upper limit of 0.05 % of the cosmic carbon in the form of coronene cation could be derived (Ehrenfreund *et al.*, 1995).

The second transition of the ovalene ion has recently been measured (Ehrenfreund *et al.*, 1994) in solid neon and displays a strong band at 9780 Å. In this wavelength range the telluric absorption can already be well corrected. In Figure 2 we show a series of telluric corrected spectra centred at 9780 Å. Surprisingly, no strong DIBs are present between 9650 and 9850 Å. Weak bands at 9672, 9783, 9798 Å, which appear only in a composite spectrum combining the 4 reddened stars between 0.54 and 1.28 and scaling to $E_{(B-V)}$=4 may be weak DIBs with approximate equivalent widths of 30 mÅ per unit reddening. The band at 9783 Å could be the weak counterpart of the ovalene cation, leading to an upper limit estimate of only 0.005 % of the cosmic carbon in the form of this particular ion. An explanation of the low abundance of these PAH cations has been outlined in Ehrenfreund et al. (1995).

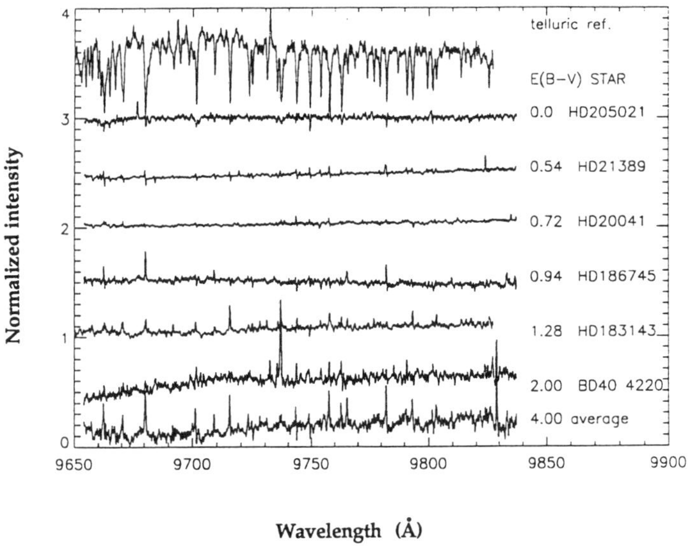

Fig. 2. Sequence of reduced (normalized to unity) program stars with increasing reddening $E_{(B-V)}$. Each observed stellar spectrum has been divided after instrumental corrections by a spectrum of a reference star of similar spectral type. No DIBs stronger than 30 mÅ per unit reddening are observed in this range. An average composite spectrum corresponding to $E_{(B-V)}$=4 is shown at the bottom, which displays small features at 9672 Å, 9783 Åand 9798 Å (Ehrenfreund *et al.*, 1995).

2.5. OBSERVATIONS AT 1 MICRON

In order to cover a larger wavelength range, we performed observations at medium resolution (R=9000). Figure 3 presents a survey from 9700-10200 Å. This range is relatively free of telluric absorption. However, the strong H I Paschen line at 10049 Å makes a DIB detection around this wavelength rather difficult. The spectrum shows a small overlap with the high resolution spectra (R=36000) of Figure 3 for comparison. Two new DIBs are possibly detected at 10116 and 10140 Å. In summary, we present in Table 1 all new detected DIBs between 9500-10200 Å. In this particular range the number of strong DIBs is rather small which may help in the identification of carrier molecules.

3. CONCLUSION

The identification of DIBs is a long lasting mystery, full of possible traps and illusions. In the visible range from 4000-8000 Å, there are now about 150 DIBs known with equivalent width above 10 mÅ/$E_{(B-V)}$. This means on average one DIB per 20 Å bin. The comparison of laboratory spectra of carbon-containing species and astronomical observations of diffuse interstellar bands showed very interesting results in recent years. In order to achieve a reliable identification of a specific carrier molecule (excluding random matches due to the vast amount of diffuse bands), only gas phase or rare gas matrix spectra should be used for comparison with astronomical data. Candidate molecules which show several strong transitions in the DIB range will allow to study better consistence in band position and intensity, intercorrelation within

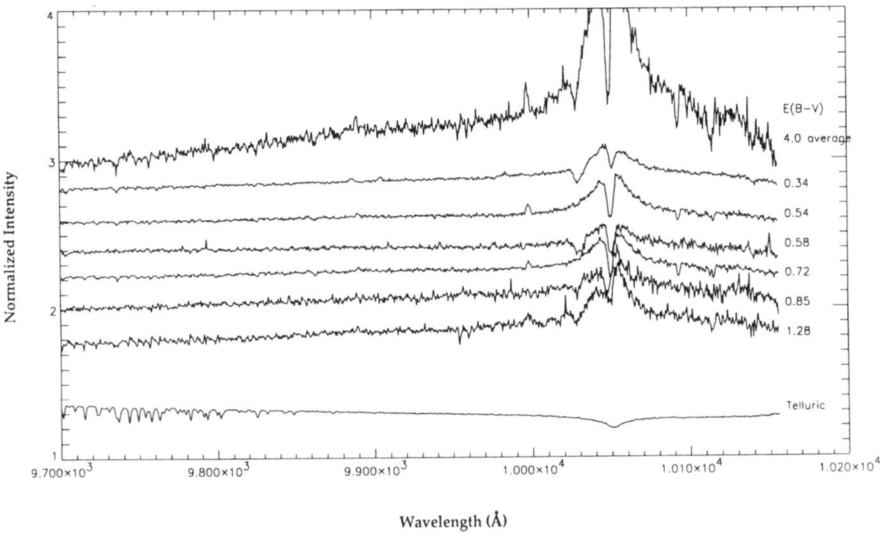

Fig. 3. Sequence of reduced spectra (normalized to unity) between 9700-10200 Å of program stars with increasing reddening $E_{(B-V)}$. An average composite spectrum corresponding to $E_{(B-V)}=4$ is shown at the bottom, which displays possible new DIBs at 10116 Å and 10140 Å. The two strong bands at 10030 (He I) and 10093 Å are of stellar origin. These observations were taken at medium resolution R=9000. In order to identify weak DIBs in this range high resolution is definitely required.

the bands, and help to evaluate experimental shifts and broadening. Outside the strong telluric absorption a direct identification of DIB carriers will be more easier in the infrared, because only a few strong bands are present.

Our conclusions can be summarized as follows:

- Infrared DIBs are particularly useful for the identification of large molecules such as PAHs and fullerenes in the interstellar medium.

- New detectors, instruments and techniques are now available for a careful infrared DIB survey, building on previous searches and on an optimised observing strategy.

- The identification of DIBs must go beyond the art of numerology (in the visible as in the near infrared) and requires a good understanding of experimental measurements as well as statistical evaluation of astronomical spectra.

- A search for near-infrared DIBs of PAHs (as coronene and ovalene cations) was conducted leading to a small upper limit abundance for these pericondensed species.

- Two new diffuse interstellar bands consistent with laboratory measurements of $C_{60}{}^+$ have been detected. This result provides strong evidence for the existence of the C_{60} cation as the largest molecule ever detected in interstellar space and likely represents the first direct identification of diffusc intcrstcllar bands. The estimated amount of cosmic carbon in $C_{60}{}^+$ (up to 0.9 %) indicates that fullerenes might play an important catalytic role in the diffuse interstellar medium. A search for other fullerene compounds is underway.

TABLE 1

New diffuse interstellar bands between 9500-10200 Å

Wavelength(Å)	FWHM(Å)	$W/E_{(B-V)}$(Å mag^{-1})
9577	2.5-4	0.370
9632	2.4-4	0.610
9672	1.5	0.03
9783	2	0.03
9798	1.8	0.03
10116	3	0.270
(10140	3-4	0.170)

ACKNOWLEDGEMENTS. We thank L. d'Hendecourt, P. Jenniskens and X. Désert for discussion and the Observatoire Haute Provence for their support of this project. BHF is on leave from CNRS. PE is a recipient of a fellowship of the European Community ERBCHBICT940939.

REFERENCES

Bendale, R.D., Stanton, J.F., and Zerner, M.C. 1992, *Chem. Phys. Lett.*, **194**, 467.
Ehrenfreund, P., d'Hendecourt, L., Verstraete, L., Léger, A., Schmidt, W., and Defourneau, D. 1992 *Astr. Ap.*, **259**, 257.
Ehrenfreund, P., Foing, B., d'Hendecourt, L., Jenniskens, P., and Désert, X. 1995, *Astr. Ap.*, in press.
Foing, B.H. and Ehrenfreund, P. 1994, *Nature*, **369**, 298.
Fulara, J., Lessen, D., Freivogel, P., and Maier, J.P. 1993, *Nature*, **366**, 439.
Gasyna, Z., Andrews, L., and Schatz, P.N. 1992, *J. Phys. Chem.*, **96**, 1525.
Goeres, A. and Sedlmayr, E. 1992, *Astr. Ap.*, **265**, 216.
Gredel, R. and Münch, G. 1986, *Astr. Ap.*, **154**, 336.
d'Hendecourt, L., Fostiropoulos, K., and Léger, A. 1992, in C. Joblin thesis, Univ. Paris VII.
Herbig, G.H. 1975, *Ap. J.*, **196**, 129.
Herbig, G.H. and Leka, K.D. 1991, *Ap. J.*, **382**, 193.
Jenniskens, P. 1992, *Organic Matter in Interstellar Extinction*, PhD Thesis Leiden University, The Netherlands.
Jenniskens, P. and Désert, F.-X. 1994, *Astr. Ap. Suppl. Ser.*, **106**, *A survey of diffuse interstellar bands (3800-8680 Å)*, 39.
Joblin, C., Maillard, J.P., d'Hendecourt, L., and Léger, A. 1990, *Nature*, **346**, 729.
Kato, T. *et al.* 1991, *Chem. Phys. Lett.*, **180**, 446.
Kraetschmer, W., Lamb, L., Fostiropoulos, K., and Huffman, D.R. 1990, *Nature*, **347**, 354.
Kroto, H., Heath, J.R., O'Brien, S.C., Curl, R.F., and Smalley, R.E. 1985, *Nature*, **318**, 162.
Kroto, H. and Jura, M. 1992, *Astr. Ap.*, **263**, 275.
Leach, S. in *Polycyclic Aromatic Hydrocarbons*, 1987 (eds Leger, A., d'Hendecourt, L., and Boccara, N.). (Reidel, Dordrecht) 99-129.
Léger, A. and d'Hendecourt, L. 1985, *Astr. Ap.*, **146**, 81.
Léger, A., d'Hendecourt, L., Verstraete, L., and Schmidt, W. 1988, *Astr. Ap.*, **203**, 145.
Salama, F. and Allamandola, L.J. 1991, *J. Chem. Phys.*, **94**, 6964.
Salama, F. and Allamandola, L.J. 1992, *Nature*, **358**, 42.
Sanner, F., Snell, R., and Vanden Bout, P. 1978, *Ap. J.*, **226**, 460.
Spitzer, L. in *Physical Processes in the Interstellar Medium*, 1978 (Wiley, New York).
Van der Zwet, G.P. and Allamandola, L.J. 1985, *Astr. Ap.*, **146**, 76.

A SEARCH FOR POLARIZATION IN DIFFUSE INTERSTELLAR BANDS TOWARDS HD197770

A.J.ADAMSON
University of Central Lancashire, Preston PR1 2HE, UK
D.C.B.WHITTET
Rensselaer Polytechnic Institute, Troy, NY, USA

ABSTRACT.

High-precision spectropolarimetry of the reddened star HD197770 reveals the 5780Å and 5797Å diffuse interstellar features to be unpolarized, in contrast to the reported polarization of the 2175Å "graphite" bump in this line of sight. This result rules out a connection between these DIBs' carrier and the small, presumably carbonaceous, grains responsible for the 2175Å absorption.

1. INTRODUCTION

1.1. THE DIBS AND INTERSTELLAR REDDENING

The diffuse interstellar bands may be seen as fine structure in the interstellar extinction curve. Apart from their intrinsic interest, properties such as their profiles and range of widths would be useful diagnostics of the composition and physical state of interstellar grains if any of them were ever established as arising in grain material.

Evidence for a solid-state carrier is, however, not convincing. Higher reddening generally produces stronger DIBs, but for most bands this correlation is loose; only the strong, narrow feature at 5797Å has been shown to correlate as well with E(B-V) as expected for a band carried by the grains which are responsible for the visual extinction (Chlewicki *et al.* 1983). Correlations between the bands and the 2175Å "graphite" bump are in general no better than those between the bands and E(B-V); in both cases there is evidence of environmental scatter in the correlations.

1.2. SPECTROPOLARIMETRY

Interstellar extinction is generally accompanied by polarization, at a rate of $\leq 9\%$ per magnitude of E(B-V); this inequality can be understood in terms of variations in the alignment of grains along the line of sight. An observation of polarization in a diffuse band would strongly implicate the polarizing grains as harbouring *that* DIB's carriers.

All attempts to detect polarization structure in the DIBs have failed (Martin & Angel 1974,1975, Fahlmann & Walker 1975, Adamson & Whittet 1992); these results

A. G. G. M. Tielens and T. P. Snow (eds.), The Diffuse Interstellar Bands, 73–79.

TABLE 1
Existing high-resolution spectropolarimetry

Reference	λÅ (1)	p% (2)	σ% (3)	Δp% (4)	$\frac{\Delta p}{\sigma}$ (5)	Star (6)
Martin & Angel 1974	4430	5.76	0.05	0.2	4	183143
	5780	5.95	0.15	0.5	3	183143
Martin & Angel 1975	4430	3.28	0.013	0.09	7	21389
	6284	3.2	0.02	0.45	22	21389
Fahlmann & Walker 1975	6284	6.0	0.2	0.9	5	183143
Adamson & Whittet 1992	5780	6.15	0.06	0.9	15	183143
	5797	6.15	0.06	0.3	5	183143

NOTES
(1) Band Wavelength in Å
(2) Continuum polarization level
(3) Continuum rms polarization
(4) Polarization excess predicted across a grain-based band
(5) Predicted excess compared to observed rms limit
(6) Star observed (HD)

are summarized in Table 1. So, although one can no longer rule out a solid-state carrier on the basis of band profiles (e.g., Shapiro & Holcomb 1986), the *observations* argue against the hypothesis that the grains responsible for the visual extinction also carry the DIBs. Continued work of this nature is driven by two facts: (i) only a few percent of the known diffuse features have been studied, and (ii) CCD-equipped spectrometers on 4-m telescopes now give high resolution and permit measurements of narrower bands than was previously possible.

In a previous paper (Adamson & Whittet 1992) we searched for polarization in the 5797 DIB, which as mentioned above shows the best correlation with interstellar reddening. On the basis of convincingly negative results (Table 1), we concluded that the DIBs cannot be identified with silicate grains (which are aligned) and that small carbonaceous grains are the most probable solid-state carriers. If we accept the usual interpretation of the 2175Å absorption feature, then it provides a handle on the alignment state of the small carbonaceous grains. HD197770 is remarkable in being the only star out of 12 so far studied in the satellite ultraviolet to show evidence for excess polarization associated with the 2175Å feature (Clayton *et al.* 1992; Somerville *et al.* 1994). This result may indicate a dependence on environment of the efficiency with which the carrier grains are aligned. The line of sight to HD197770 is therefore a particularly appropriate one in which to search for polarization associated with the visible diffuse bands: a positive detection in any DIB would greatly strengthen the link between its carrier and the 2175Å feature, and would thus support a carrier in small carbonaceous grains. Here, we present spectropolarimetry of HD197770 in a spectral region that includes the strong DIBs at 5780 and 5797Å .

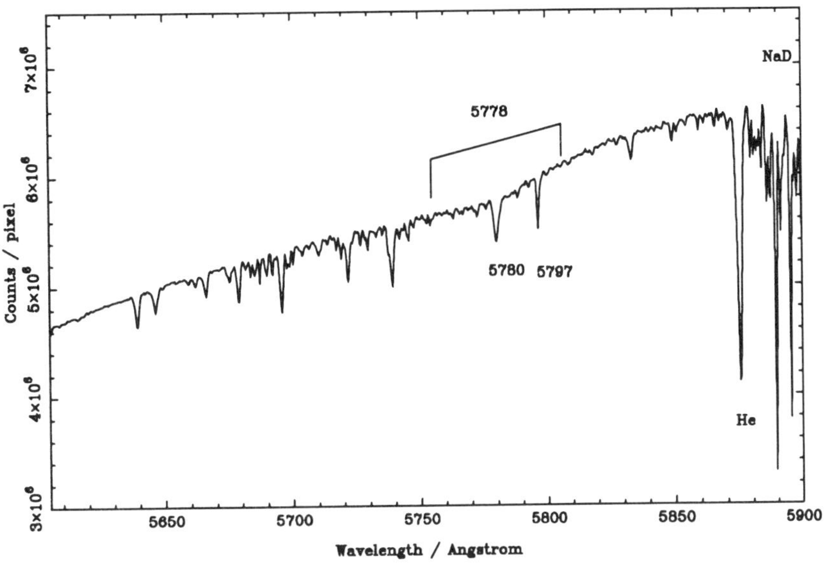

Fig. 1. The count spectrum of HD197770 between 5700Å and the Sodium D lines. This spectrum was derived by summing all available counts in the O and E beam spectra. Counts are between 3 and 7 million per pixel; statistical signal to noise is greater than 2000 throughout. Note the prominent 5778Å band, and photospheric contamination between 5640 and 5750Å.

2. OBSERVATIONS

HD197770 (spectral type B2III, E(B-V)$\simeq$0.57) was observed on the night of 6-7 September, 1993, with the ISIS spectrograph on the William Herschel Telescope, La Palma. The observations were carried out by René Rutten under the WHT service program. The R1200R (1200 line/mm) grating was used, providing 0.4Å pixels and unsmoothed spectral resolution of 0.8Å on the detector (an EEV CCD). 400Å were covered, centered on the 5797Å band. Observing conditions were good, but the source was observed at a fairly high airmass (sec(z)$\simeq$2). No polarized standard was observed; ISIS spectropolarimetry is systematically accurate to about a tenth of a polarization percent (e.g. Adamson & Whittet 1992).

A section of the count spectrum is shown in Figure 1[1]. This star appears to have slightly anomalous DIB strengths. Equivalent widths of the narrow bands at 5780 and 5797Å are 0.17 and 0.07Å respectively. By comparison with HD183143 (Herbig's 1975 diffuse-medium standard) both are a factor of 2.5 weaker than expected on the basis of the reddening quoted above. In contrast, the broad feature at 5778Å has an equivalent width of 1.56Å , stronger by a factor of 2 than normal.

Reduction from raw CCD frames to the polarization spectrum proceeded as de-

[1]Since this spectrum has not been ratioed to that of an unreddened standard star of similar spectral type, identifications of weak bands are prone to stellar contamination.

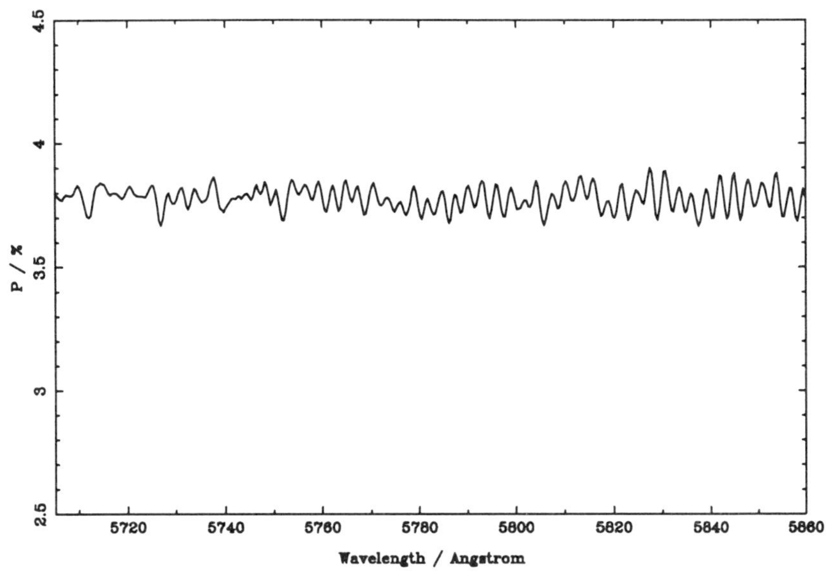

Fig. 2. Raw polarization spectrum of HD197770, showing the region of the major DIBs between 5640 and 5800Å .

scribed by Tinbergen & Rutten (1990), using sky-subtraction methods described by Adamson & Whittet (1992) and 1.5-pixel smoothing of the Stokes Q(λ) and U(λ) spectra before addition in quadrature to produce P(λ).

A substantial ripple is apparent in the resulting P(λ) spectrum (Figure 2). As discussed by Tinbergen (1993), this may be an optical effect arising in the cement layer of the Savart plate analyser employed in the ISIS polarimetry module. Thirty percent of the rms error in the polarization is produced by this ripple; uncorrected, it would limit the precision ultimately available with this instrument. However, the ripple is *periodic* (a section of its power spectrum is shown in Figure 3a). Editing of the real and imaginary parts of the polarization spectrum in the region of 0.27 Nyquist units substantially removes the high-frequency component of the ripple, producing the final P(λ) spectrum (Figure 3b). Continuum polarization in this spectrum is consistent with broadband measurements of this star. Clayton *et al.* quote p_{max} $\simeq$3.83% at λ_{max} $\simeq$5000Å ; we measure 3.75% at 5700Å , 700Å from the polarization peak.

The polarization ripple noted by Adamson & Whittet (1992) in HD183143 does not respond to the technique described above; it is possible that the difference relates to observing techniques used in these two service runs. HD 183143 was observed at a variety of positions in the spectrograph slit (and hence used various CCD rows and optical paths), whereas HD197770 was observed at essentially a single location on the detector.

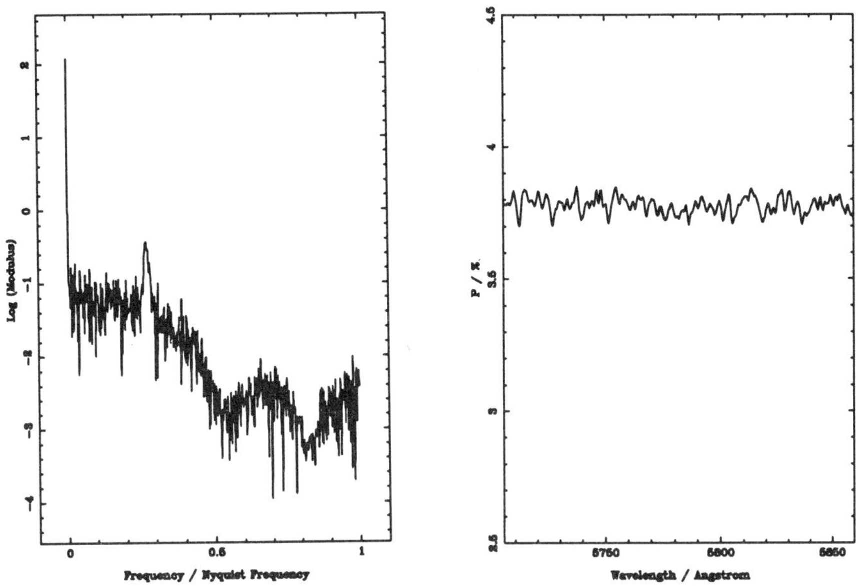

Fig. 3. (a, left) Power spectrum of the ripple. The peak at 0.27 Nyquist units was removed from the data to produce the final P(λ) spectrum (b, right).

3. LIMITS ON DIB POLARIZATION

There is no measurable polarization structure in either the 5780 or 5797 Å diffuse features in this line of sight. The region shown in Figure 3b is typical of the entire spectrum and has rms polarization 0.03%. We adopt this as an upper limit on the excess polarization allowed by these data. It compares poorly with the prediction of the solid grain model by Martin & Angel (1974), which is compared to the observations in Figure 4. The upper and lower bounds of the model envelope correspond to values of f=1.0 and 1.8 in the formula

$$\frac{\Delta p}{p} = f\frac{\Delta \tau}{\tau}, \tag{1}$$

accounting for a variety of grain geometry and composition. Here p and τ are the continuum polarization and optical depth respectively, and Δp and $\Delta\tau$ are the predicted increments in these quantities across the absorption feature. The mean prediction lies above the observed polarization by 8 and 6 times the rms in 5780Å and 5797Å respectively.

4. CONCLUSIONS

Modern instrumentation can produce spectropolarimetry with sufficient resolution and signal-to-noise to address the question of polarization structure within DIBs of a wide range of strength and widths. We have described observations of the "polarized

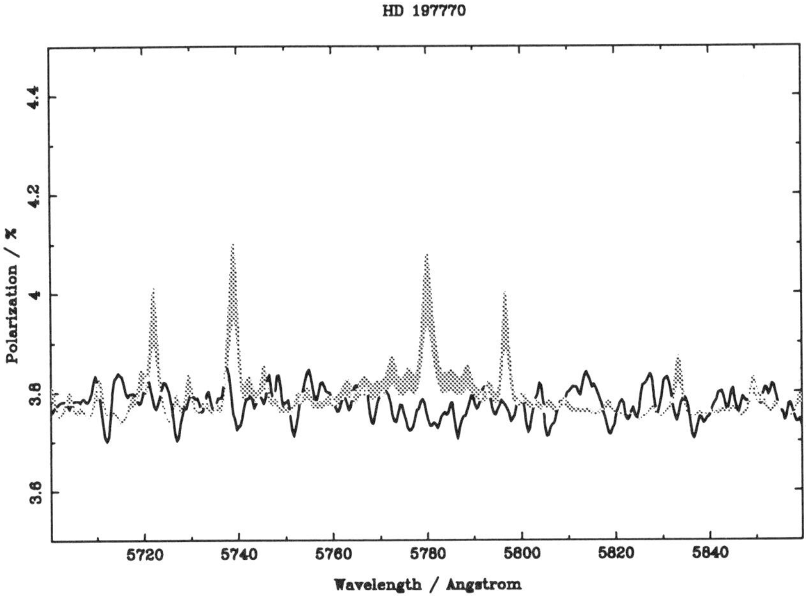

Fig. 4. Linear Polarization in HD197770. Thin line: Observed P(λ). Shaded envelope: the grain model of Martin & Angel. Peaks at 5720 and 5740 Å are photospheric artefacts.

bump" star HD197770 taken with the WHT ISIS spectrometer. The diffuse bands at 5780 and 5797Å are unpolarized in this star, to levels well below the polarization predicted by the Martin & Angel model of bands arising in aligned grains. It is also likely that the broad 5778Å feature is unpolarized. Non-detection of two of the strongest DIBs confirms our earlier result for HD183143 (Adamson & Whittet 1992) that their carriers do not reside in the aligned component of interstellar dust. If the polarized 2175Å bump reported by Clayton *et al.* (1992) is real[2], then the present results finally decouple the bump carriers from those of the DIBs. Solid state carriers thus seem increasingly implausible, requiring a grain population distinct from both large "classical" grains and small carbonaceous grains.

We finally consider whether spectropolarimetry is relevant to carriers such as free-flying PAHs. These molecules can align (by the Barnett effect) in a magnetic field. If the transitions responsible for the DIBs were intrinsically polarized, then the DIBs would themselves be either polarized or depolarized depending upon orientation. However, in normal interstellar conditions, disalignment due to angular momentum changes during the emission of photons in the "Unidentified IR" emission bands is much more efficient than magnetic alignment (Rouan *et al.* 1992). The UIR bands represent reprocessing of UV photons from the interstellar radiation field; if the *same* PAHs are responsible for the UIRs, then a PAH model for the DIBs predicts a lack of polarization structure, consistent with results presented here and in previous papers for several DIBs.

---

[2]We note that the bump polarization was reported to be present in two separate observations with the WUPPE instrument; this strengthens the above conclusion.

ACKNOWLEDGEMENTS.

We are indebted to René Rutten for obtaining the data presented here under the RGO service program, and to Adrian Webster for useful comments on small molecular carriers. Data reduction employed STARLINK software at the University of Central Lancashire STARLINK node. The William Herschell Telescope is run by the Royal Greenwich Observatory at the Spanish Roque de los Muchachos Observatory in LaPalma. A.J.A. is funded by the STARLINK project and the University of Central Lancashire.

REFERENCES

Adamson, A.J. and Whittet, D.C.B. 1992, *Astrophys. J. Letters*, **398**, L69.
Clayton, G.C., Anderson, C.M., Magalhaes, A.M., Code, A.D., Nordsieck, K.M., Meade, M.R., Wolff, M.J., Babler, B., Bjorkmann, K.S., Schulte-Ladbeck, R.E., Taylor, M.J., and Whitney, B.A. 1992, *Astrophys. J.*, **385**, L53.
Chlewicki, G., van der Zwet, G.P., van Ijzendoorn, J., Greenberg, J.M., and Alvarez, P.P. 1986, *Astrophys. J.*, **305**, 455.
Fahlmann, G.G. and Walker, G.A.H. 1975, *Astrophys. J.*, **200**, 22.
Herbig, G.H. 1975, *Astrophys. J.*, **196**, 129.
Martin, P.G. and Angel, J.R.P. 1974, *Astrophys. J.*, **188**, 517.
Martin, P.G. and Angel, J.R.P. 1975, *Astrophys. J.*, **195**, 379.
Rouan, D., Leger, A., Omont, A., and Giard, M. 1992, *Astron. Astrophys.*, **253**, 498.
Shapiro, S.L. and Holcomb, K. 1986, *Astrophys. J.*, **310**, 872.
Somerville, W.B., Allen, R.G., Carnochan, D.J., He, L., McNally, D., Martin, P.G., Morgan, D.H., Nandy, K., Walsh, J.R., Whittet, D.C.B., Wilson, R., and Wolff, M.J. 1994, *Astrophys. J. Letters*, **427**, L47.
Tinbergen, J. and Rutten, R. 1990, RGO/LPO internal note.
Tinbergen, J. 1993, *Spectrum (Royal Observatories Newsletter)*, **1**, 26.

CHAPTER II

Environmental Behavior of the Diffuse Interstellar Bands

COLLECTIVE BEHAVIOUR OF THE DIFFUSE INTERSTELLAR BANDS

W. B. SOMERVILLE
Department of Physics and Astronomy, University College London, Gower Street, London WC1E 6BT, UK

ABSTRACT. This is a review of the properties of the DIBs in relation to the different interstellar environments in which their carriers are located. It starts with a general survey of conditions in different regions of the interstellar medium. Comparisons of DIBs with each other show that there are considerable variations in relative strengths of absorption, but also have led to the concept of 'families' that vary together. The relation of DIB strengths to optical extinction gives a measure of the abundances of the carriers relative to the total amount of interstellar matter as well as to the dust grains. DIB strengths may be compared with gas-phase abundances of atoms and molecules to study their immediate relation to cloud conditions such as density. It is found that the strength of the strongest DIB, $\lambda4430$, does not vary with the cloud density, whereas for other DIBs studied the strength tends to decrease with density and depth into dark clouds, more rapidly for $\lambda5780$ than for $\lambda5797$.

1. INTRODUCTION

The diffuse interstellar bands (DIBs) are absorption features produced as starlight passes through interstellar matter. Although we often speak of "The Interstellar Medium", conditions of density, temperature, ionization and chemical equilibrium vary considerably from place to place and it is convenient to classify different regions of interstellar space as though they were different types of object. An outline of such a classification is shown in Table 1. It should be noted, however, that properties such as temperature and density actually vary continuously—there are no sharp boundaries between "clouds" and "intercloud space" and the different regions blend into each other. Interstellar matter accounts for around 10% of the known mass in the disks of spiral galaxies such as our own galaxy, 25% in irregular galaxies such as the Magellanic Clouds, and less than 0.1% in elliptical galaxies.

Because the DIBs are seen as absorptions in starlight, those observed are in diffuse clouds—these contain small column densities of material and we can see stars through them—and the outer layers of dark clouds. The dark clouds are too opaque for stars to be seen right through them but we can study stars embedded in the clouds, not too far below the surface. The stars used are *early-type*, that is they are hot, luminous and (because such stars have short lifetimes) young. There are several reasons why these stars are used. They have simple intrinsic line spectra, so interstellar absorptions can easily be studied; they are bright, so they can be seen at great distances through a long path-length of interstellar matter; and being young they are still associated with

A. G. G. M. Tielens and T. P. Snow (eds.), The Diffuse Interstellar Bands, 83–95.

TABLE 1
Characteristics of interstellar regions

Region	State of Hydrogen	Other Notable Constituents	Temperature T (K)	Density n (cm^{-3})
Coronal gas $\approx 50\%$ of ISM by volume	H^+	O^{5+}	$10^5 - 10^6$	~ 0.01
Diffuse nebulae (H II regions)	H^+	other ions	$\sim 10^4$	$10^2 - 10^3$
Intercloud medium $\approx 40\%$ of ISM by volume	H	C^+	$\sim 10^4$	~ 0.1
Diffuse clouds	H, H_2	C^+, CO	$50 - 100$	$10 - 10^2$
Dark clouds (molecular clouds)	H_2	many molecules	$10 - 50$	$10^3 - 10^7$
Giant molecular clouds $\sim 10^5$ solar masses	H_2	CO	~ 10	~ 600

the interstellar clouds from which they formed. If a cooler star, for example one like the Sun, lay behind a diffuse cloud its spectrum would contain DIB absorptions but it would be very difficult to study them. Typical values of parameters for diffuse clouds are in Table 1. Detailed models for certain individual interstellar clouds, including those in front of ζ Per and ζ Oph, have been discussed by van Dishoeck and Black (1986), fitting the models to all available observational data. Temperature decreases from 100—200 K at the surface to some tens of K in the core of the cloud, while density increases typically from 100 to several hundred hydrogen nuclei per cm^{-3}.

In addition to DIBs, the stellar spectra contain interstellar atomic and molecular absorption lines in the optical and especially the ultraviolet, from which column densities may be deduced. Relative to hydrogen, which is always dominantly in the gas phase, the heavier elements have lower abundances than the 'cosmic' values found for example in the solar system; they are depleted from the gas by incorporation into the grains. The amount of depletion is correlated with the mean gas density (Jenkins, Savage and Spitzer, 1986), suggesting that atoms tend to adhere to grains in denser regions and to be removed in regions of lower density. Depletions can be used to put some general constraints on the possible grain composition (Whittet, 1984). Note that astronomers tend to speak of all atomic species heavier than helium as *metals*.

Less than 1% of the mass of interstellar material is in the dust grains. Observationally, it is found that the gas and dust are everywhere well mixed. The dust produces interstellar extinction, a mixture of scattering and absorption, also known as interstellar reddening because more blue light is removed and the transmitted starlight looks redder. In the optical, broad-band magnitudes such as B and V are used; the amount of extinction in the V-band is written as A_V magnitudes and the *colour excess*

$$E(B-V) = (B-V)_{\text{observed}} - (B-V)_{\text{intrinsic}} = A_B - A_V \tag{1}$$

gives a measure of the total amount of dust in the line of sight to the star. The ratio $R = A_V/E(B-V)$, known as *the ratio of total to selective extinction*, is a measure of the mean grain size. In the general interstellar medium $R = 3.1$, but in local dark cloud regions it can take larger values.

In the optical, the extinction shows a steady rise to shorter wavelengths that is almost linear. There is no structure, apart from the DIBs. The shape of the curve is very much the same in different directions and can be fitted by grains of a variety of composition (e. g., Spitzer, 1968). The ultraviolet extinction curve shows more dramatic behaviour and larger variations. It can be regarded as containing three components, the linear rise to shorter wavelength, the 2200 Å absorption 'bump' (the central wavelength is 217.5 nm), and the non-linear far-UV rise (e. g., Carnochan, 1986; Fitzpatrick and Massa, 1990), although more-detailed parameterizations have sometimes been used. Relative to the optical extinction, these three components vary in strength between different sight-lines, independently of each other. The far-UV rise continues down to the shortest wavelength studied, 925 Å (Snow, Allen and Polidan, 1990), a result that must be kept in mind when attempting to interpret the extinction curve. A much-quoted empirical fit to the data (Mathis, Rumpl and Nordsieck, 1977) involves a mixture of graphite and silicates with a particular size distribution, the bump being attributed to graphite, but this fit is by no means unique.

The ultraviolet radiation field, from hot stars, plays a major role in interstellar processes. Outside the H II regions around hot stars, only photons of energy less than the hydrogen Lyman limit, 13.6 eV, are present. From direct observations of the ultraviolet sky, a general interstellar radiation field can be estimated for use in models (Draine, 1978; Gondhalekar, Phillips and Wilson, 1980). Inside the denser clouds, there is shielding from this radiation and few interstellar photons penetrate the molecular clouds; however, ionization by cosmic ray particles can happen anywhere, followed by photorecombination so that the chemistry still can be influenced by radiation, even in the densest clouds (Sorrell, 1992).

There has been some discussion of the possible presence of DIB carriers in circumstellar (as distinct from interstellar) environments. Some detections of DIB absorption have been reported (Le Bértre and Lequeux, 1993) but there is always great difficulty in establishing just where the absorption is formed along the line of sight (Seab, this volume). The recognition that emission lines observed in the spectra of the Red Rectangle nebula (Scarrott *et al.*, 1992) and the shell of the R CrB type variable star V854 Cen (Rao and Lambert, 1993) can be related to particular DIBs indicates that at least some DIB carriers are present in circumstellar regions. These observations also give the strongest evidence yet available that the carriers are molecular, at any rate for the DIBs in question.

In astronomical spectroscopy, the strength of an absorption line is often expressed as an *equivalent width* W, a measure of total absorption without regard to the line profile. The DIBs are all unsaturated absorptions; in this case, W is directly proportional to the column density N of absorbers in the line of sight from the star. In the case of a broad, shallow feature, it can be more accurate to use not W but the central depth A_c of the absorption, expressed as a percentage of the continuum intensity on either side of the feature. A_c has been used particularly for $\lambda 4430$.

2. DIB RELATIONSHIPS

We may summarize the different possible relationships for individual DIBs as follows:

With each other:

this introduces the concept of families.

With the gas:

the mean line-of-sight hydrogen abundance $< n_H >$ (the column density of H atoms plus twice the column density of H_2 molecules, divided by the distance to the star); Spitzer (1985) has shown that $< n_H >$ is a good measure of the mixture of interstellar clouds along the line of sight and so it measures the mean cloud density.

depletions of heavy elements—also related to the mean cloud density.

molecular abundances (CN, CH, CH^+, ...)—related through chemical processes to the cloud density, the chemical environment (including the radiation field), and—especially for CH^+—perhaps to processing in shock fronts.

With the grains:

A_V or $E(B - V)$—the total column density of material.

UV grain properties—grain size distribution and carrier of λ2200.

polarization—composition, shape and magnetic properties of grains.

IR grain properties—dust temperature and carriers of IR spectral features.

With depth into dark clouds:

cloud density and shielding from UV radiation field.

With galactic location:

regional variations.

differences between individual star clusters and associations.

differences between individual galaxies.

In circumstellar shells:

relation to type of star.

3. DIFFUSE BAND FAMILIES

It has been known from some of the earliest observations that the relative strengths of diffuse bands can show considerable variations between different lines of sight. This is seen, for instance, in the large dataset for λ5780, λ5797 and λ6284 given by Merrill *et al.* (1937). But it was not possible to do much more than simply record this information, until the concept of families—groups of DIBs that vary together—was introduced almost simultaneously by three groups of workers (Table 2). The three classifications agree well, except that Josafatsson and Snow rather surprisingly found λ5780 and λ5797 to be well correlated. All three used relatively small samples of stars. The fact that different families exist is thus more clearly established than the precise membership of any one family—it can be easy to demonstrate that two absorptions differ in relative strength, but to verify that they always vary together would require large datasets. The importance of the families is to suggest DIBs that belong to the same chemical carrier, or at least to carriers that tend to co-exist.

TABLE 2
Diffuse band families

λ	Ch86	KW87	JS87	BP89
2200	-	2	2?	*u*
4430	-	1	-	-
5705	-	2?	3	-
5778	-	2?	2	-
5780	II	2	1	*c*
5795	Not I	-	-	-
5797	I	3	1	?
5844	II?	2?	2	-
5849	I?	3	1	-
6180	-	1	-	-
6196	II	2	-	*c*
6203	II	2	-	?
6234	II?	-	-	-
6269	II?	2	-	*u*
6284	II	2	-	*c*
6376	-	3	-	-
6379	-	3?	-	-
6614	-	3?	-	-

Ch86: Chlewicki *et al.* (1986)—13 stars
KW87: Krełowski and Walker (1987)—4 stars
JS87: Josafatsson and Snow (1987)—59 stars
BP89: Benvenuti and Porceddu (1989)—26 stars
c: a correlated family
u: uncorrelated with any other feature

Further work has attempted to refine the membership of families. The ultraviolet bump λ2200 features in Table 2 as associated with KW87 family 2, but Seab and Snow (1984) found a slight positive correlation of λ2200 with λ4430 and none with λ5780 or λ6284, while BP89 found that both λ2200 and λ6269 are poorly correlated, if at all, with any of the other DIBs they observed. All that BP89 could say about λ5797 and λ6203 is that they do not correlate together. Porceddu, Benvenuti and Krełowski (1991) showed that λ6196, λ6203 and λ6205 vary independently, the first two having previously (Table 2) been attributed to the same family. λ5797 and λ6614 were tentatively assigned to the same KW87 family, but Fossey and Somerville (1994) have found sightlines where their relative strengths differ considerably. On the positive side, Westerlund and Krelowski (1989) found a star whose DIB spectrum contains members of KW87 family 2 but not of family 3. Le Bértre (1990) detected λ4430, λ5780 and λ6284 but not λ5797 in the spectrum of the companion to the carbon star CS 776. The observation of DIB-related emission lines, mentioned earlier, has a further significance: the emissions all are related to DIBs in the KW87 family 3, giving some confirmation of the reality of that grouping.

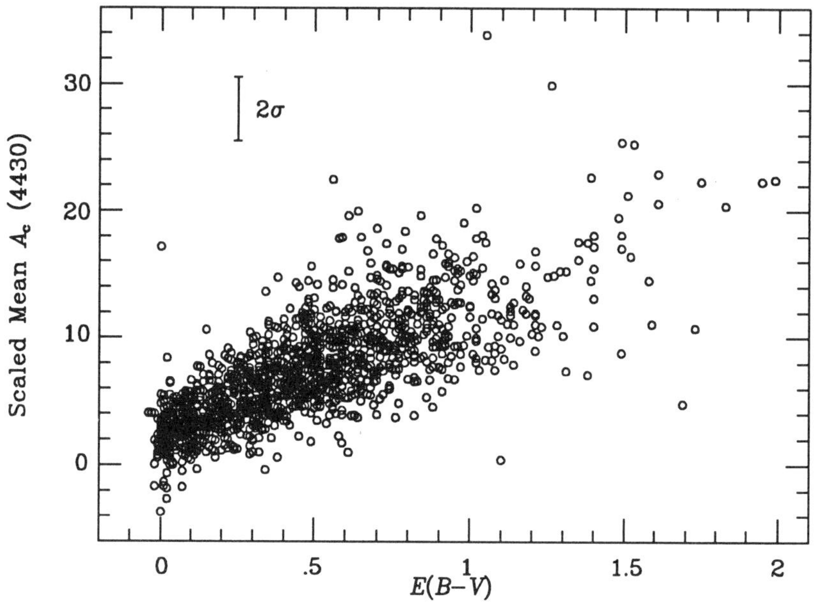

Fig. 1. A compilation of λ4430 data for 1500 stars (Somerville, 1989). Each point represents the mean scaled value of central depth A_c for one star. The 2σ bar is derived from the scatter of measurements for stars observed more than once and gives an internal estimate of a characteristic observational error for the entire dataset.

4. RELATION TO DUST PROPERTIES AND CLOUD DENSITY

We concentrate on λ4430, for which historically there is a much larger dataset of observations available than for any other of the DIBs. It is the strongest of the absorptions and has full width at half depth 20 Å (Herbig, 1975). Almost all of the results are the same for the small number of other bands that have been studied, principally λ5780, λ5797, λ6284, λ6614 and a few others in the yellow-red spectral region. Figure 1 (Somerville, 1989) shows a gathering together of data for λ4430 in relation to $E(B-V)$ that illustrates several general properties. It has been prepared along similar lines to the earlier compilation by Snow, York and Welty (1977) but with a larger dataset. Data from different observers have been scaled to put them onto a common basis, eliminating systematic bias between observers.

Several points may be noted about the data in Figure 1. Firstly, there is a strong linear correlation between A_c and $E(B-V)$, extending down to small values. While the more stable interstellar molecules such as H_2 and CO correlate very well with $E(B-V)$, their column densities fall off for small values (Savage *et al.*, 1977) because at low densities a molecule re-forms slowly after dissociation. We may therefore say that, at low densities, λ4430 is related more closely to $E(B-V)$ than is any known molecule. This result appears to hold for all DIBs studied; however, the recent data of Jenniskens, Ehrenfreund and Foing (1994) show a slight falling-off for λ5780 at very small values. The correlation with $E(B-V)$ is not perfect; the scatter about the mean in Figure 1 is significantly greater than the 2σ bar which represents a characteristic observational error for the dataset. Thus, the λ4430 strength is related closely but not perfectly to the dust grains that produce optical extinction. This result is found also for the other DIBs that have been studied.

In Figure 1, there is a non-zero intercept on the A_c axis. It was shown by Blades and Somerville (1977, 1981) that this is produced by the inclusion in the line profile of weak stellar lines, not resolved at low spectral resolution. This will affect essentially all the points in the dataset on average depending on spectral type, so that the non-zero intercept should effectively be subtracted from all the points and the whole plot moved down. Figure 1 is based primarily on older observations with photographic plates and early, low-resolution photoelectric detectors. With modern high-resolution spectrographs and CCD detectors, no non-zero intercept of this sort is found (e.g., λ5780 as observed by Herbig, 1993). It should be noted, however, that there have been few recent observations of λ4430—no doubt because with high-resolution spectrographs its width exceeds the wavelength range of many CCDs; λ4430 can occupy an entire echelle order. The older λ4430 data still often are used for comparison, frequently through the compilation of Snow *et al.* (1977).

The presence of an intercept has sometimes led analysis astray. For two quantities both of which depend on the amount of interstellar matter there will always be some correlation even if they are not physically related. It is necessary to remove the column-density dependence that they share and in effect look at the amount of scatter about it. A technique often used is to divide through by $E(B-V)$ to obtain an 'absorption per unit reddening'. This works well—unless there is a non-zero intercept, in which case as $E(B-V) \rightarrow 0$ the ratio goes hyperbolically to infinity suggesting the fictitious presence of a dependence. Somerville (1988) emphasised this and showed that a negative correlation of $A_c(4430)$ with $< n_H >$ and metal depletion reported previously is false. It will always be dangerous to work with ratios, particularly for the older DIB data, without considering possible effects of intercepts. Instead, the departure from the mean regression line in a diagram such as Figure 1 gives an essentially column-density-independent measure. Using this, Somerville found that the strength of λ4430 absorption does not vary with quantities that depend on cloud density: λ4430 is independent of density. This result does not hold for other DIBs examined, such as λ5780 and λ6284, which do decrease in strength as $< n_H >$ and metal depletion increase. This is shown very clearly in the analysis by Fossey (1990) of observations by McNally *et al.* (1987) and has been found also by other workers; it appears that λ5780 may be affected more in these situations than other DIBs such as λ5797 (Fossey and Somerville, 1994). So there are distinct differences between λ4430, and λ5780, and other DIBs studied; these are all in different families (Table 2).

Referring again to Figure 1, it is seen that there are about as many points anomalously above the distribution there as below it. There has always tended to be more emphasis on the sightlines where DIBs are anomalously weak, for the good observational reason that if something is 100% weaker than normal it isn't there at all, which is easily seen, whereas if it is 100% stronger this is not obvious without comparing it with other stars to see just what is 'normal'. Both types of anomaly have been reported over the years. Wampler (1966) discussed a selection of individual stars where λ4430 is weak compared with the general relation that he found. Similar results for λ5780 and λ5797 in the spectra of stars in CMa R1 and of dust-embedded stars were reported by Whittet and Blades (1980) and Baines and Whittet (1983); there is a suggestion in the latter paper that λ5797 is relatively weaker than λ5780. Wampler

(1966) also considered λ4430 for stars in open clusters, finding several clusters where the band is weak, but that in NGC 2244 it is anomalously strong. There are O stars in NGC 2244, so the strong λ4430 is not the result of a weaker UV radiation field. More recently, Krełowski *et al.* (1987) have compared λ4430 in the Per OB1 and Cep OB3 associations, finding a clear difference, the results for Cep OB3 being anomalously high. Another example is HD 97950, the central multiple star in the cluster NGC 3603, studied by Walborn (1977) and Somerville and Blades (1980), which shows exceptionally strong λ4430. In a poster paper at this meeting, Porceddu & Benvenuti (1994) show that for the high-latitude star HD 25137 both λ5780 and λ5797 are about as strong as for a normal star of twice the extinction. We may note that these bands do not always behave anomalously together.

As reported in other papers at this meeting (Martin, 1994; Martin *et al.*, 1994; Adamson and Whittet, 1994), there is no evidence for any structure in the interstellar polarization in the region of DIBs or of the λ2200 bump. A different comparison that can be made is with the value of λ_{max}, which is related to the mean grain size and tends to increase with R and with density, although with some scatter (Clayton and Mathis, 1988; Whittet *et al.*, 1994). In unpublished work, I have found that there is no relation between λ4430 anomalies and λ_{max}. This supports the conclusion that λ4430 is not related to cloud density.

One caveat about this work is that anomalies found in a comparison of DIB strength with $E(B-V)$ may arise from anomalous behaviour of $E(B-V)$ rather than of the DIB itself. It would be better to use the total extinction A_V, but $E(B-V)$ is used because it is readily available for very many stars whereas A_V is not. However, the evidence suggests that this may not matter much. If $R = A_V/E(B-V)$ were larger than its normal value, then a DIB with strength proportional to A_V would appear strong relative to $E(B-V)$. But it is found that DIBs anomalously *weak* relative to $E(B-V)$ tend to go with anomalously *large* values of R (e. g., Adamson, Whittet and Duley, 1991) which is in the wrong direction: such a DIB would look even more anomalously weak if compared with A_V. As discussed above, λ4430 anomalies are not related to λ_{max} and therefore probably not to R either. Further, the result that many DIBs can be both strong and weak relative to extinction, in different local regions, suggests that there is not a simple relation such that the DIB strength is everywhere inversely related to density or to R.

To summarize this discussion of the relation of DIBs to $E(B-V)$, it appears that all the DIBs studied (only a few of the strongest) have the following properties:

- Correlate well with total column density as measured by $E(B-V)$
  —better at low column densities than molecules such as H_2
  —but not perfectly
- There is no non-zero intercept
- Can be anomalously strong as well as weak
- There is no relation to interstellar polarization

Of those studied, 4430 alone has the property

- Does not vary with cloud density

whereas the other DIBs studied tend to become weaker as cloud density increases.

The larger grains responsible for optical extinction are ruled out as DIB carriers by the lack of polarization structure (Herbig, 1975; Smith, Snow and York, 1977).

The relation of DIB strengths to the far-ultraviolet extinction concerns the possibility that the carriers are among the small-size grain component, in which case a strong positive correlation would be expected. Results over the years, however, have shown either a null result or a weak negative correlation. Seab and Snow (1984) found that $\lambda4430$, $\lambda5780$ and $\lambda5797$ show no relation to the far-UV rise. Negative correlations have been found for $\lambda4430$ (Witt, Bohlin and Stecher, 1983, for the $E(1250 - V)$ colour excess; Krełowski *et al.*, 1987, for $E(15-18)$) and for yellow-red DIBs (Désert, Jenniskens and Dennefeld, 1994). These results indicate that the carriers tend not to co-exist with the small grains. Witt *et al.* (1983) suggested that DIB carriers may be related to volatile grain mantles and be weak when grain mantles evaporate to produce more grains of small size. Désert *et al.* (1994) discuss how their results may be understood in terms of PAH models.

The 4430 Å diffuse band has been detected in the spectra of stars in the two Magellanic Clouds, our nearest-neighbour galaxies: in the LMC by Houziaux, Nandy and Morgan (1980) and Hutchings (1980); and in the SMC by Hutchings and by Houziaux, Nandy and Morgan (1985). The $\lambda4430$ strength has been shown to be significantly weaker in relation to $E(B-V)$ than is normal in the Galaxy, for several SMC stars by Nandy, Morgan and Houziaux (1990) and for a LMC cluster by Heydari-Malayeri *et al.* (1993). This result may well be related to the different extinction laws in the Clouds, compared with the general galactic one—$\lambda2200$ is weak and the far-UV rise very strong, especially so in the SMC. Vladilo *et al.* (1987) detected yellow-red DIBs in the spectrum of the LMC supernova 1987A. These also appear to be weak in relation to extinction.

5. DARK CLOUDS AND THRESHOLD EFFECTS

Jenniskens *et al.* (1994) have recently published results for stars in the Orion dark cloud. When compared with general field stars, it is seen that for large $E(B-V)$ the DIB strength per unit $E(B-V)$ falls off with increasing column density; this means, with increasing distance into the cloud. For very small $E(B-V)$, the opposite trend is seen: the strength per $E(B-V)$ falls off with decreasing column density. These results are found for the four DIBs $\lambda6177$, $\lambda6196$, $\lambda6269$ and $\lambda6284$ studied. Now, in their diagrams not only have the data been divided through by $E(B-V)$ but also they are plotted logarithmically. There is no non-zero intercept effect here, and the trends shown are real, but I suggest that plotting in this way can conceal the physical reality of the situation. Figure 2 shows the data of Jenniskens *et al.* for $\lambda6284$, plotted more straightforwardly. It is seen that the DIB strength increases with $E(B-V)$ to start with and then shows no further increase with distance into the cloud. This behaviour is somewhat obscured by dividing through by $E(B-V)$. In this situation, to consider a 'DIB strength per unit $E(B-V)$' is not appropriate because an unknown amount of the reddening comes from diffuse material in front of the dark cloud. The same general variation with depth into the cloud is found for the data of Adamson *et al.* (1991), who studied seven DIBs in the Taurus dark cloud region, and for $\lambda5797$ and $\lambda5780$ in Cyg OB2 (Chlewicki *et al.*, 1986).

Physically, the DIB carrier is present in the general diffuse material and is destroyed with increasing depth into the cloud—it is converted into something else. For example, if the carrier is an ion then with depth into the cloud the density increases and there is shielding from interstellar radiation, so that the neutral molecule forms. Other model carriers can also be understood to change with depth. Thinking about

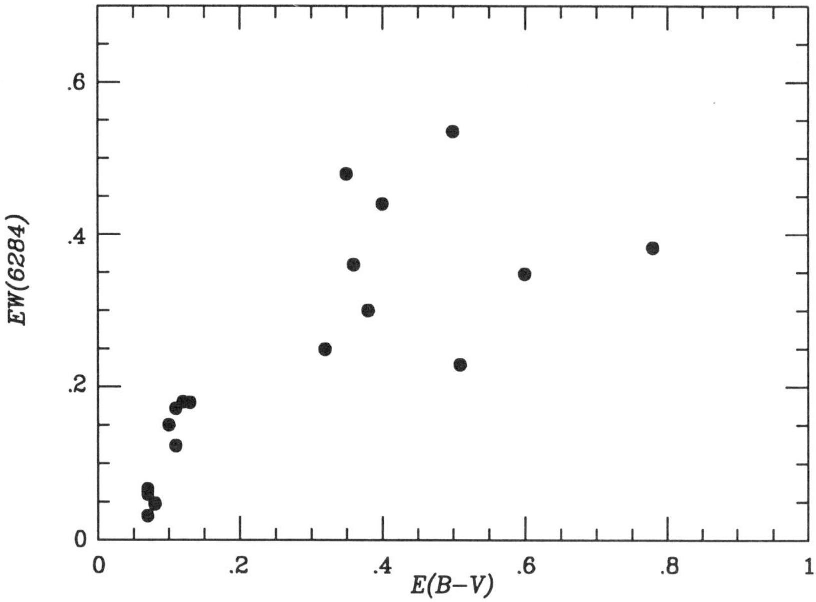

Fig. 2. Data for λ6284 from Jenniskens *et al.* (1994), replotted to show the relationships more clearly.

what it is that the carrier turns into, we might expect to see certain other DIBs *increase* in strength with depth into the cloud. These could be known DIBs that are normally weak, or they could be totally new ones. This has not yet been looked for, as far as I know, but with modern echelle spectrographs and large CCDs a sufficiently large wavelength range can be covered with good resolution and signal-to-noise to make such a search worthwhile.

For small $E(B-V)$ in Figure 2, there is a linear distribution, cutting the axis at a threshold value of $E(B-V)$. The other DIBs in the dataset of Jenniskens *et al.* show the same behaviour. If the effect is real, it means that (at least in this Orion region) the carrier is destroyed also if there is too small a column density—if the density becomes too low, or if there is inadequate shielding. For an ionic carrier, in such a region it could become doubly ionized.

McIntosh and Webster (1992) have shown that in the data of Adamson *et al.* (1991) the ratio of the strengths of the KW87 families 2 and 3 varies with R, which is a good measure of the depth into the cloud. This means that the carriers of these two families are both destroyed with increasing density (or with shielding from the radiation) but at a faster rate for family 2. By showing that the two families behave differently, this result also confirms the membership of the families (except that some of the bands involved are weak and their inclusion makes little difference).

The first observations reporting the decline of DIB strength with depth into dark clouds were by Snow and Cohen (1974), for λ4430 and λ5780 in the ρ Oph cloud.

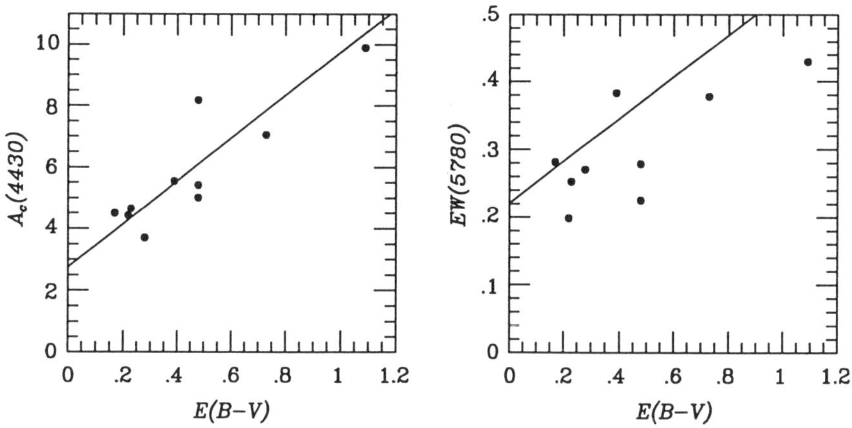

Fig. 3. The data points of Snow and Cohen (1974) related to the regression lines of Snow *et al.* (1977), for λ4430 and λ5780.

Both were shown to fall off with increasing extinction, compared with field stars. However, the fit to the field stars was constrained to pass through the origin. In Figure 3 it is seen that the λ5780 data lie below the unconstrained regression line for the dataset of Snow *et al.* (1977), in agreement with the results of later workers, but for λ4430 the data are well fitted by the line from Snow *et al.* Thus, for λ4430 there is no relative change with depth into the cloud, consistent with the result discussed above that λ4430 does not depend on cloud density.

6. RELATION TO MOLECULES

The clearest demonstration of a relation of DIB strengths to molecules is in the work by Krelowski *et al.* (1992) who give plots of spectra showing that the molecules CN, CH and CH^+ are present in sightlines where λ5797 is strong compared with λ5780 (which they call the ζ family of stars) and not in those where λ5797 is weak compared with λ5780 (the σ family). Note that 'family' here refers to stars and not to families of DIBs. In a poster paper at this meeting, Szczerba *et al.* (1994) took this work further by adding the results that the λ5797/λ5780 ratio is smallest for Be stars which lack the λ2200 feature, and that the blue DIBs λ4726, λ4763 and λ4780 have similar strengths in both families of stars.

A trend showing essentially the same result, that the CH abundance is greater where λ5780 is weaker compared with λ5797, is seen in Figure 5 of Fossey and Somerville (1994), who also show that the λ5780 abundance increases with the rotational temperature of H_2. Similar plots are given by Fossey (1990). Both the molecular abundance and the H_2 temperature may be interpreted as measures of cloud density and these results confirm the previous conclusions that the DIB strengths decline as density increases, more rapidly for λ5780 (family 2) than for λ5797 (family 3).

REFERENCES

Adamson, A. J., and Whittet, D. C. B. 1994, in *The Diffuse Interstellar Bands*, ed. A. G. G. M. Tielens and T. P. Snow, (Dordrecht: Kluwer), in press.
Adamson, A. J., Whittet, D. C. B., and Duley, W. W. 1991, *M. N. R. A. S.*, **252**, 234.
Baines, D. W. T., and Whittet, D. C. B. 1983, *M. N. R. A. S.*, **203**, 419.
Benvenuti, P., and Porceddu, I. 1989, *Astr. Ap.*, **223**, 329.
Blades, J. C., and Somerville, W. B. 1977, *M. N. R. A. S.*, **181**, 769.
——. 1981, *M. N. R. A. S.*, **197**, 543.
Carnochan, D. J. 1986, *M. N. R. A. S.*, **219**, 903.
Chlewicki, G., van der Zwet, G. P., van Ijzendoorn, L. J., and Greenberg, J. M. 1986, *Ap. J.*, **305**, 455.
Clayton, G. C., and Mathis, J. S. 1988, *Ap. J.*, **327**, 911.
Désert, F.-X., Jenniskens, P., and Dennefeld, M. 1994, in *The Diffuse Interstellar Bands*, ed. A. G. G. M. Tielens and T. P. Snow, (Dordrecht: Kluwer), in press.
Draine, B. T. 1978, *Ap. J. Suppl.*, **36**, 595.
Fitzpatrick, E. L., and Massa, D. 1990, *Ap. J. Suppl.*, **72**, 163.
Fossey, S. J. 1990, PhD thesis, University of London.
Fossey, S. J., and Somerville, W. B. 1994, in *The Diffuse Interstellar Bands*, ed. A. G. G. M. Tielens and T. P. Snow, (Dordrecht: Kluwer), in press.
Gondhalekar, P. M., Phillips, A. P., and Wilson, R. 1980, *Astr. Ap.*, **85**, 272.
Herbig, G. H. 1975, *Ap. J.*, **196**, 129.
——. 1993, *Ap. J.*, **407**, 142.
Heydari-Malayeri, M., Grebel, E. K., Melnick, J., and Jorda, L. 1993, *Astr. Ap.*, **278**, 11.
Houziaux, L., Nandy, K., and Morgan, D. H. 1980, *Astr. Ap.*, **84**, 377.
——. 1985, *M. N. R. A. S.*, **215**, 5P.
Hutchings, J. B. 1980, *Pub. A. S. P.*, **92**, 592.
Jenniskens, P., Ehrenfreund, P., and Foing, B. 1994, *Astr. Ap.*, **281**, 517.
Jenkins, E. B., Savage, B. D., and Spitzer, L. 1986, *Ap. J.*, **301**, 355.
Josafatsson, K., and Snow, T. P. 1987, *Ap. J.*, **319**, 436.
Krełowski, J., Snow, T. P., Seab, C. G., and Papaj, J. 1992, *M. N. R. A. S.*, **258**, 693.
Krełowski, J., and Walker, G. A. H. 1987, *Ap. J.*, **312**, 860.
Krełowski, J., Walker, G. A. H., Grieve, G. R., and Hill, G. M. 1987, *Ap. J.*, **316**, 449.
Le Bertre, T. 1990, *Astr. Ap.*, **236**, 472.
Le Bertre, T., and Lequeux, J. 1993, *Astr. Ap.*, **274**, 909.
McIntosh, A., and Webster, A. 1992, *M. N. R. A. S.*, **255**, 37P.
McNally, D., Ashfield, M., Baines, D. W. T., Fossey, S., Rees, P. C. T., Somerville, W. B., and Whittet, D. C. B. 1987, in *IAU Symposium 120, Astrochemistry*, ed M. S. Vardya & S. P. Tarafdar (Dordrecht: Reidel), p. 321.
Martin, P. G. 1994, in *The Diffuse Interstellar Bands*, ed. A. G. G. M. Tielens and T. P. Snow, (Dordrecht: Kluwer), in press.
Martin, P. G., Somerville, W. B., McNally, D., Whittet, D. C. B., Allen, R. G., Walsh, J. R. and Wolff, M. J. 1994, in *The Diffuse Interstellar Bands*, ed. A. G. G. M. Tielens and T. P. Snow, (Dordrecht: Kluwer), in press.
Mathis, J. S., Rumpl, W., and Nordsieck, K. H. 1977, *Ap. J.*, **217**, 425.
Merrill, P. W., Sanford, R. F., Wilson, O. C., and Burwell, C. G. 1937, *Ap. J.*, **86**, 274.
Nandy, K., Morgan, D. H., and Houziaux, L. 1990, *M. N. R. A. S.*, **245**, 318.
Porceddu, I., Benvenuti, P., and Krełowski, J. 1991, *Astr. Ap.*, **248**, 188.
Porceddu, I., and Benvenuti, P. 1994, in *The Diffuse Interstellar Bands Contributed Papers*, ed. A. G. G. M. Tielens and T. P. Snow, NASA-CP, in press.
Rao, N. K., and Lambert, D. L. 1993, *M. N. R. A. S.*, **263**, L27.
Savage, B. D., Bohlin, R. C., Drake, J. F., and Budich, W. 1977, *Ap. J.*, **216**, 291.
Scarrott, S. M., Watkin, S., Miles, J. R., and Sarre, P. J. 1992, *M. N. R. A. S.*, **255**, 11P.
Seab, C. G. 1994, in *The Diffuse Interstellar Bands*, ed. A. G. G. M. Tielens and T. P. Snow, (Dordrecht: Kluwer), in press.
Seab, C. G., and Snow, T. P. 1984, *Ap. J.*, **277**, 200.
Smith, W. H., Snow, T. P., and York, D. G. 1977, *Ap. J.*, **218**, 124.
Snow, T. P., Allen, M. M., and Polidan, R. S. 1990, *Ap. J.*, **359**, L23.
Snow, T. P., and Cohen, J. G. 1974, *Ap. J.*, **194**, 313.
Snow, T. P., York, D. G., and Welty, D. E. 1977, *A. J.*, **82**, 113.
Somerville, W. B. 1988, *M. N. R. A. S.*, **234**, 655.
——. 1989, in *Interstellar Dust: Contributed Papers*, ed. A. G. G. M. Tielens and L. J. Allamandola, NASA CP-3036, p. 77.
Somerville, W. B., and Blades, J. C. 1980, *M. N. R. A. S.*, **192**, 719.
Sorrell, W. H. 1992, *Comments Ap.*, **16**, 123.
Spitzer, L. 1968, *Diffuse Matter in Space* (New York: Interscience), p. 68.
——. 1985, *Ap. J.*, **290**, L21.

Szczerba, R., Krełowski, J., Walker, G. A. H., Kennelly, E. T., and Sneden, C. 1994, in *The Diffuse Interstellar Bands Contributed Papers*, ed. A. G. G. M. Tielens and T. P. Snow, NASA-CP, in press.
van Dishoeck, E. F., and Black, J. H. 1986, *Ap. J. Suppl.*, **62**, 109.
Vladilo, G., Crivellari, L., Molaro, P., and Beckman, J. E. 1987, *Astr. Ap.*, **182**, L59.
Walborn, N. R. 1977, *Pub. A. S. P.*, **89**, 765.
Wampler, E. J. 1966, *Ap. J.*, **144**, 921.
Westerlund, B. E., and Krełowski, J. 1989, *Astr. Ap.*, **218**, 216.
Whittet, D. C. B. 1984, *M. N. R. A. S.*, **210**, 479.
Whittet, D. C. B. and Blades, J. C. 1980, *M. N. R. A. S.*, **190**, 41P.
Whittet, D. C. B., Gerakines, P. A., Carkner, A. L., Hough, J. H., Martin, P. G., Prusti, T., and Kilkenny, D. 1994, *M. N. R. A. S.*, **268**, 1.
Witt, A. N., Bohlin, R. C., and Stecher, T. P. 1983, *Ap. J.*, **267**, L47.

DIBS AND UV EXTINCTION

F.-X. DÉSERT[1], P. JENNISKENS[2]
[1] *Inst. d'Astroph. Spatiale, Bat 121, Université Paris XI, F-91405 Orsay*
[2] *NASA/Ames Research Center, Mail Stop 239-4, Moffett Field, CA 94035*

ABSTRACT. The search for the missing link between diffuse interstellar bands (DIB) and UV continuum extinction is reviewed. Indeed, there are significant correlations between DIB strength and the shape of the UV extinction curve, which put constraints on both the nature of the DIB carrier and the origin of certain UV extinction features.

1. INTRODUCTION

Both leading schools of thought on the origin of diffuse interstellar bands, impurities in grains (e.g. Herbig 1975, Duley 1982, Shapiro & Holcomb 1986, Krełowski *et al.* 1987) or isolated molecules (e.g. Smith *et al.* 1977, Danks & Lambert 1976) predict some correlation between DIB characteristics and the features in the UV extinction curve. The shape of the extinction curve is determined by the relative abundance of small grains in the ensemble of all interstellar dust, reflects the presence of, the similarly elusive, carriers of the 2175 Å bump and far-UV non-linear rise, and affects the spectrum of the local radiation field of ionising photons.

Over the years, there has been a steady stream of publications dedicated to finding relationships between UV extinction and DIB absorption. Weak trends and outright negative results have been reported, but only recently it has become clear which parts of the extinction curves relate to DIB strength behaviour. These results are discussed in the context of current models of UV extinction and DIB absorption.

2. THE UV EXTINCTION CURVE

The extinction curve k(x) is conveniently expressed in terms of colour excess (E):

$$E_{\lambda-V} = (m_\lambda - m_V) - (m_\lambda - m_V)_o \tag{1}$$

These parameters refer to actually observed magnitudes at given wavelength and the index $)_o$ refers to the colour of an unreddened star of the same spectral type. The extinction curve in this representation is ($x = 1/\lambda$):

$$k(x) = E_{\lambda-V}/E_{B-V} \tag{2}$$

A. G. G. M. Tielens and T. P. Snow (eds.), The Diffuse Interstellar Bands, 97–103.

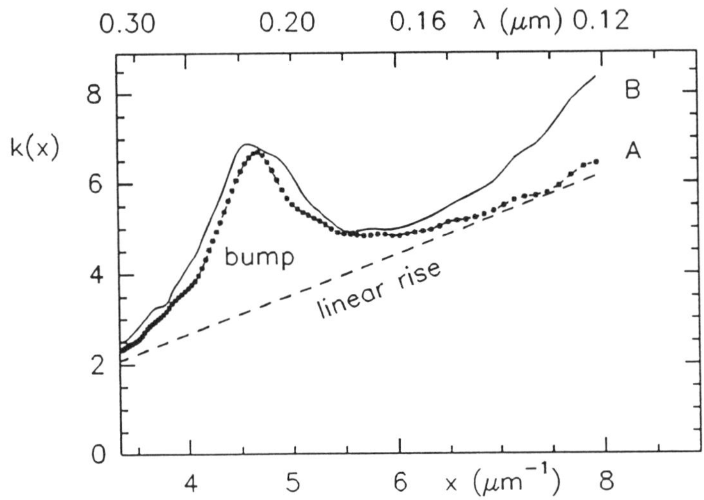

Fig. 1. Examples of extinction curves with high and low far-UV non-linear rise from the dataset of Aiello *et al.* (1988). HD190603 (A) has c4 = 0.04 and HD61827 (B) has c4=0.83. Both have the same amount of linear rise (from: Jenniskens *et al.* 1992).

All UV extinction curves are well described by only five free parameters: the bump width γ, position x_o and height c_3/γ^2, the slope of the linear rise c_2, and the amount of far-UV non-linear rise c_4 (Fitzpatrick & Massa 1988):

$$k(x) = c1 + c2x + c3D(x, x_o, \gamma) + c4f(x) \tag{3}$$

$$D(x, x_o, \gamma) = \frac{x^2}{(x^2 - x_o^2)^2 + \gamma^2 x^2}$$

$$f(x) = 0.5392(x - 5.9)^2 + 0.0564(x - 5.9)^3 + ... \quad (5.9 < x < 8.0 \mu m)$$

The shape of the bump D(x, x_o, γ) and the far-UV non-linear rise $f(x)$ does not vary from one line of sight to another (Fitzpatrick & Massa 1986, Greenberg & Chlewicki 1983). Far-UV non-linear rise, bump, and linear rise are independent components. However, there is a weak positive correlation between bump position and far-UV non-linear rise and a weak anti-correlation between bump strength and far-UV non-linear rise (Fitzpatrick & Massa 1988, Jenniskens & Greenberg 1993). Fig. 1 shows two examples of extinction curves that have similar bump strength and linear rise but varying amounts of far-UV non-linear rise.

The shape of the continuum extinction at long wavelengths in the red and near-infrared is constant (Jones & Hyland 1980, Clayton & Mathis 1988, Whittet 1989, Cardelli *et al.* 1989, Chini & Wargau 1990). Changes at visible and blue wavelengths (in the ratio of total to selective extinction in the visual) are related to the onset of the linear rise (e.g. Jenniskens 1994). Correlations of DIB strength with E_{B-V} are beyond the scope of this paper.

3. CORRELATION WITH DIFFUSE INTERSTELLAR BANDS

Early studies looked at the absolute amount of DIB absorption and UV extinction curve colours before normalisation to unit reddening (or unit extinction in the visual). Wu (1972) used *OAO-2* extinction measurements at 2175, 3500, and 1200 Å in combination with high resolution (0.23 Å) echelle spectra of DIBs λ4430, λ5780, and λ5797 in 66 lines of sight. Using $E_{2175-3500}$ as a measure of bump strength and E_{1200-V} as a measure of rise in the extinction curve, Wu found no correlation with E_{1200-V} and that λ5780 and λ5797 correlated better with E_{B-V} than with $E_{2175-3500}$. Correlations involving λ4430 were worse than those of the other DIBs.

Nandy and Thompson (1975) compared the non-normalised depth (Ac) of the 2175 Å bump with the depth of λ4430. Contrary to Wu (1972), they found a relatively good correlation (0.93 +/- 0.01) for 60 low reddened ($E_{B-V} < 0.3$) stars, now thought to be mainly due to the relative constant abundance of DIBs and a relatively constant shape of the UV extinction curve in the diffuse medium.

Dorschner, Friedemann, and Gurtler (1977) considered the equivalent width (W) of the 2200 Å bump instead of only the peak depth. They found that W_{4430} is less well correlated with W_{2200} than with E_{B-V}.

Herbig (1975) noted that the correlation of DIB strength with colours becomes worse when colours are considered above the turning point in the extinction curve, that is where the linear rise starts to contribute significantly to the total extinction.

Snow, York, and Welty (1977) compiled a list of DIB strengths, which was applied in correlation studies long after (e.g. Danks 1980, Seab & Snow 1984). However, it was soon realised that there were relatively large errors in the listed equivalent widths. The limitation of correlation studies was the accuracy of the DIB measurements and not the accuracy of the UV extinction curves. Schmidt (1978) went into considerable effort to improve the DIB measurements making a clear distinction between λ5780 and λ5778, the separation of which was discussed in detail by Herbig (1975). Schmidt found again a poor correlation between λ5780 and the total extinction at 1500 Å . Also, the correlation with bump strength was poor.

Wu, York, and Snow (1981) used *ANS* data of 110 stars in combination with DIBs listed in Snow, York, and Welty (1977) and subsets of these data thought to be most accurate. *ANS* passbands are at 1549 (149), 1799 (149), 2200 (200), 2493 (150) and 3294 (101) Å. Using the depth of the bump, they found that in their sample Ac_{4430} is as well correlated with Ac_{2175} as with E_{B-V}. The data included more heavily reddened stars than Nandy & Thompson (1975). Small differences in correlation coefficients were attributed to inclusions of background (= linear rise) extinction into the bump component. They also confirmed that the correlation of λ4430 with extinction at 2500 Å and shorter wavelengths is progressively less good.

The observers thus far did not look at band strength relative to E_{B-V}. Witt, Bohlin, and Stecher (1983) first compared normalised DIB strengths from the data compiled by Snow, York, and Welty (1977) with *IUE* extinction curves. The parameters k(2175 Å), k(2160 Å), k(1250 Å) were considered, which are proportional to the bump, the linear rise, and a sum of far-UV non-linear rise plus linear rise, respectively. The data concentrated on lines of sight with anomalous Ac_{4430}/E_{B-V}. They found a weak positive correlation between Ac_{4430}/E_{B-V} and normalised strength of the 2175 extinction bump. A weak negative correlation was found with the sum of linear and non-linear rise at 1250 Å.

Seab and Snow (1984) normalised *IUE* data of 50 stars to E_{B-V} and also found a weak positive correlation between the depth of λ4430 and the normalised bump strength. No other correlations were found. Their normalised procedure, however, included a non-zero intercept in the relation of Ac_{4430} with E_{B-V}. This non-zero intercept is now thought to be mainly a result of contamination by stellar lines in the list compiled by Snow, York, and Welty (Somerville 1988).

Josafatsson & Snow (1987) first applied modern solid state detectors to this problem. New DIB measurements of 37 stars (6-14 stars per DIB) were compared to the bump in the extinction curve as measured by *IUE*. They also included a non-zero intercept in the normalisation. It was found that in particular W_{5778} showed a good correlation with the bump strength, but this good correlation was not significant given the small sample size, 6 stars only for λ5778. Other diffuse bands showed no correlations (N=6 to 14). The far-UV rise was not mentioned.

The study of Krełowski *et al.* (1987) differs from that of Wu *et al.* (1981) in that solid state measurements of λ4430 were considered in relation to *ANS* data and the DIB measurements were normalised. Again, excess absorption (including non-zero intercept) in lines of sight to the Per OB1 and Cep OB3 associations showed an anti-correlation with $E_{1500-1800}$. This *ANS* colour excess relates to the far-UV non-linear rise, but is also sensitive to CIV contamination and contains a significant component of the linear rise. The correlation disappeared when stars of other associations at higher lattitude were added. We now believe this was because of adding lines of sight with deviating linear rise.

Krełowski (1989) and Krełowski & Wegner (1989) insisted on the difference between σ Sco and ζ Oph type of extinction curves as to reflect different behavior of DIBs. The difference between these two types of extinction curves is that σ Sco has a low linear rise and low non-linear rise, while ζ Oph has higher linear rise and higher non-linear rise. The difference goes back to the weak anti-correlations of early finds.

Benvenuti & Porceddu (1989) pursued the possible family dependence of the correlation of DIBs with the bump by correlating λ5780, λ5797, λ6196, λ6203, λ6270 and λ6284 of 26 southern hemisphere stars with *IUE* UV extinction curves. Principal component analysis was applied to the dataset after normalising to unit reddening. The first eigenvalue was found at 55% confidence level and five variables were needed to increase this value to 95%. These values reflect the family behavior, because the normalised intensity of the bump was poorly if not at all correlated with any of the observed DIBs.

Recently, a number of lines of sight with extreme UV extinction curves have been studied by amongst others Porceddu & Benvenuti (1990) and Snow & Seab (1991). Extreme weakness of DIBs per unit reddening (and weakness of the bump) has been related to high densities in molecular clouds (e.g. Snow & Cohen 1974, Joseph *et al.* 1986).

Finally, Désert, Jenniskens, and Dennefeld (1992) considered independently the far-UV non-linear rise and the linear rise and also considered the width and position of the bump in relation to new DIB measurements. Fig. 2 shows an example of the correlation found: there is a clear anti-correlation of far-UV non-linear rise with DIB extinction but no correlation with linear rise. The anti-correlation may be stronger

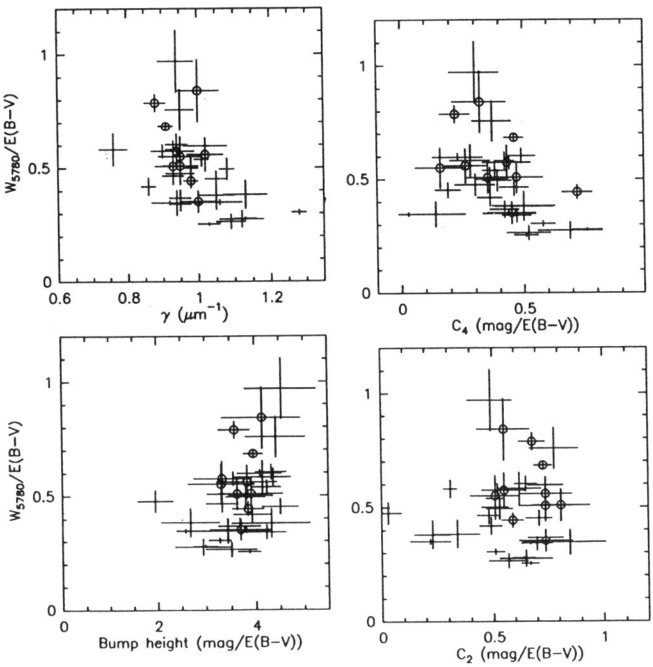

Fig. 2. Correlation diagram between the equivalent width of λ5780 per unit reddening and the four main UV extinction curve parameters defined in section 2.

for the broad and very broad DIBs like λ5780 and λ4430 than the narrow DIBs like λ5797 and λ6196. There is a similar anti-correlation with bump width and a weak correlation with bump height.

4. DISCUSSION

Wu (1972) interpreted the results in the fashion of that time in terms of impurities in grains or pre-ionisation in molecules. Some evidence for this was a blue wing found on the λ5780 profile, which now is thought to be due to the narrow spectral range covered (44 Å) and a baseline that did not account for the presence of the broad λ5778. From the lack of a a good correlation of DIBs with bump and far-UV rise, it was concluded that the far-UV rise and DIBs do not have a common carrier and DIBs are not due to the bump carrier.

Witt *et al.* (1983) interpreted the weak anti-correlation of λ4430 per unit E_{B-V} versus far-UV rise (= non-linear rise) in terms of a DIB carrier in volatile grain mantles of small grains. Evaporation of such mantles may decrease λ4430 and increase the far-UV extinction rise at the same time. The absence of λ4430 in some lines of sight suggested that λ4430 be produced in volatile grain mantles and not in the grain substance of any component of the size spectrum of grains.

Krełowski *et al.* (1987) interpreted the anti-correlation along similar lines, except they thought that a strong λ4430 was due to favourably aligned cores in a magnetic field and the bump carrier was part of large (larger than 0.22 micron) cores. However, such large grains, of course, have gray extinction in the ultraviolet (no bump, no far-UV non-linear rise) and there is no sign of polarisation in the DIBs or the bump that would imply that they are part of large grains (Martin and Angel 1975).

Désert *et al.* (1992), in the fashion of this time, interpret the anti-correlation in terms of the big molecule (PAH) hypothesis. The far-UV non-linear rise is a pure absorption feature and correlates positively with the abundance of CH, which itself is proportional to the abundance of molecular hydrogen (Jenniskens *et al.* 1992). Therefore, the far-UV non-linear rise can be due to neutral large molecules, for example due to $\sigma \rightarrow \sigma^*$ transitions. However, neutral PAHs of sizes less than about 60 carbon atoms have no strong absorptions in the visible and probably do not account for DIBs. Ionised PAHs do. If DIBs are due to some ionised PAH species and the far-UV non-linear rise is due to neutral PAHs, a low degree of PAH ionisation will result in relatively weak DIBs and a large far-UV non-linear rise. Alternatively, dehydrogenation of PAHs (i.e. in terms of radicals) rather than ionisation might be responsible for the DIBs and the absence of a far-UV non-linear rise where H_2 dissociating UV photons are abundant.

5. CONCLUSIONS

The parameterisation scheme of Fitzpatrick & Massa (Eq. 3) has helped in finding some relevant UV parameters for DIB correlation studies. The DIB strength is related to the bump and the far-UV non-linear rise, two components that show weak internal correlations.

- The far-UV non-linear rise and bump width are clearly anti-correlated with the DIB strength.
- A positive correlation with the 2175 Å bump does exist but is weaker.
- There is no correlation of DIB strength with linear rise or the ratio of total to selective extinction in the visual.

A possible interpretation of these conclusions is that the far-UV non-linear rise is due to neutral PAHs, while DIBs are due to ionised PAHs. Weak positive correlations with bump strength could reflect variations in PAH abundances. The DIBs are not related to the small grains ($< 0.05\mu$m) responsible for the linear rise.

ACKNOWLEDGEMENTS. We thank M. Dennefeld of I.A.P. for his support of this work. We are also indebted to D. Gillet and coworkers of O.H.P., J. M. Greenberg and H. Habing of Leiden Observatory, and B. Barsella of Pisa Observatory.

REFERENCES

Aiello, S., Barsella, B., Chlewicki, G., Greenberg, J.M., Patriarchi, P., Perinotto, M. 1988, *Astr. Ap. Suppl. Ser.*, **73**, 195.
Benvenuti, P. and Porceddu, I. 1989, *Astr. Ap.*, **223**, 329.
Cardelli, J.A., Clayton, G.C., and Mathis, J.S. 1989, *Ap. J.*, **345**, 245.
Chini, R. and Wargau, W.F. 1990, *Astr. Ap.*, **227**, 213.

Clayton, G.C. and Mathis, J.S. 1988, *Ap. J.*, **327**, 911.
Danks, A.C. and Lambert, D.L. 1976, *MNRAS*, **174**, 571.
Danks, A.C. 1980, *PASP*, **92**, 52.
Désert, F.-X., Jenniskens, P., and Dennefeld, M. 1992, in *Organic Matter in Interstellar Extinction*, PhD Thesis Leiden University, p. 85.
Dorschner, J., Friedemann, C., and Gurtler, J. 1977, *Astr. Ap.*, **58**, 201.
Duley, W.W. 1982, *Astroph. & Space Sci.*, **88**, 501.
Fitzpatrick, E.L. and Massa, D. 1986, *Ap. J.*, **307**, 286.
Fitzpatrick, E.L. and Massa, D. 1988, *Ap. J.*, **328**, 734.
Greenberg, J.M. and Chlewicki, G. 1983, *Ap. J.*, **272**, 563.
Herbig, G.H. 1975, *Ap. J.*, **196**, 129.
Jenniskens, P., Ehrenfreund, P., and Désert, F.-X. 1992, *Astr. Ap.*, **265**, L1.
Jenniskens, P. and Greenberg, J.M. 1993, *Astr. Ap.*, **274**, 439.
Jenniskens, P. 1994, *Astr. Ap.*, **284**, 227.
Jones, T.L. and Hyland, A.R. 1980, *MNRAS*, **192**, 359.
Josafatsson, K. and Snow, T.P. 1987, *Ap. J.*, **319**, 436.
Joseph, C.L., Snow, T.P., Seab, C.G., and Crutcher, R.M. 1986, *Ap. J.*, **309**, 771.
Krełowski, J., Walker, G.A.H., Grieve, G.R., and Hill, G.M. 1987, *Ap. J.*, **316**, 449.
Krełowski, J. and Wegner, W. 1989, *Astron. Nach.*, **310**, 281.
Krełowski, J. 1989, in *Interstellar Dust*, IAU-Symp. 135, Allamandola & Tielens eds., p. 67.
Martin, P.G. and Angel, J.R.P. 1975, *Ap. J.*, **195**, 379.
Nandy, K. and Thompson, G.I. 1975, *MNRAS*, **173**, 237.
Porceddu, I. and Benvenuti, P. 1991, in *Evolution in Astrophysics*, ESA SP-310, p. 495.
Seab, C.G., Snow, T.P., and Joseph, C.L. 1981, *Ap. J.*, **246**, 788.
Schmidt, E.G. 1978, *Ap. J.*, **223**, 458.
Shapiro, P.R. and Holcomb, K.A. 1986, *Ap. J.*, **305**, 433.
Smith, W.H., Snow, T.P., and York, D.G. 1977, *Ap. J.*, **218**, 124.
Snow, T.P. and Cohen, J.G. 1974, *Ap. J.*, **194**, 313.
Snow, T.P., York, D.G., and Welty, D.E. 1977, *Astron. J.*, **82**, 113.
Snow, T.P. and Seab, C.G. 1991, *Ap. J.*, **382**, 189.
Somerville, W.B. 1988, in *Interstellar Dust: Contributed Papers*, NASA Conf. Publ. 3036, Tielens & Allamandola eds., p. 77.
Witt, A.N., Bohlin, R.C., and Stecher, T.P. 1983, *Ap. J.*, **267**, L47.
Whittet, D.C.B. 1989, in *Interstellar Dust*, IAU-Symp. 135, Allamandola & Tielens eds., p. 445.
Wu, C.-C. 1972, *Ap. J.*, **178**, 681.
Wu, C.-C., York, D.G., and Snow, T.P. 1981, *Astron. J.*, **86**, 755.

THE ENVIRONMENT DEPENDENCE OF DIFFUSE INTERSTELLAR BANDS

P. EHRENFREUND [1] , P. JENNISKENS [2]
[1] *Leiden Observatory, PO Box 9513, 2300 RA Leiden, The Netherlands*
[2] *NASA/Ames Research Center, Mail Stop 239-4, Moffett Field, CA 94035-1000, USA*

ABSTRACT. Diffuse interstellar bands (DIBs) show different behaviour according to their environment. For some DIBs a decrease in strength is observed in the direction of dense clouds and towards H II regions. We have studied these dependencies in Orion for two DIBs in particular, the narrow well defined DIB at 6196 Å and the strong DIB at 6284 Å. A key characteristic of interstellar environment is the physical state of hydrogen in the line of sight, which can be molecular (H_2), atomic (H I), or ionized (H II). Usually, there is a range of physical conditions along the line of sight. We find that there is a good correlation of DIB strength with atomic hydrogen column density. The correlation of DIB strength with reddening (which is proportional to all matter in line of sight) breaks down for $E_{(B-V)} < 0.08$ where molecular hydrogen is absent and some hydrogen is thought to be ionized. Aparently, DIBs (both λ6196 and λ6284) are absent or much weaker in the H II medium. In lines of sight towards M42, where hydrogen is mainly atomic, λ6196 is weak but λ6284 is as strong (or stronger) as elsewhere. Finally, DIBs are indeed weaker per unit reddening (or unit extinction) for lines of sight that penetrate a dense cloud, but λ6196 tends to weaken less quickly with increasing reddening than λ6284. We interpret the results in terms of the influence of UV photons, and conclude that λ6196 is both ionized and double ionized (destroyed) by less energetic photons than λ6284. This is consistent with a molecular origin for the DIBs.

1. INTRODUCTION

Currently more than 150 DIBs are known and the number is still increasing (Herbig 1975, Herbig & Leka 1991, Jenniskens & Désert 1994). Among the many proposed carriers are gas phase molecules, dust grains and impurities embedded in grains (Smith *et al.* 1977, van der Zwet 1987). Most recent attention has been given to the hypothesis that polycyclic aromatic hydrocarbons (PAHs) may be responsible for the DIBs (Léger & d'Hendecourt 1985, van der Zwet & Allamandola 1985). In order to have strong absorption in the visual, PAHs need to be ionized (Salama & Allamandola 1992a,b, Ehrenfreund *et al.* 1992).

Diffuse bands show different behaviour according to their environment (e.g. in dark clouds and H II regions). Snow & Cohen (1974) showed that the strength of

A. G. G. M. Tielens and T. P. Snow (eds.), The Diffuse Interstellar Bands, 105–112.

the DIBs per unit reddening ($W/E_{(B-V)}$, where W is the equivalent width) decreases with increasing reddening $E_{(B-V)}$ in the Ophiuchus dark cloud. From a similar result found in the Taurus dark cloud, Adamson *et al.* (1991) concluded that the relative weakness of the DIBs in dark clouds is driven by changing grain properties.

We have studied the relative DIB intensities of 4 DIBs: λ6284, λ6177, λ6196 and λ6269 in 22 lines of sight towards Orion and GMC 214-13. Orion provides a more extreme case of UV radiation field than Ophiuchus and offers the possibility to study low reddened lines of sight, as many stars are found in front of the molecular cloud complex. The lines of sight in Orion can be grouped in 3 different types of environment: the low reddened stars in the Orion OBIa association, the medium reddened stars in the Orion Nebula and the highly reddened stars in the outskirts of the Orion giant molecular cloud (GMC). As indicators for line of sight conditions we used reddening, $E_{(B-V)}$, total extinction in the visible A_V, and column densities of hydrogen in atomic, molecular and ionized form.

2. OBSERVATIONS

The data were obtained with the 1.52 m f/27.6 coudé telescope at the Observatoire Haute Provence, France. The spectra are recorded by a 2048 diode linear Reticon array which is read by two 1024px CCDs. The resolving power of the spectrograph Aurelie was set to R = 15000 or $\Delta\lambda = 0.3$ Å, using a 1200 lines/mm grating with a blaze angle at 5000 Å and a dispersion of 8.1 Å/mm.

3. RESULTS

3.1. $E_{(B-V)}$ DEPENDENCE

The stars in Orion show a decline in DIB strength per unit reddening similar to those found for ρ Oph (Snow & Cohen 1974) and Taurus (Adamson *et al.* 1991). The decrease with reddening has roughly an $E^{-1}_{(B-V)}$ dependence, see Fig. 1. This relationship is followed most closely by λ6177 and λ6269. Towards the Orion nebula, λ6284 is stronger while λ6196 is somewhat weaker than expected from this dependence. According to Table 1, the first group contains stars HD34989 to HD36861 which are part of a large association of O and B stars. The stars HD37020-23 and HD37061 are ionisation sources for the H II regions M42 and M43, which form the Orion nebula. The three stars at highest reddening are HD38023, HD38563A and HD39557. These stars are found embedded in outer regions of the Orion GMC, far away from the active regions M42 and NGC2023, although HD38563A and HD38023 are associated with reflection nebulosity. Most, if not all, of the reddening in these lines of sight is due to matter that is part of the GMC. Because Orion is nearby, 500 pc, and far below the galactic plane, 180 pc, there is little foreground reddening. HD34085 has a reddening of $E_{(B-V)} = 0.00$.

3.2. A_V DEPENDENCE

It is better to consider DIB strength per unit amount of extinction in the visual, A_V, in order to characterize the total amount of matter in the line of sight. Fig. 2 shows the strength of the λ6284 and the weak, but well defined, λ6196 in units of W/A_V

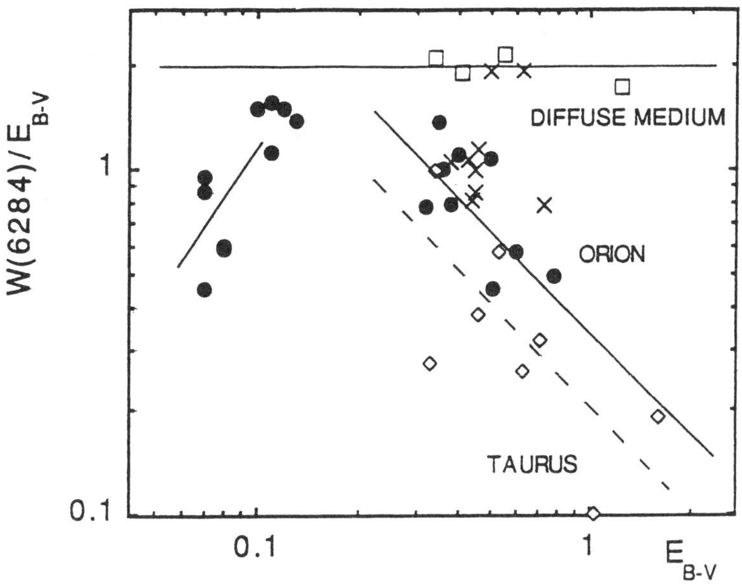

Fig. 1. DIB strength $W_{6284}/E_{(B-V)}$ for lines of sight passing the Orion Giant Molecular Cloud (filled circles) and the Taurus dark cloud (diamonds). Diffuse medium values are given by crosses and by open squares (HD183143, HD21389 and the stars in GMC214-13). An $E^{-1}_{(B-V)}$ dependence is fitted to the Taurus data.

versus A_V. Symbols are as in Fig. 1. Crosses show the diffuse medium lines of sight. The solid line is a fit of an A_V^{-1} dependence to the Taurus data. Again, the band at λ6284 is about a factor of two stronger in Orion as compared to Taurus. More precisely, the highly reddened stars are about a factor of 1.5 stronger, the stars in the Orion Nebula as much as a factor 2.6. There are also some significant differences between λ6284 and λ 6269. λ6284 is typically more reduced in strength at $A_V = 1$ than λ6196 when compared to the diffuse medium standards.

3.3. DIB STRENGTH AT LOW REDDENING

A number of Orion stars have low reddening ($0.07 < E_{(B-V)} < 0.2$). In all stars the strong DIB at 6284 Å is detected. There is a weakening of $W_{6284}/E_{(B-V)}$ for lines of sight with $E_{(B-V)} < 0.1$ compared to the typical diffuse medium values (Fig.1). The data do not allow us to infer whether or not other DIBs show a similar behaviour.

The strong λ5780 is also well studied at low reddening. This DIB correlates well with λ6284 (Jenniskens & Desert 1993). Fig. 3 shows log-log plots of observational data for 34 lines of sight (Savage *et al.* 1977). Below $E_{(B-V)} = 0.08$ molecular hydrogen is effectively dissociated due to lack of self-shielding. Also λ5780, like λ6284 (not shown), tends to weaken for $E_{(B-V)}$ less than or equal 0.08 mag, although the decrease is not as rapid as that of molecular hydrogen. The total neutral gas column density N_H ($N_H = N(H\ I) + 2\ N(H_2)$) decreases too for $E_{(B-V)} < 0.08$ mag. According to Bohlin *et al.* (1978), the decrease of N_H at low reddening is probably because hydrogen is partially in ionized form. N_H includes only the neutral atoms and molecules and does not count the ionized atoms. The equivalent width of λ 5780 follows N_H closely down to $N_H = 10^{20}$ cm^{-2}. The equivalent width of the λ6284 measured for ten stars in Orion shows a similar good correlation with N_H (Jenniskens *et al.* 1994).

TABLE 1
Equivalent width per unit reddening

HD	$E_{(B-V)}$	6177	6196	6269	6284	rms
34085	0.00(1)	-.-	-.-	-.-	-.-	1
34989	0.13(S)	1.2(3)	0.06(3)	0.15(6)	1.38(30)	4
35149	0.11(S)	0.7(4)	0.06(4)	0.07(4)	1.12(15)	4
36486	0.07(S)	-.-	0.08(4)	0.04(2)	0.95(30)	1
36822	0.11(S)	-.-	0.10(4)	0.06(3)	1.56(30)	1
36861	0.12(S)	0.8(4)	0.046(3)	0.10(3)	1.50(30)	1
37020	0.40(56)	0.29 (-)	<0.008	0.023(-)	1.10 (-)	6
37022	0.35(4)	0.5 (2)	<0.007	0.04(1)	1.37(20)	3
37023	0.36(6)	0.4(2)	<0.012	0.023(12)	1.0(2)	6
37061	0.50(4)	0.41(5)	0.012(6)	0.03(1)	1.07(11)	5
38023	0.51(4)	0.2(2)	<0.014	0.05(4)	0.45(10)	13
38563A	0.78(10)	<0.2	< 0.018	0.025(12)	0.49(17)	18
39557	0.60(7)	0.38(16)	0.037(16)	0.08(3)	0.58(10)	9
37043	0.07(S)	no data	0.04(4)	<0.002	0.45(24)	1
37128A	0.08(S)	-.-	-.-	-.-	0.59(23)	2
37490	0.10(2)	0.5(3)	0.01(2)	0.09(3)	1.5(2)	2
37742	0.08(S)	no data	<0.14	0.07(6)	0.6(2)	7
37903	0.38(6)	0.51(11)	0.035(12)	0.024(8)	0.79(15)	4
38087	0.32(6)	0.2(2)	no data	0.04(3)	0.78(20)	11
38771	0.07(2)	-.-	-.-	-.-	0.86(30)	2
42050	0.34(6)	1.4(3)	0.07(3)	0.20(6)	2.1(4)	8
46060	0.41(60)	1.04(-)	0.052(-)	0.14(-)	1.9 (-)	11
21389	0.55(6)	0.83(8)	0.069(7)	0.18(2)	2.15(22)	1
183143	1.24(8)	0.93(6)	0.067(6)	0.16(1)	1.73(12)	2
Dif.:	0.49(12)	0.50(17)	0.051(12)	0.13(3)	1.2(4)	-

"rms" is the root mean square noise per pixel of 0.1 Å in units of 10^{-3} Å. The error estimates in parenthesis are in units of the smallest significant number given and take into account the uncertainty in E_{B-V}. Values of reddening are from Savage *et al.* (1977) (marked S) and otherwise derived from photometry listed in the *Hipparchos Input Catalogue*. "Dif" refers to the diffuse medium.

Assuming that reddening is proportional to the total column density of all matter, we conclude that the DIB carriers of λ5780 and λ6284 are confined to the neutral

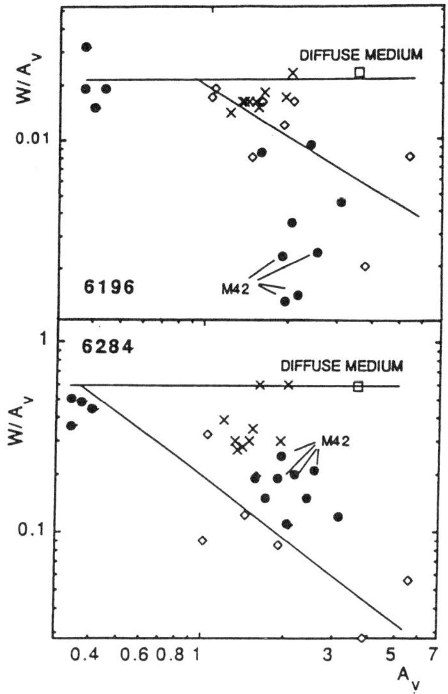

Fig. 2. The strength of 6284 and 6196 Å bands, in units of W/A_V, versus A_V. Symbols as in Fig. 1. An A_V^{-1} dependence is fitted to the Taurus data. Note that the stars M42 have somewhat stronger 6284 and weaker 6196 Å bands compared to an A_V dependence fitted to other stars in Orion.

gas and are weakened in low reddened lines of sight where a significant fraction of hydrogen is in ionized form.

4. DISCUSSION

In general there is good correlation between DIB equivalent width and reddening but a weakening in dark clouds is observed. The trend is not due to mere column density differences, like reddening or the visual extinction. The DIB strength is not decreasing towards GMC214-13 or HD183143. Therefore the determining factor for the decrease of DIB strength seems to be the attenuated UV field in dark clouds. The UV field might have a destructive role also. The observed deficiency for lines of sight with $E_{(B-V)} < 0.08$ mag is strongly linked to whether hydrogen is in neutral or ionized form, which argues for a destructive role of UV photons with energy higher than 13.6 eV. The specific behaviour of the strong DIB at 6284 Å and the weak DIB at 6196 Å in Orion is illustrated in Fig. 4. Using the state of hydrogen as indicator for environmental conditions and the visual extinction A_V, we can summarize as follows: In regions with molecular hydrogen, the DIBs at λ6196 and λ6284 are both present. The destructive role of UV is observed for λ6196 in regions where atomic hydrogen dominates. Here the λ6196 is depleted and the carrier of t his band might therefore be destroyed by photons less than 13.6 eV. The DIB at λ6284 is resistant

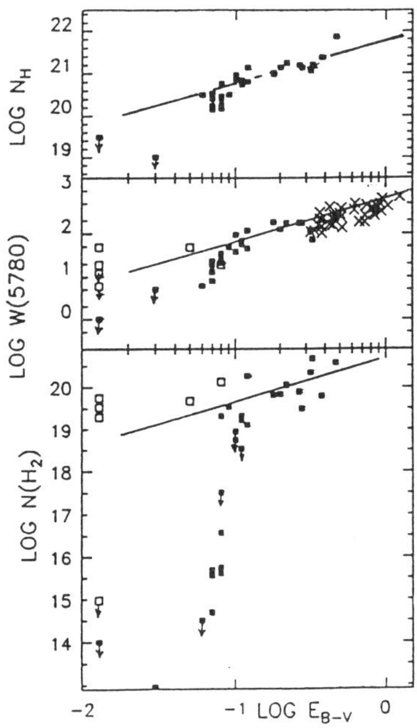

Fig. 3. DIB strength as a function of H I and H_2 column densities. Data from Snell & Van den Bout (1981), Meyer (1983), Federman *et al.* (1984), Josafatsson & Snow (1987) and Benvenuti & Porceddu (1989). Crosses are the DIBs reported in Desert *et al.* (1994). The data on atomic and molecular gas column densities are from Savage *et al.* (1977). Open symbols mark lines of sight for which no atomic hydrogen column densities is given. The solid lines have slope +1.

against these photons and actually enhanced relative to an A_V^{-1} dependence in the same environment. A depletion of both carriers (for λ6196 and λ6284) is observed and linked to the presence of ionized hydrogen in the line of sight. Furthermore both carriers show a decrease in dark clouds at $A_V = 5$. However λ6284 is already more reduced in strength at $A_V = 1$ than λ6196, which confirms that the carrier of λ6284 is ionized and destroyed by higher energy photons than the λ6196 carrier. These data support strongly a large molecular carrier in an ionized state. Ionized polycyclic aromatic hydrocarbons (PAHs) are therefore excellent candidates for the carriers of the diffuse interstellar bands (Léger & d'Hendecourt 1985, van der Zwet & Allamandola 1985, Salama & Allamandola 1992a,b).

The abundance of PAH monocations is a complex balance of ionisation by photons of energy > 7 eV, recombination and destruction. The ionisation degree of PAHs is more or less proportional to the ratio of the interstellar radiation field to the electron density (Omont 1986). In the diffuse medium 10-40% are thought to be ionized (Verstraete *et al.* 1990, Jenniskens *et al.* 1992), while this fraction is about 1 for H II regions. Destruction of PAHs probably involves doubly charged PAHs (Leach 1987). Leach argues that nascent PAH monocations of any size formed by photoionization in H I regions will preferentially relax by infrared emission. Those that have less than 30-40 carbon atoms will tend to relax by photofragmentation, after absorption of a second UV photon of sufficient energy, rather than by infrared radiative relaxation (Jochims *et al.* 1994). If the difference between the first and second ionisation

Fig. 4. Environmental dependence of the DIBs at 6284 and 6196 Å in Orion

potential ΔI is larger than 13.6 eV, the molecules should survive in all environments where neutral hydrogen is found. If less than 13.6 eV, the DIBs are destroyed in photodissociation regions. The DIB carriers of λ6284 and λ5780 have likely a ΔI larger than 13.6 eV, since they are found in regions with neutral hydrogen, but are absent when hydrogen is ionized. In contrast, the carrier of λ6196 is already weak in regions with neutral hydrogen. New measurements of the ionization potentials of 21 PAHs show that ΔI lies below 13.6 eV for many species (Tobita *et al.* 1994). These PAHs can probably be excluded as carriers for the strong DIBs at 6284 and 5780 Å.

ACKNOWLEDGEMENTS. We thank B.H. Foing, F.X. Désert and the staff of O.H.P. for their support of this project and X. Tielens and T. Millar for discussion. Special thanks goes to Vincent Icke for graphic design. PE is a recipient of a fellowship of the European Community ERBCHBICT940939. PJ holds an National Research Council ARC Research Associateship.

REFERENCES

Adamson, A.J., Whittet, D.C.B., and Duley, W.W. 1991, *MNRAS*, **252**, 234.
Benvenuti, P. and Porceddu, I. 1989, *Astr. Ap*, **223**, 329.
Bohlin, R.C., Savage, B.D., and Drake J.F. 1978, *Ap. J.*, **224**, 132.
Désert, F.X., Jenniskens, P., and Dennefeld, M. 1994, *Astr. Ap.*, submitted.
Ehrenfreund, P., d'Hendecourt, L., Verstraete, L., Léger, A., Schmidt, W., and Defourneau. D. 1992, *Astr. Ap.*, **259**, 257.
Federman, S.R., Kumar, C.K., and Vanden Bout, P.A. 1984, *Ap. J.*, **282**, 485.
Herbig, G.H. 1975, *Ap. J.*, **196**, 129.
Herbig, G.H. and Leka, K.D. 1991, *Ap. J.*, **382**, 193.
Jenniskens, P., Ehrenfreund, P., and Désert, X. 1992, *Astr. Ap.*, **265**, L1.
Jenniskens, P. and Désert, F.-X. 1993, *Astr. Ap.*, **274**, 465.
Jenniskens, P., Ehrenfreund, E., and Foing, B. 1994, *Astr. Ap.*, **281**, 517.
Jenniskens, P. and Désert, F.-X. 1994, *Astr. Ap. Suppl. Ser.*, *A survey of diffuse interstellar bands (3800-8680 Å)*, in press.
Jochims, H.W., Ruehl, E., Baumgaertl, H., Tobita, S., and Leach. S., 1994, *Ap. J.*, **420**, 307.
Josafatsson, K. and Snow, T.P. 1987, *Ap. J.*, **319**, 434.
Leach S. 1987, Polycyclic Aromatic Hydrocarbons and Astrophysics, eds. Léger A. & d'Hendecourt L. (Reidel Publishing Company), 99.
Léger, A. and d'Hendecourt, L. 1985, *Astr. Ap.*, **146**, 81.
Meyer, D.M. 1983, *Ap. J.*, **266**, L51.

Omont, A. 1986, *Astr. Ap.*, **164**, 159.
Salama, F. and Allamandola, L.J. 1992a, *Ap. J.*, **395**, 301.
Salama, F. and Allamandola, L.J. 1992b, *Nature*, **358**, 42; *Ap. J.*, **395**, 301.
Savage, B.D., Bohlin, R.C., Drake, J.F., and Budich. W. 1977, *Ap. J.*, **216**, 291.
Smith, W.H., Snow, T.P., and York, D.G. 1977, *Ap. J.*, **218**, 124.
Snell, R.L. and Vanden Bout, P.A. 1981, *Ap. J.*, **244**, 844.
Snow, T.P. and Cohen, J.G. 1974, *Ap. J.*, **194**, 313.
Tobita, S., Leach, S., Jochims, H.W., Ruehl, E., Illenberger, E., and Baumgaertel, H. 1994, *Canadian J. Physics*, in press.
Van der Zwet, G.P. and Allamandola, L.J. 1985, *Astr. Ap.*, **146**, 76.
Van der Zwet, G.P. 1987, Polycyclic Aromatic Hydrocarbons and Astrophysics, eds. Léger A. & d'Hendecourt L., (Reidel Publishing Company), 351.
Verstraete, L., Léger, A., d'Hendecourt, L., Dutuit, O., and Defourneau, D. 1990, *Astr. Ap.*, **237**, 436.

DIFFUSE INTERSTELLAR BANDS IN THE YOUNG GALACTIC CLUSTERS M16 AND M17

W. PFAU* AND TH. HENNING**
* *University Observatory, Jena, Germany*
** *Research Group of the Max Planck Society, Jena, Germany*

ABSTRACT. The closely related galactic clusters M16 and M17 were selected for spectroscopy around the DIBs $\lambda\lambda 5780/5797$ and $\lambda 6203$ for the anomalous reddening law prevailing therein. It is evident that the band strengths differ markedly *between* and that there is no increase of DIB strength *within* the clusters.

1. INTRODUCTION

In the past, the main strategy to get insight into the problem of the DIB origin was correlation analysis. This is hampered by the fact, however, that in interstellar space most quantities are correlated with each other, so that it is sometimes difficult to draw unambiguous conclusions. Another way to tackle the question of DIB absorbers may be to make use of the fact that in some regions the DIBs show deviant strengths as compared to other bands or interstellar quantities. A typical example is the large cosmic scatter in the relation between band strength and most of the colour excesses. It seems to be promising to try to trace the sometimes exceptional behaviour of the DIBs back to striking pecularities along the respective lines of sight. In the present paper, we investigate the DIB strengths in two galactic clusters – namely M16 and M17 – which were selected for the presence of anomalous extinction therein.

In the following section, a short characterization of the two clusters is given. Emphasis is on their outstanding properties, worth to be considered in connection with the DIB origin.

2. DESCRIPTION OF THE CLUSTERS

The two star-forming regions M16 and M17 are located in the Sagittarius spiral arm in one of the large OB associations in Serpens (Ser OB1 or Ser OB2). They appear to be embedded in a common huge CO interstellar cloud at about 2 kpc distance (V_{LSR}=20 kms^{-1}). This is in accordance with photometric distance determinations to the clusters and results in a separation in the plane of the sky of only about 90 pc.

2.1. M16 (NGC 6611)

The young galactic cluster M16 ($l = 17.0°, b = +0.8°$; age about $5\,10^6$ yrs) is associated with a bright H II region (see, e.g., McBreen *et al.* 1982), e.g. It is an evolved,

A. G. G. M. Tielens and T. P. Snow (eds.), The Diffuse Interstellar Bands, 113–119.

density-bounded H II region embedded within a massive molecular cloud. Optical observations indicate the presence of large turbulent as well as systematic motions between ionized and neutral gas. The total stellar luminosity of M16 is $L_\star \approx 1\,10^7\ L_\odot$ (Graf 1993) and the flux of Lyman continuum photons from the ionizing stars $N_{c,\star} \geq 1\,10^{52}$ phot s^{-1}, much more than demanded for the ionization of the nebula.

First evidence for the presence of an abnormal extinction law in M16 came from Neckel & Chini (1981). Later on Chini & Krügel (1983) published JHKL photometry of early-type stars within M16 and showed the ratio of total to selective extinction to be $R=A_V/E(B-V) = 4.8 \pm 0.2$. The deviant extinction law is clearly seen in their two-colour diagram, where normal foreground stars and the dust-embedded cluster stars populate different reddening lines. Chini & Wargau (1990) characterized the abnormal extinction law as deviating in the short-wavelength region. They then fitted the extinction by a silicate/graphite mixture of dust particles to the observed curves and explained the special reddening law in M16 as due to the presence of somewhat larger silicates and considerably larger (factor of 2) graphite spheres.

The stars observed here are located within an 9' x 5' area at the northern border between ionized and neutral gas. There are three categories of stars (notation after Walker 1961): (*i*) W455 is a true *foreground star*. It is located about halfway to M16 and definitely not related to the cluster. (*ii*) W227 and W400 are members of the cluster, but located in its outer part and mainly influenced by the normal reddening law. In the following we call them *surface stars*. (*iii*) W175, W205, W228, W231, W246, W280, W305, and W339 are *cluster stars* from deep within the cluster, some of them known as ionizing sources of the nebula. The extinction A_V in this group is between 3.6 and 6.9 mag.

2.2. *M17 (NGC6618, THE OMEGA NEBULA)*

M17 ($l = 15.1°, b = -0.7°$) is a giant H II region associated with the molecular cloud core M17SW. As in M16, recently formed luminous stars exert pressure to the adjacent molecular cloud material and initiate macroturbulent motions and ongoing star-formation. The whole complex has an integrated total infrared luminosity which is transformed from stars with altogether $6\,10^6\,L_\odot$. A warm and clumpy boundary layer of large optical depth in the visible separates the regions of ionized and molecular gas. Responsible for the H II region are about 10 stars with a total Lyman continuum flux $N_{c,\star} \geq 2.7\,10^{50}$ phot s^{-1}, close to the flux required for ionization of the nebula (Felli *et al.* 1984).

The stars observed by us are within the 4' x 4' field where most of the ionizing stars are found. Our stars show extinction A_V of between 1.9 and 5.6 mag (Chini *et al.* 1980) and $R=A_V/E(B-V) = 4.9 \pm 0.3$ (Chini & Krügel 1983), very similar to M16. From their distances and locations in the two-colour diagram the stars C6, C8, C11, and C13 are either foreground or surface stars (maximum extinction A_V=2.3 mag), whereas C2, C3, C16, and C45 definitely belong to the complex (notation after Chini *et al.* 1980).

3. OBSERVATIONS

Spectra of 11 and 9 stars in M16 and M17, resp., were obtained with the coudé spectrograph of the 2.2-m telescope of the German-Spanish Astronomical Centre on

Calar Alto (Spain). The CCD frames covered the regions 5735 to 5865 Å and 6140 to 6260 Å with spectral resolutions of 45 000 and 15 000, resp. The V magnitudes of the stars range between 8.5 and 13.5 mag. and the signal-to-noise ratios in the continuum are between 300 and 10.

According to the list by Herbig (1975) the following features are covered:

DIB	λ	5778	DIB	λ	6177 (♯1)
		5780 (♯2)			6195 (♯2)
		5797 (♯3)			6203 (♯2)
		5844			6269 (♯2)
		5849 (♯3)			

In brackets the membership of the DIBs in one of the three families introduced by Chlewicki *et al.* (1986) and further outlined by Krełowski & Walker (1987) is indicated if known. DIB λ6203 is usually considered to be a double feature blended with an interstellar band at 6205 Å. Our equivalent widths refer to the whole asymmetric band. In addition to the prominent DIBs listed above, at least some of the weak bands identified by Krełowski & Sneden (1993) are also clearly present in our spectra.

Further details of the observations, stellar data, and the DIB equivalent widths will be given elsewhere (Pfau *et al.* 1994).

4. DISCUSSION OF THE DIB STRENGTHS

Inspection of the spectra reveals two remarkable results:

1. In both clusters, the spectra of objects, either foreground, surface, or cluster stars, differ significantly. This is despite the fact that foreground and surface stars are influenced by the normal reddening law, and the two clusters are not so far from each other in space.
2. Despite measurable differences in extinction, the DIB spectra are quite similar for foreground and surface stars, as well as for cluster stars.

Whereas the average DIB spectrum of stars in the direction of M16 resembles the ζ Ophiuchi-type, we find σ Scorpii-type in M17. Typical examples of individual or averaged spectra are given in Figs. 1 to 3 for both wavelength regions.

Typical of the ζ Ophiuchi-type DIB spectra are about equal depths of $\lambda\lambda$5780 and 5797 and a moderately weaker λ5849 band. In case of the σ Scorpii-type, λ5797 and related to it λ5849 have much smaller depths than λ5780. Looking at the families, family ♯2 with its members $\lambda\lambda$5780, 6195, and 6269 is about equally strong in the M16 and M17 regions, whereas ♯3 with $\lambda\lambda$5797 and 5849 is definitely weaker in the direction to M17. From our spectra we doubt that DIB λ6203 really belongs to family ♯2, as was found by Chlewicki *et al.* (1986) and Krełowski & Walker (1987). In M16 and M17 it behaves different from the other long-wavelength members of ♯2, namely $\lambda\lambda$6195 and 6269, and is definitely stronger in M16. We have to take into consideration, however, that the difference might be due to different blending by DIB λ6205.

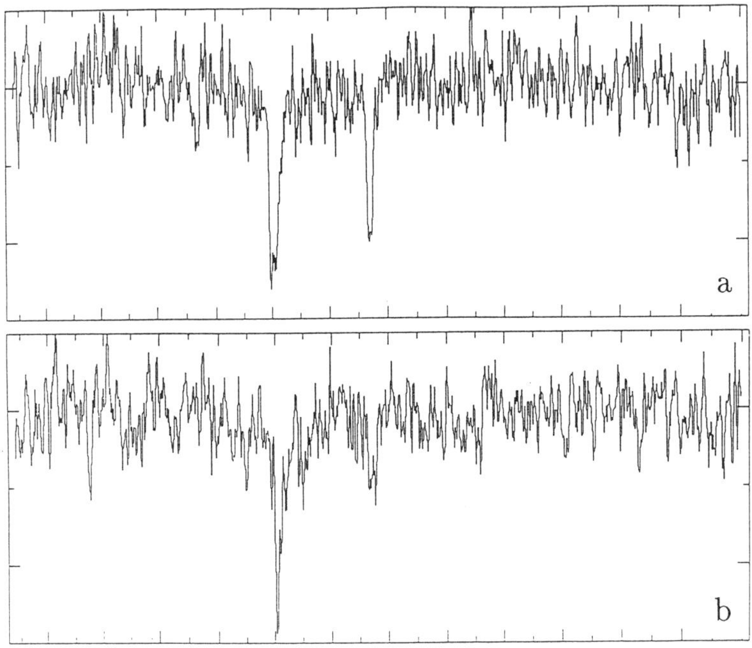

Fig. 1. Unsmoothed spectra in the λλ5780/5797 spectral region for stars W227 (E(B-V)=0.96 mag) **(a)** and C6 (0.44 mag) **(b)** in front of M16 and M17, resp. The strongest line is DIB λ5780 and the distance between larger tic marks is 10 Å in wavelength and 0.2 in relative intensity.

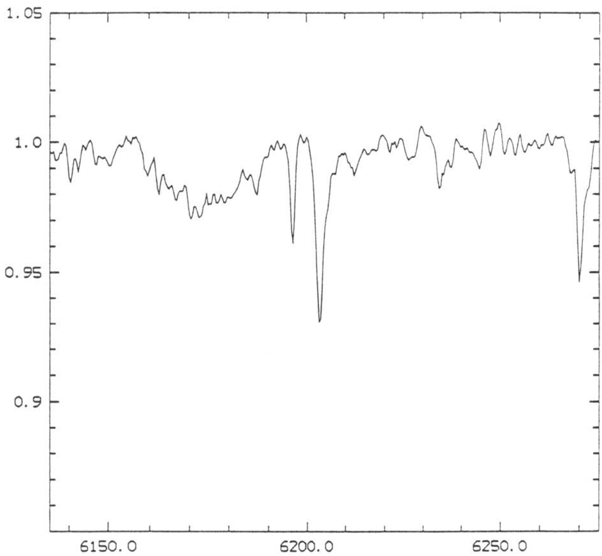

Fig. 2. Averaged spectra in the λ6200 region of stars W205 and W280 in M16 (E(B-V)=0.81 and 0.93 mag, resp.). The abscissa is wavelength in Å and the ordinate relative continuum intensity.

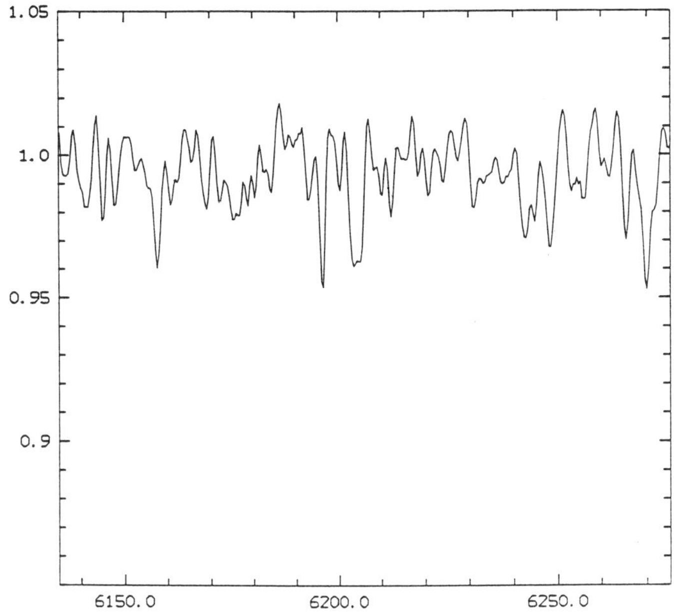

Fig. 3. Averaged spectra in the λ6200 region of stars C6, C8, and C11 in M17 (E(B-V)=0.44, 0.41, and 0.66 mag, resp.). The meaning of the axes is as in Fig. 2.

The difference in the DIB spectra in both regions is further illustrated by Fig. 4, where the equivalent widths of the DIBs λ5780 and λ5797 are compared to each other. In a similar presentation Wallerstein & Cardelli (1987, Fig. 2 in their paper) outlined the region, where the λ5780 versus λ5797 data points of the large majority of stars are located. Whereas the stars in and at the surface of M16 populate this part of the diagram, too, objects from the M17 region clearly fall to the side of relatively weak DIB λ5797. For two other star-forming regions – the ρ Ophiuchi and the Chamaeleon clouds – and for some isolated stars associated with nebulosity, Wallerstein & Cardelli (1987) found DIB λ5797 to be *generally* weak relative to DIB λ5780. From our Fig. 4 this is true only for M17 and, despite their close relation in space, definitely not for M16. There is at least one other example that the conditions for the formation of these bands might change over small linear scales: HD 21389 with a DIB spectrum of the σ Scorpii-type and, at only 19 pc distance from it, HD 21291 with an o Persei-type spectrum (Josafatsson & Snow, 1987).

In both clusters, no dependence on reddening E(B-V) of the DIB strengths for the bands investigated here is indicated. Despite a range of ΔE(B-V)$\simeq$ 0.4 and 0.8 mag *within* M16 and M17, resp., with the exception of W339 the equivalent widths keep constant to within better than $\pm$0.15 Å. As an example this is illustrated in Fig. 5 for the combined λ6203 data in M16 and M17. Stars ascribed above to the surface of the clusters, namely W227, W400, and C11, are in reddening as well as in band strength comparable to cluster stars.

Similarly, C6 and C8 in M17 are true foreground objects (Chini *et al.* 1980) and the increase of the equivalent width with reddening, EW(λ6203)/E(B-V) $\approx$ 0.48 Å mag^{-1} as indicated in Fig. 5, means the growth of DIB $\lambda\lambda$6203/6205 with reddening in the

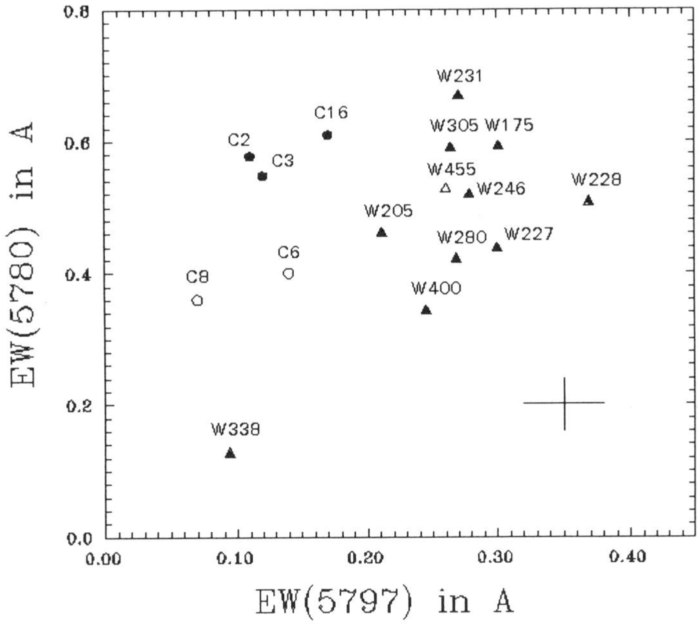

Fig. 4. Equivalent widths of DIB λ5780 versus λ5797 for stars in the fields of M16 (triangles) and M17 (circles). Stars in front are indicated by open symbols. The error bars are calculated for an estimated signal-to-noise ratio of 20 from the formula by Josafatsson & Snow (1987).

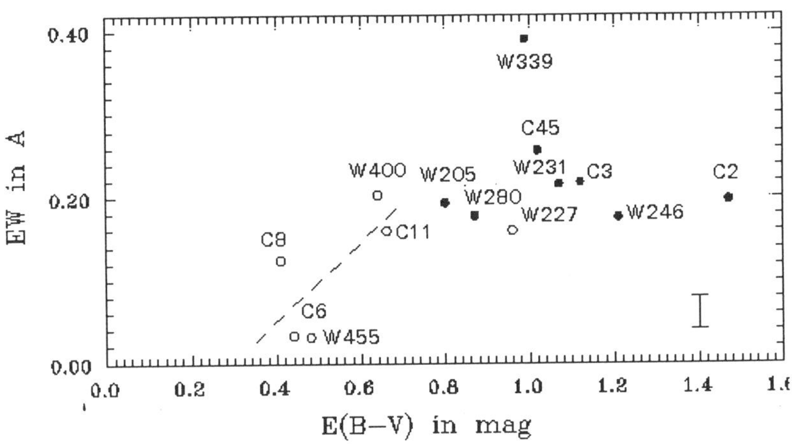

Fig. 5. Equivalent widths for DIB λλ6203/6205 in M16 and M17. Open and closed circles refer to foreground and surface and to cluster stars, resp., where the division is made as in Subsects. 2.1 and 2.2. The characteristic standard deviation is calculated for a typical signal-to-noise ratio of 50 from the formula by Josafatsson & Snow (1987).

diffuse medium along the lines of sight.

The result is that none of the DIBs observed here, show a significant increase in strength within the cluster regions. The physical conditions prevailing there are obviously not favourable for band formation. The near to constant strengths of DIBs found here for M16 and M17, confirm earlier results that the band growth is inhibited in denser regions. For dark clouds this was demonstrated again by Adamson *et al.* (1991). These authors suggest that in the Taurus dark cloud, which shows only marginal foreground reddening, DIB formation is restricted to a thin surface layer. Another example is formed by the medium and heavily reddened stars in Orion (Jenniskens *et al.* 1994). As in dark clouds, modelling the anomalous reddening law in M16, point to larger than normal grains (Chini & Wargau 1990). In dense regions of interstellar matter, such as dark clouds and star-forming complexes with their embedded denser clumps, some special or all the strong effects which density and radiation field exert on the DIB carriers are obviously hostile to the formation of DIBs. Due to large-scale motions and turbulence in such environments, the carriers additionally experience a changing history which may add to the explanation of differing band strengths, even between close lines of sight.

ACKNOWLEDGEMENTS. We gratefully acknowledge the support with the observations and part of the data reduction by Dr. J. Solf, Heidelberg.

References

Adamson, A.J., Whittet, D.C.B., and Duley, W.W. 1991, *MNRAS*, **252**, 234.

Chlewicki, G., van der Zwet, G.P., van Ijzendoorn, L.J., and Greenberg, M. 1986, *Ap. J.*, **305**, 455.

Chini, R., Elsässer, H., and Neckel, Th. 1980, *Astr. Ap.*, **91**, 186.

Chini, R. and Krügel, E. 1983, *Astr. Ap.*, **117**, 289.

Chini, R. and Wargau, W.F. 1990, *Astr. Ap.*, **227**, 213.

Felli, M., Churchwell, E., and Massi, M. 1984, *Astr. Ap.*, **136**, 53.

Graf, O. 1993, unpublished Jena diploma thesis.

Herbig, G.H. 1975, *Ap. J.*, **196**, 129.

Jenniskens, P., Ehrenfreund, P., and Foing, B. 1994, *Astr. Ap.*, **281**, 517.

Krełowski, J. and Sneden, C. 1993, *P.A.S.P.*, **105**, 1141.

Krełowski, J. and Walker, G.A.H. 1987, *Ap. J.*, **312**, 860.

McBreen, B., Fazio, G.G., and Jaffe, D.T. 1982, *Ap. J.*, **254**, 126.

Neckel, T. and Chini, R. 1981, *Astr. Ap. Suppl. Ser.*, **45**, 451.

Pfau, W., Henning, Th., and Solf, J. 1994, *in preparation for Astr. Ap.*

Josafatsson, K. and Snow, T.P. 1987, *Ap. J.*, **319**, 436.

Walker, M.F. 1961, *Ap. J.*, **133**, 438.

Wallerstein, G. and Cardelli, J.A. 1987, *A. J.*, **93**, 1522.

THE INTERSTELLAR SPACE MOSAIC: A USER INTERFACE FOR ACCESSING ISM DATA WORLD-WIDE DISTRIBUTED

I. Porceddu, S. Corda [(1)]
F. Pasian, R. Smareglia [(2)]
[(1)] *Cagliari Astronomical Observatory, Cagliari, I*
[(2)] *Trieste Astronomical Observatory, Trieste, I*

ABSTRACT. Data retrieval systems are an essential tool for scientists. Up to a few years ago, user interaction was usually fairly poor, and very often different from one Hardware/Software environment to another. Since one year or less, several portable and user-friendly interactive tools for Network Information Retrieval (NIR) gave us a new and exciting way of interacting with the data in general, and with the astrophysical data in particular. The huge amount of data of astrophysical interest can now be accessed with less pain than ever. The Interstellar Space Mosaic is aimed to be a first attempt to allow transparent access to interstellar medium data in the new NIR scenario, trying to take full advantage of the new tools. This paper will describe the logic structure of this "distributed database of archives", the data that will become soon available thru it, the advantages (many) and traps (very few) that this new tool is expected to provide.

1. INTRODUCTION

The rapid expansion of the Internet framework is not so unbelievable. Many of us became experienced with the harsh astronomical environment, where we need a continuous technology upgrade, with a not so rare lack of awareness of the powerful tools we can deal with. The *World Wide Web* "virus" originated at *CERN*, and very soon contaminated the lost people of Internet. But most of us kept it on a latent state: it was only the *Mosaic* fever which gave evidence of the need we had to use in a more active way our computer network. The availability of fairly cheap personal workstations allows us to access the information distributed world-wide in a fast and easy mode: and all this comes with no royalties, neither license agreement forms nor bureaucracy at large! But the real lane to heaven is the public domain software, such as *Mosaic*, *Gopher*, and *Veronica*. The huge amount of extremely useful, quite useful, sometime useful and completely useless available software tools thru the Internet archives is a real key for understanding the scenario.

And there is a challenge: trying to overcome the nice way of interacting with those graph tools, concentrating only on the practical application to scientific information managing and browsing. We believe these graphic data browsers will be able to

A. G. G. M. Tielens and T. P. Snow (eds.), The Diffuse Interstellar Bands, 121–128.

improve the retrieval of information, cutting down the amount of time we normally lose when, e.g., looking for target parameters, existing measurements, etc.

The NIR (Network Information Retrieval) procedures can greatly improve our "interaction" style with data. Here we present the prototype application of the WWW philosophy applied to the DIBs problem: a hypertextual database made up from a set of archives containing data and spectra of astrophysical interest related to the interstellar medium, with a subsection dedicated to DIB information. The ISM project could be a good test application of how we would like to use the WWW scenario: reliable applications for a daily use of NIR systems.

2. ACCESSING INFORMATION OF ASTROPHYSICAL INTEREST

Most of us have been using the ULDA archive as well as the wider collection of data which are accessible thru the Starcat interface, as well as the NASA ADS and RECON facilities. The on-line ULDA/USSP (Wamsteker *et al.* 1989) facility is a good example of a powerful tool used for a long time to do ultraviolet astronomy, with a simple and not so bad user interface. Now we can have access to a huge number of catalogues, databases, archives, datasets, data juke-boxes and more. But quite often the hot data which are published as microfiches or on some exotic medium and/or format within research papers are not easily accessible or are not accessible at all. The granted on-line access to most of the libraries which maintain astrophysical information can be seen as a useful way of improving our opportunities.

As for the diffuse interstellar bands, an increasing amount of laboratory, theoretical and observational studies were carried out during the last ten years in order to solve this multifaceted scientific challenge. But most of this effort is very often not completely exploited: the huge amount of laboratory and observational spectra which were included within published reports are not easily retrieved at large. Most of the data which were published to date are of great interest, but they are seldom disclosed or simply they are not fully evaluated. This is mostly true when looking at several atlases of interstellar extinction curves, which were obtained from the *IUE* and *TD-1* satellites. They could be easily browsed on-line taking advantage of a *Mosaic* or *Gopher* system instead of sleeping in a microfiche-like format. This could be a great advantage especially for those institutes which have not enough money support, but can easily be part of the Internet network. Therefore the NIR tools can also be seen as a very *democratic* tool which could boost or simply aid research, especially in developing countries.

3. THE DIBS ARCHIVE AT THE CAGLIARI ASTRONOMICAL OBSERVATORY

Since 1986 the Astrophysics branch of the Cagliari Observatory has been studying diffuse interstellar bands, one of the many open questions about the chemical composition and physical behaviour of the interstellar environments. The observational activity, carried out by I. Porceddu and P. Benvenuti, led to generate a database of DIB spectra, as well as of Na D-lines and molecular features. The main features of this database are summarized in Table 1.

Probably the most relevant features of this database, besides the number of selected targets, are the homogeneous source of the spectra and the standard calibration procedures which have been adopted in order to reduce the whole set of CES data.

TABLE 1
Main features of our DIBs spectra

Number of stars observed	greater than 160
Limit Magnitude	V = 8
Specific Targets	Be stars, Dark clouds, High Latitude Molecular Clouds Low Reddening stars
Total Number of Spectra	600 (six hundred)
Telescope/Instrument	ESO CES fed by the 1.4m CAT telescope
Detector	CCD RCA having a 15μ pixel size
Resolving power	ranging from 50,000 to 100,000
Spectral Resolution	ranging from 0.05 to 0.09 Å
DIB features spectral ranges	5487, 5705, 5780, 5844, 6010, 6196, 6284, 6614 Na D-lines for most of the targets

We can therefore compare the spectra and be very confident about them. The calibration steps were carried out with a selected subset of ESO-MIDAS commands, as well as most of the spectra parameters, concerning the profile of the diffuse bands.

3.1. *THE DIBS DATABASE STRUCTURE*

The database is intended to provide the widest range of information which relates to each target and useful to characterize its interstellar environment. A wide spread of cross-correlations between the parameters extracted from the analysis of each diffuse band spectrum and several interstellar indicators have been carried out, highlighting useful relationships or stressing lack of correlations between them. Therefore the presence of such parameters within the DIBs database would be interesting. The basic set of data which are proposed to the users include:

- targets' full coordinates: equatorial and galactic position, visual magnitude, interstellar reddening, colour indexes, rotational velocity;
- each diffuse band profile will be characterized by:
    1. equivalent width;
    2. peak wavelength;
    3. relative depth of the band;
- number of intervening clouds as reported from the sodium lines analysis (where available);
- environment characterization;

- bibliographic references, with a link to WISE systems;
- satellite-based data availability for UV and IR ranges;

This "information sheet" for each spectrum of each star is the minimum requirement for the database. More information will be added to this first implementation, whether the database, as it is expected, will be acknowledged as of common interest.

4. THE INTERSTELLAR SPACE MOSAIC

The newest network information retrieval tools give us a powerful user interface for data browsing and retrieval. Therefore we decided, besides to submit the complete DIBs dataset for publication, to let our data be accessible on-line thru the network.

4.1. THE ISM FRAMEWORK

The basic tools which have been selected for building the ISM are known and widely used worldwide. Following the scheme proposed in Pasian & Smareglia (1994), three main blocks contribute to the ISM:

- **Internet** and **WWW**, the **W**orld **W**ide **W**eb. These two logic structures constitute the framework where the information can be accessed. The wide acceptance in the international scientific community of the WWW as an information retrieval mechanism is changing day by day the way users access the data retrieval. This fast evolving philosophy leads to a different and more homogeneous design of interfaces to archives and databases;
- **NCSA Mosaic**. It is the top hypertextual interface which is widely used by most of the scientific Institutitions: it can manage information in graphic format, images, as well as animations and sounds. This hypermedia capability is unique: no other interface can currently challenge Mosaic as flexibility and user friendliness, making Mosaic a *de facto* standard. Furthermore, Mosaic is a copyrighted public domain software package you can use free of charge in a no profit environment;
- **SQL database**. Allbase, an SQL software package for database management licensed by Hewlett-Packard, is used as first guess. This is the section of the ISM project where software development is actually needed, basically to create a proper link to the Mosaic interface.

4.2. THE ISM TABLE OF CONTENTS

The components of the ISM project are the data archives it will make available to the astronomical community. The nice interaction and the painless feedback of its query forms will allow to access, right now, to three archives:

- the DIBs database, described in the previous section;

- the Aiello *et al.* (1988) *Atlas of the wavelength dependence of ultraviolet extinction in the Galaxy*. It presents a collection of 115 extinction curves derived from low dispersion *IUE* spectra. The spectra have been processed with the use of techniques designed to reduce the effects of random noise and the influence of residual spectral features due to classification mismatch. The extinction curves are presented with normalization to E(B-V) = 1 and the FUV portion (below 170 nm) is also shown for E(13-17) = 1. The atlas includes examples of extinction originating in the diffuse medium and several major nebulae and dense clouds.

- the Papaj *et al.* (1991) *Atlas of extinction curves derived from ultraviolet spectra of the TD-1 satellite*. It is a collection of 166 extinction curves derived from the published low-resolution spectra acquired with the aid of the spectrometer on board the *TD-1* satellite. The system of standards applied is constructed with the aid of a special procedure allowing the possible effects of spectral mismatch to be avoided and making possible the derivation of extinction curves even in cases of very small E(B-V).

Figures 1 and 2 show the hardcopies of the ISM homepage prototype structure, which would interface to the database server.

4.3. THE ISM ARCHITECTURE

According to the WWW philosophy of a distributed information environment, each archive, database, atlas or collection of information will be accessed at the owner's institute: only a "cross referencing table" will be continuously updated in a single Internet gateway, the "Interstellar Space Mosaic" primary front-end system. In the client/server scheme, the client side runs at the user's computer and allows access to the Mosaic interface, while there are two levels of servers: the primary server accesses the database where the description of the data are stored, and the secondary ones will be located where each archive is physically stored. The user interface allows access to the underlying structure of the archive: a catalogue of the on-line databases which are accessible thru the network, managed by a commercial SQL-based DBMS (DataBase Management System). The interface itself provides several functions:

- selection of a specific database (access to the IUE extincton curves);
- selection of a single target;
- selection of a class of targets (Be stars, O stars, a diffuse band, etc.);
- browsing of each database by specifying query constraints thru forms to be filled by the users;
- display of the selected spectra.

This *β-test* release of the ISM project will not allow to retrieve data, but simply to visualize them.

Fig. 1. The homepage for interstellar space mosaic.

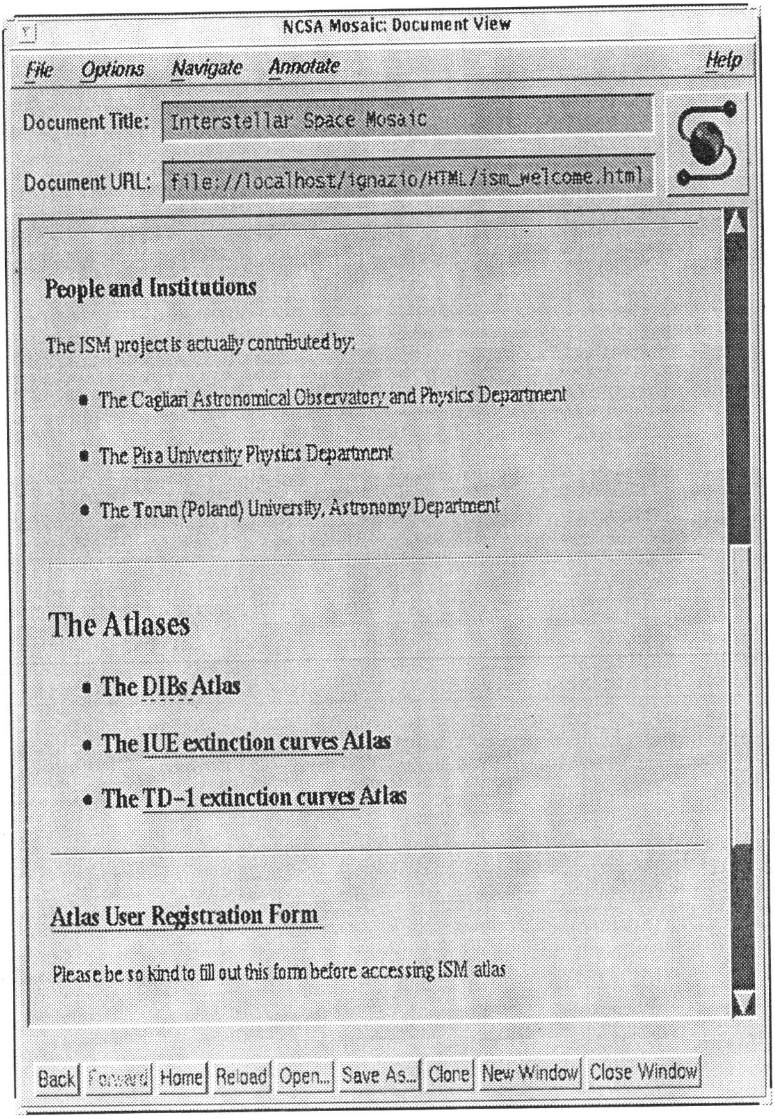

Fig. 2. The homepage for interstellar space mosaic.

5. CONCLUSION

Querying data of astrophysical interest thru WWW and Mosaic is becoming common practice. We are confident about the advantages that such a *philosophy* will bring to us, cutting down the amount of waste time which is spent looking for information. This is the approach we plan to follow for the ISM project. The project itself is open, and it is expected to grow up when new archives will join us in the future. Comments, questions, suggestions and upcoming data are kindly requested.

REFERENCES

Aiello, S., Barsella, B., Chlewicki, G., Greenberg, J.M., and Patriarchi P. 1988, *Astr. Ap. Suppl. Ser.*, **73**, 195.
Papaj, J., Krelowski, J., and Wegner, W. 1991, *M. N. R. A. S.*, **252**, 403.
Pasian, F. and Smareglia, S. 1994, in *New Tools for Network Information Retrieval*, I. Porceddu and S. Corda, eds., O.A. Cagliari.
Wamsteker, W., Driessen, C., Muñoz, J.R., Hassall, B.J.M., Pasian, F., Barylak, M., Russo, G., Egret, D., Murray, J., Talavera, A., and Heck, A. 1989, *Astr. Ap. Suppl. Ser.*, **79**, 230.

DIFFUSE CIRCUMSTELLAR BANDS

C. GREGORY SEAB*
University of New Orleans

ABSTRACT.

A review of searches for diffuse circumstellar bands is presented. There have been several recent articles reporting the detection of diffuse bands in the circumstellar environment. The most important instances are reviewed and discussed, with the conclusion that there remain enough ambiguities in the interpretation of the data to leave the matter uncertain. The largest problem area is in distinguishing the circumstellar material from foreground interstellar material. Because of this uncertainty, the reported detections cannot put hard constraints on the various proposed diffuse band carriers. However, the non-dectection of bands in targets with circumstellar material can and does constrain the carrier theories. The detection of PAHs in targets without diffuse bands implies that neutral PAHs cannot be the carriers. Observations of bands in dark clouds are also reviewed briefly, showing that the bands are weaker in these high density environments. Together, the circumstellar and dark clouds observations present a challenge that must be met by any successful carrier theory.

1. INTRODUCTION

The conventional wisdom is that there *are* no diffuse bands in the circumstellar environment. Herbig (1966) is credited with first observing this absence, although Snow and Wallerstein (1974) first searched explicitly for circumstellar bands. Savage and Mathis (1979) mention that they are absent from the circumstellar environment. Any bands found in the material around stars would have to be labeled diffuse *circumstellar* bands (DCBs) rather than interstellar bands. There have been several such detections reported in the literature; these will be discussed below.

Searches for DCBs are motivated by the strong impact an unambiguous detection would have on diffuse band theories. The reason for this impact is that the circumstellar environment presents an inherently simpler locale for interpreting the observations. Not only would more of the parameters of density, temperature, and chemical composition be known, but the history of the material would be certain: it has no history, but originates only a short time before in the outflow from the star. This is in strong contrast to the always uncertain history of clouds in the interstellar medium, where one does not know if the cloud has been recently processed through a dense molecular cloud, subjected to intense radiation fields, shocked, or otherwise abused in the past. Furthermore, typical interstellar clouds must represent

*resident for Summer 1994 at CASA, University of Colorado, under NASA JOVE program grant to the University of New Orleans

A. G. G. M. Tielens and T. P. Snow (eds.), The Diffuse Interstellar Bands, 129–142.

a mixture of materials from various different sources. Similarities in the chemical composition and dust properties (in the diffuse clouds, at least) are indicative of a well-mixed interstellar medium (ISM). There is direct evidence of this mixing in the isotopic anomalies found in meteoritic material, indicating a proto-solar cloud with dust grains originating in very different sources to produce, for example, Xe-HL as well as Xe-S anomalies (*cf.* review by Ott 1993).

In contrast, the circumstellar environment presents a relatively pristine, known set of circumstances for interpreting observations. Diffuse bands, whether present or absent, can be compared in this environment to the presence of proposed diffuse band carriers such as polycyclic aromatic hydrocarbons (PAHs), ionized PAHs, hydrogenated hydrocarbons, fullerenes or fulleranes, long-chain carbon molecules, and other carriers (for references, see various articles in this volume). Thus an unambiguous circumstellar detection of diffuse bands would necessarily present a critical test for diffuse band theories. Even a clear non-detection of DIBs can be a challenge for the theories, as will be seen below.

Section II below will discuss in some detail the various reported detections of DCBs, together with arguments for and against the circumstellar nature of the bands. Section III will discuss some of the non-detections of bands in circumstellar clouds, and how these negative findings impact on the various theories for the DIB carriers. Section IV extends the discussion to consider diffuse bands in dark clouds. Finally, Section V summarizes the present state of knowledge about the occurence of diffuse bands in circumstellar material with some remarks on the transition between circumstellar and interstellar environments.

2. REPORTED CIRCUMSTELLAR DETECTIONS

Since the initial negative search for diffuse bands in the circumstellar environment (Snow and Wallerstein 1972), there have been a handful of reported detections. However, all of these detections must overcome an inherent and serious problem: it is always difficult to demonstrate that the detected feature is circumstellar rather than interstellar. Since the DIBs seem to be ubiquitous in interstellar material, then any foreground reddening will produce a DIB feature in the absorption spectrum of the object. Although the strength of the bands generally correlates with various interstellar parameters such as E_{B-V}, there is still a large range of variation in the strengths, as seen, for example, in the compilation of Snow, York and Welty (1977). Some of this variation is attributed to the nature of that compilation from heterogeneous sources and generally low-resolution photographic data, but even modern results (e. g. Herbig 1993) using high S/N data shows a large scatter. Because of this scatter, it is generally not possible to correct for the foreground band strength with any degree of confidence. This assumes it is even possible to derive a reliable foreground reddening for the object. A simple detection of a diffuse band towards an object known to have a circumstellar dust shell is thus inadequate to demonstrate the presence of a DCB.

The detection of the DCB can be further confused by the spectrum of the underlying object itself. By definition, the targets for DCB searches have to be mass-losing objects such as Mira variables, post-AGB stars, some planetary nebulae, and Wolf-Rayet stars, and novae. For the cool type stars, the stellar lines can obscure the diffuse bands; although this can be corrected by comparison with a similar spectral type star without the circumstellar material, the quality of the data is reduced. Emission lines from planetary nebulae and WR stars can also be a problem. The ideal situation would be a mass-losing object with a hot star companion within or just behind the circumstellar nebula to provide a background object for measuring the DCB, and with another hot star nearby to help determine the foreground band strength. Even in this ideal case, the hot star companion might be suspected of altering the circumstellar environment enough to limit the generality of the results. Alternately, a time variable source that shows distinct changes in the band strength could also provide a strong detection. We will see that some of these requirements are met in some observations, but never quite enough to provide a strong, unambiguous detection. We will discuss the strongest cases for detected DCBs below on an individual basis.

2.1. CS 776

Le Bértre (1990, hereafter LB90; *cf.* Le Bértre and Lequeux 1993, hereafter LBL93) reports the detection of the diffuse bands at 4430, 5780, and 6284 Å in the circumstellar material around CS 776. This is a carbon star undergoing mass loss at the rate of about $5 \times 10^{-5} M_{\odot} yr^{-1}$ (LB90), with evidence of circumstellar *CO* bands, 3 μm absorption (Ridgway, Carbon and Hall 1978), and SiC-rich dust from the *IRAS* spectrum (*IRAS* Science Team 1986). It also has a hotter companion of spectral type A3 and uncertain luminosity class (probably V-III). Le Bértre comments that CS 776 and its circumstellar dust shell have properties similar to those of more familiar carbon mira variables, except for a lack of periodicity in its variations.

Diffuse bands are visible in the spectra of the companion and in the composite spectrum of both stars. The spectrum of CS 776 is complex in the long wavelength regime where the carbon star dominates, so that it is difficult to determine a continuum level. The A3 companion, however, can be compared to a similar spectral type to cancel the stellar lines, leaving a fairly clean spectrum showing the diffuse bands and Na D lines. Le Bértre gives the equivalent widths of the 4430, 5780, 5797, and 6284 Å bands as 1.39, 0.85, $\leq$ 0.15 and 0.83 Å, respectively, where I have used the measurements from LBL93 for all but the 6284 Å band. This latter paper also gives higher resolution spectra of the star; these are reproduced in Figure 1.

Clearly the diffuse bands are present in the spectra. The critical question is whether the bands are circumstellar or interstellar. LB90 offers two bits of evidence pointing toward a circumstellar origin. First, the star lies in a region of low reddening near galactic longitude 240° (Lucke 1978). Second, the CO emission in that direction (apart from a circumstellar component) is at larger radial velocities than the star (May, Murphy and Thaddeus 1988), and hence comes from larger distances. On the basis of these facts, LB90 and LBL93 suggest that the diffuse bands seen towards CS 776 are circumstellar.

These arguments are suggestive, but not compelling. They offer only indirect evidence for the circumstellar nature of the detected bands. The interstellar medium is so clumpy in other directions that it is quite possible that an interstellar cloud could be in this line of sight, even though it is within Lucke's window. There are some indirect arguments suggesting that the bands are interstellar rather than circumstellar. The fact that the spectrum of CS 776 is so similar to that of other carbon Miras that don't show diffuse bands argues against the bands being circumstellar: if they are produced in the carbon star winds for CS 776, then why only in this star and not in BS 3233, which is otherwise similar? The Na I D lines are also seen in absorption toward the A3 companion star. The ionization potential of Na I is only 5.139 ev, corresponding to about 2413 Å. According to the *IUE Atlas of Stellar Spectra* (Wu

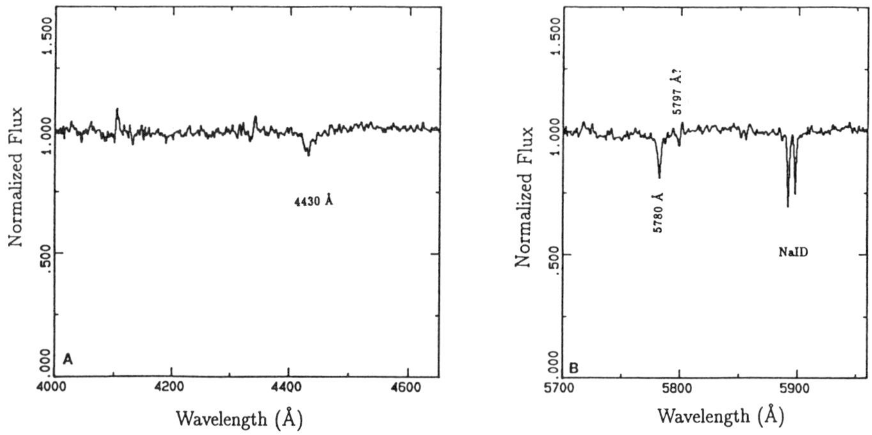

Fig. 1. Diffuse bands in the A3 companion to CS 776. The spectra have been divided by the spectra of the A3 star BS 3233 to remove stellar lines (from LBL93.)

et al. 1983), the cutoff for UV emission from an A3 star is at about 1700 Å, so that one might expect the sodium to be largely in the form of Na II for the circumstellar nebula (assuming that the A3 star is a companion to CS 776 and not merely a line-of-sight coincidence; the distance estimates given by Le Bertre are consistent with a double star system). If the A3 star ionizes the Na I in the system, then the observed line strength must be interstellar. Herbig (1993) presents a correlation between Na I column density and the equivalent width of the 5780 Å band. Although LBL93 do not present measurements on the sodium D lines, estimates made from their figure indicate a column density of about logN(NaI) $\simeq$ 13.18. For this amount of sodium, the 5780 diffuse band is somewhat strong, but not inconsistent with the scatter in the log [W(5780)] *vs.* log [N(Na I)] relationship given by Herbig. The implication is that the observed band strength is likely interstellar rather than circumstellar.

Neither of these arguments is compelling for an interstellar locale for the diffuse bands, but they tend to weaken the argument for the circumstellar case. In favor of the circumstellar interpretation is the strength of the band compared to the E_{B-V} value. Although the equivalent width for 5780 reported by LBL93 appears somewhat strong compared to a crude measurement from their published spectrum, it is still very high for the estimated E_{B-V} of 0.53. LB90 estimates an optical depth of 0.4 at μm; scaling this with the dust extinction given by Mathis (1990) suggests a circumstellar extinction of about 0.31 for 'normal' dust, leaving only 0.22 magnitudes of E_{B-V} for the interstellar component. However, this should be considered a lower limit, since circumstellar dust is often observed to be larger than normal, with R_V up to 12.6 reported by Tjin A Djie and Thé (1978). If the 0.22 magnitude figure is correct, then the band strengths would have to be extremely high in the interstellar component to avoid having at least some of the band strength in the circumstellar nebula.

We conclude that the data for CS 776 is suggestive of the presence of circumstellar diffuse bands, but the evidence is not compelling.

2.2. *IRAS 21282+5050*

Cohen and Jones (1987) report strongly enhanced diffuse bands in the line of sight toward this O7(f)-[WC 11] planetary nebula nucleus. The enhanced bands are at $\lambda\lambda$ 5797, 6196, 6203, 6283, and 6613 Å. This star also exhibits the infrared emission lines characteristic of PAHs (Cohen, Tielens, and Allamandola 1985).

In this case, there is no way to reliably distinguish between circumstellar reddening and interstellar reddening. Cohen and Jones's argument relies on the finding that the band strengths are enhanced in this line of sight; they suggest that the enhancement might be connected to the presence of PAHs around the star. LBL93 have remeasured some of the diffuse bands toward this star. Their measurements are generally somewhat smaller than the Cohen and Jones values, but not inconsistent with them, with the exception of the 5797 Å band. The two sets of authors disagree about the proportion of the blended feature to be ascribed to the 5801 lines of C IV; Le Bértre and Lequeux consequently find the 5797 Å diffuse band much weaker than Cohen and Jones. To compound the matter, Cohen and Jones find the diffuse bands very strong in comparison to HD 2905 from Krelowski and Walker (1987), while LBL93 find them consistent with the diffuse band strengths toward HD 183143 (Herbig 1975). The latter procedure is more correct, although it would be better still to compare the band strengths to the general correlations with other parameters for a number of lines of sight. Herbig (1993) includes a modern plot of W_λ *vs.* E_{B-V} for the 5780 Å band. On their plot, the IRAS 21282+5050 equivalent width of 0.90 (Le Bértre and Lequeux) to 1.0 Å (Cohen and Jones) *vs.* $E_{B-V} \simeq 1.7$ puts the star in the class of HD 183143 and Cyg OB2 #10 as having an unusually strong 5780 band, but not without precedent. The same comment applies to the 6283 Å band when compared to modern measurements toward 11 stars by Herbig and Leka (1991), although we note that the values compiled by Snow, York and Welty (1977) are generally much lower for this band.

The conclusion is that the diffuse bands toward IRAS 21282+5050 are strong, but not without precedent in the interstellar medium. Although there is good evidence for circumstellar material around this star, there is no direct evidence for assigning the E_{B-V} reddening or the diffuse bands to this material as opposed to the interstellar foreground material.

2.3. *NGC 7027*

Pritchet and Grillmair (1984) find apparent diffuse band absorption in the line of sight to the planetary nebula NGC 7027. Although the features may be real (see Figure 2), it is almost impossible to accurately evaluate their strength due to the many emission features surrounding them. Pritchet and Grillmair identify Hg I lines at 5770 and 5791 Å; Le Bértre and Lequeux (1992) further identify emission from the $(n-5)$ series of He II, plus C IV emission at 5802 Å. The diffuse bands appear in gaps between these emission lines, so that the placement of the continuum level for equivalent width measurements is highly uncertain. The 6283 Å line does appear clearly in a somewhat smoother part of the spectrum in the data from Le Bértre and Lequeux. Pritchet and Grillmair suggest that the diffuse bands may arise from an older shell formed by mass loss from the precursor of NGC 7027. Lacking additional evidence on the foreground reddening and extinction within this shell, their suggestion must remain just that — a suggestion, but not a compelling argument.

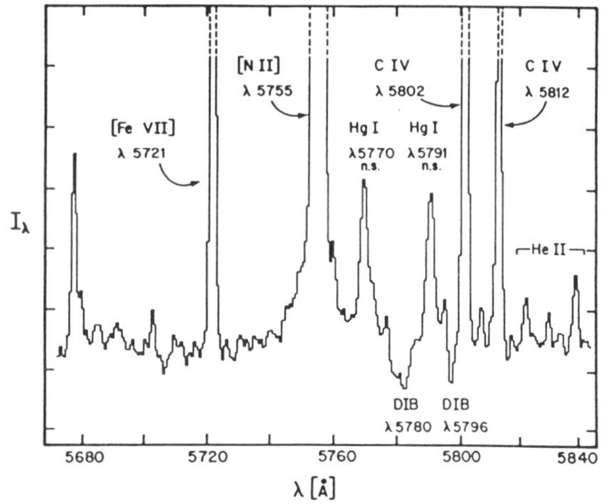

Fig. 2. Spectrum of NCG 7027 showing the possible detection of the diffuse bands (from Pritchet and Grillmair 1984.)

2.4. OTHER DETECTIONS

LBL93 report results of a search for diffuse bands in the lines of sight toward 19 mass-losing stars, divided between *C*-rich, *O*-rich, and *N*-rich sources. They find diffuse bands toward 14 of these stars, including all of their *O*-rich and *N*-rich targets, but only 6 of 11 *C*-rich targets. Targets include Wolf-Rayet stars, one planetary nebula, and several early-type supergiants. The selection criterium for these targets is not mentioned, beyond the fact that all are known to be surrounded by circumstellar matter. It is not clear for any of these sources whether the detected diffuse bands arise from this circumstellar matter or from foreground gas and dust. Given this fundamental uncertainty, no clear conclusions are possible from the data presented, although there is a suggestion of a correlation with an overabundance of nitrogen (LBL93).

There are two other reports of circumstellar bands in the literature. Heydari-Malayeri *et al.* (1993) report a time-varying 4430 Å band in the spectrum of the LMC star HDE 269828. If the time variation in the band were true, it would definitely indicate a circumstellar origin for the band. However, a close examination of the spectra presented in their Figure 3 strongly suggests that the origin of the 'variation' is in the changing strengths of the adjacent He I absorption lines at 4387 and 4471 Å. The large changes in strength of these and other lines in the spectra makes the placement of the continuum highly uncertain, and provides the most likely explanation of the apparent change in band strength, rather than a circumstellar origin.

Tüg and Schmidt-Kaler (1981) report a contribution to the central depth A_c of the 4430 Å band from circumstellar material around early B-type Ia supergiants, based on a slightly higher regression line for these stars as compared to the average. However, their analysis is based on the tabulation from Snow, York and Welty (1977), which is compiled from a number of different sources, mostly from photographic spectra, and suffers from a large degree of uncertainty. The very small difference in the regression lines (0.2% in the y-intercept) that they find is not significant in the face of the inherent errors in a compilation from diverse sources such as this.

2.5. DISCUSSION OF REPORTED DETECTIONS

The most intriguing question raised by these searches for circumstellar diffuse bands is why there seem to be no diffuse bands detected toward stars that show strong evidence of the infrared emission features normally attributed to PAHs. We caution that this is a weak finding, since it includes targets such as NGC 7027 and IRAS 21282+5050 which show some evidence of diffuse bands, but where the bands are more likely interstellar rather than circumstellar. However, the trend is strongly supported by the case of HR 4049, and will be discussed in the next section.

LBL93 suggest that nitrogen enhancement might be related to the formation of diffuse bands, and that carbon enrichment is not required. In the absence of more definitive data on the separation between circumstellar and interstellar material, it is hard to evaluate the significance of this finding.

All of these reported detections of circumstellar diffuse bands suffer from the serious difficulty in distinguishing circumstellar features from interstellar features. For this reason, the results cannot be interpreted as definitive, unambiguous detections of DCBs, and therefore cannot as yet seriously constrain the theories. On the other hand, non-detections of diffuse bands toward stars with circumstellar medium are definitive: if a band is absent in the line of sight, then it is absent from both the interstellar and circumstellar matter. Non-detections, then, can more easily provide hard constraints on the nature of the carriers. The next section deals with these non-detections.

3. NON-DETECTIONS OF CIRCUMSTELLAR BANDS

One expects to see diffuse absorption bands everywhere in the interstellar medium. It can be a challenge to find lines of sight where there are no diffuse bands. However, there are two approaches to the problem. First and most obvious is to look towards candidate stars with little or no interstellar medium in the line of sight. HR 4049 falls into this category. The second method is to look at sources where the circumstellar material varies significantly on short time scales, such as in novae and certain other classes of mass-losing objects. Examples from both these types are given below.

3.1. HR 4049

HR 4049 (= HD 89353) is a high galactic latitude ($b = +23^o$), C-rich, low surface gravity Pop II star recently reclassified as a low-mass post-AGB object by Waelkens *et al.* (1991). It exhibits a circumstellar dust shell with properties that vary with the orbital phase of the companion star, as well as with the variability of the star itself (Waelkens *et al.* 1991). Waters *et al.* (1989) report the presence of infrared emission bands at 7 – 9 μm and 11.3 μm, indicating the presence of PAHs. They also report the absence of the 2175 Å extinction bump and the diffuse bands at 5780 and 6614 Å. Their optical spectra are reproduced in Figure 3. Note that the weak feature near the 5780 Å band location is interpreted as interstellar on the basis of a +55 km s^{-1} velocity shift from the rest frame of the star. They also place an upper limit of $E_{B-V} < 0.10$ on the interstellar foreground reddening, consistent with the observed strength of the interstellar 5780 Å feature. The circumstellar extinction is in the range $0.19 < E_{B-V} < 0.29$, depending on the exact value of the foreground reddening.

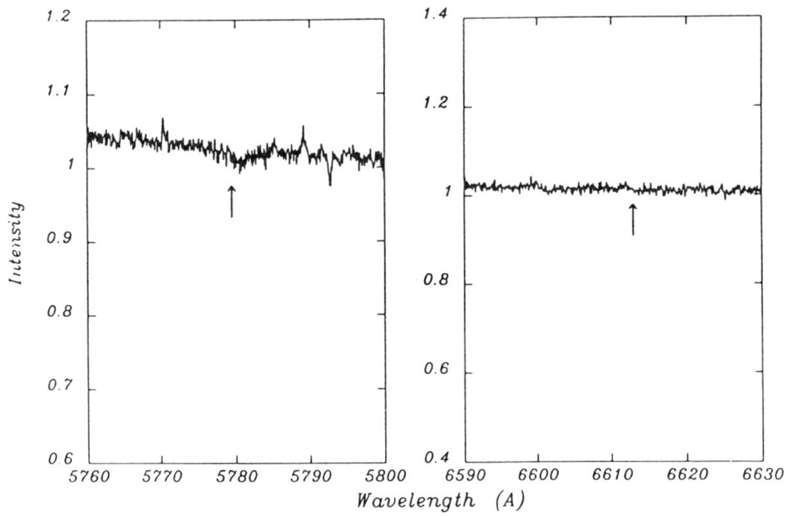

Fig. 3. The locations of diffuse bands toward HR 4049. Note the apparent feature just longward of 5780 is shifted too far away from the rest frame of the star to be circumstellar (from Waters *et al.* 1989.)

This star provides a clear case in which there is circumstellar material that is deficient in diffuse bands, although there is a substantial quantity of dust and PAHs. The presence of the PAHs is consistent with the *C*-rich environment of the stellar wind. Cleary, the PAHs which produce the infrared bands seen in this star are not the carriers of the diffuse bands. Neither are they related to the otherwise ubiquitous 2175 Å bump normally seen in the interstellar extinction curve. Apparently the dust formed in the outflow from this star is of a different character than the 'average' interstellar dust population. This conclusion agrees with prior finding for new-born dust around stars, e.g., α Sco (Buss and Snow 1988; Seab and Snow 1989), where the grains are unusually large and also lacking in the 2175 Å bump.

3.2. HD 5999

This is a young variable star (also know as HD 144668 and V856 Sco) of spectral class A 7 III, with a circumstellar dust shell studied by Andersen, Gahm and Krełowski (1982). The variation amounts to 1.5 magnitudes, and seems to be due to changing amounts of circumstellar dust, since the spectral type of the star remains constant. Andersen *et al.* report on an optical spectrum of the star taken near minimum light, when the dust shell was at its thickest. They find no evidence of diffuse bands, setting an upper limit of 0.200 Å equivalent width for the 4430 Å band, and a limit of 0.080 Å in its change from the maximum light phase. These numbers are well below the expected value based on the average correlation with $E_{B-V} \simeq 0.5$; Tüg and Schmidt-Kaler's (1981) correlation give an expected value of about 1.3 Å. It is also interesting that the dust grains formed during the minimum light episodes in this star appear to be much larger than normal, with R_V as high as 12.6 (Tjin A Djie and Thé 1978), with some indications of anomalous UV extinction (Thé *et al.* 1981) and a missing 2175 Å bump (Hack *et al.* 1980).

3.3. NOVA CAS 1993

Nova Cas 1993 was first observed on December 7, 1993 (IAU Circ. No. 5902). Shore *et al.* (1994) report spectroscopic observations of the nova's evolution in time. In early 1994 the brightness of the source dropped by a factor of 30, coincident with grain formation in the expanding envelope. At this time there were no diffuse bands visible (Shore 1994) in the spectrum but there was some evidence of weak PAH emission at 3.4 μm. In addition, the UV extinction was flat from 1200 – 2900 Å, indicative of large sized grains ($> 0.2\mu$m). The evidence points to an expanding, dust-forming, cooling shell, wherein one might expect to see molecules, but with no evidence of diffuse bands.

3.4. HD 213985

This is another post-AGB star, studied by Buss, Lamers and Snow (1989). It has an anomalous UV extinction curve, with a strong, broad UV bump shifted to longer wavelengths than normal. It is *C*-rich, although less so than HR 4049. In contrast to that star, however, HD 213985 lacks the infrared emission lines usually attributed to PAHs (Geballe *et al.* 1989). It also has no sign of the 6614 Å diffuse band (Buss *et al.* 1989).

3.5. DISCUSSION OF NON-DETECTIONS

Unlike the reported cases of detections of circumstellar bands, there is no ambiguity in these cases of non-detections. Clearly, the bands are *not* present in the circumstellar environments of these stars. These cases show circumstellar material either through the detection of emission lines from PAHs, or through IR excesses coupled with UV grain extinction, or from both.

The fact that the diffuse bands are absent from sites of new-formed dust says that the band carriers are distinct from this type of dust. This provides a weak constraint on dust theories such as that of the embedded cavity model of Shapiro and Holcomb (1986a, b). It cannot give a strong constraint because the dust in these circumstellar environments is so clearly anomalous. It is possible that the sort of processing that turns pristine circumstellar dust into the average mixture of interstellar dust could also alter the grains in such a way as to produce the diffuse bands. For example, the impurity centers postulated by Shapiro and Holcomb might be implanted in the grains by cosmic ray impacts or by processing through interstellar shocks. However, we note that there are other reasons for questioning the embedded cavity model; Seab and Snow (1994) show that the sort of profile changes expected with increasing grain size according to the Shapiro and Holcomb (1986b) theory are not seen in the ρ Oph cloud. The argument is somewhat stronger against the porphyrin model of Johnson (1994), since the matrix-embedded porphyrin molecule would more reasonably be formed at the time of the initial grain condensation, rather than by grain processing at later times. Neither of these models can be definitely ruled out, but the observations do provide a strong challenge to the grain theories to not only devise a band carrier model, but also to explain why it is not effective in the circumstellar environment.

This same challenge applies to the other theories for the diffuse band carriers. In particular, the neutral PAHs in circumstellar environments cannot be the band carriers themselves. Since these are largely cool stars (HR 4049 is B9, however), and the PAHs are expected to have ionization potentials in the range of 5 – 9 ev, then they are most likely neutral in these regions. This is not a significant constraint, since the lack of observable neutral-PAH structure in the UV extinction curve has lead to the suggestion of ionized PAHs as the band carriers anyway (van der Zwet and Allamandola 1985; Lèger and d'Hendecourt 1985; Crawford, Tielens and Allamandola 1985). Moreover, the diffuse bands are observed to be greatly weakened or absent from the cores of interstellar clouds (Snow and Cohen 1974; Adamson, Whittet and Duley 1991), where the PAHs also might be expected to be neutral. However, Snow *et al.* (1995) present evidence of circumstellar clouds hot enough to have ionized PAHs, but still lacking diffuse bands. This finding argues against even ionized PAHs as the diffuse band carriers.

Other molecular carriers of the diffuse bands perhaps have the advantage of not being so easily recognizable in the circumstellar environment as the PAHs. Thus the observations of circumstellar environments cannot make definitive statements about the partially-hydrogenated hydrocarbons suggested by Fulara *et al.* (1994), the methyl-pyrene of Léger, d'Hendecourt and Défourneau (1994), or the ionized naphthalene suggested by Salama and Allamandola (1992), because these are harder to positively identify in the circumstellar material. However, all of these candidates must still meet the strong challenge of being absent, or at least inactive, in the environment of the stellar winds, but yet omnipresent in the general ISM. This challenge in general is met by models depending on particular ionization levels or hydrogenation states of the carriers to produce the diffuse bands.

4. DIFFUSE BANDS IN DENSE CLOUDS

We will briefly discuss the observations of diffuse bands in dense clouds, as these produce constraints on the carriers related to those in circumstellar environments, and thus help to present a more rounded picture of the challenges to be met by carrier theories.

The basic observation that the strength of diffuse bands is weaker in dark clouds was explored by Snow and Cohen (1974). More recently, Adamson, Whittet and Duley (1991) have made a study of the Taurus dark cloud, and Seab and Snow (1994) of the ρ Oph cloud. These later findings confirm the weakness of the bands in denser environments. Snow and Seab (1991) show that the 5780 Å band toward the Taurus cloud star HD 29647 is particularly weak and narrow; so narrow, in fact, that tends to rule out aborption centers in solid grains as the band carriers, since these require larger bandwidths than are observed toward this star. This is a single extreme case of a weak band in a dark cloud, but it exemplifies the generally observed trend.

Adamson *et al.* (1991) go a step further in suggesting that the bands may be produced entirely on the surfaces of the clouds, with no contribution from the cloud interior. Figure 4 (reproduced from their Figure 7) clearly shows that the ratio $W_\lambda(5780)/E_{B-V}$, identified with the 'production efficiency' of the band, declines steadily with depth in both the Taurus and ρ Oph clouds. They interpret this result in terms of a model with the bands arising from the partial hydrogenation of unspecified hydrocarbons (Webster 1993a,b), with the fulleranes suggested as the most likely species involved. In this model, the diffuse bands can be suppressed either from having either too much or too little hydrogenation. This feature gives their theory more flexibility than the ionized PAH hypothesis, since the second ionization potential of the PAHs is unreasonably high for the interstellar medium. In dark clouds,

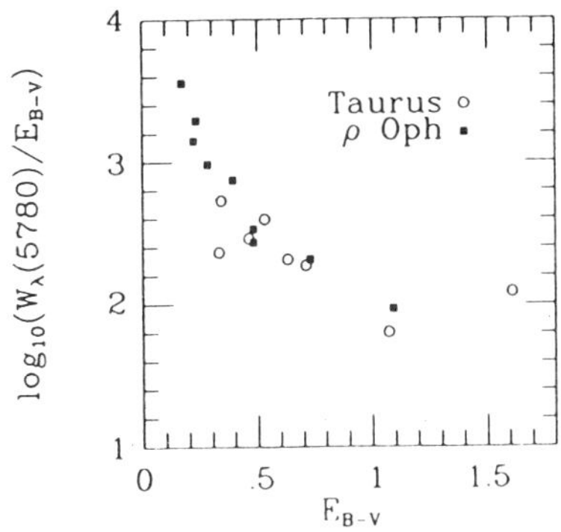

Fig. 4. The decrease in production efficiency of the 5780 Å band with depth in the Taurus and ρ Oph clouds (from Adamson *et al.* 1991.)

the higher densities lead to greater degrees of hydrogenation of the molecules with the result that the diffuse bands are suppressed.

McIntosh and Webster (1992, 1993) also attempt to deduce the relative degree of hydrogenation for the different families of diffuse bands. In observations of the Taurus cloud, McIntosh and Webster (1992) found that the Family 2 bands (Krelowski and Walker 1987) (includes the 5797 and 5850 Å bands) become weaker relative to the Family 3 bands (includes the 5780, 6284, and other bands) as the value of R_V increases. Taking R_V as an indication of cloud density, they interpret this effect as meaning that the 5797 band prefers lower densities and therefore has a smaller degree of hydrogenation than the 5780 band. This is in agreement with their investigation of diffuse bands at higher galactic latitudes (McIntosh and Webster 1993), where the 5797 Å band again is found to prefer lower densities. Although this sort of agreement is encouraging, it is directly contradicted by the recent findings of Seab and Snow (1994), who show that the opposite is true for stars in the ρ Oph cloud: the ratio of the 5780 to 5797 band strengths decreases with λ_{max}, a parameter well-correlated with R_V. Thus the straightforward interpretation of the change in relative strengths of these bands in terms of differing degrees of hydrogenation (Webster 1993a,b) will not work here. As in almost all diffuse band work, there must be another significant parameter influencing the results.

5. SUMMARY AND CONCLUSIONS

The search for circumstellar diffuse bands has a potentially large payoff in terms of significant constraints on the proposed carriers of the interstellar diffuse bands. The constraints are evident both from the negative and positive results. There is clear evidence for the absence of diffuse bands in circumstellar nebulae where the presence of strong infrared emission bands indicates the presence of PAHs. Therefore one can confidently conclude that the neutral PAHs are not the diffuse band carriers; if the PAHs are related to the carriers, it can only be the ionized PAHs responsible. This is an important and possibly testable conclusion, since one can search for diffuse bands in regions where ionized PAHs would be expected. A negative finding in those regions would essentially rule out the PAHs as the band carriers altogether. This test is most easily applied to the PAHs because they are detectable around stars from their infrared emission features. The other types of proposed carriers must pass the same stringent test: if they are formed in the winds from stars, where dust grains, PAHs, and presumably other types of large molecules are formed, then one must explain why the molecules or grains do not produce the bands in this environment (at least around stars like HR 4049 and other locales with no observed bands).

The constraint becomes double-barreled when the observations of diffuse bands in dark clouds is included. Not only must the carriers be suppressed in the outflows from stars, but they must also be weakened in the dense, cold cores of clouds. The combination of these two constraints favors molecular models over grain models for the bands. Methods of suppressing the bands in grains by accreting mantles on grains would tend to give broader band profiles, or at least some changes in the band profiles. The observations of Seab and Snow (1994) showing a remarkably constant profile at different depths in the ρ Oph cloud argues against these grain models. On the other hand, changing either the average ionization or degree of hydrogenation would weaken the band without changing its profile, as is observed. The challenge for the molecular theories is simply to have a mechanism for suppressing the bands in these locales.

An unambiguous positive detection of diffuse bands in the circumstellar environment would present an even more significant constraint on the carrier theories, since typically much more can be known or found out about these relatively small, well-studied regions. Unfortunately, even the best case for a circumstellar band, that of CS 776 (LB90, LBL93) suffers from the difficulty of clearly separating the circumstellar from the interstellar material. In fact, there is some reason to believe that the detected bands in CS 776 may be interstellar. As long as this ambiguity is there, then the reported detections of circumstellar bands cannot place any hard constraints on the theories.

At some point the circumstellar material must become interstellar. There is strong evidence that the circumstellar dust grains differ in character from the interstellar grains; just how and when they change from one nature to the other is not clear. Similarly, the circumstellar gas and dust fails to exhibit diffuse band absorption in at least some unambiguous cases, and yet the bands are nearly ubiquitous in the diffuse interstellar medium. It does not seem likely that the difference arises from merely mixing gas and dust from different sources; rather, it is more likely that there is some form of processing that happens to the material to convert it to interstellar stuff. Where, when, and how the processing occurs is open to speculation. Large grains can be sputtered and shattered down to a population of smaller grains by interstellar shocks (Tielens *et al.* 1994). Molecules are perhaps more easily processed by aging in the harsh environment of the interstellar medium: bathed in radiation, subjected to chemical attack by hydrogen and other species, bombarded with cosmic rays, and cycled through the various phases of the ISM. It is possible, however, that the reported detections of circumstellar bands may represent a more prompt processing mechanism, where the molecules are converted to diffuse band carriers just outside the immediate circumstellar region. For example, Prichett and Grillmair (1984) suggest that the bands seen around NGC 7027 might arise in an older ejected shell of matter; if this is the case, then it would indicate a rather prompt transition to the interstellar type of gas. Certainly the change must happen somewhere.

We conclude that searches for diffuse circumstellar bands have provided important constraints on the various theories for the band carriers. There have been several possible detections, but each suffers from some degree of ambiguity about the location of the band. Even so, the clear negative findings, especially when coupled with the dark cloud results, provide a challenge that must be met by any successful carrier theory.

ACKNOWLEDGEMENTS. I would like to thank Drs. T. P. Snow and A. G. G. M. Tielens for many, many interesting discussions about dust grains and diffuse bands. I am indebted to the NASA Conference organizers and the University of New Orleans for financial support to attend the conference, and the NASA JOVE program for support while this review was written at the University of Colorado.

REFERENCES

Adamson, A.J., Whittet, D.C.B., and Duley, W.W. 1991, *M. N. R. A. S.*, **252**, 234.
Andersen, J., Gahm, G.F., and Krelowski, J. 1982, *Astr. Ap.*, **113**, 176.
Buss, R.H. and Snow, T.P. 1988, *Ap. J.*, **335**, 331.
Buss, R.H., Lamers, J.G.L.M., and Snow, T.P. 1989, *Ap. J.*, **347**, 977.
Cohen, M. and Jones, B.F. 1987, *Ap. J. (Letters)*, **321**, L151.
Cohen, M., Tielens, A.G.G.M., and Allamandola, L.J. 1985, *Ap. J. (Letters)*, **299**, L93.
Crawford, M.K., Tielens, A.G.G.M., and Allamandola, L.J. 1985, *Ap. J. (Letters)*, **293**, L45.
Fulara, J., Lessen, D., Freivogel, P., and Maier, J.P. 1994, *Nature*, **366**, 439.
Geballe, T.R., Noll, K.S., Whittet, D.C.B., and Waters, L.B.F.M. 1989, *Ap. J. (Letters)*, **340**, L29.
Hack, M., Thé, P.S., and Tjin A Djie, H.R.E. 1980, *Proc. Second Eur. IUE Conf.*, ESA SP-157, 219.
Herbig, G.H. 1966, *Vistas in Astr.*, **8**, 109.
Herbig, G.H. 1975, *Ap. J.*, **196**, 129.
——. 1993, *Ap. J.*, **407**, 142.
Herbig, G.H. and Leka, K.D. 1991, *Ap. J.*, **382**, 193.
IRAS Science Team 1986, *Astr. Ap. Suppl. Ser.*, **65**, 607.
Johnson, F. 1994, this volume.
Krelowski, J. and Walker, G.A.H. 1987, *Ap. J.*, **312**, 860.
Léger, A., d'Hendecourt, L., and Défourneau, D. 1994, preprint.
Léger, A. and d'Hendecourt, L. 1985, *Astr. Ap.*, **146**, 81.
Le Bértre, T. 1990, *Astr. Ap.*, **236**, 472 (LB90).
Le Bértre, T., and Lequeux, J. 1992, *Astr. Ap.*, **255**, 288.
——. 1993 *Astr. Ap.*, **274**, 909 (LBL93).
Lucke, P.B. 1978, *Astr. Ap.*, **64**, 367.
Heydari-Malayeri, M., Grebel, E., Melnick, J., and Jorda, L. 1993, *Astr. Ap.*, submitted.
Mathis, J.S. 1990, *Ann. Rev. Astr. Ap.*, **28**, 37.
May, J., Murphy, D.C., and Thaddeus, P. 1988, *Astr. Ap. Suppl. Ser.*, **73**, 51.
McIntosh, A. and Webster, A. 1992, *M. N. R. A. S.*, **255**, 37.
——. 1993, *M. N. R. A. S.*, **261**, L13.
Ott, U. 1993, *Nature*, **364**, 25.
Pritchet, C.J. and Grillmair, C.J. 1984, *Pub. A. S. P.*, **96**, 349.
Ridgway, S.T., Carbon, D.F., and Hall, D.N.B. 1978, *Ap. J.*, **225**, 138.
Salama, F. and Allamandola, L.J. 1992, *Ap. J.*, **595**, 501.
Savage, B.D. and Mathis, J.S. 1979, *Ann. Rev. Astr. Ap.*, **17**, 73.
Seab, C.G. and Snow, T.P. 1989, *Ap. J.*, **347**, 479.
——. 1994, *Ap. J.*, submitted.
Shapiro, P.R. and Holcomb, K.A. 1986a, *Ap. J.*, **305**, 433.
——. 1986b, **310**, 872.
Shore, S.N. 1994, private communication.
Shore, S.N., Starrfield, S., Gonzales-Biestra, R., Hanschildt, P.H., and Sonneborn, G. 1994, *Nature*, submitted.
Snow, T.P. 1991, *Ap. J.*, **401**, 775.
Snow, T.P. and Cohen, J.G. 1974, *Ap. J.*, **194**, 313.
Snow, T.P. and Seab, C.G. 1991, *Ap. J.*, **382**, 189.
Snow, T.P. and Wallerstein, G. 1972, *Pub. A. S. P.*, **84**, 492.
Snow, T.P., Bakes, E.L.O., Buss, R.H., and Seab, C.G. 1995, *Astr. Ap.*, in press.
Snow, T.P., York, D.G., and Welty, D.E. 1977, *A. J.*, **82**, 113.
Tüg, H. and Schmidt-Kaler, Th. 1981, *Astr. Ap.*, **94**, 16.

Thé, P.S., Tjin A Djie, H.R.E., Bakker, R., Bastiaansen, P.A., Burger, M., Cassatella, A., Fredga, K., Gahm, G., Liseau, R., Smyth, M.J., Viotti, R., Wamsteker, W., and Zeuge, W. 1981, *Astr. Ap. Suppl. Ser.*, **44**, 451.
Tielens, A.G.G.M., Hollenbach, D.J., McKee, C.F., and Seab, C.G. 1994, *Ap. J.*, in press.
Tjin A Djie, H.R.E. and Thé, P.S. 1978, *Astr. Ap.*, **70**, 311.
van der Zwet, G.P. and Allamandola, L.J. 1985, *Astr. Ap.*, **146**, 76.
Waelkens, C., Lamers, H.J.G.L.M., Waters, L.B.F.M., Rufener, F., Trams, N.R., Le Bértre, T., Ferlet, R., and Vidal-Madjar, A. 1991, *Astr. Ap.*, **242**, 433.
Waters, L.B.F.M., Lamers, H.J.G.L.M., Snow, T.P., Mathlener, E., Trams, N.R., van Hoof, P.A.M., Waelkens, C., Seab, C.G., and Stanga, R. 1989, *Astr. Ap.*, **211**, 208.
Webster, A. 1993a, *M. N. R. A. S.*, **262**, 831.
——. 1993b, *M. N. R. A. S.*, **263**, 385.
Wu, C.-C., Ake, T.B., Boggess, A., Bohlin, R.C., Imhoff, C.L., Holm, A.V., Levay, Z.G., Panek, R.J., Schiffer, F.H. III, and Turnrose, B.E. 1983, *NASA IUE Newsletter*, No. 22.

RED RECTANGLE EMISSION BANDS

J.R. MILES AND P.J. SARRE
Department of Chemistry, University of Nottingham,
University Park, Nottingham NG7 2RD, UK
S.M. SCARROTT
Physics Department, University of Durham,
South Road, Durham DH1 3LE, UK

ABSTRACT.

The Red Rectangle is a bipolar nebula with a remarkable spectrum in which emission is excited by the light of the central star HD 44179. The emission spectrum contains lines of Na, Ca II, CH^+ and C_2, extended red emission (ERE) and infrared emission bands. Most striking, however, is the presence of strong unidentified optical emission bands near 5800, 5853, 6378 and 6615 Å (Schmidt *et al.* 1980, Warren-Smith *et al.* 1981), the most prominent of which appear to arise from the same carriers as a set of relatively narrow diffuse interstellar absorption bands (λλ5797, 5849, 6376/9 and 6614) (Scarrott *et al.* 1992); these diffuse bands have been classified as a family by Krełowski and Walker (1986, 1987). New high signal-to-noise spectra recorded with the Anglo-Australian telescope confirm the link between the Red Rectangle emission and diffuse absorption bands; it is found that the peak wavelengths and widths of the Red Rectangle bands alter smoothly from the hot region near the star to the cooler outer regions of the nebula and extrapolate to the corresponding diffuse band characteristics. The results provide strong evidence that the carriers of these particular diffuse bands are free molecules.

1. INTRODUCTION

One of the most serious problems limiting progress towards a solution of the diffuse band problem is that, while readily recorded in the spectra of many reddened stars, the spectra along various lines-of-sight generally have many more similarities than differences. It has therefore proved difficult to determine which bands might have a common or related origin. While differences in the relative strengths of a few bands, for example λ5780 and λ5797, are well established, classification of the bands into groups or 'families' has been put forward only recently. Most notable is the work of Krełowski and Walker (1986, 1987) who proposed three families of bands based on the variation of their strengths along different lines-of-sight. Their third family of bands, with members at 5797, 5850, 6376/9 and 6614 Å, is the most important in relation to our work. There have been a number of other studies which agree with the classification to a greater or lesser extent; a compilation of the most important features of previous work has been presented by Miles and Sarre (1993).

A. G. G. M. Tielens and T. P. Snow (eds.), The Diffuse Interstellar Bands, 143–148.

The laboratory spectroscopist has clear advantages in the assignment of spectra of unidentified species. It may be possible to record both emission and absorption spectra, at various temperatures and with control of the chemical composition. The Red Rectangle spectra discussed here are somewhat analogous to those which may be obtained in the laboratory in that it has proved possible to study the spectra of a small subset of diffuse band carriers in emission, in absorption, and over a range of temperatures. As discussed later this provides very strong evidence for a molecular origin for the carriers of these particular bands. The chemical formula(e) of the carriers remain, however, tantalisingly elusive. A further link between laboratory and astrophysical spectra is the knowledge that some carriers possess an emission spectrum; this means that laboratory experiments can be carried out to search for spectra in emission rather than absorption which is a far less sensitive approach.

2. BACKGROUND

The Red Rectangle has several peculiar characteristics including an unusual ultraviolet extinction curve (Sitko *et al.* 1981), very strong unidentified infrared emission bands (Cohen *et al.* 1975, Russell *et al.* 1978, Geballe *et al.* 1985) and extended red emission (Cohen *et al.* 1975, Greenstein and Oke 1977), the unidentified optical bands discussed here, and emission from both CH^+ (Balm and Jura 1992, Hall *et al.* 1992) and C_2 which was discovered in this study.

The optical and other properties of the central star and nebula were first examined in detail by Cohen *et al.* (1975) and the first clear indication of *molecular* bands arose from a study by Schmidt *et al.* (1980) who reported bands with peak wavelengths at 5800.0, 5825.7, 5852.9, 5884.7, 5913.7, 5939.7, 6361.2, 6378.1, 6393.7 and 6615.3 Å, which may be summarised as groups in the region of λ5800, λ6380 and λ6617. Spectra at higher resolution were reported by Warren-Smith *et al.* (1981) in which the wavelengths were measured more precisely, the band shapes were determined and new bands were discovered.

The fact that the wavelengths of the principal unidentified Red Rectangle emission bands are close to those of the diffuse bands in Family 3 of Krełowski and Walker (1986, 1987) was noted independently by Sarre (1991) and Fossey (1991); the band shapes for λ5797 and λ6614 in absorption and emission were also recognised as being qualitatively similar (Sarre 1991). A reanalysis of spectra recorded in 1980 has provided evidence in support of a link (Scarrott *et al.* 1992). In addition, high quality optical spectra of the Red Rectangle have been reported by Schmidt and Witt (1991), who examined the spatial distribution of the carriers, the changes in wavelength and width of the emission features, and the relationship of the sharp bands with ERE in particular.

3. OBSERVATIONAL ASPECTS

The data described here were recorded on the nights of 1992 January 10, 11 with the RGO spectrograph on the Anglo-Australian telescope, with the slit aligned along the NW–SE whisker at a PA of 162° and with the light from the central star obscured by a dekker. One particular advantage of detection with a CCD is that spectra can be obtained for a wide range of offsets simultaneously. Three regions, centred on wavelengths of 5400, 6000 and 6600 Å and each covering 600 Å, were recorded at a resolution of 0.64 Å. The data were reduced using standard Starlink FIGARO routines.

4. RESULTS

An overall view of the part of the spectrum obtained by summing a total of 22 pixel rows centred at offsets of 8 arc sec NW and SE of HD 44179 is shown in Figure 1. The Na lines are narrower than the molecular emission lines, confirming that the widths of the unidentified features arise from unresolved rotational structure rather than motion within the nebula. The spectra recorded in the λ5800 region for two different offsets are shown in Figure 2. It is clear that the widths of the bands increase with proximity to the central star and closer examination shows that the peak wavelengths also vary with offset. These changes are most likely attributable to the development of the rotational contour with increase in temperature, although a contribution from vibrational sequence band structure on the redward wing of the profiles also seems likely. It is clearly of importance to demonstrate quantitatively that there is a correspondence between emission and absorption features. We have been able to do this in the case of the three strongest bands, λ5797, λ5850 and λ6614. Measurements from the Red Rectangle are shown in Figure 3, which illustrates the behaviour of the peak wavelength and FWHM of the λ5797 feature in relation to the distance from the star. It is found that both the width and wavelength decrease with increasing offset and extrapolate to the 'asymptotic' values which hold for the cold diffuse cloud medium in absorption. It is worth noting that this result is independent of the origin of the wavelength shift and change in width, which may arise from a combination of the variation in population of rotational and vibrational levels with temperature (offset).

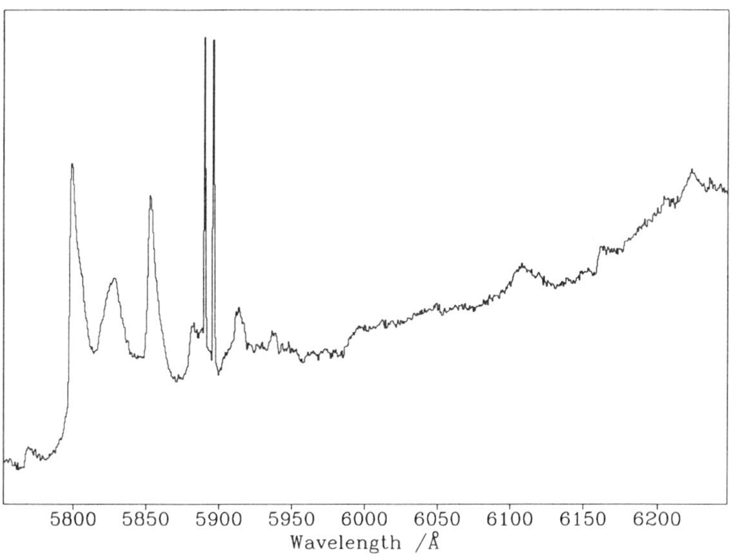

Fig. 1. Spectrum of the Red Rectangle recorded along PA = 162° (NW–SE) at an offset of 8 arc sec.

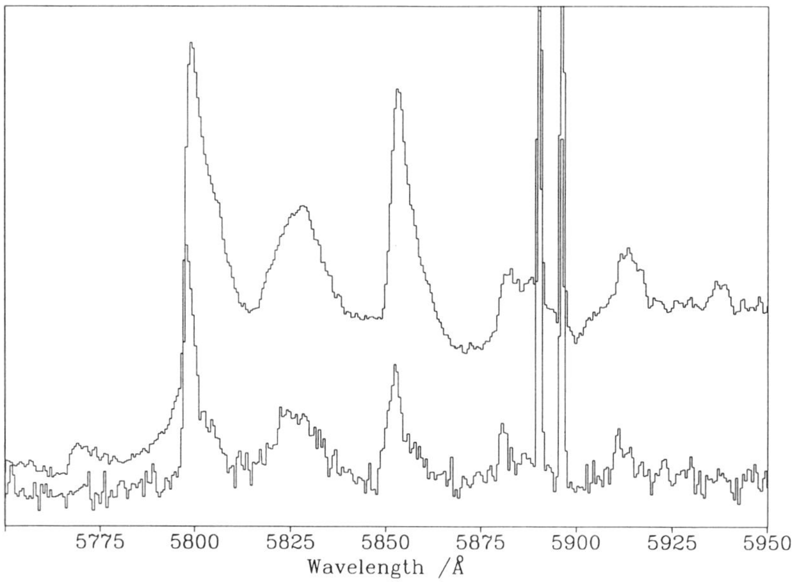

Fig. 2. Spectra of the λ5800 region at offsets of 8 (upper) and 17 (lower) arc sec, showing the changes in peak profile with increasing distance from HD 44179.

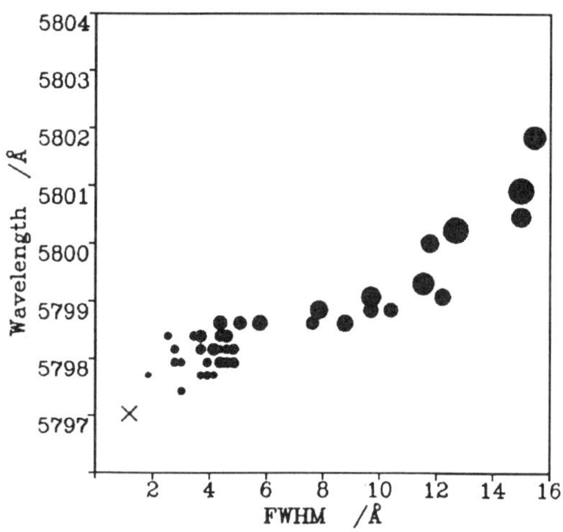

Fig. 3. The peak wavelength of the "5799" Red Rectangle feature plotted vs. the FWHM, where the radius of the circle is inversely proportional to the offset from the central star. The cross marks the values for the λ5797 diffuse band from Herbig (1975). Reproduced with permission from J. Chem. Soc Farad. Trans. **89**, 2308 (1993).

Further evidence for the emission-absorption correspondence emerges from a comparison between the new Red Rectangle data and recent high signal-to-noise diffuse absorption band studies in the $\lambda 5800$ region. Our data and the diffuse band results of a range of workers is presented in the table below. In each case the emission peak occurs to slightly longer wavelength than that of the diffuse band absorption feature.

TABLE 1
Nominal wavelengths in Å of emission features in the red rectangle and reported diffuse absorption bands

Red Rectangle (this work)	Diffuse Band	Reference
5769	5766	H, C, JD, KS
5799	5797	H
5819 (sh)	5818	JD, KS
5828	5828	JD, KS
5853	5850	H
5912	5910	K

sh = shoulder
References:
H = Herbig (1975)
C = Chlewicki *et al.* (1987)
JD = Jenniskens and Désert (1993, 1994)
KS = Krełowski and Sneden (1993)
K = Krełowski, private communication (1994)

5. DISCUSSION AND CONCLUSION

The identification of some of the diffuse absorption bands with emission features in the Red Rectangle opens a new dimension in the study of the carriers. The data provides some of the strongest evidence to date for molecular carriers for these particular bands, and under the assumption of a common carrier for the features in the $\lambda 5800$ region, a characteristic interval of *ca.* 90 cm^{-1} may be identified (from the separations of 5766–5797, 5797–5828, and 5818–5850); this might be a vibrational interval in the carrier. These emission bands have also been discovered in the spectrum of an R CrB star, V854 Cen, at minimum light (Rao and Lambert 1993). This R CrB star is unusual in exhibiting hydrogen lines, a characteristic in common with HD 44179, which may suggest a role for hydrogen in the formation of the carriers. While many of the possible deductions from Red Rectangle studies remain in the realms of speculation, we would assert that these results provide strong support for a molecular carrier, will encourage the design of laboratory experiments based on the search for emission, and justify the search for emission from other astrophysical objects.

ACKNOWLEDGEMENTS. We wish to thank PATT for the allocation of telescope time, Dr J. Bailey for his assistance with the observations, and PPARC for a

Fellowship to JRM.

REFERENCES

Balm, S.P. and Jura, M. 1992, *Astr. Ap.*, **261**, L25.
Chlewicki, G., de Groot, M.S., Van der Zwet, G.P., Greenberg, J.M., Alvarez, P.P. and Mampaso, A. 1987, *Astr. Ap.*, **173**, 131.
Cohen, M., Anderson, C.M., Cowley, A., Coyne, G.V., Fawley, W.M., Gull, T.R., Harlan, E.A., Herbig, G.H., Holden, F., Hudson, G.S., Jakoubek, R.O., Johnson, H.M., Merrill, K.M., Schiffer III, F.H., Soifer, B.T., and Zuckerman, B. 1975, *Ap. J.*, **196**, 179.
Fossey, S.J. 1991, *Nature*, **353**, 393.
Geballe, T.R., Lacy, J.H., Persson, S.E., McGregor, P.J., and Soifer, B.T. 1985, *Ap. J.*, **292**, 500.
Greenstein, J.L. and Oke, J.B. 1977, *Pub. A. S. P.*, **89**, 131.
Hall, D.I., Miles, J.R., Sarre, P.J., and Fossey, S.J. 1992, *Nature*, **358**, 629.
Herbig, G.H. 1975, *Ap. J.*, **196**, 129.
Jenniskens, P. and Désert, F.-X. 1993, *Astr. Ap.*, **274**, 465.
Jenniskens, P. and Désert, F.-X. 1994, *Astr. Ap. Suppl. Ser.*, **106**, 39.
Krełowski, J. and Walker, G.A.H. 1986, *J. R. A. S. C.*, **80**, 274.
Krełowski, J. and Walker, G.A.H. 1987, *Ap. J.*, **312**, 860.
Krełowski, J. and Sneden, C. 1993, *Pub. A. S. P.*, **105**, 1141.
Miles, J.R. and Sarre, P.J. 1993, *J. Chem. Soc. Farad. Trans.*, **89**, 2269.
Rao, N.K. and Lambert, D.L. 1993, *M. N. R. A. S.*, **263**, L27.
Russell, R.W., Soifer, B.T., and Willner, S.P. 1978, *Ap. J.*, **220**, 568.
Sarre, P.J. 1991, *Nature*, **351**, 356.
Scarrott, S.M., Watkin, S., Miles, J.R., and Sarre, P.J. 1992, *M. N. R. A. S.*, **255**, 11P.
Schmidt, G.D., Cohen, M., and Margon, B. 1980, *Ap. J.*, **239**, L133.
Schmidt, G.D. and Witt, A.N. 1991, *Ap. J.*, **383**, 698.
Sitko, M.L., Savage, B.D., and Meade, M.R. 1981, *Ap. J.*, **246**, 161.
Warren-Smith, R.F., Scarrott, S.M., and Murdin, P. 1981, *Nature*, **292**, 317.

HAC AND THE ERE

ADOLF N. WITT, *The University of Toledo*
DOUGLAS G. FURTON, *Rhode Island College*

ABSTRACT.

Extended Red Emission (ERE) results from a photoluminescence process in dusty interstellar environments which are illuminated by UV/visible radiation. ERE characteristics are a broad emission feature of 60nm < FWHM < 100nm, with locally variable peak wavelengths in the range 630nm < λ_c < 850nm, with strong intensity enhancements in H_2 photodissociation regions. Observed integrated ERE band intensities range over more than three orders of magnitude, in objects extending from dark nebulae illuminated by the interstellar radiation field, to reflection nebulae, to H II regions and planetary nebulae. In each case, the maximum ERE intensities are roughly proportional to the estimated intensity of the local UV/visible radiation field. The ERE efficiency is about 1 to 10%, if the ERE producing cross section is equal to the total dust cross section.

The variability of ERE characteristics with position within a given object and from object to object provides dramatic evidence for highly effective processing of interstellar dust materials by interaction with gas and radiation in the interstellar environment. We summarize the astronomical evidence for the suggestion that ERE is a result of photoluminescence by hydrogenated amorphous carbon (HAC), that the intensity of the ERE is a strong function of the hydrogenation of the HAC, and that the peak wavelength of the ERE band is strongly dependent upon the radiation processing of the HAC material. Recent laboratory experiments involving the processing of HAC with atomic hydrogen and UV irradiation provide strong support for these conclusions.

We advance a proposal that sputtering of HAC mantles of grains in the interstellar medium could provide a population of complex carbonaceous molecules which might be carriers of diffuse interstellar bands.

1. INTRODUCTION

Hydrogenated amorphous carbon (HAC) was invoked as an interstellar dust component by Duley (1985) in order to explain the (then) unique broad red emission band in the spectrum of the Red Rectangle (HD 44179) (Schmidt, Cohen, and Margon 1980), noting the close similarity between that band and the photoluminescence spectra of HAC observed in the laboratory by Watanabe *et al.* (1982). Since then, the same emission band has now been observed in numerous sources, ranging from high-$|b|$ dark clouds, through reflection and planetary nebulae, to H II regions. It is generally referred to generically as extended red emission (ERE). If HAC is responsible for ERE, HAC must be an abundant dust component throughout the Galaxy.

A. G. G. M. Tielens and T. P. Snow (eds.), The Diffuse Interstellar Bands, 149–155.

The potential significance of HAC for the problem of understanding the origin of the diffuse interstellar bands rests on the notion that HAC mantles on dust grains could well be the source, through sputtering in the diffuse interstellar medium, of complex hydrocarbon molecules and fragments, now widely suspected as DIB absorbers. This idea is strengthened by several observations and arguments:

1. The amount of cosmic carbon not in the gas phase (Sofia, Cardelli, and Savage 1994; Lacy *et al.* 1994) and available for depletion appears to be smaller than assumed by most current models. This constraint, coupled with possible HAC production methods (Jenniskens *et al.* 1993) and requirements imposed by the high efficiency of the interstellar ERE process, demands that HAC resides on the surfaces of grains.

2. There is strong observational evidence for active processing of the ERE carrier in interstellar environments. In particular, ERE is remarkably enhanced in H_2 photodissociation zones, and the ERE band is broadened and shifted toward longer wavelengths in the presence of strong radiation fields. Laboratory experiments with HAC (Furton and Witt 1993) have reproduced these phenomena under simulated conditions.

3. A set of sharp emission bands in the Red Rectangle ($\lambda\lambda$5799, 5855, 6380, 6615) appears to be related to DIBs at $\lambda\lambda$5797, 5850, 6376/9 and 6614 (Fossey 1990; Scarrott *et al.* 1992; Miles and Sarre, this volume). Schmidt and Witt (1991) reported that the structure of these bands changes with distance from the central star, the band centers shifting toward the blue with increasing offset and thus approaching the DIB wavelengths. Moreover, the sharp bands are concentrated in the cone-shaped walls of a bipolar outflow through a disk bright in the ERE band. A possible interpretation is that the narrow-band emitters are produced by erosion of the HAC dust thought responsible for the ERE in the Red Rectangle.

4. The DIBs are a phenomenon mainly of the diffuse interstellar medium (Herbig 1993), where molecules other than simple diatomic species are unlikely to form. If the DIB absorbers are complex molecules, the most likely path for them to get into the diffuse medium is as integral parts of dust grains. The fairly close correlation between DIB strength and reddening measured by E(B-V) supports this possibility.

5. The proposed process of deriving the DIB absorbers by sputtering off HAC mantles requires an efficient re-forming of HAC mantles in molecular clouds during the time that grains reside in the molecular cloud environment. A recent experimental study by Jenniskens (1993) provides important details on the conversion of ice mantles of cosmic composition into HAC-like materials via photolysis. The overall recycling process of HAC in interstellar space has been discussed at length by Jones, Duley, and Williams (1990).

In the following, we will summarize the observational characteristics of the ERE, correlations of the ERE with environmental parameters and radiation from other emitters, and finally the relevant laboratory results.

2. OBSERVATIONAL CHARACTERISTICS OF THE ERE

The ERE consists of a broad, featureless emission band with an onset near 540 nm and a tail of continuous emission extending well into the near-IR. The peak of the emission band occurs anywhere from 640 nm, as in the Red Rectangle (Schmidt, Cohen, and Margon 1980), to a range of 650 nm-700 nm in reflection nebulae (Witt and Boroson 1990), to values of 680 nm to 780 nm in planetary nebulae (Furton and Witt 1990, 1992), to values beyond 800 nm in the Orion Nebula (Perrin and Sivan 1992). Even within a given object, such as the Red Rectangle (Witt and Boroson 1990) or the Orion Nebula (Perrin and Sivan 1992), the peak wavelength of the ERE band is found to shift, occurring at longer wavelengths the smaller the distance to the illuminating source. We interpret this as resulting from the narrowing of the bandgap in the emitting material due to increased irradiation or heating.

The observed ERE band intensities found in different sources are roughly proportional to the intensity of the illuminating radiation field over several orders of magnitude, suggesting that under ideal conditions the photon conversion efficiency of the ERE process is about the same everywhere. If the incident radiation from 91nm-500nm is considered as excitation source, a photon conversion efficiency in the range of 1-10% can be estimated, with uncertainties mainly due to projection and differential extinction effects within the sources.

The detection of the ERE band in the spectrum of a given nebula is complicated by the fact that the low-contrast band must be extracted from among other, usually much stronger, nebular spectral components, e.g. the scattered continuum in reflection nebulae, the recombination continuum and line spectrum of planetary nebulae, and a combination of all three in H II regions. Nevertheless, it has been possible to detect significant spatial variability of the ERE intensity within given objects (Witt and Schild 1988; Witt and Malin 1989; Witt and Boroson 1990, Furton and Witt 1990). Regions of strongly enhanced ERE are generally associated with H_2 photodissociation zones. It is likely that rehydrogenation of HAC by the dense, warm atomic hydrogen in such regions is responsible for the locally enhanced ERE (Furton and Witt 1993).

Strict upper limits have also been established against the presence of ERE in certain other reflection nebulae, e.g. the Pleiades nebulae, (Witt and Boroson 1990) and in O-rich planetary nebulae (Furton and Witt 1992). We believe, the environment, through radiation exposure, has rendered the photoluminescence inactive in the former case, while the likely absence of carbon dust is the explanation in the latter case. On the other hand, the positive detection of ERE in C-rich planetaries (Furton and Witt 1992) is one of the strongest observational clues that carbon dust is responsible for the ERE.

The source of excitation of the ERE needs to meet only one requirement: abundant photons with energies in excess of the optical band gap of HAC of $E_{opt} \simeq 2.3$ eV. In most objects where ERE is seen, the illuminating stars are moderately hot; this property influences the ERE intensity more through the greater overall luminosity rather than the hardness of the spectrum. Only in sources where the ERE is locally enhanced in H_2 photodissociation fronts does the hardness of the spectrum come into play: photons with $\lambda < 110$ nm are needed for H_2 dissociation. Some of the most efficient ERE production is found in high-$|b|$ dark clouds, which appear to be illuminated solely by the galactic interstellar radiation field (Chlewicki and Laureijs 1987; Guhathakurta and Tyson 1989).

3. CORRELATIONS OF THE ERE WITH OTHER EMISSIONS

The high degree of spatial variability and sensitivity to the local environment of the ERE process invite the search for correlations with other emissions, with the expectation that the exact conditions favoring maximum ERE efficiency can be deduced.

3.1. CORRELATION WITH SHARP EMISSION FEATURES NEAR 580 NM

The sharp emission features near 580nm (Schmidt and Witt 1991) have been seen so far only in the Red Rectangle in conjunction with ERE, although with distinctly different spatial distribution. The sharp features have also been reported in the spectrum of the R CrB star V854 (Rao and Lambert 1993), but without any sign of ERE. Also, sensitive searches for the sharp features in several strong ERE filaments in NGC 2023 (Schmidt and Witt, unpublished) have not revealed their presence. None of the ERE spectra of Witt and Boroson (1990), with the exception of the Red Rectangle, exhibit obvious signs of the sharp features. There is thus no evidence for any general correlation between these two emission mechanisms.

3.2. UV FLUORESCENCE OF H_2 AND ERE

Noncoherent resonant fluorescence of H_2 at $\lambda < 165$ nm is the most directly visible sign of ongoing photodissociation of H_2 (Stecher and Williams 1967). Witt *et al.* (1989) demonstrated the simultaneous presence of ERE and H_2 fluorescence in IC 63, a small molecular cloud illuminated externally by Gamma Cas (e.g. Jansen *et al.* 1994). A neighboring nebula, IC 59, also illuminated by Gamma Cas, displays neither emission. A number of ERE filaments in NGC 2023 and NGC 7023 (Witt, Bohlin, and Stecher, unpublished) have been observed with IUE, and they all exhibit H_2 fluorescence. A close correlation appears strongly indicated, but problems due to the large extinction at $\lambda < 165$ nm in internally illuminated nebulae and the lack of a suitable observing capability will prohibit its detailed observational study for some time.

3.3. NEAR-IR FLUORESCENCE OF H_2 AND ERE

Near-IR transitions between the rotational-vibrational energy levels of the electronic ground state of H_2 can occur both as a result of thermal or shock excitation or as a by-product of the UV fluorescence discussed above, which can populate higher levels within the ground state. Near-IR luminescence from H_2 is therefore not a reliable indicator of H_2 photodissociation, although a study of the intensity ratios among near-IR transitions may decide among the excitation mechanisms (Burton 1992). Gatley *et al.* (1987) provided a low-resolution (beam 19.6") map of the v = 1 - 0 S(1) line emission of H_2, in which there is a good correlation with the distribution of the ERE as found by Witt and Schild (1988) and by Witt and Malin (1989), the differences in resolution disregarded. However, Gatley *et al.* show a bright region of H_2 emission about 130" NNE of the central star, for which no ERE counterpart has been detected. Interestingly, they also find no 3.28μm emission counterpart at this location.

To understand the correlations between these emissions, observations at the highest (and comparable) resolutions must be obtained. Examples of such studies which focus upon the detailed structure of individual dissociation fronts, are the work by Burton *et al.* (1989) on the NGC 2023 ERE filament 80"S of the central star and on the Orion Bar, the work by Tielens et al. (1993) on the Orion Bar, and the work by Roche, Aitken, and Smith (1994) on NGC 2023 and NGC 1333. In collaboration with K. Sellgren and C. Rogers, we are also involved in detailed studies of dissociation fronts in NGC 7023. The emerging picture shows ERE to be closely correlated with H_2 photodissociation, with ERE being spatially somewhat more diffuse. This could be due to scattering, given the optical depth difference. The idea that ERE is intensified by hydrogenation, which requires atomic rather than molecular hydrogen, is most strongly supported by these observations.

An excellent correlation between ERE and the interface between ionized and molecular gas has also been demonstrated in the planetary nebula NGC 7027 through the high-resolution observations of the ERE by Furton and Witt (1990) and the CO, H_2, H II, and 3.28μm maps by Graham *et al.* (1993).

3.4. THE 3.28μm-BAND, THE UIBS, AND THE ERE

In general, sources exhibiting the strong near-IR unidentified emission features (UIBs) at 3.28, 6.2, 7.8, 8.6 and 11.3μm, such as the Red Rectangle, NGC 7027, NGC 2023 and NGC 7023, are also strong sources of ERE. Wherever the spatial distribution can be analyzed, the UIB emission appears to occur near photo-dissociation fronts but slightly shifted toward the source of excitation with respect to the peak H_2 emission, as best exemplified in the studies of the Orion Bar by Tielens *et al.* (1993) and the study of NGC 7027 by Graham *et al.* (1993), while the ERE and the H_2 emission appear to coincide. A case in point is the unsuccessful attempt by Perrin and Sivan (1992) to correlate the intensities of the ERE and the 3.28μm emission in the Orion nebula. They conclude from this that the 3.28μm emission band and the ERE are caused by different carriers, which agrees with our interpretation which attributes the ERE to HAC grain mantles, while the UIBs are more likely the result of molecules like PAHs (e.g. Geballe *et al.* 1989).

4. RESULTS FROM THE LABORATORY

Material candidates advanced to explain the ERE phenomenon must be expected to exhibit the following characteristics:

1. they must be carbon-based;
2. they must photoluminesce in a broad featureless band at $\lambda > 540$ nm;
3. the wavelength of peak emission must shift toward longer wavelengths in response to irradiation;
4. exposure to warm, atomic hydrogen must result in greater photoluminescence efficiency;
5. the overall photon conversion efficiency must be high to match the observed efficiency of 1-10%, which is derived assuming that the ERE cross-section equals the total extinction cross-section of the grains.

Watanabe *et al.* (1982) published photoluminescence spectra of HAC deposited on substrates with a range of temperatures, resulting in films with varying H/C ratios. The onset of the luminescence occurs around 550 nm and the peak emission is seen between 640nm and 730 nm, in close match to the data for reflection nebulae of Witt and Boroson (1990). The photoluminescence yield is strongly dependent on the hydrogen content, which can range from hydrogen atom fractions of 0.6 to 0.2 in various HACs (Angus and Hayman 1988). The hydrogen atomic fraction in HAC is strongly correlated with the sp^3/sp^2 bonding ratio, and the photoluminescence efficiency is high when both are near their maximum values.

Because the degree of variability among HACs is extremely high, depending on substrate temperature and material, deposition technique, feed gas composition, and subsequent processing, no two HAC samples produced in different laboratories are identical. Consequently, freshly deposited HAC produced from ethylene by chemical vapor deposition in our laboratory (Furton and Witt 1993) exhibits slightly blue-shifted spectra relative to the Watanabe *et al.* results. UV irradiation with doses corresponding to 1 to 10 years in the radiation field in a photodissociation zone near a B star leads to a significant reduction in the photoluminescence yield, accompanied by a shift in the emission peak by about 40 nm to the red. Our experiments showed further that the total yield can be restored almost fully by exposure of the irradiated HAC sample to atomic hydrogen, with the peak emission remaining at the redshifted value. Further irradiation following rehydrogenation leads to an additional redshift of the emission peak without change in yield. We conclude from this that the wide range of ERE spectra observed in different sources are the result of various degrees of environmental processing, in which UV irradiation and hydrogenation with atomic hydrogen are the primary agents.

Another variant of HAC is the so-called quenched carbonaceous composite (QCC) of Sakata *et al.* (1983). This material was first advanced to explain the 220 nm UV extinction hump (Sakata *et al.* 1983), as well as the UIBs (Sakata *et al.* 1984, 1987, 1990). Recently, Sakata *et al.* (1992) reported typical HAC photoluminescence behavior exhibited by QCC, with a series of spectra shifted according to the temperature of the substrate during deposition. QCC deposited in the range -50°C to -20°C on quartz provides an excellent match to the ERE spectrum of NGC 7023.

The HAC (and QCC) photoluminescence is intense even at room temperature. It has been pointed out (Wdowiak 1994, private communication) and been confirmed in our experiments that the photoluminescence yield increases substantially at cryogenic temperatures to still higher values.

It is noteworthy that the fullerene C_{60} also exhibits fluorescence (Kim *et al.* 1992) and Webster (1993) has pointed out the close similarity of the C_{60} spectrum of Kim *et al.*, and the ERE spectrum of NGC 2327, which happens to be one of the most red-shifted ERE spectra among reflection nebulae. A concern is that the efficiency of photoluminescence of C_{60} is about two orders of magnitude lower than that of HAC under similar conditions.

5. CONCLUSIONS

A hydrocarbon solid, falling in composition and structure somewhere in the large family of HAC materials (Angus and Hayman 1988), is a prominent dust component found in many astronomical interstellar environments in the Galaxy. This material is strongly affected and easily modified by environmental factors such as UV radiation and the density and temperature of atomic hydrogen gas, to the extent that enhanced ERE luminescence from this material can be used as an effective tracer of the locations of H_2 photodissociation fronts in space. Since the processing is external, it is the surface layers of dust grains where this HAC material is present. We have suggested that sputtering of hydrocarbon fragments off these dust surfaces in the diffuse interstellar medium can lead to a population of hydrocarbon molecules in space which may be responsible for the diffuse interstellar bands.

ACKNOWLEDGEMENTS.

This work was supported by NASA through Laboratory Astrophysics Grant NAGW-2794.

REFERENCES

Angus, J.C. and Hayman, C.C. 1988, *Science*, **241**, 913.
Burton, M.G. 1992, *Austr. J. Phys.*, **45**, 463.
Burton, M.G. *et al.* 1989, *NASA Conf. Publ.*, **3036**, 87.
Chlewicki, G. and Laureijs, R. 1987, in *Polycyclic Aromatic Hydrocarbons and Astrophysics*, eds. A. Leger, L. d'Hendecourt, and N. Boccara (Dordrecht; Reidel), p. 335.
Duley, W.W. 1985, *M. N. R. A. S.*, **215**, 259.
Fossey, S.F. 1990, *Ph.D. Thesis*, Univ. College, London.
Furton, D.G. and Witt, A.N. 1990, *Ap. J.*, **364**, L45.
Furton, D.G. and Witt, A.N. 1992, *Ap. J.*, **386**, 587.
Furton, D.G. and Witt, A.N. 1993, *Ap. J.*, **415**, L51.
Herbig, G.H. 1993, *Ap. J.*, **407**, 142.
Gatley, I. *et al.* 1987, *Ap. J.*, **318**, L73.
Geballe, T.R. *et al.* 1989, *Ap. J.*, **341**, 278.
Guhathakurta, P. and Tyson, J.A. 1989, *Ap. J.*, **346**, 773.
Jansen, D.J., van Dishoeck, E.F., and Black, J.H. 1994, *Astr. Ap.*, **282**, 605.
Jenniskens, P. 1993, *Astr. Ap.*, **274**, 653.
Jenniskens, P. *et al.* 1993, *Astr. Ap.*, **273**, 583.
Jones, A.P., Duley, W.W., and Williams, D.A. 1990, *Quart. J. R. A. S.*, **31**, 567.
Kim, D., Lee, M., Suh, Y.D., and Kim, S.K. 1992, *J. Am. Chem. Soc.*, **114**, 4429.
Lacy, J.H., Knacke, R., Geballe, T.R., and Tokunaga, A.T. 1994, *Ap. J.*, **428**, L69.
Perrin, J.-M. and Sivan, J.P. 1993, *Astr. Ap.*, **255**, 271.
Rao, N.K. and Lambert, D.L. 1993, *M. N. R. A. S.*, **263**, L27.
Roche, P.F., Aitken, D.K., and Smith, C.H. 1994, *M. N. R. A. S.*, in press.
Sakata *et al.* 1983, *Nature*, **301**, 493.
Sakata *et al.* 1984, *Ap. J.*, **287**, L51.
Sakata *et al.* 1987, *Ap. J.*, **320**, L63.
Sakata *et al.* 1990, *Ap. J.*, **353**, 543.
Sakate *et al.* 1992, *Ap. J.*, **393**, L83.
Scarrott, S.M., Watkin, S., Miles, J.R., and Sarre, P.J. 1992, *M. N. R. A. S.*, **255**, 11.
Schmidt, G.D., Cohen, M., and Margon, B. 1980, *Ap. J.*, **239**, L133.
Schmidt, G.D. and Witt, A.N. 1991, *Ap. J.*, **383**, 698.
Sofia, U.J., Cardelli, J.A., and Savage, B.D. 1994, *Ap. J.*, (Aug 1), in press .
Stecher, T.P. and Williams, D.A. 1967, *Ap. J.*, **149**, L29.
Tielens, A.G.G.M. *et al.* 1993, *Science*, **262**, 86.
Watanabe, I., Hasegawa, S., and Kurata, Y. 1982, *Jpn. J. Appl. Phys.*, **21**, 856.
Webster, A. 1993, *M. N. R. A. S.*, **264**, L1.
Witt, A.N. and Boroson, T.A. 1990, *Ap. J.*, **355**, 182.
Witt, A.N. and Malin, D.F. 1989, *Ap. J.*, **347**, L25.
Witt, A.N. and Schild, R.E. 1988, *Ap. J.*, **325**, 837.
Witt, A.N. Stecher, T.P., Boroson, T.A., and Bohlin, R.C. 1989, *Ap. J.*, **336**, L21.

VISIBLE LUMINESCENCE FROM PAH CATIONS AND THE DIFFUSE BANDS IN EMISSION

C. JOBLIN, F. SALAMA, AND L. ALLAMANDOLA
NASA Ames Research Center, Space Science Division, Moffett Field, CA 94035-1000

ABSTRACT. The cations of polycylic aromatic hydrocarbons (PAHs) constitute plausible candidates to account for some of the diffuse interstellar bands. As it has been recently shown that some of these bands are also observed in emission, we have searched for electronic luminescence from PAH cations. We report here measurements on the perylene cation ($C_{20}H_{12}^+$) and on the phenanthrene cation ($C_{14}H_{10}^+$) isolated in Ne matrices, showing that these molecules indeed give rise to fluorescent emission. Thus PAH cations should also be considered as possible carriers of the DIBs in emission.

1. INTRODUCTION

Optical emission bands have been observed by Schmidt *et al.* (1980) in the Red Rectangle and by Rao and Lambert (1993) in the R CrB star V854 Cent. The correspondence of some of these bands with diffuse interstellar bands (DIBs), which are observed in absorption, has been clearly established by Scarrott *et al.* (1992). These observations put new constraints on the DIB carriers. Scarrott *et al.* concluded that the carriers of these DIBs are gas-phase molecules.

Among the possible carriers, the polycyclic aromatic hydrocarbons (PAHs) constitute plausible candidates. Indeed, these molecules could contain about 10% of the cosmic carbon and their electronic spectra exhibit strong bands in the visible when the molecules are large or ionized (Van der Zwet and Allamandola 1985; Léger and d'Hendecourt 1985; Crawford, Tielens, and Allamandola 1985). Recent laboratory measurements have provided encouraging results to assign some of the DIBs with the electronic transitions of PAH cations. Salama and Allamandola (1991, 1992a) have shown that most of the bands of the naphthalene cation ($C_{10}H_8^+$) fall at DIB positions, although they cannot account for the relative band intensities. The same authors (1992b) proposed that a derivative of the pyrene cation could explain the strongest DIB, that at 4430 Å. Léger, d'Hendecourt, and Défourneau (1994; see also d'Hendecourt and Léger in these proceedings) showed that the carrier could possibly be the methylene-pyrene cation ($C_{16}H_9^+ - CH_2$) which would then account for the strong 4430 Å DIB and for four weaker DIBs. All the above experiments are performed using the technique of matrix isolation spectroscopy which is generally used to stabilize and study short-lived reactive species. Neon matrices at 4 K are chosen

A. G. G. M. Tielens and T. P. Snow (eds.), The Diffuse Interstellar Bands, 157–163.

because they give the weakest interactions with the reactive species. The measured spectrum is very close to the spectrum of cold isolated molecules as in the gas phase of the interstellar medium. However, the final identification must await gas-phase measurements. These experiments are very difficult and only a few indirect results have been obtained up to now for ions. Neon matrices are thus very useful and important as the only alternative currently available to study many complex molecular candidates of astrophysical interest.

The detection of DIBs in emission implies that some carriers, after the absorption of a UV-visible photon, can relax by electronic emission. In this paper, we present some preliminary results of experiments performed to study the electronic luminescence of PAH ions and provide a test to their relevance as potential carriers for these bands. Except for benzene and some of its derivatives (Braltbard et al. 1983; also Bondybey and Miller 1983 for review), no such studies on aromatic ions have been published. The experimental set-up and procedure are first described. The fluorescence spectra of perylene and phenanthrene cations isolated in Ne matrices are then reported. Finally, the relevance of these experiments for the DIB identification is discussed.

2. EXPERIMENTAL PROCEDURE

A mixture of neon and the PAH of interest, evaporated from an oven, is deposited on the cold (4 K) saphire window of a liquid helium transfer cryostat. The sample is then irradiated by VUV photons (mostly Lyα) from an H_2 discharge lamp. Ions formed in-situ in the matrix are probed by absorption measurements using spectral lamps and a UV-Visible-near IR spectrograph (see Salama, Joblin, and Allamandola 1994 for further description). The light collected both in the absorption and emission configurations, is guided by a UV optical fiber to the entrance slit of a 0.5 m monochromator. Only the longest dimension of the CCD detector is used for spectroscopy measurements and binning is performed in the other dimension to increase the S/N ratio. For emission measurements, the excitation source and the detection system axes are 90° apart with the sample window positioned at about 30° with respect to the excitation source (not at 45° to avoid direct reflected light on the detector). The excitation source is a quartz tungsten halogen lamp which provides a stable output from 320 to 2500 nm. Radiation from this lamp is dispersed with a monochromator for broad band excitation (typically 25 nm FWHM). Emission spectra are measured at a resolution of about 1 nm.

3. RESULTS

We present here the preliminary results obtained for the perylene ($C_{20}H_{12}^+$) and the phenanthrene ($C_{14}H_{10}^+$) cations. Upon the electronic excitation of the neutral molecules, strong emission in the blue or in the green is observed for the perylene and phenanthrene samples respectively. This signal has been used to optimize the optical alignment of the system. According to Birks (1970), the $S_1 \rightarrow S_0$ fluorescence dominates the emission of neutral perylene whereas it is the phosphorescence $T_1 \rightarrow T_0$ in the case of phenanthrene. The fluorescence spectrum of perylene has been extensively studied in jet-cooled experiments (Wittmeyer and Topp 1993, 1989, Fourmann *et al.* 1985, Schwartz and Topp 1984, and references therein). The phosphorescence spectrum of neutral phenanthrene has been measured in solution in various solvents (see Birks 1970). In the following, we will focus only on the emission of the cations of these molecules for which no result has been published up to now.

The absorption spectrum of the perylene cation has been measured in Ar matrices by Szczepanski, Chapo and Vala (1993). The absorption spectrum in Ne matrices is shown as the lower trace in Fig. 1. It exhibits a very strong band at about 525 nm associated with the $D_0 \rightarrow D_5$ transition. The first transition ($D_0 \rightarrow D_1$) is weak and its fundamental (0–0) is located at 794.3±0.1 nm. By tuning the excitation wavelength to the strong 525 nm band, a set of emission bands are observed to the red side of the first absorption system (Fig. 2). The two strongest emission bands are located at 794.3±0.2 nm and 817.4±0.2 nm. Weaker features are also observed at longer wavelengths, at about 842.4, 870.1, 877.7, 892.6, 908.2 and 938.5 nm. A similar emission spectrum is obtained by exciting in the band around 720 nm which is identified as the $D_0 \rightarrow D_3$ transition by Negri and Zgierski (1994).

To check that the perylene cation is the carrier of this emission, we have measured the intensity variation of the strongest emission band at 794.3 nm as a function of the excitation wavelength (fluorescence excitation spectrum). The peak intensity of the 794.3 nm band is plotted as a function of the excitation wavelength in the top trace of Fig. 1. The excitation bands are about 25 nm wide at half maximum. The measured excitation spectrum correlates well with the absorption spectrum which strongly supports the perylene cation as the common carrier. Note, however, that we have not attempted to correct the emission spectrum for the instrumental response of the excitation and detection optics. This will be done in future calibration experiments. Within the precision of the measurements, the 794.3 nm emission band is very close to the first electronic transition . Thus, we assign it to the fluorescent emission associated with the 0–0 transition from the D_1 to the D_0 state. The other emission bands at lower energies probably arise from transitions between the bottom of the D_1 state to various excited vibrational levels of the ground state, D_0. For example, the second strong band at 817.4 nm corresponds to a vibrational level at 357±2 cm^{-1}. More work has to be done to identify these weaker features which seem to involve fundamental modes at about 1100, 1200, 1390 and 1580 cm^{-1}. Selection rules require that the bands originating from the zero level of the D_1 state should involve only fully symmetric vibrations. The observed values are in good agreement with the calculated frequencies of the a_g modes (which are fully symmetric) for the D_0 state of the perylene cation. Negri and Zgierski (1994), in particular, calculated frequencies at 363, 1120, 1188, 1362 and 1557 cm^{-1}. Finally two other weak bands in the emission spectrum at 719 and 1935 cm^{-1} are assigned to combination modes, (357×2) and (357+1579) respectively.

The absorption spectrum of the phenanthrene cation isolated in Ne matrices has been recently reported by Salama, Joblin, and Allamandola (1994) for the 200 nm to 1 μm range. The first transition, $D_0 \rightarrow D_1$, is expected to be weak and to fall in the infrared. This position is estimated to be around 4.3 μm from the photoelectron spectrum (Boschi, Clar, and Schmidt 1974), and calculated at 2.7 μm in quantum chemistry calculations (Parisel, Berthier, and Ellinger 1992). As this transition is out of the range of the CCD detector, we have then searched for emission near the origin of the strong $D_0 \rightarrow D_2$ absorption band system which is shown in Fig. 3. When exciting in the absorption bands at 856 nm and at shorter wavelengths, emission around 900 nm has been clearly observed in one experiment (Fig. 4). Although this emission is in some way related to the presence of the phenanthrene cation as indicated by the correlation between the emitted and absorbed energies (Fig. 3), we are not presently able to identify its origin.

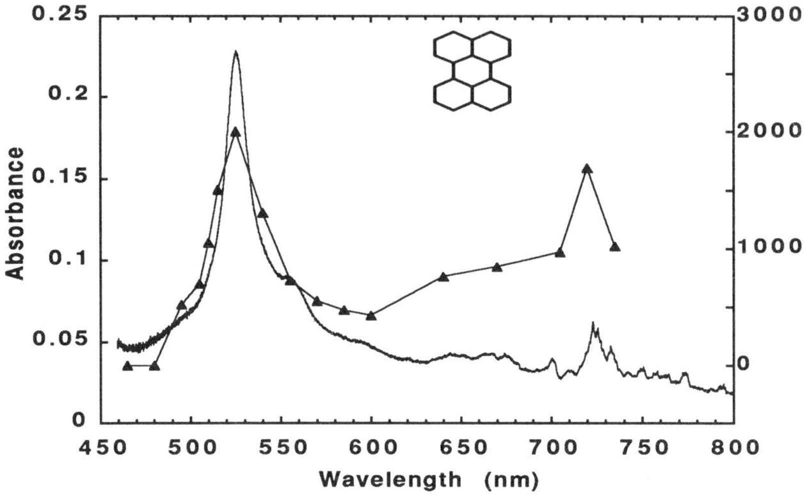

Fig. 1. The visible absorption spectrum of the perylene cation ($C_{20}H_{12}^{+}$) measured in Ne matrices is represented by the lower trace. The triangles of the upper trace represent the fluorescence excitation spectrum. This spectrum has been obtained by plotting the peak intensity of the strongest emission band at 794.3 nm (see Fig. 2) versus the central wavelength of the scanning 25 nm wide excitation band. Note: The absorption spectrum appears positive in these plots as it is expressed in absorbance units.

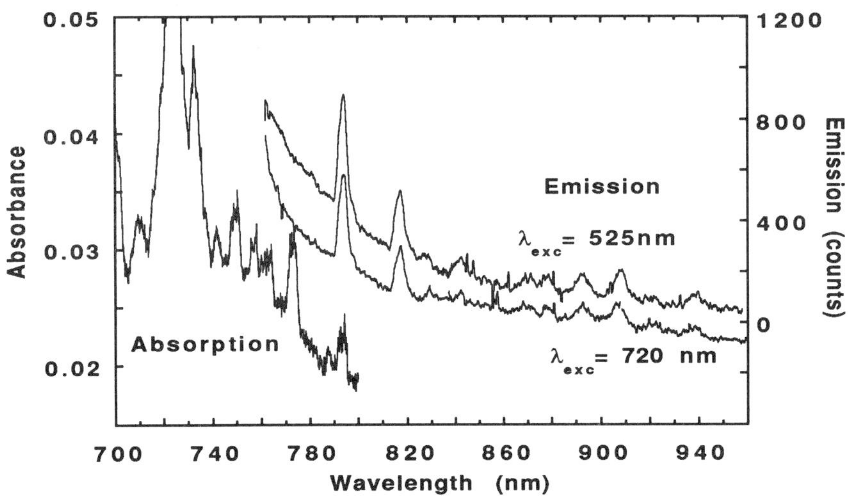

Fig. 2. The two upper traces represent the fluorescence spectrum of the perylene cation induced by pumping in the two strongest visible absorption bands at 525 and 720 nm. The first emission band at 794.3 nm coincides with the first absorption band (see lower trace). This supports the assignment of the observed emission to fluorescence from the bottom of the D_1 electronic state. See also note in Fig. 1 caption.

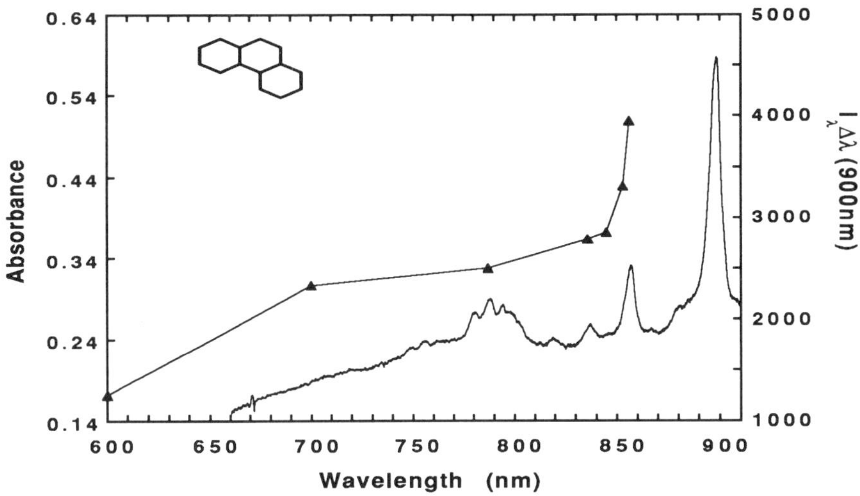

Fig. 3. The absorption spectrum of the phenanthrene cation ($C_{14}H_{10}^{+}$) measured in Ne matrices is represented by the lower trace (see Salama, Joblin, and Allamandola 1994). The triangles of the upper trace represent the fluorescence excitation spectrum. This spectrum has been obtained by plotting the integrated intensity (counts.nm) of the broad emission structure observed around 900 nm (Fig. 4) versus the central wavelength of the scanning 25 nm wide excitation band. See also note in Fig. 1 caption.

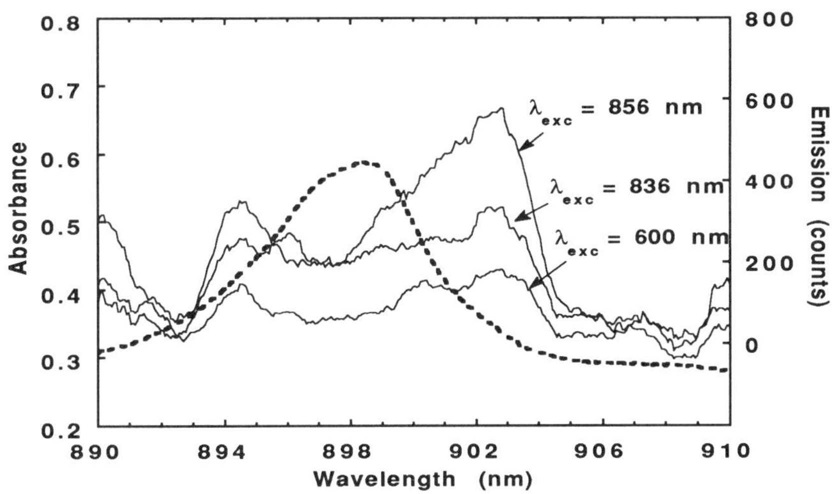

Fig. 4. The strong 898.3 nm absorption band of the phenanthrene cation, corresponding to the origin of the $D_0 \rightarrow D_2$ transition, is represented by the dashed curve. The other curves show the strongest emission which has been measured in that region and for different excitation bands centered at 600, 836, and 856 nm, respectively. See also note in Fig. 1 caption.

First, it is not clear whether the observed structure is due to a broad emission band which is reabsorbed, producing a double-bumped emission (self absorption), or if this structure is the sum of various contributions. Also the fact that some emission, although weaker, is observed when exciting at 600 nm, is puzzling since there is no distinct absorption band of the ion in that region. Finally, we do not yet understand what made this particular phenanthrene experiment successful. The sample was prepared in a similar way as for other experiments in which no or only weak emission is observed. One possibility would be that the emission shown in Fig. 4 arises from non-isolated molecules which emit at the same position as for the isolated ion, or from exciton-like behavior where some of the absorbed energy is transferred to the isolated ions.

4. DISCUSSION

These measurements are the first evidence of electronic fluorescence from PAH cations. In the case of the perylene cation, a fluorescence spectrum is observed corresponding to emission from the D_1 state (Fig. 2). The well-known mirror effect between the fluorescence and the absorption spectrum is observed, with the highest energy fluorescence band falling at the position of the lowest energy absorption band. Such a mirror effect is usually observed in the condensed phase due to a rapid relaxation of the excess vibrational energy in the host matrices prior to emission. This could also explain why fluorescence arising from the phenanthrene cation is not clearly observed, as in that case we have searched for emission near the $D_2 \rightarrow D_0$ transition, not the $D_1 \rightarrow D_0$ transition. Note, however, that fluorescence from upper levels is observed in the condensed phase, for example the $S_2 \rightarrow S_0$ fluorescence of neutral coronene has been detected in solution (Ho, Babbitt & Topp 1987). Our experimental set-up favors the detection of emission from the perylene cation versus that from the phenanthrene cation. The perylene cation can easily be excited in its strong 525 nm band while the emission occurs in the readily accessible 794 nm region. For the phenanthrene cation, the strongest absorption band falls at 898.3±0.1 nm precisely where the emission is expected. Clean excitation in the strong 898.3 nm band is then out of the question with a broad-band excitation source.

5. ASTROPHYSICAL IMPLICATIONS

The fact that some DIBs are observed both in absorption and emission implies that the carrier(s) can relax radiatively. For the ions, electronic fluorescence is likely. Neutral species can relax by fluorescence from the singlet and phosphorescence from the triplet states. For instance, d'Hendecourt *et al.* (1986) have proposed that the extended red emission measured in the Red Rectangle (continuum plus sharp features) could be due to the phosphorescence of the large neutral PAH hexabenzocoronene. *Our measurements show that PAH cations can also relax by electronic fluorescence and should be considered as possible carriers of the DIBs in emission.* The fluorescence quantum yields associated with the cations have yet to be determined. In our experiments in neon matrices, the quantum yields are lower for cations than for neutrals. For gas-phase benzene cations, Braltbard *et al.* (1983) have reported a yield lower than 4×10^{-5}. This does not rule out PAH cations as carriers of the emission bands observed in the Red Rectangle. Indeed, even for low fluorescence yields, these features could be observable due to abundance of the carriers, excitation conditions, or unique source geometry.

6. CONCLUSION

We have shown that luminescence can be induced from PAH cations. This is of astrophysical interest as these molecular ions have been proposed to be candidates for several DIBs. As some DIBs have also been seen in emission, it is important to determine whether or not this is possible from PAH cations as well. The positive results reported here which have been obtained in neon matrices should be followed-up by jet experiments where molecules are truly isolated as in the ISM. Under these conditions, emission can be measured from excited electronic states as there is no competing faster relaxation with the surroundings.

REFERENCES

Birks, J. B. 1970, *Photophysics of Aromatic Molecules*, (London: Wiley).
Bondybey, V.E. and Miller, T.A., 1983, in *Molecular Ions: Spectroscopy, Structure and Chemistry*, eds. T. A. Miller and V. E. Bondybey, (North-Holland Publishing Company).
Boshi, R., Clar, E., and Schmidt, W. 1974, *J. Chem. Phys.*, **60**, 4406.
Braltbart, O., Castellucci, E., Dujardin, G., and Leach, S. 1983, *J. Phys. Chem.*, **87**, 4799.
Crawford, M.K., Tielens, A.G.G.M., and Allamandola, L.J. 1985, *Ap. J.*, **293**, L45.
Fourmann, B., Jouvet, C., Tramer, A., Le Bars, J.M., and Millie, Ph. 1985, *Chem. Phys.*, **92**, 25.
d'Hendecourt, L.B., Léger, A., Olofsson, G., and Schmidt, W. 1986, *Astr. Ap.*, **170**, 91.
Ho, C.J., Babbitt, R.J., and Topp, M.R. 1987, *J. Phys. Chem.*, **91**, 5599.
Léger, A. and d'Hendecourt, L.B. 1985, *Astr. Ap.*, **146**, 81.
Léger, A., d'Hendecourt, L.B., and Défourneau, D. 1994, *Astr. Ap.Lett.*, in press.
Negri, F. and Zgierski, M.Z. 1994, *J. Chem. Phys.*, **100**, 1387.
Parisel, O., Berthier, G., and Ellinger, Y. 1992, *Astr. Ap.*, **266**, L1.
Rao, N.K. and Lambert, D.L. 1993, *M. N. R. A. S.*, **263**, L27.
Salama, F. and Allamandola, L.J. 1991, *J. Chem. Phys.*, **94**, 6964.
Salama, F. and Allamandola, L.J. 1992a, *Ap. J.*, **395**, 301.
Salama, F. and Allamandola, L.J. 1992b, *Nature*, **358**, 42.
Salama, F., Joblin, C., and Allamandola, L.J. 1994, *J. Chem. Phys.*, in press.
Scarrott, S.M., Watkin, S., Miles, J.R., and Sarre, P.J. 1992, *M. N. R. A. S.*, **255**, 11P.
Schmidt, G.D., Cohen, M., and Margon, B. 1980, *Ap. J.*, **239**, L133.
Schwartz, S.A. and Topp, M.R. 1984, *Chem. Phys.*, **86**, 245.
Szczepanski, J., Chapo, C., and Vala, M. 1993, *Chem. Phys. Lett.*, **205**, 434.
Van der Zwet, G.P. and Allamandola, L.J. 1985, *Astr. Ap.*, **146**, 76.
Wittmeyer, S.A. and Topp, M.R. 1989, *Chem. Phys. Lett.*, **163**, 261.
Wittmeyer, S.A. and Topp, M.R. 1993, *J. Phys. Chem.*, **97**, 8718.

CARRIER OF THE EMISSION λ5797 DIB IN THE RED RECTANGLE: SUPRATHERMALLY ROTATING PAH?

D. ROUAN *Observatoire de Paris-Meudon - CNRS-, 92195 Meudon, France*
A. LÉGER *Institut d'Astrophysique Spatiale - CNRS - bat 121, 91405 Orsay, France*

ABSTRACT.

Assuming that free PAHs are the carriers of the λ5797 DIB observed in emission in the Red Rectangle, we investigate the physics of the rotation of such a molecule in order to interpret: *i)* a band structure indicating a very high rotational temperature (450 to 150 K, much higher than the gas temperature) and *ii)* a decrease of this temperature with distance from the star as r^{-1}. We consider several processes that exchange angular momentum with a free PAH, and we conclude that only an efficient rocket effect (H-atom ejection) leads to the suprathermal rotation required by *i)* and that its competition with radio emission can explain the power-law behaviour *ii)*. Using observational constraints for the gas and radiation densities in the Red Rectangle, we build a self-consistent model that leads to a good fit of the observed rotational temperature variation. It implies rather small PAH size (10 - 30 carbons).

1. INTRODUCTION

Recently, superb spectroscopic observations of the Red Rectangle (RR) have shown: a) an unambiguous identification of several emission features with known interstellar DIBs (Scarrott *et al.*, 1992); b) a redshifted and asymmetric band profile, well explained by the rotational band profile of an electronic transition with a variation of the rotational constant between initial and final electronic states ($B_f \neq B_i$, Schmidt & Witt 1991); c) a deduced rotational temperature decreasing with distance from the star as $T_{rot}(r) \propto r^{-1}$ from 6 to 18 arc-sec.

One can show that change in the rotational constant induces a shift in frequency: $\Delta(h\nu) = T_{rot} \times \Delta B/B$. Since $\Delta B/B$ can hardly be larger than 3%, the data of Scarrot *et al.* would then indicate a rotational temperature in the RR larger than 450 K at 6 arc-sec of the central star! What does lead to such a high temperature and govern the radial dependence of rotational temperature? In order to answer those questions, we built a model that assumes that the carrier of the DIB is a free PAH molecule (van der Zwet & Allamandola 1985; Léger and d'Hendecourt 1985; Crawford, Tielens & Allamandola 1985). We first show that thermal equilibrium with the gas cannot explain this behaviour. Then we investigate the local conditions of density and radiation in the Red Rectangle and the different processes that exchange angular momentum with a PAH in order to build a self-consistent model.

A. G. G. M. Tielens and T. P. Snow (eds.), The Diffuse Interstellar Bands, 165–172.

2. TEMPERATURE OF THE CIRCUMSTELLAR GAS

The first idea to explain the observed PAH rotational temperature would be that the molecule is in collisional equilibrium with the gas, this one being hot, and its temperature decreasing with the distance to the star. To test this hypothesis, we consider two classical mechanisms for the heating of the gas:collisions with the grains and photo-electric effect on molecules such as PAHs.

If heating of the gas is dominated by collisions with grains, the gas temperature is probably much lower than 50 K and does not follow a r^{-1} law, for the following reasons:

a) $T_g \ll T_d$ as long as $n_H < 10^{6-8}\ cm^{-3}$ (e.g. Spitzer 1978) because cooling rate by C^+ is more efficient than heating by grains, and a maximum grain temperature (no extinction, low-emissivity graphite) would be $T_{Graphite} = 110K$ at $\theta = 10$";
b) scaling the gas temperature deduced in IRC+10216 (Keady, Hall & Ridgway 1984) to the situation of the RR leads to $T_d = 50K$ and $T_g = 10K$ at a radius corresponding to $\theta = 10$" in the RR;
c) assuming no extinction in the nebula (see below) and $Q_{abs} \propto \lambda^{-1}$, then $T_d \propto r^{-2/5}$, a behaviour which is very different from the r^{-1} law observed in the RR.

On the other hand, if heating of the gas is dominated by photoelectric effect on aromatic molecules (d'Hendecourt & Léger 1987), one shows that, whatever the degree of ionisation of the medium is, the gas temperature is independent of distance to the star: this is because the gas density and the radiation density both follow r^{-2} laws. Once again, this result is in conflict with the r^{-1} law observed in the RR.

A strong conclusion is thus that **the gas temperature is probably much lower than the observed DIB rotational temperature and does not follow the r^{-1} law**. However, one cannot exclude that the gas is heated by other mechanisms, specific of the mass-loss process (compression in shocks?) and measurements of the actual gas temperature in the Red Rectangle by any technique would be extremely useful.

3. BUILDING-UP OF THE ANGULAR MOMENTUM J

To explain a suprathermal rotation of a PAH molecule, we have to consider several processes leading to exchange of angular momentum with it. For each process (collision, emission of a photon, etc.), the variation of the angular momentum J can be split-up, from a statistical point of view, into: a) a symmetric random walk with step δJ and b) a systematic effect with step ΔJ (Rouan *et al.* 1992). Both depends generally on J. At equilibrium, the most probable value $\overline{J}$ is such that $dJ/dt = 0$ i.e. $(\sum_i \tau_i^{-1} \Delta J)_{\overline{J}} = 0$. We have thus to establish, for each process i, its rate τ_i^{-1} and the expression of the systematic term ΔJ_i, with a peculiar attention to the particle size dependence, i.e. the number of carbon atoms N_C in the molecule skeleton.

4. PHYSICS OF THE ROTATION. I: EXCHANGES WITH PHOTONS

Visible photons: The only efficient mechanism for exchange of angular momentum with visible photons could have been the so-called streaming effect (Purcell 1969). One can show that the condition for such an effect to take place is that the pressure of radiation is of the order of the gas pressure, which is far to be fullfilled in the Red Rectangle.

IR photons: A burst of emission of IR photons follows, during a few seconds, the transient heating by absorption of a UV photon by a PAH. The characteristics of the corresponding exchange of angular momentum are:

i) $\delta(angularmomentum) = \pm\hbar$ per IR photon $\Rightarrow \quad \delta J = \pm 1$

ii) There is an asymmetry of probabilities between $\delta J = 1$ and $\delta J = -1$, to which corresponds a systematic effect that reads (see Rouan *et al.* 1992):

$$\Delta J_{IR} = \frac{1}{J_{IR}}\left(\frac{J_{IR}}{J} - \frac{J}{J_{IR}}\right) \quad (1) \quad \text{with} \quad BJ_{IR} = kT_{IR} \approx h\langle\nu_{IR}\rangle/6.$$

Taking into account the different parameters dependence with size leads to $J_{IR} \propto N_{at}^{5/6}$

iii) The process is self-regulating (see equation 1), and would lead, if alone, to the value $J = J_{IR}$ at equilibrium.

iv) The rate is $\tau_{IR}^{-1} = \tau_{UV}^{-1} \times \frac{\lambda_{IR}}{\lambda_{UV}}$, i.e. $\tau_{IR}^{-1} \propto (\frac{N_c}{80})^{4/3}$ when taking into account the dependence of T_{peak} and PAH UV absorption with number of atoms.

One must note that the return effect to $J = J_{IR}$ is tight (see Fig. 3 of Rouan *et al.* 1992) and may *clamp the rotational temperature to* $T_{IR} \approx 100K$ *if the UV density is high enough* to give a large number of re-emitted IR photons.

Radio Photons: This corresponds to the spontaneous emission of radio purely rotational photons. This supposes that there is a static dipole moment $\mu \neq 0$ in the plane of the molecule, a likely possibility for that is the presence of heretero-atoms in the aromatic molecule. The characteristics of the corresponding exchange of angular momentum are:

i) $\Delta(angularmomentum) = -\hbar \quad \Rightarrow \quad \Delta J = -1;$

ii) The rate is given by the Einstein coefficient $A_{J\to J-1}$:

$$A_{J\to J-1} \quad \propto \quad B^3\,\mu^2\,J^3 \quad \propto \quad N_c^{-6}\,\mu^2\,J^3$$

We note, for a given μ, the *extremely fast dependence of* T_{rot} *vs* N_c *that will severely limit reaching high temperatures for small particles* because they will cool efficiently by radio emission.

5. PHYSICS OF THE ROTATION. II: EXCHANGES WITH GAS

Three processes have been considered:

Friction by collisions with the gas: inelastic or elastic collisions give a frictional effect that slow down the rotation. With a simple geometrical model one finds the following dependences:

- $(\Delta J)_{friction} \; = \; -Jm_H a^2/I \quad \propto \quad N_c^{-1} J$

- Rate: $\tau_{friction}^{-1} = n_H v_H \pi a^2 \quad \propto \quad N_c \, n_H \, T_g^{1/2} \; s^{-1}$

Random walk: because $J = |\vec{J}|$, it must be positive and tends to increase indefinitely as $N_{coll}^{1/2}$. This is equivalent to a systematic increase ΔJ:

$(\Delta J)_{random} = [(N+1)^{1/2} - N^{1/2}] \times \delta J \approx (\delta J)^2 J^{-1}$. One shows that the dependence is then: $(\Delta J)_{random} \propto \quad T_g \; N_c \; J^{-1}$. The rate is the same as for friction. One checks that $(\Delta J)_{random} = (\Delta J)_{friction}$ implies $T_{rot} = T_{gas}$, a result expected when exchange of angular momentum is dominated by collisions.

Rocket effect: this is an unusual effect that we introduce, in the spirit of the work of Purcell (1979) for grains; it corresponds to a non-symmetric ejection of hydrogen atoms or molecules from some sites on the PAH molecule, under the action of either a temperature spike (photo-thermo-dissociation) or of the energy of recombination of H_2. It is analog to the rocket effect introduced in big grains to explain the Davis-Greenstein alignment on magnetic field. For large grains, such an effect assumes the existence of hetero-atoms or pentagonal cycles or radicals in the PAH skeleton in order to have peculiar sites. We define the quantity α_{asym} which is the relative number of "asymmetric" ejections versus the total number of ejections ($\alpha_{asym} < 1$). The energy of ejection is a fraction of the energy available and we assume $.1 \, eV \; < \; E_{ej} \; < \; 1 \, eV$. One shows that $(\Delta J)_{rocket} \; = \; \alpha_{asym} (m_H a^2 E_{ej})^{1/2} \quad \propto \quad \alpha_{asym} \times N_c^{1/2} \; E_{ej}^{1/2}$.

We assumed that the rate of this process is nothing but the inelastic collision rate (chemisorption of H atom), that we assume to be $\beta \tau_{friction}^{-1}$, since replenishment in H atoms is required before any ejection; β is the fraction of ionized PAHs.

6. A COHERENT MODEL OF DUST AND PHOTONS IN THE RED RECTANGLE

Since several of the previous processes depend on the gas and/or radiation density, we need a fair evaluation of those two quantities along the whiskers of the RR where the rotational temperature is measured. Flux and density are related through the intensity of the scattered light, so that a simple coherent model can been built, constrained only by the observation of this scattered light. The following assumptions are made:

- the density follows the classical law $n_H \propto r^{-2}$ characteristic of a stationary mass-loss;
- two cases are examined for the central exciting star: a 10 000 K B9-A0 (Cohen *et al.* 1975) and a 7500 K star (more recent spectroscopic derivation by Waelkens *et al.* 1992);
- a Kurucz (1979) model is used for the spectrum of the stellar light;
- the nebular spectrum must be purely scattered light, so we consider only the 3400Å – 4200Å range to avoid ERE (Warren-Smith, Scarrott & Murdin 1981);
- the grain model is a mixture of Graphite and Silicate (Draine & Lee 1984), with $n(a) \propto a^{-3.5}$ (MRN), but a_{max} left as a free parameter;
- we assume that extinction occurs *only* close to the star, but we do not derive this extinction from photometry on the star because it can be different in the RR

disc and towards the biconical nebula;

- we assume optically thin dust along the line of sight.

Mie computation of $< Q_{ext} >$ and $< \phi(\theta) \times Q_{sca} >$ is performed, where $\phi(\theta)$ is the phase function of scattering. a_{max} is adjusted so that a unique C can be found for all wavelengths in the expression of the ratio of scattered to stellar light:

$$F_{Neb}/F_{*} \propto \langle \phi(\theta) \times Q_{sca} \rangle exp(-C\langle Q_{ext} \rangle)$$

$\tau_{ext} = C\langle Q_{ext} \rangle$ can then be derived and, in turn, the quantities $F_{Neb}(r)$ and $n_H(r)$ which are then fully consistent.

The best fit is found for the following conditions:

- $T_* = 7500$ K ($T_* = 10000$ K corresponds to too large an extinction);
- a_{max} ranging from 0.05 to 0.1, (note that for those values of a_{max}, τ_{ext} remains almost constant);
- typically $A_B = 2.2$, $A_V = 1.7$, a value consistent with the bolometric to visible ratio that would give a maximum opacity $A_v = 2.7$;
- UV Flux: $U_{UV} = 30\ (\theta/10")^{-2} eV cm^{-3}$;
- gas density: $n_H = 40\ 10^3 (\theta/10")^{-2}\ cm^{-3}$;
- gas temperature: $T_g \propto r^{-0.7}$ (Goldreich & Scoville 1976).

7. RESULTS OF THE MODEL

Taking into account all the processes, we solve now the equation for $\bar{J}$ which reads:

$$-a_{radio} J^3 - (b_{IR} + b_{friction}) J + c_{rocket} + (d_{IR} + d_{coll})/J = 0$$

with: $a_{radio} \propto N_c^{-6}, b_{IR} \propto N_c^{-1/3} U_{UV}, b_{friction} \propto n_H T_g^{1/2}, c_{rocket} \propto N_c^{3/2} n_H T_g^{1/2} \alpha_{asym} E_{ej}^{1/2}, d_{IR} \propto N_C^{4/3} U_{UV}, d_{coll} \propto N_c^2 n_H T_g^{3/2}$.

The free parameters (within certain limits) are: N_c, $\alpha_{asym}\ E_{ej}^{1/2}$, β, μ and T_g.

Different conditions and values of the parameters have been explored. We show in Fig. 1 the observed radial dependence of the rotational temperature and the one found with our model for a set of parameters giving a good fit ($\alpha_{asym}\ E_{ej}^{1/2} = 1.eV^{1/2}$, $A_v = 2.0$, $T_g = 10.(\frac{\theta}{10"})^{-.7}\ K$, $\mu = 3.0 Debye$). The main behaviours of the solution can be summarized as follows:

1. In the fitting procedure, the radial dependence of the temperature is a sensitive criterion for parameters adjustment, while the absolute temperature is not as sensitive since $\Delta B/B$ can generally be adjusted within an acceptable range.
2. The Rocket effect is the only process that gives high enough temperatures, provided that $\alpha_{asym}\ E_{ej}^{1/2}\ >\ 0.3 \times 0.3 \times 1.eV^{1/2}$. We show in Fig. 2-c the effect of a low value of α_{asym}: high enough temperatures cannot be reached.
3. The Radio cooling is the only process that gives an efficient decrease with distance, because pure IR and pure rocket/friction both give a temperature independent from distance to the star. Since radio cooling is only efficient for small particle, we conclude that *the carrier of the λ5797 DIB must be a rather small PAH ($N_c \approx 20$) to explain the observed r^{-1} decrease of temperature.* In addition, a rather

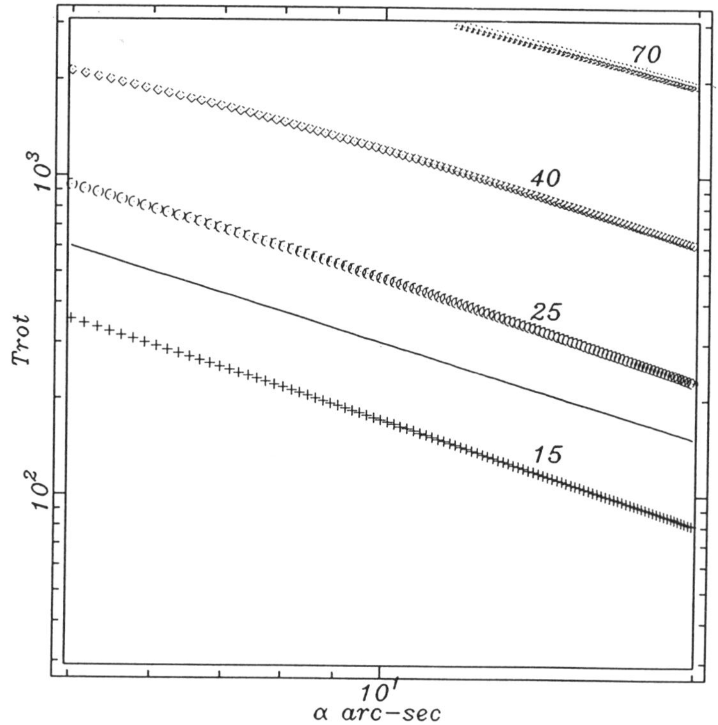

Fig. 1. Rotational temperature as a function of the angular distance to the central star. The plain line is the observed r^{-1} law; the other curves correspond to the model, for different sizes of the PAH: they are labelled with the number of carbons. Conditions are: $\alpha_{asym}\ E_{ej}^{1/2} = .3eV^{1/2}$, $A_v = 2.0$, $T_g = 10(\frac{\theta}{10''})^{-.7}K$, $\mu = 3.0 Debye$. We note the good agreement between observed and theoretical profiles for PAH of ≈ 20 carbons.

high magnetic moment is required: for instance, with $\mu = 1$ *Debye* one cannot find an acceptable fit of the slope, as shown on Fig. 2-a.

4. Decrease of T_{gas} with radius helps for fitting the -1 slope but this process alone is not sufficient, as already shown (see Fig. 2-d).
5. **The *UV* intensity must not be too large** because IR emission would then clamp the temperature to $T_{IR} \approx 150K$ (Rouan *et al.* 1992), as illustrated on Fig 2-b, where $A_v = 1.0$. On the other hand, too small a UV density leads to a slope too steep.
6. Fits are found only in a quite restricted range of parameters.

We emphasize the fact that we use *gas and radiation densities fully consistent with the observed scattered light from the Red Rectangle.*

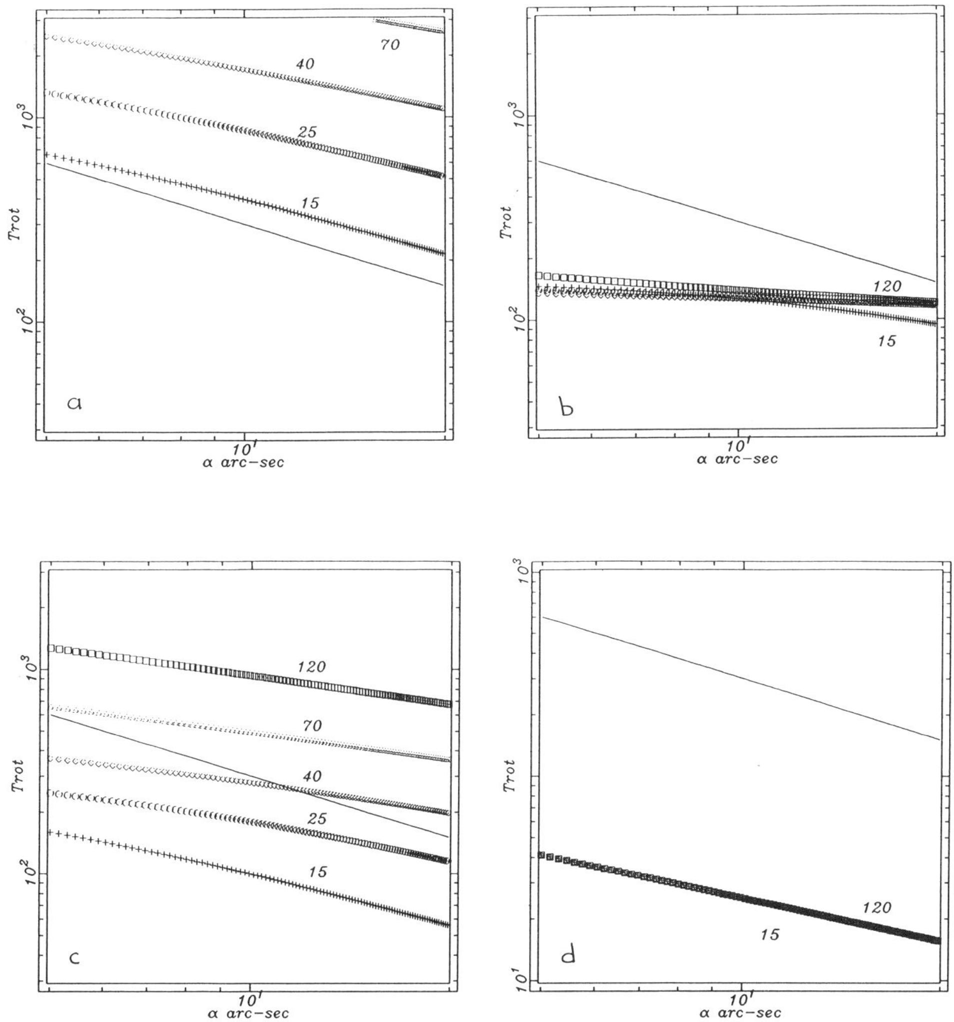

Fig. 2. a) Same as Fig. 1, but with $\mu = 1.0 Debye$: there is no acceptable fit whatever the PAH size is. b) same as Fig. 1, but with $A_v = 1.0$; IR emission imposes the temperature, far from the observed one. c) Same as Fig. 1, but for $\alpha_{asym} E_{ej}^{1/2} = .1 eV^{1/2}$: no satisfactory fit can be found on the slopes. d) No other processes than collisions with the gas are considered: the rotational temperature is equal to the gas temperature.

8. CONCLUSION

In order to interpret the very high rotational temperature of the λ5797 DIB observed in emission in the Red Rectangle *and* its dependence with radial distance to the star, we have explored a model that assumes that the carrier of the DIB is a free PAH molecule. We have shown that, *if* there is no gas heating mechanisms other than grain collisions and photo-electric effect, the gas temperature is much too low to explain the implied molecule rotational temperature. A self-consistent model of the density and radiation density structure of the Red Rectangle is used to compute the rotational temperature and its radial dependence, taking into account the processes exchanging angular momentum with a PAH. We show that a very efficient **rocket effect** based on the ejection of H atoms or molecules from peculiar sites of the molecule, **is needed to get the fast rotation**. We show that among all the processes the emission of radio photons is the only one which can explain the decrease as r^{-1} of the temperature, constraining a **small size for the carrier of the DIB** that would feature typically **15 to 30 atoms**. Finally we want to point out the rather narrow range for each parameter that can give a satisfactory fit to the observational data.

ACKNOWLEDGEMENTS. We thank P. Thaddeus for a fruitful discussion during the workshop.

REFERENCES

Cohen, M. *et al.* 1975, *Ap. J.*, **196**, 179.
Draine, B.T. and Lee, H.M. 1984, *Ap. J.*, **285**, 89.
Goldreich and Scoville 1976, *Ap. J.*, **205**, 144.
d'Hendecourt, L.B. and Léger, A. 1987, *A&A*, **180**, L9.
Keady, Hall and Ridgway 1988, *Ap. J.*, **326**, 832.
Kurucz, R.L. 1979, *Ap. J.*, **40**, 1.
Léger, A. and d'Hendecourt, L. 1985, *Astr. Ap.*, **146**, 81.
Purcell, E.M. 1969, *Physica*, **41**, 100.
Purcell, E.M. 1979, *Ap. J.*, **231**, 404.
Rouan, D., Léger, A., Omont, A., and Giard, M. 1992, *Astr. Ap.*, **253**, 498.
Scarrott, S.M., Watkin, S., Miles, J.R., and Sarre, P.J. 1992, *M. N. R. A. S.*, **255**, 11.
Schmidt, G.D. and Witt, A.N. 1991, *Ap. J.*, **383**, 698.
Waelkens, *et al.* 1992, *Astr. Ap.*, **256**, L15.
Warren-Smith, R.F., Scarrott, S.M., and Murdin, P. 1981, *Nature*, **292**, 317.

CHAPTER III

Laboratory Studies of Proposed Carriers

LABORATORY STUDIES OF DIB CARRIERS

L.J. ALLAMANDOLA
NASA Ames Research Center, MS 245-6
Moffett Field, CA 94035-1000

ABSTRACT. Spectroscopic studies of the following potential diffuse interstellar band (DIB) carriers are reviewed: unspecified organics, carbon chains, polycyclic aromatic hydrocarbons (PAHs), fullerenes and derivatives, as well as porphyrins and related material. An assessment of each is given, along with suggestions for further experimental studies needed to fully test each candidate. Of the experimental techniques in common use, matrix isolation spectroscopy with neon matrices is the most appropriate for the DIBs. The low vapor pressure and high reactivity of these materials preclude gas phase studies on many of these species. At this point, given the type and quality of published data available, carbon chains and PAHs are the most promising candidates for a number of the DIBs.

1. INTRODUCTION

Identifying the carriers of the diffuse interstellar bands (DIBs) has become the classic spectroscopic problem of the 20th Century. Since their discovery in the early 1900's the DIBs have successfully challenged spectroscopists, astronomers and physicists. During this time so many suggestions have been made, experiments carried out, and theories proposed that a thorough review would more than fill this volume. The same holds true for a summary of experiments on species which might be relevant to the DIB problem. Here, the focus will rather be placed on a review of the experiments which were specifically carried out to address the DIB question, with the further restriction that the types of materials discussed represent what are currently thought to be promising candidates. Even with this, the scope is large. Many researchers active in this area have been asked for relevant references and cross references in order to be as thorough as possible. Nontheless, some important contributions have almost certainly been overlooked.

The experimental challenge was succinctly stated by Johnson (1970) twenty five years ago, "... one not only has to match 25 diffuse interstellar lines as far as wavelengths are concerned, but also as far as intensity. In addition, the... interstellar line widths vary from 40 Å to 1 Å...and are invariant to 0.1 Å..." Since that time the number of DIBs has grown to well over 100, their relative intensities have been shown to vary from one line-of-sight to another, and some DIBs seem to go together in loosely connected families (Herbig 1975; Jenniskens and Désert 1994; Krełowski and Walker 1987). Wavelength invariance has been taken to indicate that the carriers cannot reside in or on dust particles. This is because particle size, shape and composition influence peak position and profile, and it is difficult to imagine that

A. G. G. M. Tielens and T. P. Snow (eds.), The Diffuse Interstellar Bands, 175–198.

the grains along all lines-of-sight are exactly the same. Thus, while this is used to further limit the work reviewed here to molecular sized species in the gas phase, one must keep in mind that any sweeping conclusions based on global DIB observational properties are probably invalid. This is intrinsic to the DIB problem. Quite often generalizations are based on studies of only a handful of the strongest DIBs. There is good reason for focusing on these, as most DIBs are weak and inherently difficult to measure. Nonetheless, these strong DIBs are certainly not representative of the entire collection along all lines-of-sight.

The criteria which must be met for a particular material to be accepted as a DIB carrier are that its visible and near-infrared absorption features match the known DIBs in wavelength, bandwidth and relative intensities, while not possessing additional features that are absent in the interstellar spectra. There are no species which satisfy these criteria to date, there are only candidates. Experimentalists are charged with measuring the spectra of candidate materials which can be compared with the DIBs. Ultimately, one strives to obtain spectra which can be directly compared to the DIB spectrum. Ideally, this requires measuring the absorption spectra of various candidates in the gas phase at interstellar temperatures and pressures. This also includes determining the oscillator strength for each transition, as this is needed to derive the column density necessary to produce the observed equivalent width. Practically, since many of the promising species are difficult to work with, these goals are extremely challenging, if not impossible, given current experimental capabilities.

These issues are discussed in this review. First, experimental techniques are summarized, pointing out the inherent advantages and disadvantages of each approach. Next, specific examples are given of species which are currently considered to be promising as contributors to the diffuse interstellar bands. Finally, an assessment of these candidates as potential DIB carriers is given along with suggestions for the experiments needed to provide additional tests for species.

2. EXPERIMENTAL TECHNIQUES

On face value, measuring DIB candidate absorption spectra is simple enough. One needs a visible light source, a monochromator, and a detector. This equipment is readily available. The experimental challenge often lies in sample preparation and, during the time required to measure it's spectrum, maintenance. This is because the most promising candidate materials are either refractory or, due to the harsh interstellar radiation environment, short lived under most terrestrial conditions. Thus, the experimentalist must be able to suspend individual refractory molecules in a manner which minimizes or eliminates any perturbation from the surrounding medium, or prepare and store reactive species unperturbed for periods long enough to permit the measurement of their spectra.

The advantages and disadvantages of the sample preparation techniques which are in common spectroscopic use are summarized in Table 1. The first three, while straightforward, are not appropriate for measuring the spectra of isolated, individual species. In all three - ground powders, liquid solutions, or pellets (transparent solids containing suspensions of the material of interest) - interactions with the host material can severely perturb the transitions in the guest. These perturbations cause shifts in band positions, intensities and profiles. Thus, while these methods are useful as an initial screen to identify promising candidates, they cannot be used in one-on-one comparisons with the DIBs. The last two methods - matrix isolation and gas phase spectroscopy - provide data which is more appropriate for the DIB issue.

Table 1
Advantages and disadvantages of sample preparation techniques

TECHNIQUE	ADVANTAGES	DISADVANTAGES
Ground Powders	straightforward	not isolated
Liquid Solutions	straightforward	interactions
Solid Pellets		
salt	straightforward	matrix interactions
glass	somewhat involved	matrix interactions
Inert Gas Matrices		
argon	less matrix interaction low temperature	involved rotation bands suppressed
neon	minor matrix interaction low temperature	very involved rotation bands suppressed
Gas Phase	no interaction	very involved very difficult to get sufficient population $T > ISM$: hot bands (eliminated in jets) Spectrum deduced indirectly

Matrix Isolation Spectroscopy and the DIBs- Matrix isolation spectroscopy, developed by Professor Pimentel and his colleagues in the 1950's, is the technique in which the species of interest is suspended in a frozen inert gas matrix. This is ideally suited to the challenges provided by both refractory and reactive DIB candidates. Briefly, the technique consists of simultaneously codepositing the species of interest with an overabundance of inert gas such as neon, argon, or krypton onto a cold window which is suspended in a vacuum chamber. The vacuum should lie in the high- to ultra-high range (10^{-7} – 10^{-8} mbar). Such high vacuum conditions, as well as very pure starting materials (>99.5% for the guest, 99.999% for the host), are essential to ensure that all the spectral features obtained can be confidently attributed to the target species. This is especially true in the cases where the species is a reactive intermediate produced by some form of energetic alteration of the starting materials. Here processes such as irradiation or electron bombardment can produce many new and exotic species. It is often a daunting task to identify the precise chemical identity of the species in such matrices under the best of conditions. Thus, while often glossed over, it is essential that great care be given to the mundane task of sample preparation. The deposition rate must be slow and the window temperature quite low to ensure that the gases freeze out quickly, without allowing diffusion to occur. The rapid formation of a rigid matrix is required since any diffusion will result in the production and trapping of aggregates, as well as isolated species; or promote reactions between reactive species such as ions and radicals, resulting in the loss of the target species and the production of unwanted reaction products. Both processes produce spurious spectral features. The maximum window temperature acceptable is determined by the inert gas, with temperatures below about 20 K sufficient for argon and below about 5 K required for neon. Since 10 K is the minimum temperature mechanical coolers can currently achieve with reasonable cooling powers, argon matrices are commonly employed. However, due to the fact that the band shifts from gas phase values are greater in argon matrices than neon matrices, neon is required for the DIB problem, a point discussed further below. Thus, liquid helium coolant (and all the care it entails) is required to produce the matrices which are most appropriate for the DIB issue. Once the matrix is produced, absorption spectra are measured using customary techniques. These data are then compared to the DIBs.

The key question here is, "How relevant is neon matrix data for the DIB issue?" Neon matrices perturb transitions, resulting in band shifting and broadening. Relative band strengths may also be slightly affected. The material is frozen in a solid, thus eliminating the rovibrational structure inherent in a gas phase spectrum, and trapping in different sites can produce spurious spectral features. Thus, one wonders if neon matrix spectra are at all relevant. They are very relevant. These matrices are used because it is exceedingly difficult (if not impossible) to prepare and study the gas phase spectra of many serious DIB candidate materials (see below) and, of all known sample preparation techniques, the spectra of materials suspended in solid neon have been shown to provide the best wavelength matches to the bands in the gas phase spectra of the same species. Thus, at the very least, neon matrices provide a means by which many candidate materials, some exceedingly difficult to work with, can be screened.

The situation is actually much better, and serious DIB candidates can be identified using neon matrix isolation spectroscopy (See, for example, the articles by Maier, Salama, and d'Hendecourt elsewhere in this volume). In neon, the shifts are very small, broadening is not severe, the 0-0 band position is often readily discernable, and site splitting is generally not a problem. While gas phase profiles are not duplicated in neon, one can calculate reasonably accurate rotational band envelopes. The tables of shifts between the gas phase and neon matrix frequencies for a particular ion transition listed by Bondybey and Miller (1983) illustrate this point. The ions listed include diacetylene, hexatriene, the aromatic molecule benzene (C_6H_6) and many halogen, hydroxy, and alkyl substituted benzenes. Less than 0.5% frequency shifts ($((\nu_g - \nu_{ne})/\nu_g)$ x 100) are reported for the majority of bands listed (15 out of 26). Eight of the remaining nine are shifted by less than 1.0%, with the largest occurring in species containing chlorine and bromine. For the transitions having less than a 0.5% shift, the average difference between the neon matrix and gas phase values is 66 cm^{-1}, with the neon data always shifted to lower frequencies. For those less than 1%, the average is 94 cm^{-1}. Therefore, the criterion for a material to be seriously considered as a plausible DIB candidate is straightforward. The band frequencies of the species, when isolated in a neon matrix, must fall within 1% that of a DIB frequency. A smaller shift is required for neutral species for the same confidence level. Bondybey and Miller (1983) also show that the upper state lifetimes for a particular ionic species isolated in neon are very close to the gas phase values, allowing one to deduce gas phase band widths from neon matrix data as well. The reason that neon is particularly useful, especially for ionized species, arises from its 24.6 eV ionization potential as compared to the 15.76 eV ionization potential for argon. Much less charge transfer between guest and host can take place in neon, resulting in much weaker interactions (Bondybey and Miller, 1983).

Gas Phase Spectroscopy and the DIBs- Gas phase absorption spectroscopy of many suitable candidate materials remains in the wishful thinking category. While it is often expressed that the final test awaits gas phase data on the species of interest taken under the appropriate conditions, the extreme difficulties associated with attaining this obvious goal are not at all appreciated by many in this field. There are several severe stumbling blocks which must be overcome. While it is exciting to realize that this type of data may become available on a few systems during the next few years, unless there is a significant experimental breakthrough, it is unlikely that many systems will be so studied. Here, the challenges have several sources.

Although molecular beam techniques employing jets can generate cold samples in a collision-free environment, ideal conditions for mimicking the interstellar medium, these jets generally do not contain enough material to permit an absorption spectrum measurement over a narrow spectral region, let alone across the entire UV-Visible-near IR range. Most jet experiments take advantage of the extreme sensitivity of the laser-induced-fluorescence technique, which provides high resolution emission spectra over a very narrow spectral range. In order to maintain sufficient column densities of the individual molecular constituents of the refractory material of interest to permit the measurement of an absorption spectrum over a broad spectral range requires sample cell temperatures significantly higher than relevant for the ISM. Thus, low lying states are populated and the absorption spectrum must be corrected appropriately. As the molecular spectroscopic parameters required to accurately and precisely carry out such an analysis are poorly known for many of the species of interest, simplifying a high temperature spectrum will introduce uncertainties in band positions, profiles, and relative intensities. These difficulties apply to reactive species as well, which carry the additional burden of sample maintenance. Even for refractory species which do not need very high temperatures, sample generation often requires the sudden input of energy and the species are not prepared in their ground state, once again requiring spectral deconvolution. Thus, while it seems reasonable to presume that this type of data will be obtained for a few DIB candidates within several years, it is not likely that spectra of the large number of species that must be measured to adequately address the more than 100 DIBs will be available for some time to come. See Boissel (1995, elsewhere in this volume) for the description of a novel technique which may ultimately yield moderate resolution spectra of volatile, reactive DIB candidates. (The author hopes he is overly pessimistic here, and that this critique stimulates a few spectroscopists to prove him wrong.)

Consequently, neon matrix isolation spectroscopy, in spite of it's limitations, remains the most promising technique to use. It offers the ability to survey many possibilities in a reasonably short period of time providing data that is directly applicable to the DIB problem.

Apart from the spectroscopic studies discussed here, additional experimental work is also called for. Properties such as ionization potentials and cross sections, photodissociation rates and cross sections, electron affinities, electron recombination rates and so forth are required for the most promising candidates. With this information one can ascertain which forms are most likely to be present in various regions of the diffuse interstellar medium. Since the spectrum of a species changes with ionization state, these data give much needed supplemental insight into the DIB problem. For example, knowledge of these properties enables one to determine the fraction of a given species that will be in its cationic (+), anionic (-), and neutral forms, and which species are likely to survive interstellar conditions. Leach (1995, elsewhere in this volume) has reviewed the available data on this subject.

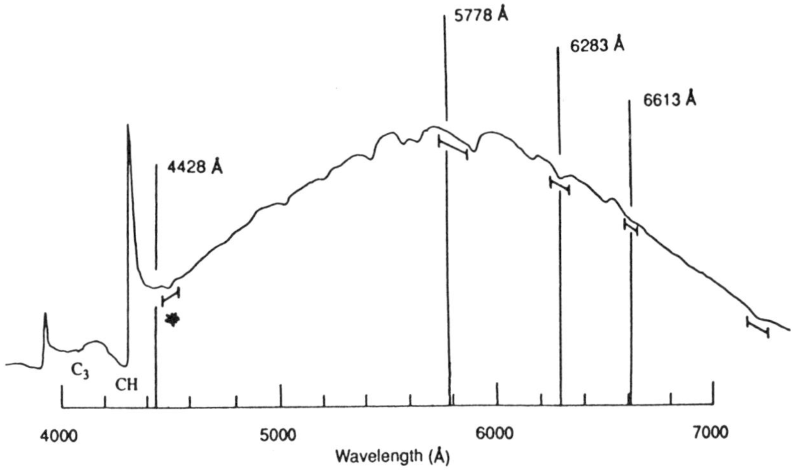

Fig. 1. Single beam absorption spectrum of an argon matrix containing the discharge products from an Ar:CH_4 (200:1) mixture which has been irradiated with ultraviolet light. Vertical lines indicate the positions of 4 of the strongest DIBs. The broad laboratory absorption features proposed as possible DIB counterparts are also indicated (Wdowiak, Lee, and Beegle 1993). Table 2 lists the peak absorption wavelengths of the DIB candidates.

Table 2
Absorptions produced in an Ar:CH_4 (200:1) discharge compared to the DIBs[b]

DIB		CORRELATED LABORATORY FEATURE			
λ (Å)	cm^{-1}	Feature	λ (Å)	cm^{-1}	$\frac{\nu_{DIB} - \nu_{Lab}}{\nu_{DIB}} \times 100\%$[a]
4430	22573	a	4500	22220	+1.57
4754, 4763	21025, 20989	g	4755	21030	−0.02, −0.20
5535	18062	f	5587	17900	+0.90
5705	17523	d	5687	17583	−0.34
5779	17304	c	5788	17277	+0.16
6269	15952	b	6271	15946	+0.03
6283	15916	b	6279	15926	−0.06
6613	15122	e	6601	15150	−0.19
6992	14302	IR	6882	14530	−1.59
7223	13845	IR	7194	13900	−0.40

[a] Attributed to matrix shifts.
[b] Features labeled a, g... are discussed in the original paper (Wdowiak 1980).

3. SPECIFIC EXAMPLES OF DIB CANDIDATES

The following materials are currently considered as potential DIB carriers.

- Unspecified Organics
- Carbon Chains
- Polycyclic Aromatic Hydrocarbons
- Fullerenes and Their Derivatives
- Porphyn/Porphyrins

All contain the cosmically abundant element carbon. Since these are molecules, not grains, the goal has been to obtain spectra of the isolated species. As will become clear in the following discussion, while some may be more attractive than others as potential contributors to the DIBs, none satisfy all the criteria summarized by Snow near the end of this book. Each will now be discussed in turn.

3.1. UNSPECIFIED ORGANICS

In 1980 Wdowiak produced candidates for the carriers of several DIBs in argon matrix experiments. Samples were prepared by passing an Ar:CH_4 (200:1) mixture through an electric discharge and condensing the products onto a sapphire window (~12K). In addition to features arising from several well known species such as C_2^-, CNN, and CH, a few new bands were evident as shown in Figure 1. Several (see Fig. 1 and Table 2) fell within reasonable matrix shifts of DIBs. The bands were photostable under ultraviolet radiation suggesting that the carriers could survive the interstellar radiation field. While several fall near the strong DIBs at 5778, 6283, and 6613 Å, the strongest, a broad absorption band near 4500 Å, is particularly interesting. It becomes evident only after CH, C_2^- and C_3 bands are strong, suggesting it arises from a species containing several carbon atoms. This is supported by the observation that a similar band is produced in pure carbon vapor deposition experiments (next section). These experiments have been repeated and subjected to further analysis as described by Wdowiak *et al.* (1994).

3.2. CARBON CHAINS

Krätschmer (1986) investigated the possibility that carbon chains might be responsible for the DIBs by preparing argon matrices containing small carbon molecules and atoms. These matrices were produced by depositing a stream of argon onto a cold sapphire window (10 K) simultaneously with carbon vapor generated from a resistively heated carbon rod. The ratio of argon to trapped species (100 to 1000) was controlled by adjusting the relative deposition rates. This produced an argon matrix containing small carbon species such as C, C_2, and C_3. As shown in the lower trace of Figure 2, these absorb below about 4200Å. By warming the matrix slightly, these smaller species can diffuse and react with each other producing larger carbon molecules (Figure 2). These often absorb at longer wavelengths. For example, the strong 4100 Å absorption assigned to C_3 decreases while bands at 3110, 3480, 3940, 4470 Å and longer wavelengths grow in. That at 4470Å was tentatively assigned to C_7. Tentative assignments of some of the longer wavelength bands are shown in Figure 3. The bands seem uncorrelated, indicating that each arises from a different molecule.

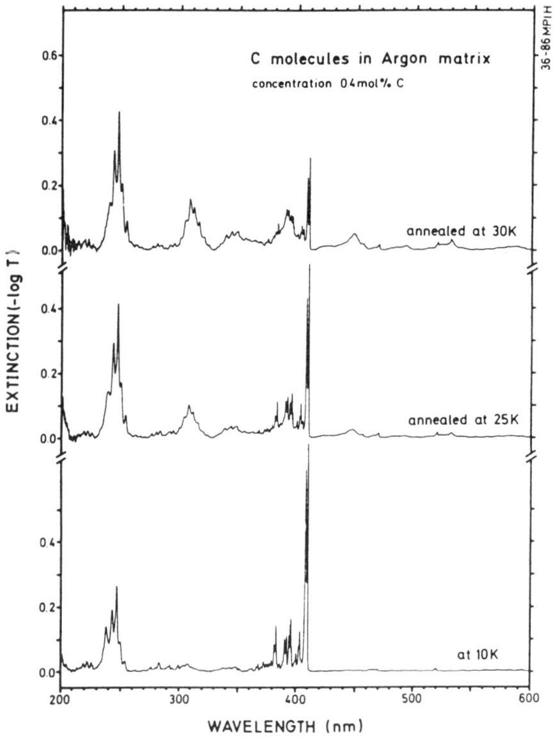

Fig. 2. The UV-Vis spectra of carbon molecules isolated in an argon matrix. The lower trace is of a freshly deposited sample. The middle and upper spectra were taken after the matrix had been warmed to 25 and 30 K respectively. The new bands produced upon warm-up are due to the formation of larger carbon species (Krätschmer 1986).

Some of the species which produce the new features that lie in the visible could be DIB candidates. A comparison of the spectrum produced by this mixture of larger carbon molecules with the diffuse bands as listed by Herbig (1975) is shown in Figure 4. Disregarding the broadening, the pattern of bands in both spectra is remarkably similar, especially between 4400 and 5600Å. Interestingly, Krätschmer's 4470Å band shifts and appears quite similar to Wdowiak's 4500 Å feature. Although these experiments were carried out to test the suggestion made by Douglas (1977) that linear carbon chain molecules might produce some of the DIBs, IR studies of these matrices indicate that a considerable fraction of the larger carbon molecules have nonlinear structures (Krätschmer 1986). These too must contribute to the absorptions shown in Figure 4.

In 1993, Fulara *et al.* reported further laboratory evidence for highly unsaturated hydrocarbons as carriers of some of the DIBs. Their results are based on an elegant experimental technique in which the UV-Vis spectra of specific mass selected species can be measured in a neon matrix. Negative C_nH_m ions (anions), where n=2-12 and m<3 are produced in a discharge from a mixture of 10% diacetylene in argon. These were selected out and co-deposited with neon onto the matrix substrate at 5

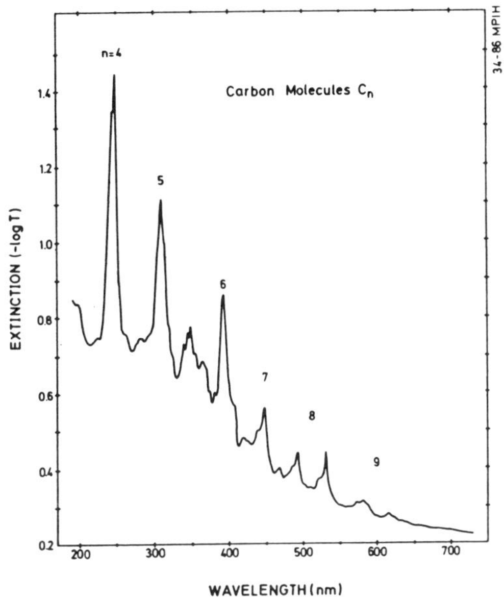

Fig. 3. The spectrum of large carbon molecules produced by annealing an argon matrix containing small carbon species to 35 K. The bands were tentatively assigned to the C_n species, C_4 through C_9 (Krätschmer and Nachtigall 1987).

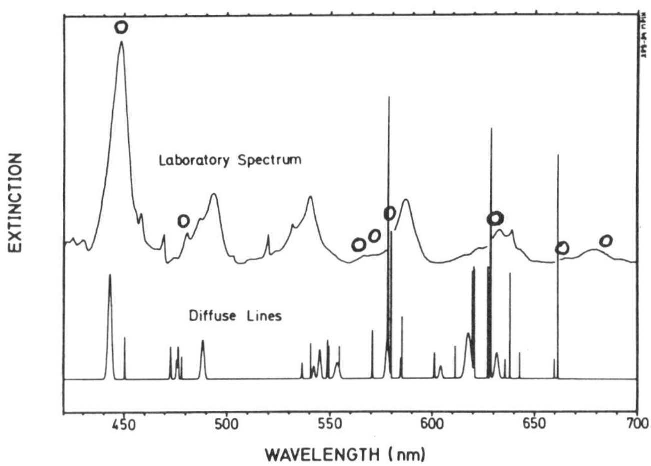

Fig. 4. Comparison of the laboratory spectrum of carbon molecules in an argon matrix with a schematic version of the DIB spectrum reported by Herbig (1975). (From Krätschmer 1986) The circles indicate the positions of the bands reported by Wdowiak (1980) in his CH_4 discharge experiments. The large number of coincidences suggests that the same carriers are responsible for many of these absorptions.

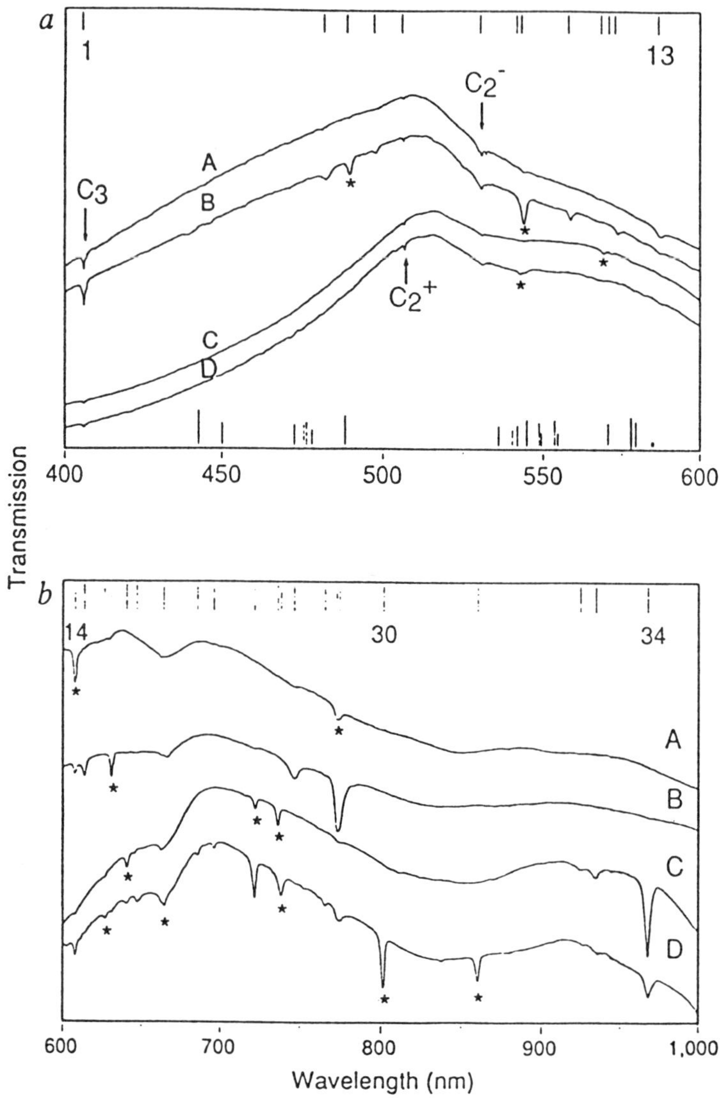

Fig. 5. The absorption spectra produced by codepositing mass-selected $C_6H_m^-$, $C_8H_m^-$, $C_{10}H_m^-$, or $C_{12}H_m^-$ and neon, with simultaneous UV irradiation (traces A, B, C, and D). The DIBs are shown schematically at the bottom of the upper figure, with height representing equivalent width (Herbig 1975; Herbig and Leka 1991). The observed absorption bands, numbered at the top of the figure, are listed in Table 3. The bands labeled with asterisks lie within 36 cm^{-1} of known DIBs (from Fulara *et al.* 1993).

K. The neon matrix was simultaneously irradiated with UV light. Figure 5 shows the absorption spectra produced by codepositing mass-selected $C_6H_m^-$, $C_8H_m^-$, $C_{10}H_m^-$, and $C_{12}H_m^-$ with neon. The intensity pattern depends on deposition conditions, indicating

Table 3
Absorptions of unsaturated hydrocarbons (Figure 5) compared to the DIBs[a].

No.		λ_{matrix} (nm)	Spectrum	λ_{DIB} (nm)
1		405.8	C_3; A, B, C, D	
2		@ 481.9	B	
3		@ 489.0	B	488.20
4		@ 497.5	B	
5		506.2	C_2^-; B, C, D	
6		@ 530.6	C_2^-; A, B, D	
7		@ 541.9	D	542.00
8		543.5	**B**, A, D	544.90
9	#	557.8	B	
10	#	@ 568.4	C	
11		570.8	C	570.51
12		572.8	B	
13		586.5	A	
14		607.7	**A**, B, C, D	608.97
15		614.1	B	
16	#	@ 627.1	D	626.97
17		@ 631.2	**B**, D	631.80
18		@ 640.7	**C**, D	639.69 } 641.26 }
19		647.2	D	
20		@ 664.6	D	666.07
21	[#]	@ 685.9	D	
22		696.0	D	
23	[#]	721.9	**D**, C	722.40
24		736.4	C	735.75
25		738.3	D	736.70
26		746.6	B	
27		765.5	D	
28		772.8	**B**, A	772.17
29		774.5	D	
30		801.9	D	802.62
31		860.7	D	862.08
32		924.6	**C**, D	
33		934.4	**C**, D	
34		967.6	**C**, D	

[a] Taken from Fulara *et al.* (1993). The numbers in the first column correspond to the vertical lines at the top of Figure 5 which indicate the position of a laboratory band in one of the spectra. A, B, C, and D indicate the particular spectrum shown in Figure 5 which have the band. Those absorptions labeled with [#] are close to bands reported by Wdowiak (1980), and those with [@] are close to bands reported by Krätschmer (1986).

that the absorptions arise from many different species. Bands marked with asterisks correlate well with DIB positions. Fulara *et al.* propose that the species which produce the absorptions are highly unsaturated hydrocarbon radicals, C_nH_m, because the anions are believed to be neutralized by the UV irradiation. On the basis of the very good agreement with DIB positions listed in Table 3, they conclude that these species with carbon numbers between 6 and 12 may be the carriers of some of the DIBs. Smaller species are readily photofragmented, while larger structures arrange themselves into cyclic forms (see also Maier, elsewhere in this volume).

It is very important to determine oscillator strengths (f-values) of the transitions in these species. Oscillator strengths are needed to calculate the column density of each proposed DIB candidate implied by the observed equivalent width. This, in turn, allows one to determine the fraction of the elemental cosmic abundance consumed by the candidate and decide whether or not it is reasonable.

3.3. POLYCYCLIC AROMATIC HYDROCARBONS

In 1968 Donn considered the possible contributions of individual polycyclic aromatic hydrocarbons (PAHs) to the interstellar extinction curve. In discussing the influence of size, he pointed out that below about 200 Å, molecular rather than macroscopic effects become important and the calculated extinction curve starts to show considerable structure. Using the spectra of neutral, compact PAHs (presumably suspended in pellets or solution) such as coronene, ovalene and hexabenzocoronene it was argued that some of the poorly understood general features of the interstellar extinction curve - such as the high UV absorption - could be accounted for. Donn and Krishna Swamy (1969) showed that a PAH mixture was consistent with the visible extinction as well. Donn (1968) also pointed out that, "...narrow band structure may...be introduced by the radical side-chains (on PAHs). Such...narrow absorption features may be the cause of the diffuse interstellar bands in a manner having some similarity to that proposed by Johnson (1967)."

Johnson (1967), after reviewing the DIB situation at the time, concluded that, "one is probably dealing with either very complex hydrocarbon molecules or simpler molecules in electronic states not ordinarily achieved under normal laboratory conditions..." In 1970, Johnson reported an extensive study of possible contributors to the DIBs. This work included measuring the spectra of over fifty *neutral* aromatic compounds suspended in various media. He concluded that "These studies do not rule out the existence of molecules of Tables IV (aromatic hydrocarbons) and V (porphyrins) entirely, however they do preclude their discovery in the *optical region* of the spectrum with the sensitivity presently available to astronomers." This conclusion is based on the well-known absorption property of neutral PAHs that the long wavelength absorption band limit is determined by the long dimension of the molecule (Platt, 1956; Clar, 1972). Thus, although all PAHs absorb strongly in the UV, only larger PAHs with more than five or six rings linearly arranged will absorb in the visible. To span the entire visible and near-infrared range, the region in which most DIBs lie, very large *neutral* PAHs are required. Neutral PAHs containing between 25 to 30 carbon atoms just start to absorb in the visible. Much higher numbers are required for absorptions in the red (e.g. see Salama *et al.* 1995 elsewhere in this volume). Consequently, since PAHs are extremely complex molecules by interstellar standards, they were not seriously considered further as DIB carriers. This conclusion that small neutral PAHs cannot contribute to the DIBs has been confirmed by the experiments of Ehrenfreund *et al.* (1992a) on the visible absorption spectra of the PAHs sublimed from coal pitch and isolated in a neon matrix. This mixture of about 150 aromatic molecules has only a few transitions in the region between 3800 and 5000 Å.

Since, on the other hand, PAH ions of all sizes absorb in the visible and near IR, this situation changed when it was recognized that PAHs might account for the interstellar infrared bands emitted from many different astronomical objects, and that a large fraction of any PAHs present in the diffuse interstellar medium would be ionized (Allamandola, Tielens, and Barker 1985; van der Zwet and Allamandola 1985; Léger and d'Hendecourt 1985; Crawford, Tielens, and Allamandola 1985). Spectra of many ionized PAHs suspended in boric acid glass, alkali halide, and freon matrices have been available for some time (Hoijtink 1959; Hoijtink, Velthorst, and Zandstra 1960; Shida and Iwata 1973). More recently argon matrix data on several ionized PAHs have also been published (Andrews, Kelsall, and Blankenship 1982; Andrews, Friedman, and Kelsall 1985; Szczepanski *et al.*, 1993a,b; Vala *et al.* 1994). Although this work provided early support for the notion that ionized PAHs could contribute to the DIB spectrum, as described above in Section 2 - Experimental Techniques, these matrices cause spectral shifts which preclude definitive comparisons with the DIBs. Thus, since little information was available concerning the spectroscopic properties of ionized PAHs in the visible and near-infrared, new experimental studies on the electronic spectra of ionized PAHs were needed. To meet this need, a major effort was initiated at several laboratories to obtain the spectra of ionized PAHs in neon matrices.

The first comparisons of the DIBs with neon matrix spectra of PAH ions appeared only recently. These data are summarized in Table 4 (see Salama *et al.* 1995 for a detailed discussion of these data). Salama and Allamandola (1991, 1992a) reported studies of the electronic absorption spectrum of the naphthalene cation ($C_{10}H_8^+$) isolated in a neon matrix. The visible portion is shown in Figure 6. Neutral naphthalene does not absorb at wavelengths longer than about 3000Å. The strongest band, that at 6742Å, falls close to the DIB at 6741Å which was first suggested by Herbig in 1975 and confirmed by Jenniskens and Désert in 1994. The second strongest naphthalene cation band falls at 6520Å. A new DIB was subsequently found at this wavelength (Table 4), although with a different strength than expected. Ehrenfreund *et al.* (1992b) measured the spectra of neutral and ionized ovalene ($C_{32}H_{14}$) and coronene ($C_{24}H_{12}$) in neon matrices as these species were known to absorb near the strong 4430Å DIB. While these spectra showed that neither neutral ovalene nor the coronene cation could account for the 4430Å DIB, as shown in Table 4, the coronene cation could be responsible for the 4595 and 9466Å DIBs. Salama and Allamandola (1992b) showed that a pyrene-like cation might well be responsible for the strong 4430Å DIB. This was based on the observation that the pyrene cation ($C_{16}H_{10}^+$) has only one strong absorption in the entire visible and near-infrared spectral range. This band, also shown in Figure 6, falls at 4435Å in an argon matrix and at 4390Å in neon. The difference between the DIB wavelength and the wavelength in a neon matrix casts doubt on the assignment to the pyrene cation itself. However, the proximity to the 4430Å DIB and simplicity of the pyrene cation spectrum suggested that a closely related molecule such as alkyl substituted pyrenes might prove to be strong candidates (Salama and Allamandola 1992b). Léger and d'Hendecourt (1995, elsewhere in this volume), show that this is indeed plausible. This point is also discussed further by Salama *et al.* (1995, elsewhere in this volume).

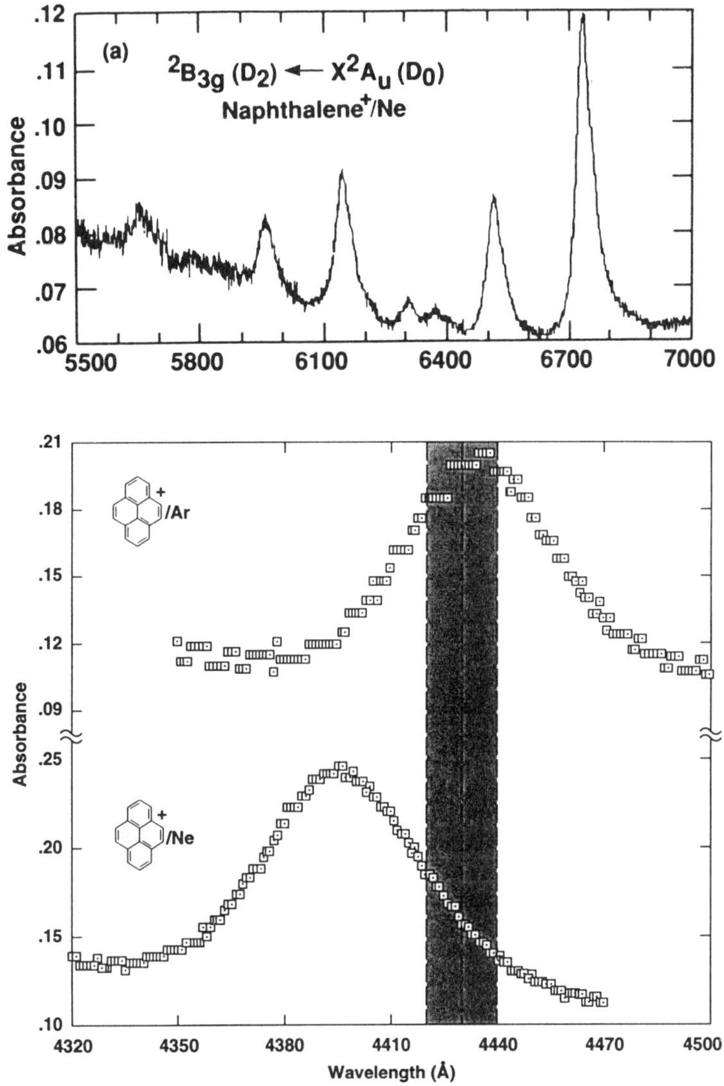

Fig. 6. Top - The visible absorption spectrum of the naphtahlene cation isolated in a neon matrix. The ion was produced by irradiating a naphthalene:neon (1:600) matrix at 12 K (Salama and Allamandola 1991, 1992a). Bottom - The strong UV absorption band of the pyrene cation isolated in neon and argon matrices compared to the position and width of the 4430Å DIB (shaded area, figure from Salama and Allamandola 1992b).

Table 4
Absorption of PAH cations compared with the DIBs (Salama *et al.* 1995)[a]

PAH^+	λ_{peak} (nm) (Neon matrix)	DIBs (nm)[b]	$\Delta\nu$(cm^{-1})
Pyrene ($C_{16}H_{10}^+$)	439.5[f] (443.0 in Ar)	442.9	175
1-Methylpyrene ($CH_3 - C_{16}H_9^+$)[c]	444.2[f]	442.9	66
4-Methylpyrene ($CH_3 - C_{16}H_9^+$)	(457.7)		
	482.8	482.4	17
	757.6	758.1	9
Naphthalene ($C_{10}H_8^+$)	674.2[f]	674.1	2
	652.0	652.0	0
Phenanthrene ($C_{14}H_{10}^+$)[d]	898.3[f]		
	856.8	857.2	5
Tetracene ($C_{18}H_{12}^+$)	864.7[f]	864.8	2
Benzo(ghi)perylene ($C_{22}H_{12}^+$)	502.2[f]	503.9 (?)	67
	758.4	758.1; 758.6	5; 4
	755.2	755.8 (?); 756.2	11; 8
	794.3	793.5 (prob.)	13
Coronene ($C_{24}H_{12}^+$)[e]	459.0[f]	459.5	24
	946.5	946.6	1

[a] Except for the 4439 Å band, the percent frequency shifts between the bands in the neon matrix and the DIBs are all much less than 0.5%. For the 4439 Å band it is 0.75%.
[b] Herbig, 1975; Sanner *et al.*, 1978; Herbig and Leka, 1991; Jenniskens and Désert, 1994; Ehrenfreund and Foing, 1994.
[c] See also Léger and d'Hendecourt, 1995.
[d] Phenanthrene data from Salama *et al.* 1994.
[e] See also Ehrenfreund *et al.*, 1992.
[f] Strongest absorption band.

The oscillator strengths determined for these species permit one to calculate the abundance of the PAH candidate required to account for the observed DIB equivalent width. For each of the PAHs listed in Table 4, only a few tenths of a percent of the cosmic carbon is implied, showing that PAHs readily satisfy the cosmic abundance constraint.

An important question related to the issue of PAHs as DIB carriers has to do with the UV spectral properties of these species. Since neutral PAHs are known to possess strong absorptions in the UV, the lack of substructure in the UV portion of the interstellar extinction curve has been considered as a criticism of the PAH hypothesis (see Donn 1995, elsewhere in this volume). There are two factors that would tend to weaken this substructure. First, the absorptions produced by a mixture of PAHs would overlap, producing only a weakly structured pseudocontinuum in a manner similar to that described by Joblin *et al.* (1992). Second, the UV band strengths of some PAH ions may be weaker than for their neutral counterparts. The UV absorption strength of the naphthalene cation appears to be weaker than for neutral naphthalene (Salama and Allamandola 1991, 1992a). This tendency has also been found for the cations of several PAHs suspended in borosilicate glasses (Lee and Wdowiak 1993). If this is a general property of ionized PAHs, the implication is that ionized interstellar PAHs might not add significant structure to the interstellar extinction curve at wavelengths shortward of about 4000 Å along lines-of-sight with strong DIBs. The ionization state of a mixture of PAHs is expected to vary in the diffuse interstellar medium (Bakes and Tielens, elsewhere in this volume). If ionized PAHs indeed contribute to the DIBs, it would be interesting to search for UV structure in those regions where DIBs are known to weaken. Since the PAH spectrum is such a sensitive function of ionization state, as the population of ionized PAHs decreases, UV absorption bands due to the neutral species may become apparent.

3.4. FULLERENES AND RELATED MATERIALS

A spell was cast over much of the astronomical community when Kroto et al. (1985) suggested that C_{60} had the structure of a soccerball and that "it may be widely distributed throughout the Universe". Kroto *et al.* also speculated that C_{60} or a derivative may be the carrier of the DIBs. As with PAHs, there is a need for the spectroscopic properties of these species taken under conditions appropriate for the interstellar medium. During the past few years some of this data has become available.

The 2000 to 7000Å spectrum of neutral C_{60} in solution or as a solid film has been reported by Krätschmer *et al.* (1990) and Leach *et al.* (1992). C_{60} absorbs most strongly at about 2150, 2600, and 3350Å, possessing very weak bands in the visible. The spectrum presented in Figure 7 shows that neutral C_{60} is not a DIB candidate. However, as with PAHs, the ions of this species absorb at longer wavelengths. Recently, the spectra of matrix isolated C_{60}^+ and C_{60}^- have become available (argon: Gasyna, Andrews, and Schatz 1992; neon: Joblin 1992; neon and argon: Fulara, Jakobi, and Maier 1993). The spectra of neon matrix isolated C_{60}^+ and C_{60}^- are shown in Figure 8. The strongest absorptions of C_{60}^+ fall at 10368 and 10435 cm^{-1} (9645 and 9583Å), while the strongest band of C_{60}^- is at 9443 cm^{-1} (10590Å). While these fall beyond the spectral region normally associated with the DIBs, recent observations reported by Foing and Ehrenfreund (1994), have revealed two possible new DIBs at 9632 and 9577Å. These lie within 14 and 7 cm^{-1} of the two strong neon matrix isolated C_{60}^+ bands. The relative band strengths are also similar to those found for matrix isolated C_{60}^+. Thus, these new DIBs may well arise from C_{60}^+ (Foing and Ehrenfreund 1994; and Foing and Ehrenfreund 1995, elsewhere in this volume). Questions remain however as the doublet structure shown in Figure 8 may arise from C_{60}^+ being trapped in two different forms (not sites), one metastable, by the matrix. Gas phase C_{60}^+ may not show this doublet structure. Since any DIB identification is strengthened by multiple band agreements, this would weaken the C_{60}^+ assignment,

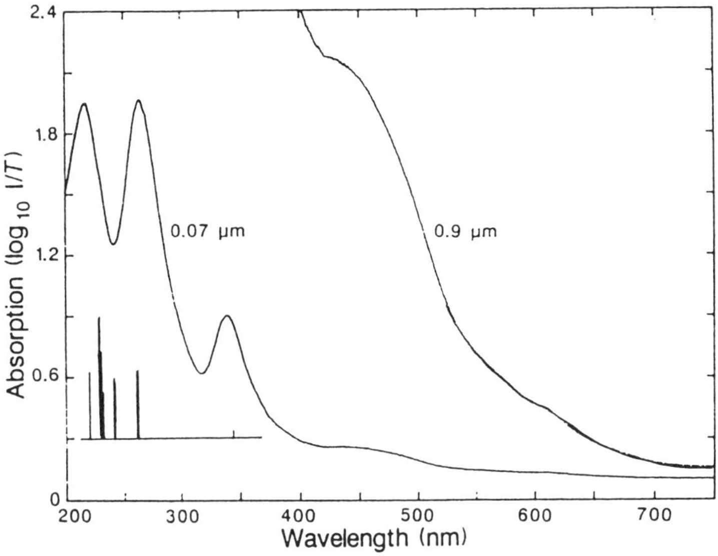

Fig. 7. The Visible-Ultraviolet absorption spectra of two thicknesses of solid, neutral C_{60}. The stick spectrum at the bottom shows the positions and relative oscillator strengths theoretically calculated for isolated C_{60}. Figure from Krätschmer *et al.* (1990).

but not eliminate it as one of the doublet components would remain. Maier (1994) has also discussed this issue.

As with the carbon chains, it is very important to determine the oscillator strengths for these transitions in order to deduce the fraction of the elemental cosmic abundance required by the candidate.

3.5. PORPHYRINS AND RELATED MATERIALS

In 1970 Johnson reported an extensive spectroscopic study of potential DIB candidates including dyes, sulfur containing species, aromatic compounds with and without hetero atoms, and porphyrins. This exhaustive work encompassed possible grain materials as well as individual molecules suspended in matrices of various sorts. Since the spectra on porphyrins looked promising, while the other spectra "did not even remotely suggest similarity to the diffuse interstellar spectra" a large number of porphyrin compounds were studied further. Since many porphyrins possess an unusually strong band (the Soret band) in the vicinity of 4430Å, the position of the strongest DIB, Johnson considered many closely related compounds. The compound, bispyridylmagnesiumtetrabenzoporphine, (MgTBP-dipyridine), with the structure shown in Figure 9, seemed particularly promising.

As shown in Figure 9, its two strongest absorptions fall close to the two strongest DIBs, namely those at 4428 and 6281Å. The bands are broad as are the two strong DIBs. Sixteen spectral coincidences with known DIBs were reported at that time, with six others added later (Johnson, Bailey, and Wegner 1973). Johnson (1991;

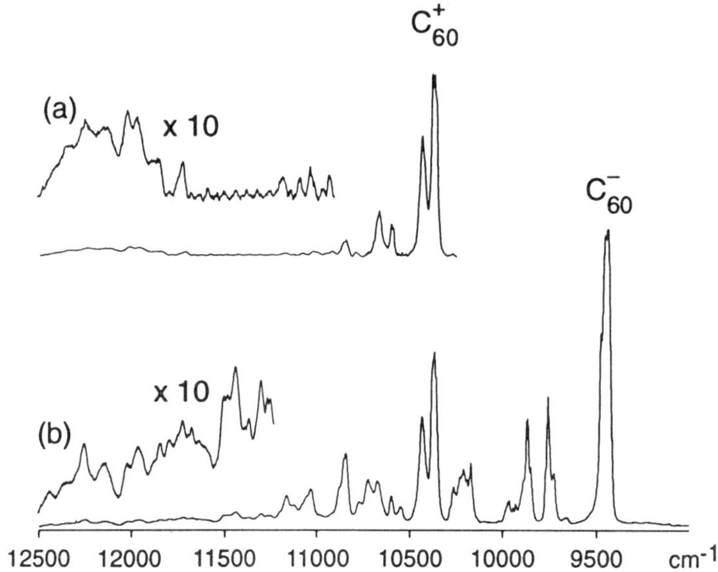

Fig. 8. The electronic absorption spectra of C_{60}^+ and C_{60}^- isolated in neon matrices. Figure from Fulara, Jakobi, and Maier (1993).

1995, elsewhere in this volume) reports that most of the DIBs can be reproduced by the spectra obtained when this compound is suspended in several different paraffin matrices.

The novel suggestion that porphyrins or closely related materials might account for the DIBs was recently furthered by Miles and Sarre (1992), when they pointed out that the strong origin band of cold, gas phase chlorin ($C_{20}H_{16}N_4$), a reduced porphyrin, falls within 1Å of the prominent DIB at 6284Å. Miles and Sarre (1993) discussed additional spectroscopic and analytical studies of this species which supports the notion of chlorin as a DIB carrier. Figure 10 which compares a schematized version of the DIB spectrum with measured and calculated chlorin spectra shows the suggestive nature of this assignment. In their thorough discussion of this suggestion, Miles and Sarre (1993) also point out that the very strong and broad Soret band which falls near 3900 Å must be searched for to test this assignment and that currently available data are of insufficient quality to settle the issue. See Hibbins *et al.* 1994, for a discussion of new, narrow DIBs detected in this region.

4. ASSESSMENTS

Here, a brief assessment of the status of each suggested material is presented, with emphasis placed on what is needed to really put each candidate to the test. Limitations are also discussed.

UNSPECIFIED ORGANICS

The specific materials responsible must be identified, or at the very least placed

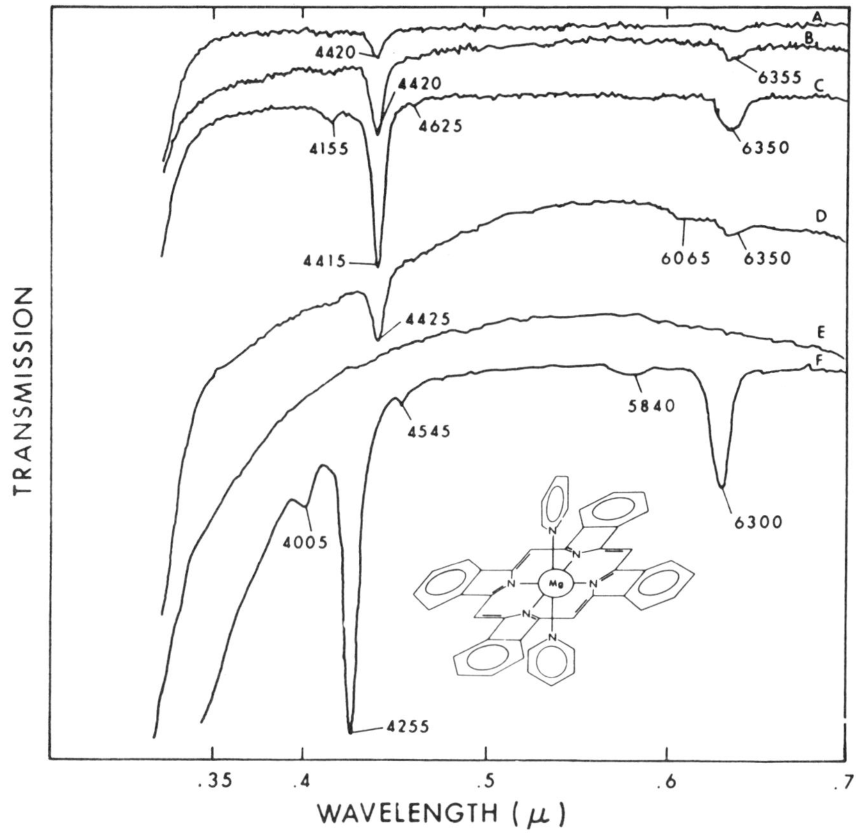

Fig. 9. The 3500 to 7000Å spectrum of Bispyridylmagnesiumtetrabenzoporphine dissolved in pyridine as a function of pathlength (A, B, C) and concentration (A and D). E and F are traces of pyridine alone, and the compound dissolved in TEA. Figure from Johnson (1970).

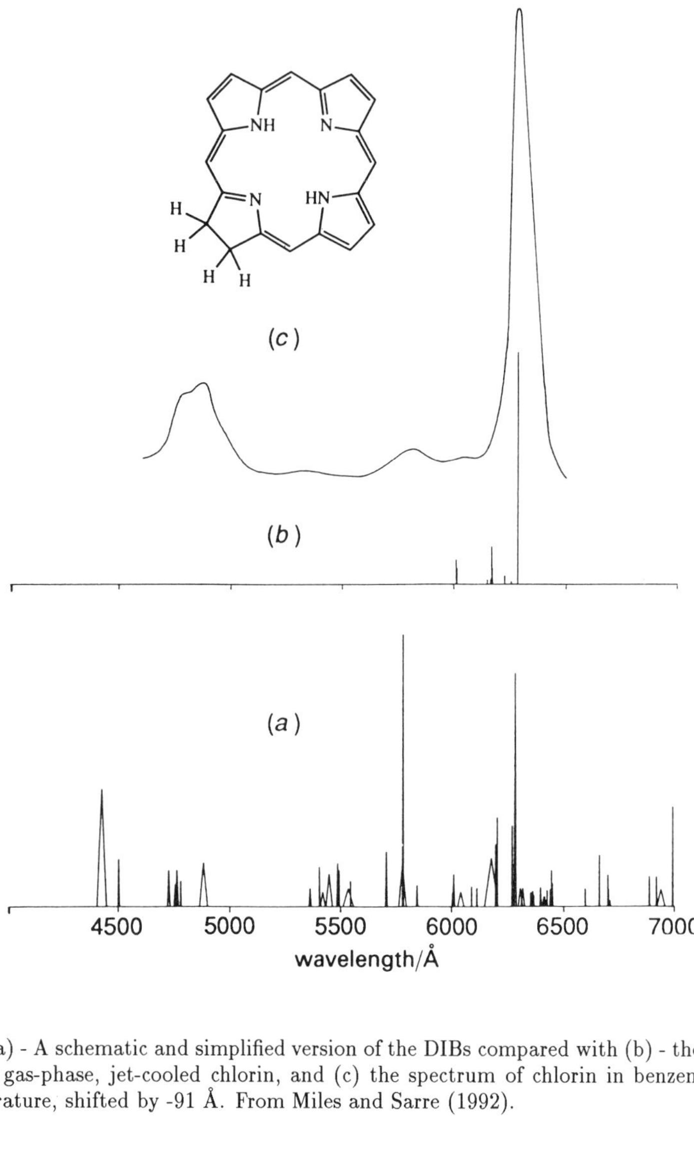

Fig. 10. (a) - A schematic and simplified version of the DIBs compared with (b) - the fluorescence spectrum of gas-phase, jet-cooled chlorin, and (c) the spectrum of chlorin in benzene solution at room temperature, shifted by -91 Å. From Miles and Sarre (1992).

in general chemical categories. To achieve this will require new experiments carried out at the low end of the ultra-high vacuum range to eliminate any questions of contamination, allowing a confident identification of the elements involved. Once this is achieved, one can then compare the properties of these materials with the properties of the other candidates. One key question at this stage is whether or not these materials are the same as, or closely related to, those generated in the carbon chain studies. Once classification has been made, studies in neon matrices will be required to see if the new bands still fall within 1% of the DIB wavelengths. If so, f-values must be determined. If all of these criteria are met, gas phase experiments, on a least a few representative samples, will be called for.

CARBON CHAINS

In view of the encouraging neon matrix results, these species look quite promising as candidates for some of the DIBs. It is important to increase the spectral library of these species and determine oscillator strengths (f-values). The most promising candidates warrant the work needed to determine the specific molecular formulae for several reasons. First, this will provide guidance for the gas phase experimental work called for to verify band positions; second, specific identifications will allow theoretical calculations on gas phase band strengths, profiles, and relative intensities; and third, the species suggested will give much needed insight into the chemical and physical processes of the interstellar medium. It presently seems that while carbon chains might account for some of the DIBs, they probably cannot account for them all since each species is likely to produce a few bands at most . This constraint appears fairly tight because small species ($\lesssim$ 6 C atoms) are not photostable under the interstellar radiation field, while larger species ($\gtrsim$ 10-12 C atoms) cyclize into aromatic moities. Even allowing for mixtures of slightly larger isomers, it is difficult to imagine such a handful of species producing all of the DIBs.

POLYCYCLIC AROMATIC HYDROCARBONS

In view of the encouraging neon matrix results, these too, look very promising as DIB candidates. It is important to increase the spectral library of neon matrix isolated PAHs, not only as positive and negative ions, but also in their radical forms. The f-values are extremely important to determine as these are used to derive column densities and there is great uncertainty in the current experimental literature on this topic. Gas phase spectroscopic work is called for on the most promising species to verify band positions and calculated profiles. While it appears that PAHs might well account for some of the DIBs, they suffer from a concern which is the opposite to that for carbon chains. There are so many PAH isomers conceivable that one must envision a mechanism which favors several tens of PAHs, while weeding many out. Unless this is the case, the cosmic carbon available must be distributed among all PAHs, lowering the column density of any specific PAH to the point where it cannot absorb strongly enough to produce a typical DIB equivalent width. Since different PAHs have widely varying stabilities, weeding out may occur during formation in circumstellar shells or in the interstellar medium itself.

C_{60} AND FRIENDS

While there is no spectroscopic evidence in support of C_{60} as an abundant interstellar species (due to the lack of structure in the 2000-3000Å portion of the extinction curve), there is intriguing agreement between the two strongest absorption bands in the spectrum of neon matrix isolated C_{60}^+ and two newly detected DIBs. To test this further requires that the gas phase spectrum of C_{60}^+ be measured and that f-values be determined. Observationally, correlation studies demonstrating that the two new DIBs are locked together are also required to test the identification. To determine if

C_{60}^- is present requires the search for a new DIB in the vicinity of 10590Å, the wavelength of its strongest absorption. In any event, neither C_{60}, C_{60}^+, nor C_{60}^- contribute to most of the known DIBs. See Kroto (1988) for a discussion of various other conceivable forms of fullerenes which may be considered for the DIBs, including metal entrapped species, and Webster (1995, elsewhere in this volume) for a discussion of the many hypothetical variations on the C_{60} theme such as hydrogen additions at various positions on the molecule which can isolate conjugated bond systems and create radical sites, providing systems analogous to polyenes such as polyethylene which can absorb in the visible.

PORPHYRINS AND RELATED MATERIALS

This intriguing suggestion, which has been proposed to account for the DIBs and UV- Visible extinction, involves the suspension of MgTBP-dipyridine in 5 types of paraffin matrices. For this suggestion to be widely accepted will require that all of the spectra be published over the entire spectral range considered. Band positions, intrinsic strengths (f-values), and relative strengths need to be tabulated as well. Although the picture involves MgTBP-dipyridine suspended in paraffin, gas phase data, especially for the very intense Soret band, is also needed. This is in addition to spectra of this compound in or on other interstellar materials known since it is difficult to imagine that this molecule would only be present in paraffin in the interstellar medium. These data are needed to fully test this DIB candidate as it would provide a tell-tale tracer for this compound prior to paraffin incorporation. In addition, the infrared spectra of the paraffin/MgTBP-dipyridine suspensions should also be published in order to provide spectral signatures with which the presumably more abundant paraffin (and possibly MgTBP-dipyridine) can be searched for. Although there is currently little data on the infrared spectral properties of the diffuse interstellar medium, the upcoming *ISO* mission should address this issue squarely. For chlorin, a complete UV-Visible spectrum in a neon matrix or in the gas-phase (including f-values) is needed to pin down the other band positions.

Finally, although it almost goes without saying, the species considered must meet a number of criteria in addition to spectral matching since there are so many DIBs that apparent wavelength agreement, while crucial, should only be viewed as the first hurdle. For example, the species must satisfy cosmic abundance constraints and they must be reasonable. The later point is important because of the seductive power of the DIB problem to spectroscopists. This has often led to the vigorous pursuit and development of unreasonable ideas. The following serves to illustrate the pitfalls of relying solely on wavelength comparisons. Some time ago the author (and one of the editors!) expended a very serious experimental and theoretical effort on the study of simple radicals such as HCO and NH_2 frozen in various matrices. The motivation was the remarkably close agreement between a number of the strongest DIBs reported by Herbig (1975) and the strongest absorption bands of these frozen radicals. The strongest absorption band wavelengths of each species are listed, followed by possible DIB counterparts in parentheses. The first DIB wavelength is from Herbig (1975), the second from Jenniskens and Désert (1994). HCO in CO: 5102Å (-,5109Å), 5330Å (5362Å, 5362Å), 5548Å (5545Å, 5545Å), 5789Å (5778Å, 5789Å), 6052Å (6042Å, 6045Å) 6352Å (6352Å, 6354Å), 6695Å (6660Å, 6699Å) [HCO data from Ewing, Thompson, and Pimentel 1960]; NH_2 in argon: 4711Å (4726Å, 4727Å), 5150Å (-,-), 5684 (5705Å, 5705Å), 6275 (6270Å, 6278Å) [NH_2 data from Robinson and McCarthy 1959)]. This work was pursued notwithstanding the fact that this "wavelength agreement" necessitated the use of different matrices - CO and Ar; both of which are absurd for the diffuse interstellar medium (matrices were rationalized for two reasons, first to shift the wavelengths into agreement since many of the gas phase values were off by more than a hundred angstroms, and second to protect these fragile species from photodissociation by the interstellar radiation field). Fortunately, we came to our senses in time.

5. CONCLUSION

Spectroscopic studies of potential diffuse interstellar band carriers are reviewed. The discussion is limited to the spectroscopic properties of those species thought to be promising by various workers in this field. These include - unspecified organics, carbon chains, polycyclic aromatic hydrocarbons (PAHs), fullerenes and derivatives, as well as porphyrins and related material. A critical assessment of the status of each material as a potential DIB candidate is given, along with suggestions for further experimental studies needed. In addition to the spectra discussed here, many other properties of the DIB candidates are required to fully test the feasibility of each.

The criteria which must be met for a particular material to be accepted as a DIB carrier are that its visible and near-infrared absorption features match the known DIBs in wavelength, bandwidth, and relative intensities, while not possessing additional features which are absent in the interstellar spectra. There are no species which presently satisfy these criteria. At this point, given the type and quality of published data available, carbon chains and PAHs are the most promising candidates to account for a number of the DIBs.

ACKNOWLEDGEMENTS. I wish to thank both Xander Tielens and Ted Snow for their extreme patience and understanding while I prepared this manuscript. I also wish to thank Farid Salama for his careful reading of the manuscript. This work was supported by NASA's Infrared, Submillimeter, and Radio Laboratory Astrophysics Program (Grant 188-44-57-01), NASA's Long Term Space Astrophysics Research Program (Grant 399-20-00-05), and NASA's Ultraviolet, Visible, and Gravitational Astrophysics Research and Analysis Program (Grant 188-41-57-41).

REFERENCES

Allamandola, L.J., Tielens, A.G.G.M., and Barker, J.R. 1985, *Ap. J. (Letters)*, **290**, L25.
Andrews, L., Friedman, R.S., and Kelsall, B.J. 1985, *J. Phys. Chem.*, **89**, 4016.
Andrews, L., Kelsall, B.J., and Blankenship, T.A. 1982, *J. Phys. Chem.*, **86**, 2916.
Bakes E. and Tielens, A.G.G.M. 1995, in *The Diffuse Interstellar Bands*, eds. A.G.G.M. Tielens and T.P. Snow (Kluwer, Dordrecht), this volume.
Boissel, P. 1995, in *The Diffuse Interstellar Bands*, eds. A.G.G.M. Tielens and T.P. Snow (Kluwer, Dordrecht), this volume.
Bondybey, V. E. and Miller, T.A. 1983, in *Molecular Ions: Spectroscopy, Structure and Chemistry*, eds. T.A. Miller and V.E. Bondybey, (North Holland Publishers), 125.
Clar, E. 1972, *The Aromatic Sextet*, (London, John Wiley and Sons).
Crawford, M.K., Tielens, A.G.G.M., and Allamandola, L.J. 1985, *Ap. J. (Letters)*, **293**, L48.
Donn, B. 1968, *Ap. J.*, **152**, L129.
Donn, B. and Krishna Swamy, K.S. 1969, *Physica*, **41**, 144.
Donn, B. 1995, in *The Diffuse Interstellar Bands*, eds. A.G.G.M. Tielens and T.P. Snow (Kluwer, Dordrecht), this volume.
Douglas, A.E. 1977, *Nature*, **269**, 130.
Ewing, G.E., Thompson, W.E., and Pimentel, G.C. 1960, *J. Chem. Phys.*, **32**, 927.

Ehrenfreund, P., d'Hendecourt, L.B., Joblin, C., and Léger, A. 1992a, *A & A*, **266**, 429.
Ehrenfreund, P., d'Hendecourt, L.B., Verstraete, L., Léger, A., Schmidt, W., and Defourneau, D. 1992b, *A & A*, **259**, 257.
Ehrenfreund, P. and Foing, B.H. 1994, (private communication).
Foing, B. and Ehrenfreund, P. 1995, in *The Diffuse Interstellar Bands*, eds. A.G.G.M. Tielens and T.P. Snow (Kluwer, Dordrecht), this volume.
Foing, B. and Ehrenfreund, P. 1994, *Nature*, **369**, 296.
Fulara, J., Jakobi, M., and Maier, J.P. 1993, *Chem. Phys. Lett.*, **211**, 227.
Fulara, J., Lessen, P., Freivogel, P., and Maier, J.P. 1993, *Nature*, **366**, 439.
Gasyna, Z., Andrews, L., and Schatz, P.N. 1992, *J. Phys. Chem.*, **96**, 1525.
Herbig, G.H. 1975, *Ap. J.*, **196**, 129.
Herbig, G.H. and Leka, K.D. 1991, *Ap. J.*, **382**, 193.
Hibbins, R.E., Miles, J.R., Sarre, P.J., and Herbig, G.H. 1994, in *The Diffuse Interstellar Bands: Contributed Papers*, eds. A.G.G.M. Tielens, (NASA CP 10144), 31.
Hoijtink, G.J. 1959, *Molecular. Phys.*, **2**, 85.
Hoijtink, G.J., Velthorst, N.H., and Zandstra, P.J. 1960, *Molecular. Phys.*, **3**, 533.
Jenniskens, P. and Désert, F.X. 1994, *A & A Supp. Ser.*, **106**, 39.
Joblin, C. 1992, Ph.D. Dissertation, University of Paris 7, 192.
Joblin, C., Léger, A., Martin, P., and Defourneau, D. 1992, in *Astrochemistry of Cosmic Phenomena*, ed. P.D. Singh (Kluwer:Dordrecht), 21.
Johnson, F. 1967, in *Interstellar Grains*, eds. J.M. Greenberg and T.P. Roark (NASA SP-140), 229.
Johnson, F.M. 1970, *Spectroscopic Studies of Interstellar Grains*, NASA CR - 1667.
Johnson, F.M., Bailey, D.T., and Wegner, P.A. 1973, in *Interstellar Dust and Related Topics*, eds. J.M. Greenberg and H.C. van de Hulst, (Reidel, Dordrecht), 317.
Johnson, F.M. 1991, *Bull. Am. Astron. Soc.*, **23**, 933.
Johnson, F.M. 1995, in *The Diffuse Interstellar Bands*, eds. A.G.G.M. Tielens and T.P. Snow (Kluwer, Dordrecht), this volume.
Krätschmer, W. 1986, *Ap.& Sp. Sci.*, **128**, 93.
Krätschmer, W., Lamb, L.D., Fostiropoulos, K., and Huffman, D.R. 1990, *Nature*, **347**, 354.
Krätschmer, W. and Nachtigall, K. 1987, in *Polycyclic Aromatic Hydrocarbons and Astrophysics*, eds. Léger, A., d'Hendecourt, L.B., and Boccara, H., 75.
Krełowski, J. and Walker, G.A.H. 1987, *Ap. J.*, **312**, 860.
Kroto, H.W. 1988, *Science*, **242**, 1139.
Kroto, H.W., Heath, J.R., O'Brien, S.C., Curl, R.F., and Smalley, R.E. 1985, *Nature*, **318**, 162.
Leach, S. 1995, in *The Diffuse Interstellar Bands*, eds. A.G.G.M. Tielens and T.P. Snow (Kluwer, Dordrecht), this volume.
Leach, S., Vervloet, M., Despres, A., Breheret, E., Hare, J.P., Dennis, T.J., Kroto, H.W., Taylor, R., and Walton, D.R.M. 1992, *Chem. Phys.*, **160**, 451.
Lee, W. and Wdowiak, T.J. 1993, *Ap. J. (Letters)*, **410**, L127.
Léger, A. and d'Hendecourt, L.B. 1985, *A & A*, **146**, 81.
Léger, A. and d'Hendecourt, L.B. 1995, in *The Diffuse Interstellar Bands*, eds. A.G.G.M. Tielens and T.P. Snow (Kluwer, Dordrecht), this volume.
Maier, J.P. 1994, *Nature*, **370**, 423.
Miles, J.R., and Sarre, P.J. 1992, *J. Chem. Soc. Faraday Trans.*, **88**, 1075.
Miles, J.R. and Sarre, P.J. 1993, *J. Chem. Soc. Faraday Trans.*, **89**, 2269.
Platt, J.R. 1956, *Ap. J.*, **123**, 486.
Robinson, G.W. and McCarthy, M. 1959, *J. Chem. Phys.*, **30**, 999.
Salama, F. and Allamandola, L.J. 1991, *J. Chem. Phys.*, **94**, 6964.
Salama, F. and Allamandola, L.J. 1992a, *Ap. J.*, **395**, 301.
Salama, F. and Allamandola, L.J. 1992b, *Nature*, **358**, 42.
Salama, F., Joblin, C., and Allamandola, L.J. 1994, *J. Chem. Phys.*, **101**, 10252.
Salama, F., Joblin, C., and Allamandola, L.J. 1995, in *The Diffuse Interstellar Bands*, eds. A.G.G.M. Tielens and T.P. Snow (Kluwer, Dordrecht), this volume.
Sanner, F., Snell, R., and Vanden Bout, P. 1978, *Ap. J.*, **226**, 460.
Shida, T. and Iwata, S. 1973, *J. Am. Chem. Soc.*, **95**, 3473.
Szczepanski, J., Chapo, C., and Vala, M. 1993a, *Chem. Phys. Lett.*, **205**, 434.
Szczepanski, J., Vala, M., Talbi, D., Parisel, O., and Ellinger, Y. 1993b, *J. Chem. Phys.*, **98**, 4494.
Vala, M., Szczepanski, J., Pauzat, F., Parisel, O., Talbi, D., and Ellinger, Y. 1994, *J. Phys. Chem.*, in press.
van der Zwet, G.P. and Allamandola, L.J. 1985, *A & A*, **146**, 76.
Webster, A. 1995, in *The Diffuse Interstellar Bands*, eds. A.G.G.M. Tielens and T.P. Snow (Kluwer, Dordrecht), this volume.
Wdowiak, T.J. 1980, *Ap. J. (Letters)*, **241**, L55.
Wdowiak, T.J., Beegle, L.W., Lee, W., and Robinson, M.S. 1994, in *The Diffuse Interstellar Bands: Contributed Papers*, eds. A.G.G.M. Tielens and T.P. Snow, (NASA CP 10144), 65.
Wdowiak, T.J., Lee,W., and Beegle, L.W. 1994 in *Molecules and Grains in Space*, ed. I. Nenner, AIP Conf. Proc. 312, (Am. Inst. Phys. Press, New York), 687.

LABORATORY STUDIES OF HIGHLY UNSATURATED CARBON CHAINS AND THE DIFFUSE INTERSTELLAR BANDS

J.P. MAIER, P. FREIVOGEL, J. FULARA AND D. LESSEN
Institut für Physikalische Chemie, Universität Basel, Klingelbergstrasse 80, CH-4056 Basel, Switzerland

Highly unsaturated hydrocarbon chains, identified by radioastronomy, are known to be important constituents of dark molecular clouds (see chapter by Thaddeus in this volume). Such species have also been proposed to play a significant role in diffuse clouds, and in fact may be related to the absorption features of the diffuse interstellar bands (DIBs) (Douglas 1977; Smith, Snow and York 1977; Bell, Feldman and Matthews 1983; Thaddeus *et al.* 1993). The article of Douglas on this topic provides an especially useful guideline, pointing out the required spectral features and underlying physical phenomena. However, direct, gas-phase absorption matches from laboratory settings are sought to completely identify the carrier of a given DIB.

In the absence of gas-phase data, the best information one can obtain on a given cation is from the measurement of their electronic spectra in neon matrices. There are now several examples of this for the aromatic hydrocarbon cations (see chapter by Allamandola in this volume). In connection with our spectroscopic studies on ions, ionic clusters and radicals, we have been developing a technique aimed at obtaining absorption spectra on highly unsaturated hydrocarbons. The approach is conceptually simple, but technically demanding; it combines the advantages of mass-spectrometry with matrix isolation (Forney, Jakobi and Maier 1989). Mass-selected ions and radicals can be embedded in neon matrices at 5K, and the electronic spectra subsequently recorded at leisure. The general features of the experimental apparatus are shown in Fig.1. The ion source which has been used for the generation of cations for the deposition of these species in 5K neon matrices has been described (Freivogel *et al.* 1994).

The success of the experimental technique is demonstrated by the initial identification of the $B^4\Sigma_u^- - X^4\Sigma_g^-$ electronic transition of mass-selected C_2^+. With this information in hand, spectroscopic characterisation in the gas-phase rapidly followed (Maier and Rösslein 1988). For purposes of DIBs comparison, the origin of the 0-0 band of the $B^4\Sigma_u^- - X^4\Sigma_g^-$ transition of C_2^+ is at 505.6nm, with an oscillator strength of 0.023(3). In this connection it should be pointed out that the electronic transitions of the cations of the polyacetylenes, $H(CC)_nH^+$ n=2-4 (Allan, Kloster-Jensen and Maier 1976) have also been measured in the gas-phase, as well as for the related species $H(CC)_2CN^+$ and $NC(CC)_nCN^+$, n=1 , 2 (see Maier 1988 for references therein). Their oscillator strengths can be obtained from the measured lifetimes and

A. G. G. M. Tielens and T. P. Snow (eds.), The Diffuse Interstellar Bands, 199–205.

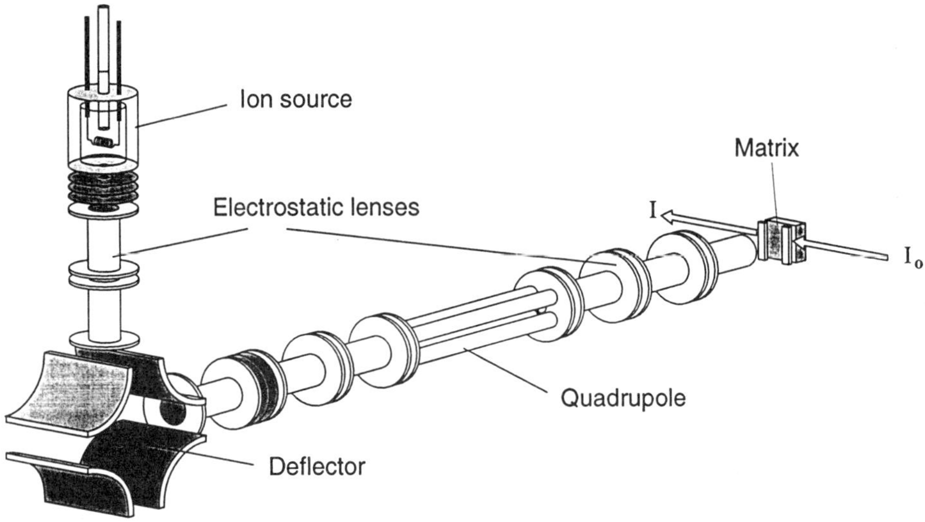

Fig. 1. Schematic of the experimental apparatus which combines the advantages of mass-selection and matrix isolation spectroscopy. Highly unsaturated carbon chains, generated in a filament-driven discharge, are extracted and transmitted to the matrix through a series of electrostatic ion optics. Prior to deposition, an ion of interest is mass-selected with a quadrupole filter. Absorption spectra of the resultant 5K neon matrix are measured by a waveguide technique.

fluorescence quantum yields (Maier and Thommen 1984). These data may be of relevance for the DIB discussions.

We have used the mass-selected approach to obtain data for astrophysical purposes for C_{60}^+ and C_{70}^+, and for their anions (Fulara, Jakobi and Maier 1993a, 1993b). The latter neon-matrix, and gas-phase data (Haufler *et al.* 1991), determined for the lowest electronic transition for C_{60} and C_{70}, show that both are shifted around $50cm^{-1}$ to the red in a neon matrix. The absorption features of the ions in the gas-phase can thus be searched for. In fact a possible DIB correlation has been reported for C_{60}^+ (Foing and Ehrenfreund 1994).

Application of the approach has focused on the mass-selection of anions of carbon chains containing dominantly one hydrogen atom, i.e. $C_{2n}H^-$ n=3-8 (Fulara et al. 1993, Freivogel, Fulara and Maier 1994). For technical reasons (i.e. sufficient anion current), the studies have been concerned with the even-number carbon chains. Species are produced in the source by introducing a $1:5$ diacetylene/helium mixture. The anions (currents in the 1-5 nA range) were codeposited with neon (in ratio $\sim 1:10^6$) to form the 5K matrices. Usually concomitant irradiation of the surface during deposition with a high pressure mercury light source was found to enhance the intensity of the absorption signals.

Absorption spectra observed in the 600-1000nm region are shown in Fig.2. This portion of the total measured wavelength region (220 nm to near infrared), shows intense peaks which can be correlated with the number of carbon atoms in the primary ion beam. The displayed region is of particurlar interest for DIB comparisons. When C_6H^- is selected, a previously unknown band system with origin at 608 nm is observed, as well as bands indicating fragment products C_2, C_2^+, C_2^- and C_3. As the size of the anion increases, so does the number of bands for a given mass-selected deposition. It is important to note that all the bands present in an absorption curve do not belong to a common carrier. Fragmentation and isomerization are possibly due to the kinetic energy of the ion beam and the concomitant irradiation during deposition. Additionally, isomers may be present within the ion beam for a given mass. Thus when C_8H^- is deposited, bands also appear in the 500-800 nm region, including the band at 608nm as a result of fragmentation. Similarly, evidence for multiple components is observed for C_nH^-, n=10-16 codepositions. For these larger systems, bands also appear in the 220-300nm region, reminiscent of aromatic molecules. The bands shift to the red with the number of carbon atoms. These are ascribed to cyclic isomers present in the mass-selected beam. It is known from both calculations and experiments that monocyclic rings are formed for the carbon species C_n, $n > 9$ (von Helden *et al.* 1993). Nonetheless, some peaks may be uniquely assigned to a given system size containing a known number of carbon atoms but an unknown number of hydrogen atoms.

Further insight into the nature of carriers responsible for bands observed in the matrix data is provided by theory. A simple but useful formalism to describe the trend in electronic transitions of linear, conjugated systems is given by the free-electron molecular orbital theory (Levine 1983). This determines the eigenvalues for π electrons by applying a one-dimensional particle-in-a-box boundary condition, and predicts that the electronic transition wavelength increases with system size. A series of systems, whose electronic origins as a function of size comprise a unique line, have similar characteristics and are called homologous.

Two homologous series may be constructed from the data on the deposition of $C_{2n}H^-$, n=3 to 8. One series included the bands at 608, 773, and 967nm and another one those at 631, 722, and 802nm. Although the exact hydrogen content for the systems is not known, mass-spectra of the ion beam suggest that the two series probably arise from one of the following linear groups: C_n, C_nH, C_nH^-, or C_n^-. Anions have to be considered only if their electron affinities are around $4eV$ (anions with lower affinities are destroyed by the mercury lamp irradiation, as is the case for C_2^-). This is indeed the case for the C_n^- species (Cheshnovsky, Pettiette and Smalley 1989) and perhaps for some of the C_nH^- ones. However, the C_nH^- species can be excluded on the basis of calculations which show that such species will not have electronic transitions in these regions (Feher and Maier 1994). The relevant transitions are $^3\Sigma - ^3\Sigma$ for C_n, and $^2\Pi - ^2\Pi$ for C_n^- and C_nH (n=even). It should also be pointed out, that with the exception of the bands which are due to the carbon chain anions C_n^-, the bands are persistent after irradiation of the matrix with medium pressure mercury lamp. Thus the C_nH and C_n species produced are photochemically stable, down to wavelengths of $\sim 200nm$.

What is striking in Fig. 2 is the overall spectral pattern, similar in some ways to the DIBs observations (Herbig 1975, 1988). For each electronic band system the origin band is dominant, with little vibrational excitation in the excited state. Furthermore, the oscillator strengths are large; estimations based on the ion current and observed absorbance, lead to values in the 0.01-0.1 range. For linear chains, the oscillator strength will increase with the number of carbon atoms, towards the infrared (Kuhn 1948). This is an appealing characteristic for DIB correlations, because the fall-off in the abundance of such species with size is counterbalanced by the rise in the oscillator strength.

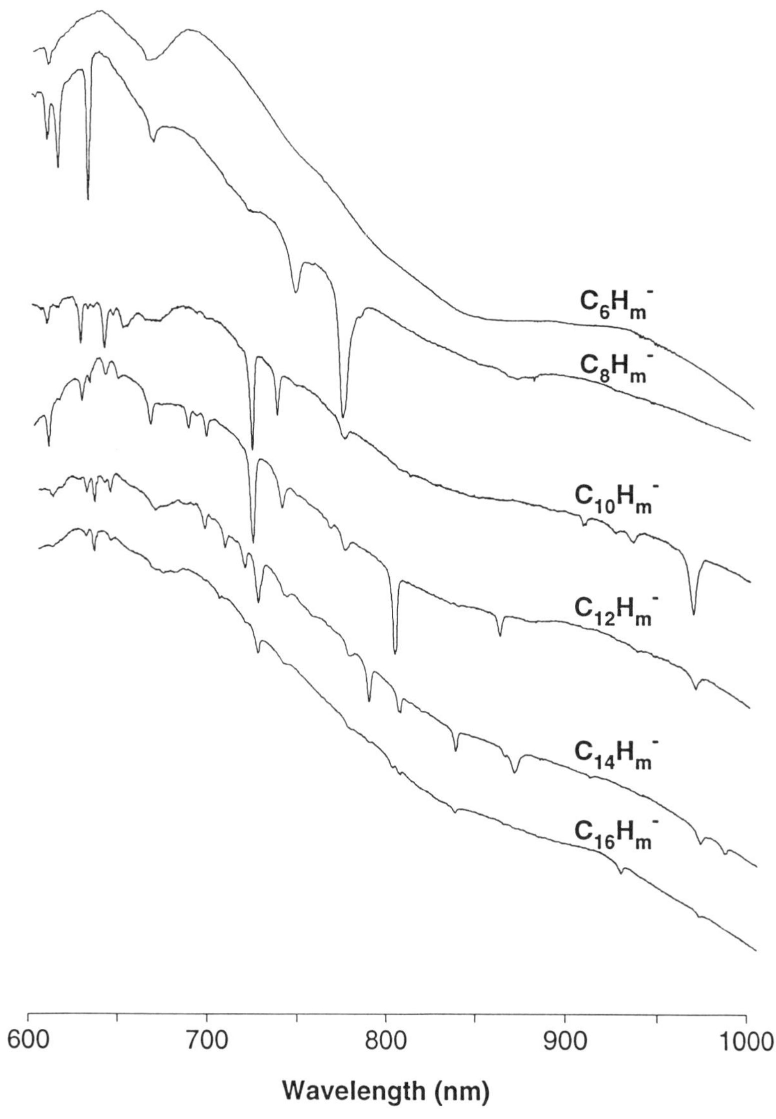

Fig. 2. Absorption spectra in the 600-1000 nm region following the mass-selected matrix-deposition of $C_{2n}H^-$ n=3-8. The spectra, displayed as raw data, have not been corrected to the transmission background which is a consequence of the optical characteristics of the matrix, light source, and detector. Each curve represents a separate experimental trial in which the annotated system was mass-selected for deposition. As the system size increases, fragmentation and isomerization increase the total number of peaks observed in a given curve. Through a process of elimation, unique peaks may be assigned to a given system size.

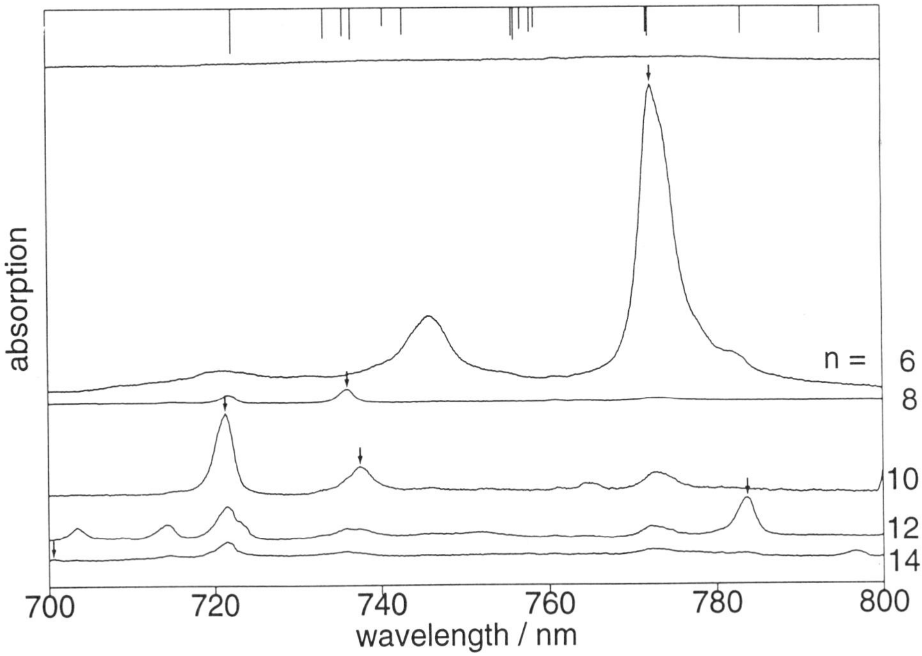

Fig. 3. Absorption spectra in the 700-800 nm region in comparison with DIB positions which are indicated with vertical bars at the top. The notation n= 6 - 16, identifies the primary ion C_nH^- selected in the deposition.

In Figs. 3 and 4 the observed bands are compared with the DIB positions (Herbig and Leka 1991). In the 600-860 nm region, 24 distinct bands are observed, of which 17 lie within $136cm^{-1}$ of the DIBs positions. The latter criterion is chosen somewhat arbitrarily; due to the lack of data on electronic transitions of hydrocarbon radicals, we chose the known gas-matrix shift ($36cm^{-1}$) of the C_3 radical as the reference (Jacox 1988). That this is reasonable is indicated by the comparison of the gas-matrix shift observed for the lowest electronic transitions of C_{60} (at $\sim 600nm$) and C_{70} ($\sim 670nm$), both of which are red shifted in the neon by $\sim 50 - 60cm^{-1}$.

With this as a basis to count the number of "coincidences" between the matrix bands and the DIBs, one can carry out the following statistical evaluation based on the methods used for element identification in stellar observations (Hartoog, Cowley and Cowley 1973). In 10^4 test runs, 17 random wavelengths (i.e. the number of apparent matches, H) were generated in the 600-860 nm range and the number found within $136cm^{-1}$ of the 54 DIBs counted. The histogram of the matches shows a maximum for 12 ($< H >$). Thus a significance parameter $S = H - < H > /\sigma$, of 1.7 is obtained (σ is the standard deviation). It should be pointed out, however, if the number of DIBs observed in this region would be doubled, the 17 matches would be random.

There are two ways to follow up these observations and suggestions. One is for astrophysical measurements to be made based on the following. For some of the origin bands for which a DIB apparently matches, at least one member of a vibrational

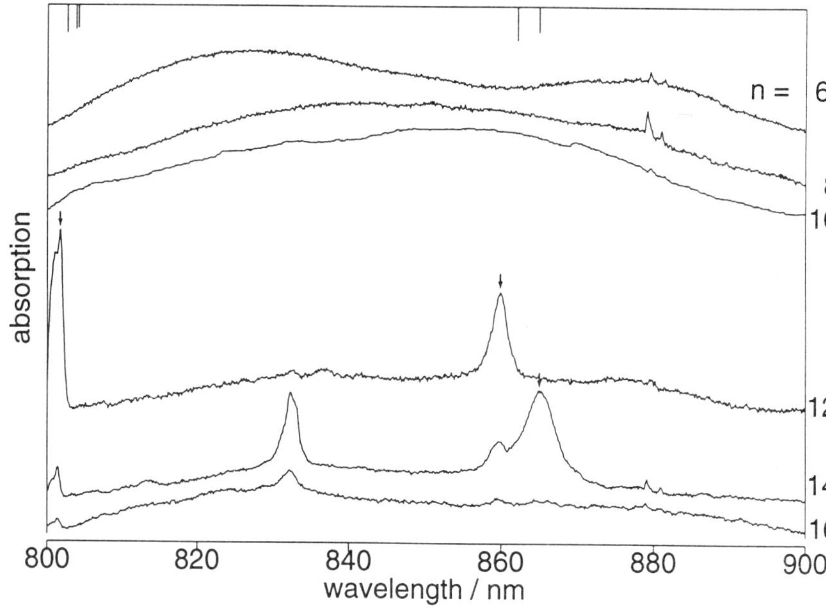

Fig. 4. Absorption spectra in the 800-900 nm region in comparison with DIB positions which are indicated with vertical bars. Several of the matrix bands match closely to those of DIB positions. The number n, identifies the C_nH^- species mass-selected for deposition.

progression in the excited state is discernible in the matrix absorption spectrum. Thus assuming that the correspondence with a DIB is real, one can predict where a further DIB should be, displaced to higher energy by the difference between the two peaks in the matrix (i.e. by the observed vibrational frequency). This reduces the uncertainty due to the change of environment from matrix to gas-phase to less than $110cm^{-1}$, as is known from a wealth of data on vibrational frequencies in neon matrices (Jacox 1988). Furthermore, the ratio of the intensities of such pairs of absorption bands in the matrix should be comparable to the ratio of the equivalent widths of the DIBs. The third criterion for this pair of DIBs, and this is now independent of the matrix data, is that their full widths at half maxima should be comparable if they belong to the same carrier. One such set of bands is located in the matrix absorption spectrum at 631.2 (origin), 573.0 and 557.8 nm, with relative intensities of 1 : 0.6 : 0.7. Another is the pair of matrix bands at 721.9 (origin) and 627.1 nm with an intensity ratio of 1 : 0.24. These would be good candidates for astrophysical studies.

The other direction to be pursued is in the laboratory, and has as goal the observation of the absorption bands in the gas-phase. Though this is undoubtedly a challenging task, the matrix observations do provide the necessary information so that this can be undertaken. Absorption measurements using sophisticated laser methods can be envisaged and hold promise.

ACKNOWLEDGEMENTS. This work has been supported by the Swiss National

Foundation (project no. 20-36153.92).

REFERENCES

Allan, M., Kloster-Jensen, E., and Maier, J.P. 1976, *Chem. Phys.*, **7**, 11.
Bell, M.B., Feldman, P.A., and Mathews, H.E. 1983, *Ap. J.*, **273**, 35.
Cheshnovsky, O., Pettiette, C.L., and Smalley, R.E. 1989, in *Ion and Cluster Ion Spectroscopy and Structure*, ed. J. P. Maier (Amsterdam: Elsevier), p. 373.
Douglas, A.E. 1977, *Nature*, **269**, 130.
Feher, M. and Maier, J.P., unpublished results.
Foing, B.H. and Ehrenfreund, P. 1994, *Nature*, **369**, 296.
Forney, D., Jakobi, M., and Maier, J.P. 1989, *J. Chem. Phys.*, **90**, 600.
Freivogel, P., Fulara, J., and Maier, J.P. 1994, *Ap. J.*, in press.
Freivogel, P., Fulara, J., Lessen, D., Forney, D., and Maier, J.P. 1994, *Chem. Phys.*, in press.
Fulara, J., Jakobi, M., and Maier, J.P. 1993a, *Chem. Phys. Letters*, **211**, 227.
Fulara, J., Jakobi, M., and Maier, J.P. 1993b, *Chem. Phys. Letters*, **206**, 203.
Fulara, J., Lessen, D., Freivogel, P., and Maier, J.P. 1993, *Nature*, **366**, 439.
Haufler, R.E., Chai, Y., Chibante, L.P.F., Fraelich, M.R., Weisman, R.B., Curl, R.F., and Smalley, R.E. 1991, *J. Chem. Phys.*, **95**, 2197.
Hartoog, M.R., Cowley, C.R., and Cowley, A.P. 1973, *Ap. J.*, **182**, 847.
Herbig, G.H. 1975, *Ap. J.*, **196**, 129.
Herbig, G.H. 1988, *Ap. J.*, **331**, 999.
Herbig, G.H. and Leka, K.D. 1991, *Ap. J.*, **382**, 193.
Jacox, M.E. 1988, *J. Phys. Chem. Ref. Data*, **17**, 269.
Kuhn, H. 1948, *Helv. Chim. Acta*, **31**, 1441.
Levine, I.N., Quantum Chemistry (Allyn and Bacon, London, 1983).
Maier, J.P. 1988, *Chem. Soc. Rev.*, **17**, 45.
Maier, J.P. and Thommen, F. 1984, in *Ions and Light*, Vol. 3 of Gas-Phase Ion Chemistry, ed. M. T. Bowers (New York: Academic Press), p. 357.
Maier, J.P. and Rösslein, M. 1988, *J. Chem. Phys.*, **88**, 4614.
Smith, W.H., Snow, T.P., and York, D.G. 1977, *Ap. J.*, **218**, 124.
Thaddeus, P., Gottlieb, C.A., Mollaaghababa, R., and Vrtilek, J.M. 1993, *J. Chem. Soc. Far. Trans.*, **89**, 2125.
von Helden, G., Kemper, P.R., Gotts, N.G., and Bowers, M.T. 1993, *Science*, **259**, 1300.

ABSORPTION SPECTROSCOPY OF NEUTRAL AND IONIZED PAHS. IMPLICATIONS FOR THE DIFFUSE INTERSTELLAR BANDS

F. SALAMA, C. JOBLIN AND L. J. ALLAMANDOLA
NASA-Ames Research Center, Space Science Division, Moffett Field, California 94035

ABSTRACT.

A detailed study of the ultraviolet-to-near infrared absorption spectra of neutral and ionized PAHs is presented. The experimental results obtained using Matrix Isolation Spectroscopy are interpreted in view of the general spectral properties of PAHs derived from theoretical modeling and compared with astronomical observations. From this comparative study, it is shown that: (i) Small neutral PAHs (less than 25 carbon atoms) absorb only in the ultraviolet and cannot contribute to the known diffuse interstellar bands although they may contribute to the extinction curve. (ii) Small ionized PAHs (less than 25 carbon atoms) absorb in the visible and near infrared close to the position of well-known diffuse interstellar bands. (iii) The absorption spectra of compact PAH ions are always clearly dominated by one single band. Moreover, (iv) the classification of the DIBs between strong and weak features can be explained, in the context of the PAH proposal, in terms of a distribution of the band carriers between compact and non-compact PAH ions. Based on all these promising results it is argued that PAH ions constitute strong candidates to explain some (if not all) of the diffuse interstellar bands. Predictions for new diffuse interstellar bands to be searched for are also presented.

1. INTRODUCTION

Polycyclic aromatic hydrocarbons (PAHs) have been postulated as an important, ubiquitous component of the interstellar medium based on their high stability against ultraviolet radiation and the close resemblance of their infrared spectral signature with the interstellar IR emission bands observed between 3.2 and 14.0 μm. For a detailed description of the PAH model, the reader is referred to the abundant literature which has been devoted to this topic (see e.g., Allamandola *et al.* 1989; Puget and Léger 1989). According to the model, PAHs are expected to be present as a mixture of free, neutral and ionized, gas phase molecules following a large size distribution which ranges from small molecules ($<$ 25 Carbon atoms) to large graphitic platelets. Thus, PAHs can be viewed as (a component of) the missing link between the gas phase and the solid phase of interstellar dust.

PAHs, in their neutral and ionized forms, have also been proposed as possible carriers for the diffuse interstellar bands (Van der Zwet and Allamandola 1985; Léger

A. G. G. M. Tielens and T. P. Snow (eds.), The Diffuse Interstellar Bands, 207–217.

and d'Hendecourt 1985; Crawford *et al.* 1985; for a review of the PAH-DIB proposal see Léger in this issue). This proposal, although attractive, was initially hampered by the lack of astrophysically relevant experimental data on PAH ions to compare with observations as well as by the limited amount and quality of the -then available- astronomical data. The situation has since greatly improved thanks to the development of new, more sensitive, detectors allowing higher resolution astronomical observations and the discovery of new, weak, DIBs from the UV down to the NIR (see, e. g., the section "DIBs observational characteristics" in this volume). The situation has also dramatically evolved on the experimental front thanks to the laboratory effort recently undertaken to provide quantitative spectroscopic data on neutral and ionized PAHs which are relevant to address astrophysical issues, i. e., data measured under conditions which come *as close as possible* to the conditions known (or expected) in the interstellar medium (See also in this volume: Canosa *et al.* Leach; Léger and d'Hendecourt. For a chronological review of the laboratory effort devoted to the DIB issue, see Allamandola in this issue).

We describe here the laboratory program initiated to determine the ultraviolet, visible and near infrared spectroscopic characteristics of neutral and ionized PAHs. The experimental work is supported by the modeling of the charge state and spectral properties of PAHs expected in the diffuse interstellar medium (Salama *et al.* 1994a). In the first part of the program, small PAHs (< 25 Carbon atoms) and their derivatives have been studied (Salama and Allamandola 1991, 1992, 1993; Salama *et al.* 1994b). This report summarizes the results obtained so far with this selected set of PAH candidates and the conclusions derived for this subclass of the PAH family. An assessment of the PAH-DIB proposal is then presented and predictions are made for new interstellar absorption bands to be searched for.

2. EXPERIMENTAL

Few experimental data are available on the spectroscopy of charged PAHs and most of the data available are not appropriate for astrophysical applications because they are taken in perturbing media (i. e., glassy or organic matrices; for a review see Shida and Iwata, 1973). The interaction between the trapped PAH ion and its environment induces strong perturbations of the energy levels which result in large shifts of the electronic transitions compared to the *ideal case* of a free, isolated PAH ion.

Matrix isolation spectroscopy (MIS) represents a particularly well suited technique for the measurement of the spectroscopic properties of minimally perturbed PAH ions by allowing to maintain PAHs truly isolated at low temperature (4 - 15 K) in an inert medium. Among the rare gases commonly used as matrix-cage material (Ne, Ar, Kr, and Xe), solid neon provides the best (i. e., less polarizable) medium for the study of quasi-unperturbed electronic spectra of ions. Until molecular jet spectroscopy techniques are developed and adapted to measure the absorption spectra of cold, isolated PAH ions in the gas phase (a difficult task considering the low vapor pressure and the high reactivity of PAHs ions), *neon-matrix isolation spectroscopy remains the best (and only) tool available to laboratory astrophysicists* to simulate the conditions of the diffuse interstellar medium.

A computer-controlled UV-Visible-NIR spectrograph, coupled to a helium-cooled cryogenic cell, is used to measure the absorption spectra of neutral and ionic PAHs in the 180 - 1060 nm range under conditions relevant to astrophysical environments. A detailed description of the experimental apparatus and technique is reported elsewhere (Salama *et al.* 1994b) and only a synopsis will be given here.

Briefly, the experimental apparatus consists of a cryogenic sample chamber which is part of a high vacuum system and consists of four ports at 90° and two gas injection ports at 45°. The cryogenic sample holder suspended at the center of the chamber is cooled down to 4.2 K by a variable temperature, liquid helium transfer cryostat. The substrate (sapphire window) can be positioned to face alternatively the spectroscopy ports, the gas injection ports, an irradiation lamp, or vacuum deposition furnace. The spectral light sources consist of a 160 - 360 nm output deuterium lamp for UV spectroscopy and a 320 - 2500 nm output tungsten filament lamp for visible-NIR spectroscopy. The computer-driven monochromator is equipped with three interchangeable gratings of variable linear dispersion blazed at 300 and 650 nm for UV and visible-NIR spectroscopy respectively. A UV fiber optic cable guides the focused light collected at the spectroscopy port of the cryogenic sample chamber to an optical fiber adaptor mounted on the entrance slit of the monochromator. The photon detector consists of a thermoelectrically cooled charge coupled device (CCD) area array detector sensitive in the spectral range 180 - 1060 nm. The CCD is mounted directly on the exit port of the monochromator and interfaced to the computer system. The ionization source is a microwave-powered, flow discharge hydrogen lamp generating photons of 10.2 eV energy (Lyman α line). The PAH sample is simultaneously condensed with the neon gas onto the cold (4.2 K) substrate. The frozen matrix is then spectroscopically analyzed. Ions are generated *in situ* from the stable precursor, via vacuum-ultraviolet photoionization.

3. RESULTS

3.1. NEUTRAL PAHS

The absorption spectra of *all* small (< 25 Carbon atoms), neutral, PAHs show transitions in the UV-near UV range (180 - 400 nm). The absorption band systems are associated with vibronic transitions between the electronic states of the molecule and can be classified, in a first approximation, into very strong (oscillator strength, f, of the order of 1), moderately strong (f of the order of 0.1) and weak (f of the order of 0.001). A representative spectrum is shown in Fig. 1 for the PAH naphthalene ($C_{10}H_8$) isolated in a neon matrix. Note that *this class of PAHs does not absorb in the visible-near IR range.* [It is also interesting to note that in the case of naphthalene, for example, the largest shift observed for the band positions, compared to the gas phase, is 0.5 nm for the strongest band at 221.6 nm. This corresponds to a 0.25 percent gas phase-to-neon matrix relative shift in energy and can be compared to the 2.5 percent shift observed in the more polarizable argon matrix].

3.2. PAH CATIONS

Vacuum ultraviolet irradiation of the neutral PAHs isolated in neon matrices produces *new spectral features* in the UV-near IR range (180 - 1060 nm). All the new features are found to be associated with the PAH cation (PAH^+) formed by direct one-photon ionization of the neutral precursor. Some representative spectra are shown in Figures 2 - 5 for the PAH cations naphthalene ($C_{10}H_8^+$), phenanthrene ($C_{14}H_{10}^+$), pyrene ($C_{16}H_{10}^+$) and benzo(ghi)perylene ($C_{22}H_{12}^+$).

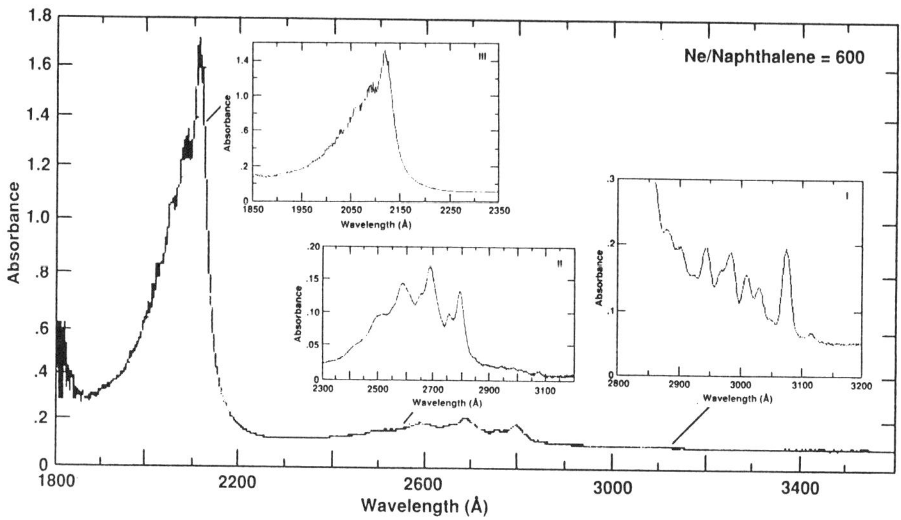

Fig. 1. Absorption spectrum of neutral naphthalene ($C_{10}H_8$) isolated in a neon matrix at 4.2 K. (From Salama and Allamandola 1991).

These spectra show many interesting features: (i) In all cases, a single band dominates the entire spectrum in the near UV-to-near IR range (the spectral range of interest for the DIB issue). (ii) In the case of non-compact PAHs (such as naphthalene and phenanthrene), the strongest ion absorption lies at the low energy end of the spectrum while it is the opposite which is observed for compact PAHs (such as pyrene and benzo(ghi)perylene). (iii) The oscillator strengths associated with the strongest ion transition are of the order of 0.001 - 0.0001 for non-compact PAHs and of the order of 0.1 - 0.01 for compact PAHs. These characteristics are verified for *all small PAH cations studied to date* and are most probably verified for larger PAH cations (see below for a general discussion of the spectral properties of neutral and ionized PAHs).

The effect of the environment on the electronic spectra of charged PAHs is even more drastic here as compared to the case of the neutral molecules. For example, the strongest pyrene cation absorption falls at 439.5 nm in a neon matrix (Salama and Allamandola 1992b) while the same band is shifted to 443.5 nm and 450.0 nm in argon (Salama and Allamandola 1992b) and organic (Shida and Iwata 1973) matrices, respectively. No gas phase data are available for comparison although the shift expected from a neon environment should be small (a fraction of a percent; in the case of the fully substituted fluorobenzene cation (a mono ring molecule) where data are available, the relative shift in energy between the gas phase and neon matrix is of the order of 0.3 percent (Bondybey and Miller 1983)). Thus, it is demonstrated again that neon matrix measurements provide the only experimental data appropriate for astrophysical applications until the absorption spectra of cold, isolated PAH ions can be measured in the gas phase.

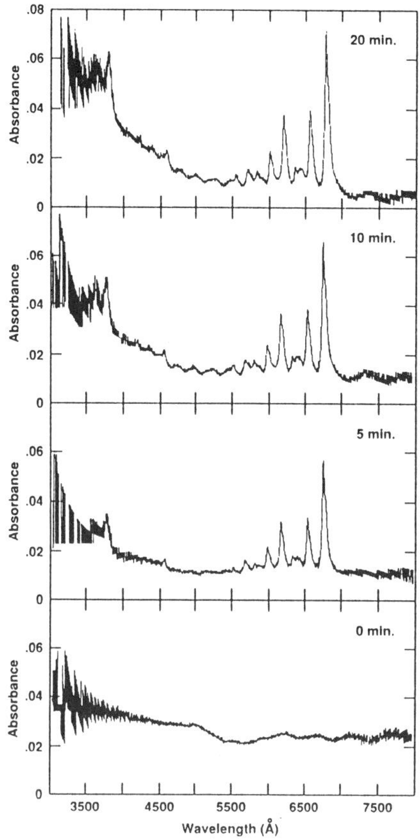

Fig. 2. Growth of the naphthalene cation ($C_{10}H_8^+$) absorption spectrum as a function of irradiation time in a neon matrix at 4.2 K. (From Salama and Allamandola 1992a).

4. DISCUSSION AND ASTROPHYSICAL IMPLICATIONS

4.1. GENERAL PROPERTIES OF NEUTRAL AND SINGLE-IONIZED PAHS

The experimental technique used here (i. e., MIS) to study isolated PAHs is a difficult and lengthy procedure when dealing with such a large family of low vapor pressure, reactive species. To be fully effective, the experimental work needs then to be supported and guided by theoretical considerations which can help reduce this large family of molecules into a series of subclasses having similar spectral properties. This approach should help build a useful database by allowing to focus the experimental studies on the representative elements of each subset.

The classification of the spectral properties of neutral PAHs has been established many years ago (for representative reviews on the subject see, e.g., the pioneering work of Platt 1949, Streitwieser 1961, and Clar 1964). No equivalent classification has been established, however, for single-ionized PAHs except for the preliminary studies of Shida and Iwata (1973) and Khan *et al.* (1993). In an attempt to fill

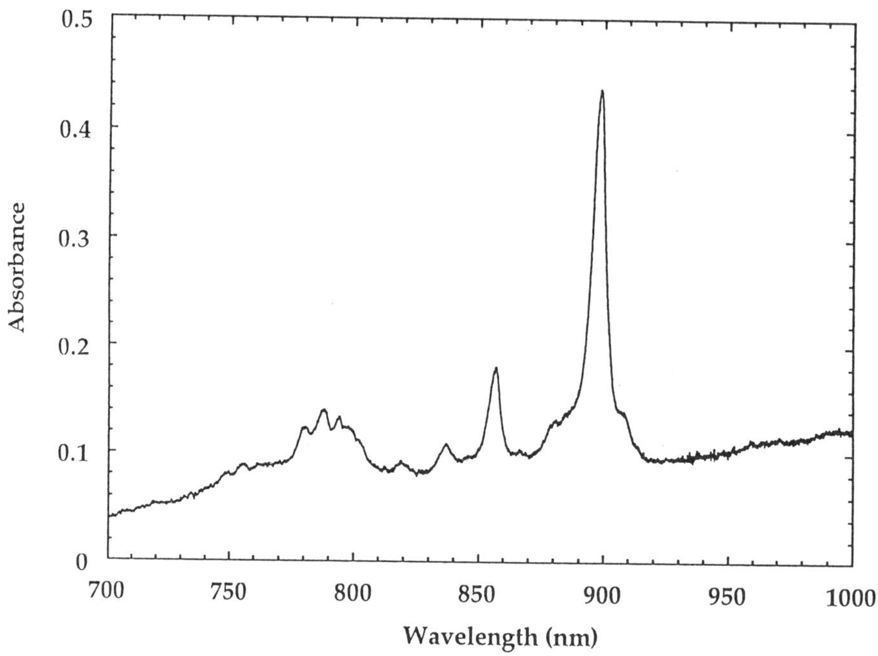

Fig. 3. The visible-near infrared absorption spectrum of the phenanthrene cation ($C_{14}H_{10}^{+}$) isolated in a neon matrix at 4.2 K. (From Salama *et al.* 1994b).

this gap, we have recently completed a detailed study of the UV-visible-near IR properties of neutral and ionized PAHs (see Salama *et al.* 1994a and references therein). The purpose of the study was to extrapolate the general spectral properties of these species towards large size molecules up to the graphitic limit and to derive significant constraints regarding astrophysical implications. The main conclusions, relevant to the present discussion, are: (i) The absorption band energies of neutral and ionized PAHs can be described as a function of the size of the molecular species. (ii) The absorption band energies of neutral PAHs shift towards lower energies (longer wavelengths) when the molecular size increases. (iii) The strongest absorption bands of *compact* PAH ions shift towards lower energies when the molecular size increases. These conclusions are illustrated in Figure 6 which shows the variation of E_p (energy of the absorption onset for neutral PAHs) and E_A (energy of the strongest absorption band for single charged compact PAHs) when the molecular size varies.

4.2. NEUTRAL PAHS

In the case of small neutral PAHs (less than 10 rings, i. e., less than 20 - 30 carbon atoms), the astrophysical implications derived from the combined experimental and theoretical study are the following: (i) *Neutral PAHs absorb only in the ultraviolet* ($\lambda <$ 390 *nm*) *and, hence, cannot contribute to the known diffuse interstellar bands.* This is illustrated by the absorption spectrum shown in Fig. 1 and by the energy-to-molecular size relation described in Fig. 6. (ii) All neutral PAHs exhibit strong UV absorption bands (oscillator strengths in the range 0.2 - 1.0) and could produce,

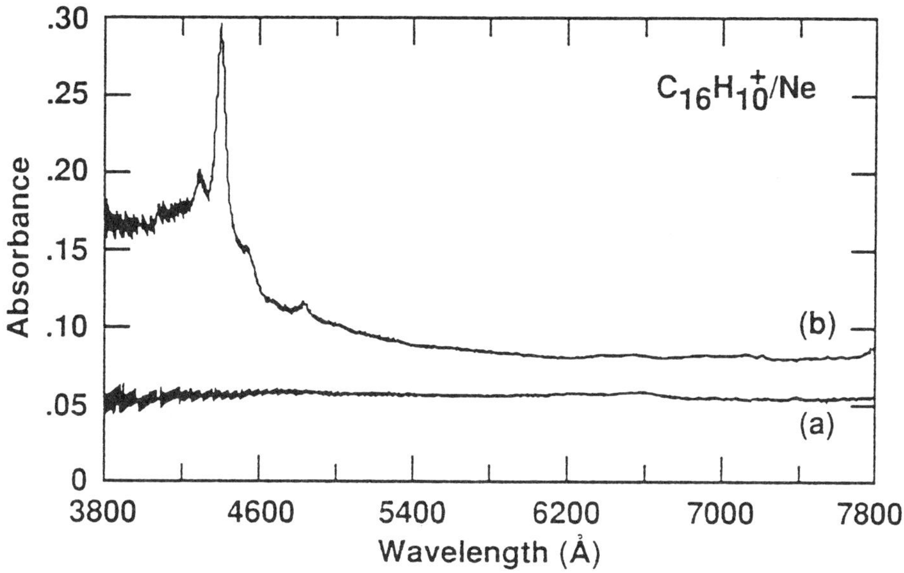

Fig. 4. The visible absorption spectra of (a) neutral and (b) ionized pyrene ($C_{16}H_{10}$) isolated in a neon matrix at 4.2 K. (Adapted from Salama and Allamandola 1992b).

if present, substructure on the UV interstellar extinction curve. A search for the stronger UV features on the interstellar extinction curve would provide a powerful test to the PAH model and would allow to establish significant limits on the abundances of *specific* PAHs.

4.3. SINGLE-IONIZED PAHS

In the case of small PAH ions (less than 10 rings, i. e., less than 20 - 30 carbon atoms), the astrophysical implications derived from the combined experimental and theoretical studies are the following: (i) Contrary to their neutral precursors, *ionized PAHs absorb in the visible and near-infrared* (see Figs. 2 - 5) *and thus could contribute to the diffuse interstellar bands.* Moreover, the energy-to-molecular size relation (Fig. 6) shows that large, compact, PAH ions (larger than about 100 rings, i. e., 250 - 300 carbon atoms) are not expected, if present, to significantly contribute to the DIBs. (ii) A comparison of the stronger PAH ion transitions, measured in neon matrices, with the known DIBs is shown in Table 1. This comparison indicates a *surprisingly good agreement in wavelength positions between the two sets* (i.e., a relative shift in energy of 0.3 percent (at the most) which corresponds to the relative shift expected between neon matrix measurements and gas phase data; see the experimental section above). (iii) Another striking result when looking to Table 1 is that *all compact PAH ion absorptions come close to strong and broad DIBs while all non-compact PAH absorptions come close to weak and narrow DIBs.* This effect could offer a plausible explanation to the recent observation that DIBs can be grouped into "families" according to the degree of correlation of their strengths and widths (Krełowski and Walker 1987; see also Jenniskens and Désert, in this volume).

There are two ways in which one can examine the possible contribution of any

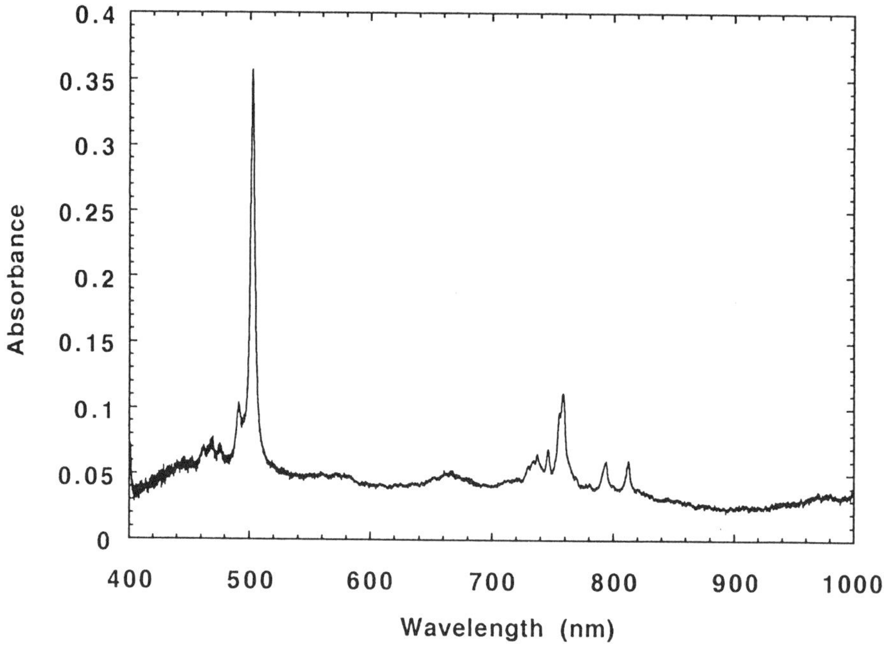

Fig. 5. Absorption spectrum of the benzo(ghi)perylene cation ($C_{22}H_{12}^{+}$) isolated in a neon matrix at 4.2 K.

given PAH ion (or any other candidate) to the DIBs. One extreme is to search for all of the ion bands, regardless of their intensity, among the DIBs. Another approach is to only require that the strongest ion band(s) be present. The former requires a higher fractional abundance among the PAHs than the latter. In *all* PAH ion spectra measured to date it is clear, however, that *one single band dominates the entire absorption spectrum in the visible-near IR range* (see section 3 and Figs. 2 - 5). Thus, we approach the problem from the point of view of "one PAH, one DIB" (i. e., by requiring that, in a first step, only the strongest absorption band of each potential PAH ion candidate be present, or searched for, in the DIB spectrum). This empirical approach is further supported by the following considerations: (i) PAHs form an extremely rich chemical family (particularly when including derivatives such as methyl and ethyl PAHs which have been recently proposed (Salama and Allamandola 1992b; 1993) and shown (Léger 1995; Léger *et al.* 1995) to be also very good candidates). (ii) A distribution of PAHs is expected in the interstellar medium. Although some arguments may be used to reduce the large, original, number of potential candidates (e. g., by a study of their spectral properties as a function of the molecular size (see the discussion above), by examining their stability as a function of the molecular structure: compact PAHs being more stable than linear or non-compact PAHs, they would be expected to survive longer in the ISM) one is still faced with very large subsets of molecules to consider. (iii) It is difficult to conceive why and how only very few, specific, PAHs would dominate any given size distribution. If we require that most of the vibronic bands of any given PAH ion must be found in the DIB spectrum (i.e., each PAH be responsible for at least 10 DIBs), this implies that a total population of 15 - 20 individual PAH ions would be sufficient to explain all the

TABLE 1

Comparison of PAH cation (PAH^+) stronger absorption bands with known DIBs

PAH^+	λ_{peak} (nm) (Neon matrix)	DIBs (nm)[a]
Pyrene ($C_{16}H_{10}^+$)	439.5* (443.0 in Ar)	442.9
1-Methylpyrene ($CH_3 - C_{16}H_9^+$)[b]	444.2*	442.9
4-Methylpyrene ($CH_3 - C_{16}H_9^+$)	(457.7)	
	482.8	482.4
	757.6	758.1
Naphthalene ($C_{10}H_8^+$)	674.2*	674.1
	652.0	652.0
Phenanthrene ($C_{14}H_{10}^+$)	898.3*	
	856.8	857.2
Tetracene ($C_{18}H_{12}^+$)	864.7*	864.8
Benzo(ghi)perylene ($C_{22}H_{12}^+$)	502.2*	503.9 (?)
	758.4	758.1; 758.6
	755.2	755.8 (?); 756.2
	794.3	793.5 (prob.)
Coronene ($C_{24}H_{12}^+$)[c]	459.0*	459.5
	946.5	946.6

*Strongest absorption band.
[a]Herbig 1975; Sanner *et al.* 1978; Herbig and Leka 1991; Jenniskens and Désert 1994; Ehrenfreund and Foing 1994.
[b]See also Léger and d'Hendecourt 1995.
[c]See also Ehrenfreund *et al.* 1992.

known DIBs. It seems more realistic, however, to expect to find one (the strongest) or two bands belonging to any given PAH candidate in the DIB spectrum. This argument is particularly valid, here, because PAH ion spectra are indeed clearly dominated by only one (or two) bands. In the case of compact PAH ions, the strongest absorption band has a large oscillator strength (of the order of 0.1) and tends to be broad (falling in the high energy range of the spectrum). In the case of non-compact PAH ions, the strongest absorption band has a small oscillator strength (of the order of 0.001) and tends to be narrower (falling in the low energy side of the spectrum). Finally, using the oscillator strengths measured in the laboratory, abundances of the order of 0.2 - 0.3 percent of the cosmic carbon are calculated for PAH ions such as naphthalene and pyrene if these ions are indeed responsible for DIB features.

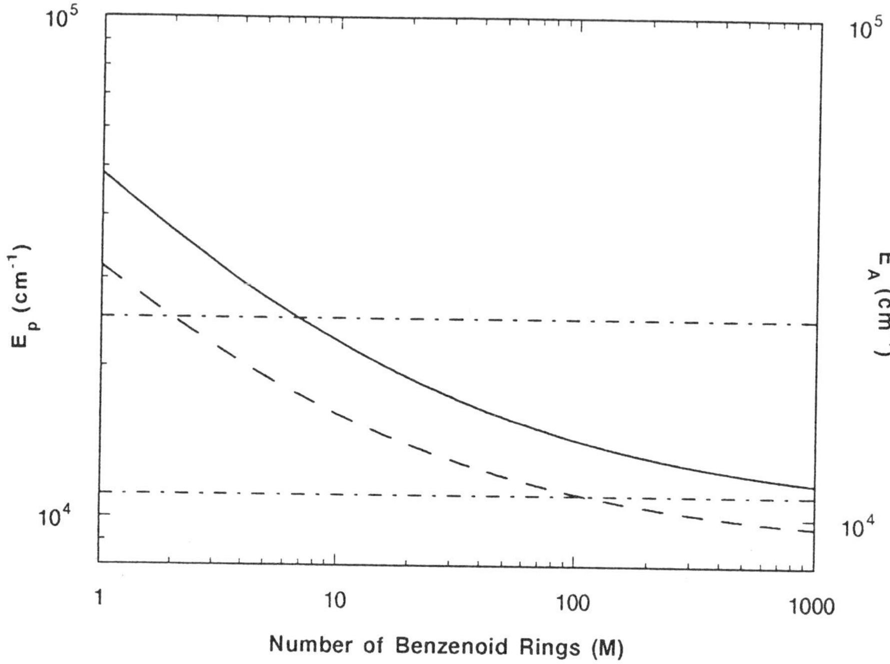

Fig. 6. PAH absorption band energy as a function of the molecular size. Top curve: E_p (Neutrals); bottom curve: E_A (Ions). The two horizontal lines define the energy range inside which DIBs are observed. (From Salama *et al.* 1994a).

5. CONCLUSION

A detailed study of the ultraviolet-to-near infrared spectroscopy of neutral and ionized PAHs has been performed. Combining the experimental results obtained for these species trapped and *isolated* in a cold (4 K), inert environment (neon) with the results of a study of their general spectral properties, the following conclusions have been reached:
(i) Matrix isolation spectroscopy is an appropriate tool to simulate the interstellar medium environment provided the low polarizability, inert gas, *neon* is chosen as the matrix material. This technique stands as *the only alternative* available to measure the absorption spectra of cold, isolated, PAH ions until reliable, adequate, gas phase experiments (such as jet spectroscopy) are developed.
(ii) Neutral and ionized PAHs absorb strongly in the UV. A search for their spectral signatures would provide a test to the PAH model and would allow to set significant limits for the abundances of specific PAHs.
(iii) Small ionized PAHs (less than 25 carbon atoms) appear to be very good candidates (if not the best candidates) to explain (some of) the diffuse interstellar bands. This comes from the combination of many factors: the recognition that DIBs are most probably due to a family of closely related, gas phase, molecular species; the high abundance of PAHs derived from interstellar infrared emission spectra; the high stability of PAHs against UV photons; the fact that their ion absorption spectra are always dominated by *one single band* which falls close to a known DIB, and the ability of the PAH model to explain the distribution of DIBs between strong and weaker

features as a direct consequence of the spectral properties related to the compact and non-compact PAH subclasses.

In conclusion, it is clear that an enormous progress has been made in just the last few (2 - 3) years in terms of the quality of the laboratory and observational data available to address the complex issue of the identity of the DIB carriers. We may, thus, be close to a better (?) understanding of the material present in the diffuse ISM. Such progress has been made possible because of all the research work preceding it and it is hoped that this study will simulate, at its turn, further experimental and observational research.

ACKNOWLEDGEMENTS. This project is supported by NASA Ultraviolet, Visible, and Gravitational Astrophysics Research and Analysis Program (Grant 188-41-57-41), NASA Long-Term Space Astrophysics Research Program (Grant 399-20-00-05) and NASA Infrared, Submillimeter, and Radio Astronomy Research and Analysis Program (Grant 188-44-57-01).

References

Allamandola, L.J. 1995, *this volume.*
Allamandola, L.J., Tielens, A.G.G.M., and Barker, J.R. 1989, *Ap. J. Suppl.*, **71**, 733.
Boissel, P. 1995, *this volume.*
Bondybey, V.E. and Miller, T.A. 1983, in *Molecular Ions: Spectroscopy, Structure and Chemistry*, eds. T. A. Miller and V. E. Bondybey, (North Holland Publishing Company), p. 138.
Canosa, A., Pasquerault, D., Parent, D., Rebrion, C., Gomet, J.C., and Rowe, B.R. 1995, *this volume.*
Clar, E. 1964, in *Polycyclic Hydrocarbons*, Academic Press: New York.
Crawford, M. K., Tielens, A. G. G. M., and Allamandola, L. J. 1985, *Ap. J.*, **293**, L45.
Ehrenfreund, P., d'Hendecourt, L., Verstraete, L., Léger, A., Schmidt, W., and Defourneau, D. 1992, *Astr. Ap.*, **259**, 257.
Ehrenfreund, P. and Foing, B.H. 1994, (private communication).
Herbig, G.H. 1975, *Ap. J.*, **196**, 129.
Herbig, G.H. and Leka, K.D. 1991, *Ap. J.*, **382**, 193.
Jenniskens, P., and Désert, X. 1994, *Astr. Ap. Suppl. Ser.*, **106**, 39.
Jenniskens, P., and Désert, X. 1995, *this volume.*
Khan, Z.H., Husain, M.M., and Haselbach, E. 1993, *Appl. Spectrosc.*, **47**, 2140.
Krełowski, J. and Walker G.A.H. 1987, *Ap. J.*, **312**, 860.
Leach, S. 1995, *this volume.*
Léger, A. 1995, *this volume.*
Léger, A. and d' Hendecourt, L.B. 1985, *Astr. Ap.*, **146**, 81.
Léger, A., d' Hendecourt, L., and défourneau, D. 1995, *Astr. Ap.*, **293**, L53.
Platt, J.R. 1949, *J. Chem. Phys.*, **17**, 484.
Puget, J.L. and Léger, A. 1989, *Ann. Rev. Astr. Ap.*, **27**, 161.
Salama, F. and Allamandola, L.J. 1991, *J. Chem. Phys.*, **94**, 6964.
Salama, F. and Allamandola, L.J. 1992a, *Ap. J.*, **395**, 301.
Salama, F. and Allamandola, L.J. 1992b, *Nature*, **358**, 42.
Salama, F. and Allamandola, L.J. 1993, *J. Chem. Soc. Faraday Trans.*, **89**, 2277.
Salama, F., Bakes, E., Allamandola, L.J., and Tielens, A.G.G.M. 1994a, *Ap. J.*, submitted.
Salama, F., Joblin, C., and Allamandola, L.J. 1994b, *J. Chem. Phys.*, in press.
Sanner, F., Snell, R., and Vanden Bout, P. 1978, *Ap. J.*, **226**, 460.
Shida, T. and Iwata, S. 1973, *J. Am. Chem. Soc.*, **95**, 3473.
Streitwieser, A. 1961, in *Molecular Orbital Theory for Organic Chemists*, Wiley, New York.
Van der Zwet, G.P. and Allamandola, L.J. 1985, *Astr. Ap.*, **146**, 76.

A CRITICAL ASSESSMENT OF THE PAH HYPOTHESIS: REVISITED

BERTRAM DONN
Howard University and the University of Virginia

Following the proposal that PAH's may be the source of the unidentified infrared emission bands, extensive laboratory spectroscopy has been carried out. These measurements have been generally consistent with that hypothesis, and interstellar PAH's have been suggested to explain the far ultraviolet extinction and the diffuse interstellar bands. However, there are several discrepancies between experiments and observations in addition to inconsistencies in the hypothesis that rule out a tendency to accept the PAH concept at this time.

The infrared emission, both continuum and bands, are highly attributed to high vibrational excitation of a molecule caused by absorption of an ultraviolet photon. It has not been shown that the resulting continuum could resemble the observed, approximate 1000K black body spectrum. The infrared bands have been compared to the PAH absorption spectrum, not an infrared fluorescent spectrum. Only the 3.3 μm band from naphthalene, excited by 193nm laser radiation, has been seen in fluorescence. Depending upon the source, a size distribution of PAH's between about 10-40 carbon atoms is required for the strong bands and several hundred carbons needed for the underlying continua (Allamandola *et al.* 1989). A definitive comparison needs to be made using the appropriate arrays and for a range of stellar radiation capable of being absorbed by each molecule.

Almost no attention has been given to the ultra-visible fluorescence although neutral PAH's are efficient fluorescent species (Felker and Zewail 1984; Amirov *et al.* 1988) with quantum yields the order of unity for measurements on isolated molecules in supersonic jets. In the diffuse interstellar medium and the outer parts of clouds, calculations indicate that a substantial fraction of PAH's will be neutral (Verstraete *et al.* 1990; Bakes and Tielens, this volume). Therefore, ultraviolet-visible fluorescence or broad absorption features are expected. It was pointed out by Donn and Krishna Swamy (1969) and Joblin *et al.* (1992) that a sufficiently broad distribution of PAH's might obscure the well defined ultraviolet bands.

However, this raised another problem. The mixture of PAH ions and derivatives would add to the complexity of the infrared bands and likely destroy the near agreement found for a simple distribution of neutral species (Szczepanski and Vala 1993). It is emphasized, again, that because of the large number of possible PAH species, strong conclusions cannot be drawn by the match of a single molecule with some observed visible or infrared features as was done in the case of the naphthalene cation (Salama and Allamandola 1992a). A similar objection applies to the suggestion that a pyrene-like ion is the cause of the ubiquitous 4428 Å band (Salama and Allamandola 1992b).

A. G. G. M. Tielens and T. P. Snow (eds.), The Diffuse Interstellar Bands, 219–220.

PAH cations have also been proposed as the source of the long unidentified diffuse interstellar bands. A difficulty that arises with such an identification are the large number of DIBs. At the Conference, Mathis pointed out that, on the average, there is one every 20 Å. The combination of a large number of DIBs and a large number of lines in the visible from an array of ions, very accurate laboratory wavelengths are required for a meaningful identification. At present, laboratory wavelengths, almost always, have been obtained in solid matrices which differ by tens of Angstroms from the gas phase values. Therefore, only a suggestive comparison can be made. In addition, a significant criterion that must be met is to account for the division of DIBs into families which vary together in relative intensity (Krełowski 1988). There are three well-recognized families which would require at least three sets of distinct sources. The same source cannot easily explain more than one family.

An argument given to support the PAH hypothesis is the expected stability of these species in the interstellar medium. Photolysis experiments on pyrene and coronene by Stief *et al.* (1968, unpublished) found that both of these molecules apparently were photodissociated by the 2537 Å Hg line at 400° C. Thermal dissociation occurred only above 800° C. As highly excited vibrational excitation is an important step in the emission process, it is important to investigate the photodissociation phenomena more thoroughly.

A kinetic mechanism for the formation of PAH's in circumstellar envelopes has been proposed by Frenklach and Feigelson (1989). Their mechanism consists of over 100 reactions including several three-body processes based on soot production in hydrocarbon flames. The process has been modified to take into account the 8-10 orders of magnitude lower pressures in space. This correction does not allow for the vibrational disequilibrium (Nuth and Donn 1981) that occurs at very low pressures and can have a significant effect on processes and rates.

Alternative explanations have been presented to account for the astrophysical phenomena attributed to PAH's and it is necessary to compare and critically examine these interpretations before accepting any one, including the PAH hypothesis. It is premature to claim that PAH's are present and indeed ubiquitous in interstellar space. Such an approach will have a chilling effect on research in this important area.

*Abstract of invited paper presented at the DIB Conference. The full report will be published elsewhere, at a later time.

REFERENCES

Allamandola, L.J., Tielens, A.G.G.M., and Barker, J.R. 1989, *Ap. J. Suppl.*, **71**, 733.
Amirov, A., Horwitz, C., and Jortner, J. 1988, *J. Chem. Phys.*, **88**, 3092.
Bakes, E.L.O. and Tielens, A.G.G.M., 1995, in *The Diffuse Interstellar Bands*, eds. A.G.G.M. Tielens and T.P. Snow (Dordrecht:Kluwer), this volume.
Donn, B. and Swamy, K. 1969, *Physica*, **41**, 13.
Joblin, C. *et al.* 1992, *Ap. J. Lett.*, **393**, L79.
Felker, M. and Zewail, T. 1984, *Chem. Phys. Lett.*, **108**, 3031.
Frenklach, M. and Feigelson, E.D. 1989, *Ap. J.*, **341**, 372.
Krełowski, J. 1988, *PASP*, **100**, 896.
Nuth, J.A. and Donn, B. 1981, *Ap. J.*, **247**, 925.
Salama, F. and Allamandola, L.J. 1992a.
Salama, F. and Allamandola, L.J. 1992b, *Nature*, **358**, 42.
Szczepanski, J. and Vala, M. 1993, *Nature*, **363**, 699.
Verstraete, L. *et al.* 1990, *Astron. Astrophys.*, **237**, 436.

PHOTOFRAGMENTATION OF ISOLATED PAH CATIONS

P. BOISSEL
Laboratoire de Photophysique Moléculaire, Orsay, France

ABSTRACT.

Photofragmentation of isolated polycyclic aromatic hydrocarbons (PAH) cations has been studied, using an ion trap and irradiation by visible light. This experiment provides indications about the selection among PAHs that can result from differences in photostability. In some cases, the fragment remaining after dissociation seems to be more resistant than the parent ion itself. In the interstellar medium, such fragments might then be more abundant than PAH cations, and should be considered in the search for DIB carriers. From a spectroscopic point of view, the exact line positions cannot be measured precisely, but results concerning the regions of absorption and the oscillator strength are obtained. These results are complementary to those given by rare gas matrix experiments.

1. INTRODUCTION

Molecular ions, specifically PAH cations, have been proposed as possible carriers of diffuse interstellar bands (Léger and d'Hendecourt 1985, van der Zwet and Allamandola 1985). Up to now, spectroscopic data concerning these species have only been obtained in condensed phase. In the latest experiments, isolation of ions in low temperature rare gas matrices has been widely used. This technique gives valuable results concerning the positions and the relative intensities of the absorption lines. However, apart from the problem of the exact influence of the environment on the lines position, ambiguities sometimes remain concerning even the exact nature of species present in the matrix. The presence of negative ions, the formation of fragments during the irradiation, or the ionisation of dimer or polymer species cannot generally be simply ruled out. Moreover, the mechanisms of ions formation in the matrix are not completely understood and uncertainties in the evaluation of the amount of ions produced leads to discrepancies in the determination of oscillator strength of the transitions. Gas phase studies are then useful to check these points.

An ion trap has been built in our laboratory, with the general aim to study medium sized ions of astrophysical interest under conditions approaching those encountered in the interstellar medium (ISM). From that point of view, the key point is that, in the trap, collisions with residual neutral species are negligible. The ions are then placed in an environment in which energy exchanges can only take place through emission or absorption of radiation. These energy exchanges, with typical time constants of the order of one second, can be followed in real time. Concerning the spectroscopy, the number of ions that can be trapped is much too low to allow direct absorption measurements, but information can be drawn from photofragmentation experiments.

A. G. G. M. Tielens and T. P. Snow (eds.), The Diffuse Interstellar Bands, 221–229.

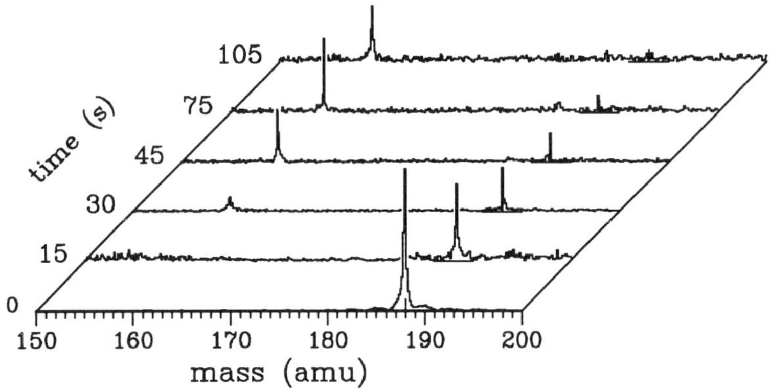

Fig. 1. Photofragmentation of perdeuterated anthracene cation: $C_{14}D_{10}^{+} -> C_{12}D_{8}^{+} + C_{2}D_{2}$.

2. EXPERIMENTAL

The ion trap is a modified version of an ionic cyclotronic resonance (ICR) cell. Ions are trapped transversely in a magnetic field. A static quadrupolar electric field prevents them from escaping along the axis. The only difference with classical ICR cells is the replacement of the four transverse plates by rods, which allows larger optical access. At any time, the content of the trap can be analysed by Fourier transform mass spectroscopy (FT-MS) with impulse excitation (McIver *et al.* 1989). The whole device, enclosed in a vacuum cell, is cooled down to temperatures between 12 and 30 K by a closed cycle helium cryostat to ensure an efficient cryopumping. In this way, trapping times longer than 10 minutes are obtained.

PAH cations are produced by laser impact on a solid pellet, located on the trap axis, behind one end plate. The neutral species evaporated by the laser shot are immediately condensed on the low temperature walls, while the ions enter the trap, guided by the magnetic field. In a large range of laser pulse intensity, the number of ions that are trapped is limited by the space charge. This saturation effect leads to reproducible initial conditions from shot to shot. Irradiation of the ions is performed using a continuous xenon arc lamp, focused on the ion cloud with fused silica lenses. In the wavelength range of interest (280-800 nm), the emission spectrum is practically that of a blackbody at 5500K. Coloured filters can be introduced to select a given spectral domain.

The result of a typical run on perdeuterated anthracene is presented on Figure 1. After the laser shot, a dead time of a few seconds is left, to ensure the decay of the initial internal energy of the ions, consecutive to their creation. An initial mass spectrum is taken (time t=0 on Figure 1) and the irradiation is switched on. Then, the evolution of the fragmentation process is followed by successive mass spectra, taken at regular time intervals. Unavoidable losses of the trap are accounted for through a normalisation of the peak heights to those of the parent ion, taken at the same instants, during runs without irradiation. As can be observed, the peak at M=188amu, corresponding to the parent ion decreases with time, while a peak at M=160amu gradually appears. A fragmentation, corresponding to the loss of a deuterated acetylene molecule, has then occurred. Another peak at M=184amu, corresponding to the loss of two D atoms can be also seen, at a much lower level.

The important point to notice, on this example, is that this fragmentation is obtained with a RG610 coloured filter, limiting the irradiation to wavelength longer than 610nm. A comparison with the absorption spectrum of anthracene cation in argon matrix (Sczepanski *et al.* 1993) show that, in this case, only the first electronic transition, with lines at 720 and 660 nm can be excited. The line positions for the deuterated ion are probably not very different. We can conclude that photons of 1.73 and 1.88eV can only be involved in the process. This energy being lower than the energy needed to break the molecular bonds, a multiphoton process has to be invoked to explain the fragmentation.

3. MECHANISM OF THE FRAGMENTATION

The multiple photon absorption can be understood from a molecular point of view. After the absorption of the first photon, a competition takes place between reemission of the energy (fluorescence) and redistribution of this energy among the vibrational degrees of freedom of the electronic ground state (internal conversion). When internal conversion occurs, the photon energy is stored in the ion, and can only be evacuated radiatively via infrared vibrational fluorescence. This process is very inefficient and the resulting cooling is slow. Then, the same electronic transition being again available, the absorption of a second photon is possible before the energy of the first one has been completely dissipated. As the vibrational energy content increases, the internal conversion becomes faster and the energy is stored more efficiently. The neat result is a gradual increase of the ion internal energy, up to the point where the dissociation rate becomes comparable to the cooling rate.

The key point of this mechanism is then a competition between heating by photon absorption and radiative cooling. A numerical simulation of the experiment has been achieved, taking into account all the elementary processes:

1. absorption of photons on each vibronic transition. The rate is assumed to be independent of the energy content.

2. infrared vibrational emission, calculated as a function of the internal energy, using a microcanonical treatment (Barker 1984), including the whole set of vibrational transitions. Standard values of the oscillator strength for neutral PAHs are used (Joblin *et al.* 1994), but an adjustable scaling factor γ is used to check the consistency of these values in the case of an ion.

3. electronic fluorescence (Poincaré fluorescence), a process that has been predicted (Léger *et al.* 1988) in the case of highly isolated species. It is due to the possibility of equilibrium between internal conversion and inverse internal conversion. In the present case, it becomes the dominant cooling path at internal energies larger than 6 eV.

4. fragmentation. Two different paths, acetylene molecule loss and hydrogen atom loss, are taken into account, using the inverse Laplace transform (ILT) formalism (Forst 1972). For the case of the perdeuterated anthracene cation, the relevant parameters are deduced from experimental results on naphthalene (Rühl *et al.* 1989).

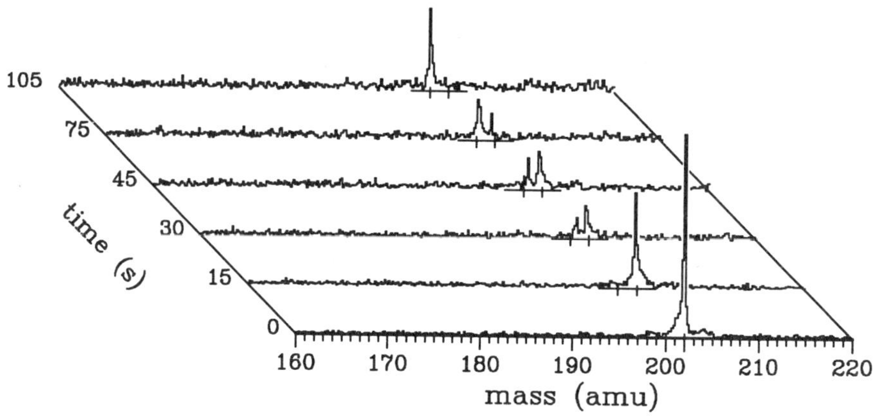

Fig. 2. Photofragmentation of the pyrene cation: $C_{16}H_{10}^{+} -> C_{16}H_8^{+} + 2H$.

4. RESULTS

4.1. FRAGMENTATION PATHS

As already mentioned, for perdeuterated anthracene, besides the fragment $C_{12}D_8^+$, corresponding to the ejection of a C_2D_2 molecule, a fragment $C_{14}D_8^+$ is also observed, which indicates the loss of two D atoms. In the case of normal anthracene, this second fragmentation path occurs with a higher branching ratio, but the acetylene loss is still dominant. A similar result was found for the naphthalene cation in another ion trap experiment (Kim and Dunbar 1980). In this case, the hydrogen loss is even completely absent.

An opposite result is found with the pyrene radical cation ($C_{16}H_{10}^+$). As can be seen on Figure 2, no fragment is found at M=176amu ($C_{14}H_8^+$), and the only species remaining after 90 seconds of irradiation, $C_{16}H_8^+$, corresponds to the loss of two hydrogen atoms. Owing to imperfections in our experiment, it is not easy to distinguish unambiguously if the two atoms come out sequentially or simultaneously in an H_2 molecule. However, a peak at M=201amu ($C_{14}H_9^+$) seems to be present in the intermediate mass spectra at 30 and 45 seconds.

These results can be compared with those of experiments using hard UV photons, probing simultaneously the ionisation and the fragmentation of the ion, in the microsecond timescale (Jochims *et al.* 1994). In these experiments, the appearance potentials (AP) of the fragmentation paths leading to the ejection of H, H_2 and C_2H_2 have been determined for different PAH cations. The AP is defined as the energy above which the fragmentation rate becomes higher than a certain threshold, $10^4 s^{-1}$ in that work.

Starting with pyrene, the fragment $C_{14}H_8^+$ $(-C_2H_2)$ exhibit an AP 1.3 eV higher than $C_{16}H_9^+$ $(-H)$. This is coherent with the absence of the former one in our ICR experiment. Moreover, the AP of $C_{16}H_8^+$ $(-H_2)$ is still higher (+2.1 eV), giving a confirmation that, in our case, this fragment is produced by the sequential ejection of two hydrogen atoms.

The case of naphthalene and anthracene is not so simple. The AP of the $(-C_2H_2)$ and $(-H)$ paths appears to be nearly the same, but the former one is slightly higher (+0.1 eV for naphthalene and +0.2 eV for anthrecene). ICR trap experiment seems to lead to contradictory results, since the $(-H)$ reaction is observed only marginally for anthracene and not detected for naphthalene. However, this apparent discrepancy can be removed by considering the difference in timescale. In the trap, the appearance of fragments is not limited by the observation time, but only by the competition with radiative cooling. The critical rate to consider for the AP is then the cooling rate i.e. $1s^{-1}$ (see section 4.2), rather than 10^4s^{-1}. The comparison of the results obtained by the two methods indicate that the slope of the energy dependence of the fragmentation rate is larger for the $(-H)$ reaction than for the $(-C_2H_2)$ one.

Another advantage of the ICR technique is that the charged species produced during a fragmentation process remain in the trap. Starting from a given parent ion, it is then possible to follow not only the first step of photofragmentation, but also the whole cascade, leading eventually to the total destruction.

In the case of the pyrene cation, for a longer irradiation, the fragment $C_{16}H_8^+$, obtained in the first step looses again two hydrogen atoms to give $C_{16}H_6^+$. The appearance of this species in the mass spectra is a confirmation that stable highly dehydrogenated PAHs ions do exist.

With the anthracene cation, a remarkable result is obtained. With an irradiation extending to the whole visible range ($\lambda > 330$ nm), the parent ion is dissociated in a few seconds, while the fragment corresponding to the loss of an acetylene molecule, exposed to the same irradiation, can be observed for several minutes, and decay only with the natural trap losses. This increased resistance may be due to two different reasons: i: absorption lines in the visible being less intense, the internal energy distribution is shifted downwards. ii: the energy barrier for the lowest dissociation channel is shifted upwards. Further studies are needed to check which of this two reason is dominant. This is an important point for the astrophysical problem. Actually, if the fragment exhibit a higher dissociation barrier, it will be also more resistant under UV flux in the interstellar medium.

4.2. COMPARISON WITH SIMULATION

Quantitative comparisons between experimental results and numerical simulation have been done in the case of the perdeuterated anthracene cation. The measurements are performed on the dominant fragmentation channel, the loss of an acetylene molecule, but both channels are considered in the calculation. Due to a non-linearity in our device, the peak heights in the mass spectra are not exactly representative of the number of ions present in our system. The fragmentation efficiency is then monitored by the inverse of the time at which the peaks of the parent and the fragment ions are the same (crossing time t_c).

With this method, we have checked the evolution of this effective fragmentation rate, $1/t_c$, versus the irradiation intensity, using the RG610 filter and glass plates as neutral attenuators. A power law dependence is found, with an exponent $\alpha = 3.7$. The same dependence comes out of the simulation, but with a slightly lower exponent $\alpha = 3.2$. It is to be noticed that the photons involved have an energy $h\nu$ around 1.8 eV. The product $\alpha h\nu$ is then in the order of 6.5 eV, which is just the energy range where the fragmentation rate becomes comparable to the radiative cooling rate. The exponent in the power law can then be considered as an indication of the number of photons needed to reach the fragmentation threshold.

A second comparison concerns the cooling rate. Using a chopper in front of the lamp, it is possible to perform the fragmentation experiment with an irradiation alternatively set on and off. For an incident photon flux ϕ, if the alternation is fast enough, the mean effective rate is simply that corresponding to the flux $\phi/2$ and is independent of the chopper period. But, when this period becomes comparable to the inverse of the cooling rate, a dissymmetry appears, owing to the power law dependence, and the mean value of the rate increase. This effect is experimentally observed for chopper period larger than 0.4 seconds. It is correctly reproduced by the simulation for a value of the scaling parameter γ comprised between 1 and 2. It can then be concluded that the cooling of an isolated ion is almost the same as that of the corresponding neutral species. In the energy range of interest, between 2 and 6 eV, the calculated decay rate of the energy varies from $0.5s^{-1}$ to $1.5s^{-1}$, this last value including the effect of electronic emission. The same order of magnitude of the cooling rate has been found by Dunbar *et al.* (1987) in a similar experiment on the naphthalene cation.

A critical input parameter in the numerical simulation is the photon absorption rate. A value of $1.5s^{-1}$ gives a result in agreement with the experiment described in Figure 1. The absorption is limited in this case to the two lines at 660 and 720nm attributed to the first allowed electronic transition by Sczepanski *et al.* (1993). A measurement of the luminous flux at these wavelength allows then a direct determination of the oscillator strength of this transition. Unfortunately, the inhomogeneity of the irradiation limits the precision and an interval of values ranging from 0.004 to 0.090 can only be obtained at present. This interval is in agreement with the value of 0.081 obtained by Sczepanski *et al.* (1993) for the same transition and confirms that the value 0.58, calculated using quantum chemistry method (Parisel *et al.* 1992) is too large.

4.3. SPECTROSCOPIC INFORMATION

The intensity of the xenon lamp used is not sufficient to allow a spectrally resolved study with the help of a monochromator. However, a crude survey can be obtained by the use of a set of long wavelength pass coloured filters. The xenon lamp has strong emission lines in the near infrared but we have checked that they have no influence on the fragmentation. Then, with a given filter, the spectrum of the radiation impinging on the ion cloud is that of a blackbody at a temperature around $5500K$, limited on the blue side by the cut-off wavelength λ_c. When λ_c is increased, the fragmentation efficiency should decrease each time an absorption region is left behind. This behaviour is apparent on Figure 3, where the fragmentation efficiency of the pyrene cation is plotted as a function of λ_c. A calibration of the intensity dependence of $1/t_c$ at $\lambda_c = 420nm$ allows the representation of the efficiency in terms of absorbed intensity. A strong decrease is indeed observed between 400 and 480 nm, indicating the presence of the absorption band.

A quantitative comparison can be made with the absorption spectrum measured in neon matrix (Salama and Allamandola 1992, 1993). Knowing this spectrum, the intensity absorbed by the ion can be calculated for each λ_c. The result, normalised at $\lambda_c = 420nm$, is reported as a dashed line on Figure 3.

On the red side of the line, the difference observed can be understood, remem-

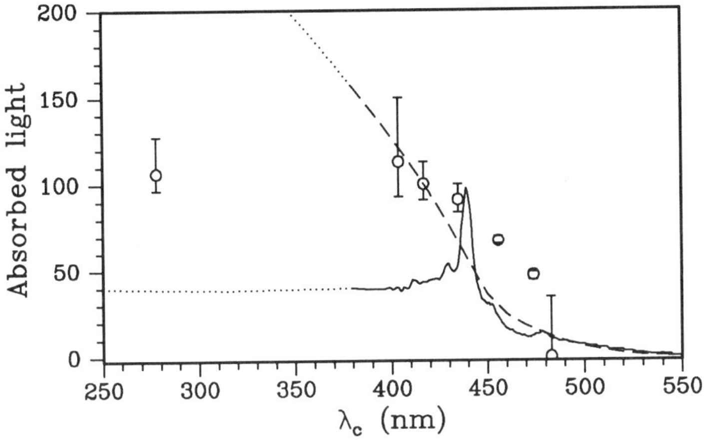

Fig. 3. Light absorption by the pyrene cation. circles: deduced from the fragmentation efficiency; dashed line: deduced from the absorption spectrum; full line: absorption spectrum in neon matrix (Salama and Allamandola 1992); dotted line: extrapolation with an UV continuum.

bering that, in our process, the internal energy of the ions increases up to 6 or 7 eV. The absorption to consider is that of an ion with non zero internal energy. The observation of a long wavelength wing is then probably due to the appearance of hot bands.

The discrepancy observed on the blue side is more important. In our measurement, in the limit of the experimental uncertainties, the fragmentation efficiency does not vary when λ_c is shifted from 280 to 400nm. This region should then exhibit no absorption. The published matrix spectrum is not drawn below 380nm, however Salama and Allamandola (1993) observed "a continuum which extend from 200 to 360 nanometers, having a weak tail out to 510 nanometers". As can be seen on Figure 3, with a reasonable estimate for this continuum (dotted line), the amount of light absorbed with a cut-off at 280 nm should be at least 3 times higher. The broad absorption feature observed in the matrix experiment is then probably not due to the pyrene cation. A possibility, which has not been considered, would be the ionisation of pre-existing pyrene dimers or polymers. This would also give a feature growing at the same rate as the main line.

5. ASTROPHYSICAL IMPLICATIONS

Concerning the photostability, the results of the present experiment cannot be directly transferred to the astrophysical problem. Although the absence of collisions mimics quite well an important characteristic of the ISM, the radiation flux involved are different. In the ISM, multiple photon processes are practically excluded and fragmentation occurs only under the influence of UV photons. Nevertheless, the present results can give some clues concerning the abundances in two ways.

First, one has to remember that ICR experiments probe the fragmentation process at its very threshold, in the energy range where it is only in competition with radiative cooling. This energy range scales up with the size of the PAH, whereas the UV flux, in many regions, is limited to 13.6eV. Then the general trends that we observe are transferable to the smallest PAHs that can survive in these regions. In this sense, the difference observed between anthracene and pyrene is coherent with the absence of lines characteristic of non compact PAHs in the IR emission spectra (Léger *et al.* 1989). Pyrene, a compact PAH, looses only hydrogen atoms that can be recovered from the surrounding medium, whereas anthracene, a non compact one, looses two carbon atoms in the acetylene molecule.

However, species smaller than the critical limit have been considered in the search for DIB carriers. Naphthalene and pyrene cations (Salama and Allamandola 1993), methyl pyrene (Léger *et al.* 1994), have been proposed, among many other ones. These hypothesis are conceivable if such species are located in regions where the UV flux is screened, or if their production is fast enough to maintain a sufficient population, despite their destruction. In this case, large differences in abundance can result from differences in photostability. Then, if the increased resistance of some of the fragments observed in our experiment is due to a higher dissociation barrier rather than to a lower absorption, these fragments might be better candidates than the parents ions. A particular attention should be paid to $C_{12}H_8^+$.(anthracene$-2H$). This fragment seems particularly stable. Its structure, probably naphtyl-acetylene ($C_{10}H_7 - C_2H^+$), is very similar to that of naphthalene, which exhibit several coincidences with DIBs. Another good candidate could be $C_{16}H_8^+$ (pyrene$-2H$), owing to the results obtained with pyrene and methyl-pyrene. More generally, concerning the dehydrogenation, fragments having lost an odd number of hydrogen atoms appears to be less stable and could be forgotten.

6. CONCLUSION

The main interest of this experiment is the possibility to follow the succession of photofragmentations undergone by a given parent ion. The results obtained can be used as guidelines in the search for new DIB carrier candidates. Fragments derived from anthracene and pyrene have to be considered.

The first photofragmentation has been used as a crude spectroscopic tool to check some results obtained in rare gas matrix experiments. An estimation of the oscillator strength of the first allowed electronic transition of the anthracene cation leads to an interval between 0.004 and 0.090, in agreement with the measurement of 0.081 by Sczepanski *et al.* (1993). Concerning the absorption regions, the position of the main lines are correctly found, but the attribution of a continuum to the pyrene cation is questioned.

ACKNOWLEDGEMENTS. This work has received financial support from the CNRS Groupe de Recherche “Physicochimie des molecules et grains interstellaires”.

REFERENCES

Barker, J.R. 1984, *J. Phys. Chem.*, **88**, 11.
Dunbar, R.C., Chen, J.Y., So, H.Y., and Asamoto, B. 1987, *J. Chem. Phys.*, **86**, 2081.
Forst, W. 1972, *J. Phys. Chem.*, **76**, 342.
Joblin, C., d'Hendecourt, L., Léger, A., and Défourneau, D. 1994, *Astr. Ap.*, **281**, 923.

Jochims, H.W., Rühl, E., Baumgartel, H., Tobita, S., and Leach, S., 1994, *Ap. J.*, **420**, 307.
Kim, S.M. and Dunbar, R.C. 1980, *J. Chem. Phys.*, **72**, 4405.
Léger, A. and d'Hendecourt, L. 1985, *Astr. Ap.*, **146**, 81.
Léger, A., Boissel, P., and d'Hendecourt, L. 1988, *Phys. Rev. Letters*, **60**, 921.
Léger, A., d'Hendecourt, L., and Défourneau, D. 1989, *Astr. Ap.*, **216**, 148.
Léger, A., d'Hendecourt, L., and Défourneau, D. 1994, to be published.
McIver, R.T. Jr., Hunter, R.L., and Baykut, G. 1989, *Rev. Sci. Instr.*, **60**, 400.
Parisel, O., Berthier, G., and Ellinger, Y. 1992, *Astr. Ap.*, **266**, L1.
Rühl, E., Price, S.D., and Leach, S. 1989, *J. Phys. Chem.*, **93**, 6312.
Salama, F. and Allamandola, L.J. 1992, *Nature*, **358**, 42.
Salama, F. and Allamandola, L.J. 1993, *J. Chem. Soc. Faraday Trans.*, **89**, 2277.
Szczepanski, J., Vala, M., Talbi, D., Parisel, O., and Ellinger, Y. 1993, *J. Chem. Phys.*, **98**, 4494.
van der Zwet, G.P. and Allamandola, L.J. 1985, *Astr. Ap.*, **146**, 76.

EXPERIMENTAL STUDY OF NAPHTHALENE AND ANTHRACENE REACTIONS

A. CANOSA, D. PASQUERAULT, D.C. PARENT*, C. REBRION, J.C. GOMET, AND B.R. ROWE
Département de Physique Atomique et Moléculaire, URA 1203 du CNRS, Université de Rennes I, Campus de Beaulieu, 35042 Rennes cedex FRANCE

ABSTRACT. The Flowing Afterglow Langmuir Probe Mass Spectrometer (FALPMS hereafter) technique has been applied to the experimental study of dissociative recombination of benzene and naphthalene cations. Measurements of the rate coefficients α have been carried out at room temperature and were found to be respectively $\alpha(C_6H_6^+) = 1 \times 10^{-6}$ cm^{3}s^{-1} and $\alpha(C_{10}H_8^+) = 3 \times 10^{-7}$ cm^{3}s^{-1}. The apparatus has also been used to determine the rate coefficients of reactions of anthracene with Ar^+, He^+ and C^+. The obtained values are found to be close to the Langevin rate. At last, the FALPMS instrument has been used to measure the rate coefficient β for the electron attachment to anthracene. A value of 1×10^{-9} cm^{3}s^{-1} has been obtained suggesting that an activation barrier could exist for this process.

1. INTRODUCTION

Polycyclic aromatic hydrocarbon (PAH) molecules and ions are thought to be present in interstellar and circumstellar environments. It is more particularly believed that they are responsible for the diffuse infrared emission bands (Léger and Puget 1984) and there are also some arguments that suggest they could be carriers of the visible diffuse interstellar bands (Van der Zwet and Allamandola 1985). A recent study led to the suggestion that small PAHs such as naphthalene or anthracene are mostly responsible for the infrared emission (Duley and Jones 1990).

Astrochemical models are now attempting to reproduce the chemistry of PAH compounds (Pineau des Forêts *et al.* 1988). Clearly, the neutralisation of PAH ions by electron recombination appears to be one of the significant processes that must be included in the models. Very few data (Abouelaziz *et al.* 1993) are presently available on this topic as well as in the field of ion-PAH reactions in general (Bohme 1992). In addition, the attachment of electrons to large molecules is believed to be efficient (Lepp and Dalgarno 1988), thus measurements of this process are also needed.

In our laboratory, several experiments have been performed at room temperature concerning reactions of naphthalene and anthracene (either ions or neutrals) using

*permanent address: Laboratoire de Physico-Chimie des Rayonnements, Université de Paris-Sud, URA 75 du CNRS, Bâtiment 350, 91405 Orsay Cedex France.

A. G. G. M. Tielens and T. P. Snow (eds.), The Diffuse Interstellar Bands, 231–238.

a flowing afterglow apparatus. In the next section the apparatus will be described briefly. Then sections 3, 4 and 5 will be dedicated to the presentation of the results we have obtained in the fields of dissociative recombination, ion-PAH reactions and electron attachment respectively. Perspectives will be presented in the last section.

2. APPARATUS

The flowing afterglow apparatus that is available in our laboratory is associated to a Langmuir probe and a mass spectrometer which can both be moved along the flow axis. As it has been widely described elsewhere (Rowe *et al.* 1992), only a brief presentation of the main features will be made in this contribution. A sketch of the apparatus can be seen in figure 1. A plasma is produced by ionising a helium buffer gas by means of a microwave discharge. This initial plasma is constituted of He^+ and He_2^+ ions, metastable helium He^M, electrons ($\approx 10^{10}$ cm^{-3}) and neutral helium. Several entry ports enable us to introduce different neutral gases into the flow, leading to the conversion of the helium plasma into a plasma containing the desired ions. The evolution of their density along the flow axis is measured with a movable mass spectrometer and the electron density is also determined with a movable Langmuir probe. Rate coefficients can then be obtained by analysing these data. One of the major problems in these experiments involving PAHs is the generally low vapor pressure of these species at room temperature. They were then introduced into a cylindrical oven which can be heated up to 600 K by means of a heating wire connected to a temperature regulation device. A helium buffer was also introduced into the oven to transport the vaporised PAH towards the 8 needle entry port G3 where the helium-PAH mixture merged with the plasma. To avoid condensation of the PAH, the stainless steel line connecting the oven to the needle entry port and the under-vacuum-torus that supports the 8 stainless steel needles were also heated. Two temperature regulation devices allowed us to maintain these to a higher temperature than it is inside the oven. The 8 needles themselves were electrically heated. By using such a technique, it has become possible to introduce large flow rates of PAH vapor (several tens of scc) into the plasma.

3. DISSOCIATIVE RECOMBINATION OF NAPHTHALENE

In this experiment, xenon was introduced into the plasma flow through the entry port G2. This led to the conversion of the helium plasma into a plasma where Xe^+ became the dominant ion at the PAH inlet position. Introducing naphthalene $C_{10}H_8$ resulted in a fast charge transfer reaction between Xe^+ and $C_{10}H_8$. This led to the production of the desired $C_{10}H_8^+$ ion which then recombined with electrons. Then, in the zone of measurements, the plasma was only constituted of $C_{10}H_8^+$, electrons and neutral helium. Neglecting diffusion and the $C_{10}H_8^+$ production term, the evolution of the $C_{10}H_8^+$ density can then be given by:

$$v\frac{d[C_{10}H_8^+]}{dz} = -\alpha[C_{10}H_8^+]n_e - k[C_{10}H_8^+][I] \tag{1}$$

where v is the flow velocity (measured by pulse modulating the microwave discharge), z the position along the flow axis, $[C_{10}H_8^+]$ the ion density, n_e the electron density, and α the dissociative recombination rate constant. The second term of the right

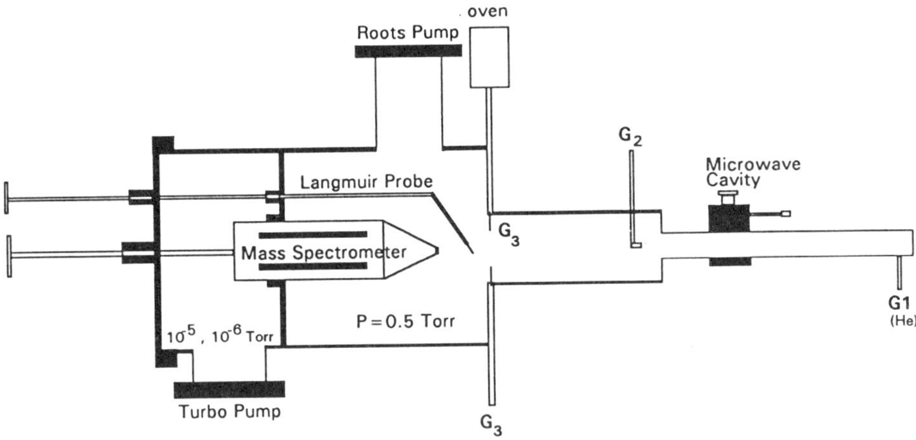

Fig. 1. Sketch of the FALPMS apparatus at the University of Rennes.

member corresponds to a reaction (at a rate coefficient k) between $C_{10}H_8^+$ and a possible impurity. Integrating equation 1 within an origine z_0 and the position z where measurements are carried out leads to the following equation for the $C_{10}H_8^+$ density at position z:

$$\ln \frac{[C_{10}H_8^+]_z}{[C_{10}H_8^+]_{z0}} = -\frac{\alpha}{v} \int_{z0}^{z} n_e dz - k\frac{[I]}{v}(z - z_0) \qquad (2)$$

If the density of impurities [I] is very low or/and if k is very small, then α can be determined by plotting the left member of equation 2 as a function of $\int_{z_0}^{z} n_e dz/v$ for different z values along the flow. This should yield a straight line plot with slope α. An alternative method of obtaining α consists in measuring the ion density ratio at a fixed position z while the initial electron density is varied by moving the microwave cavity or/and changing the microwave power. This is particularly necessary when impurities are not negligible. Indeed, in this second approach, the second term of the right member of equation 2 is a constant. In practice, both methods are used and good agreement has generally been found between the values measured in each case. Results for naphthalene ions are presented in figure 2.

One of the main interests of this method lies on the possibility of measuring several α at the same time when various terminal ions are produced. This was the case in the study that we carried out on benzene ions (Abouelaziz *et al.* 1993). Introduction of benzene in the plasma (dominated by C^+ ions in this case) led to the production of four different ions for which the rate coefficient α has been determined. These results are presented in table 1 as well as the α value that we obtained in a separate experiment for toluene ions $C_7H_8^+$.

We are now attempting to measure the rate coefficient α for the anthracene ion $C_{14}H_{10}^+$. In the frame of this work, negative anthracene ions have been detected with the mass spectrometer. Their presence was resulting from the electron attachment process:

$$C_{14}H_{10} + e \longrightarrow C_{14}H_{10}^-$$

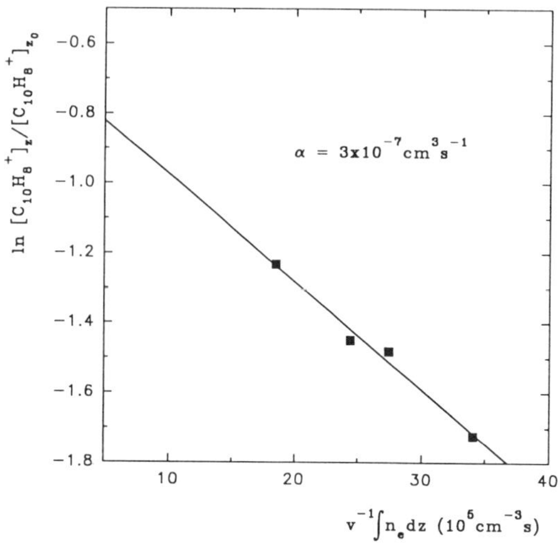

Fig. 2. Plot of decay of $C_{10}H_8^+$ at fixed distance as a function of integrated electron density n_e varied by changes in the discharge power and cavity position.

TABLE 1

Dissociative recombination rate coefficients at 300 K

Ion	Rate coefficient (cm^3 s^{-1})
$C_3H_3^+$	7×10^{-7}
$C_5H_3^+$	9×10^{-7}
$C_6H_6^+$	10×10^{-7}
$C_7H_5^+$	7×10^{-7}
$C_7H_8^+$	7×10^{-7}
$C_{10}H_8^+$	3×10^{-7}

for which we have measured the rate coefficient β_{ant} at room temperature.

4. ELECTRON ATTACHMENT TO ANTHRACENE

The rate coefficient β_{ant} for the electron attachment to anthracene was obtained by introducing in the plasma flow a mixture composed of sulfur hexafluoride SF_6 and anthracene. In order to avoid ion-ion recombination, the electron density was lowered to about 4×10^8 cm^{-3} and the SF_6 density was chosen to be around 10^{10} cm^{-3}. With such conditions, the destruction of negative ions was inhibited. It was indeed observed that the negative ion signals were increasing regularly and eventually attained a plateau when the mass spectrometer was moved to the downstream side of the flow. The evolution of the negative ion densities was then:

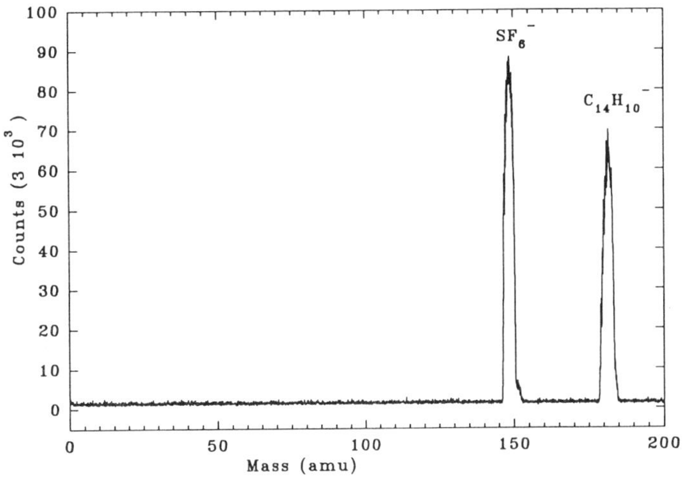

Fig. 3. A typical negative ion mass spectrum obtained when introducing SF_6 and $C_{14}H_{10}$ in the plasma flow.

$$v\frac{d[M^-]}{dz} = \beta[M]n_e \tag{3}$$

where M is either SF_6 or $C_{14}H_{10}$. Integrating over position z yields:

$$[M^-] = \beta[M]\int n_e \frac{dz}{v} \tag{4}$$

so that the ratio of the two ion signals can be written as:

$$\frac{[C_{14}H_{10}^-]}{[SF_6^-]} = -\frac{\beta_{ant}}{\beta_{SF_6}}\frac{[C_{14}H_{10}]}{[SF_6]} \tag{5}$$

The relative intensities of the two ions were measured with the mass spectrometer in the plateau zone. Figure 3 illustrates a typical mass spectrum that we obtained for given flow rate conditions.

The flow rates were precisely determined thanks to two flow controllers. One can then see that a plot $[C_{14}H_{10}^-]/[SF_6^-]$ *vs.*$[C_{14}H_{10}]/[SF_6]$ should be a straight line whose slope is equal to β_{ant}/β_{SF_6} (see figure 4). Provided that β_{SF_6} is known, it is then possible to determine the rate coefficient β_{ant}.

Although the rate coefficient for the electron attachment to sulfur hexafluoride has been measured by a variety of laboratories and techniques we have measured its value again. To do so, The electron density in the plasma was increased to about 10^9 cm^{-3}. SF_6 was introduced in the plasma at a very small flow rate. The electron density was then measured as a function of the SF_6 flow rate for two different positions of the Langmuir probe. It can then be shown that the rate coefficient β is yielded by

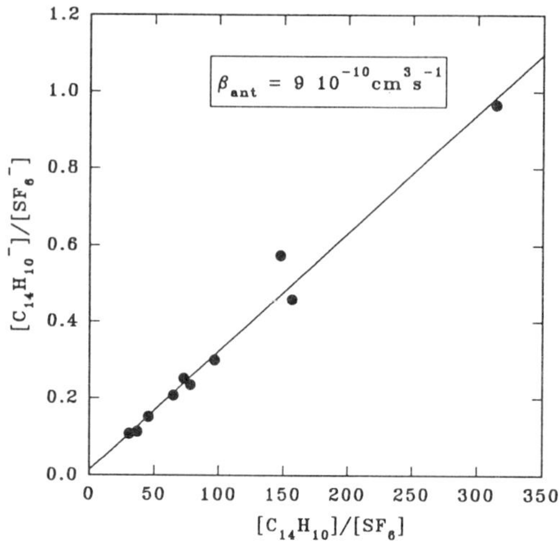

Fig. 4. Plot of $[C_{14}H_{10}^-]/[SF_6^-]$ versus $[C_{14}H_{10}]/[SF_6]$. The slope is equal to β_{ant}/β_{SF_6}.

$\beta = \Delta slope/\Delta t$ where Δt is the transit time of the flow to reach position 2 from position 1 and Δslope is the difference in slopes of the electron density decays at these two positions. By using such a method we have measured a rate coefficient of 2.8×10^{-7} cm^3s^{-1} for the electron attachment to SF_6 in very good agreement with previous studies (Smith *et al.* 1984).

Using this value and taking into account that the uncertainty on β_{SF_6} is about 15%, and that the dispersion of our measurements is also about 15%, we calculated a rate coefficient of: $\beta_{ant} = 9 \times 10^{-10}$ cm^3s^{-1}. Note that this rate coefficient has also been measured using the method we have just described for the determination of β_{SF_6}. A good agreement was found with the previous value of β_{ant} as we obtained a rate coefficient of 12×10^{-10} cm^3s^{-1} with an uncertainty of 30%. Further details concerning this study can be found in Canosa *et al.* 1994.

These results are in very good agreement with the rate coefficient that was deduced from cross section experiments by Tobita *et al.* 1992. Their estimation was indeed 9 $\times 10^{-10}$ cm^3s^{-1} at room temperature.

5. REACTIONS OF AR^+, HE^+ AND C^+ WITH ANTHRACENE

In this experiment, the desired ions (Ar^+, He^+ and C^+) were produced by introducing in the flow either argon or carbon monoxyde through the G2 entry port. As argon does not react with He^+, a mixture of Ar^+ and He^+ ions was produced in the first case whereas C^+ was the only present ion in the second case. Using the mass spectrometer and the Langmuir probe it was possible to measure the atomic ion density decrease as a function of distance for a given anthracene flow rate. This flow rate could be changed either by heating (or cooling) the oven or by changing the oven pressure by closing (or opening) a little bit the valve that insulates the oven from the experiment.

If M^+ is the atomic ion, the decrease of M^+ with position z is given by:

$$v\frac{d[M^+]}{dz} = -k[M^+][C_{14}H_{10}] \quad (6)$$

where k is the rate coefficient of the reaction of M^+ with anthracene. Figure 5 shows a typical decrease of the atomic carbon ion density. The slope of such decreases was then plotted as a function of the anthracene density leading to the determination of the rate coefficient k. Figure 6 presents the complete compilation of our study.

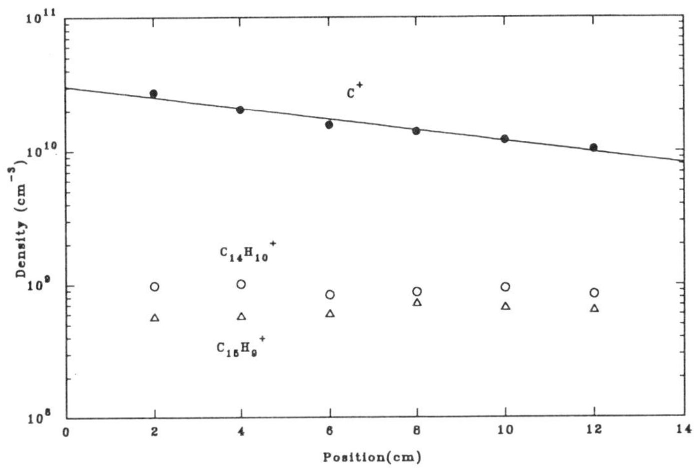

Fig. 5. The decay of C^+ ion density as a function of position for a given anthracene flow rate.

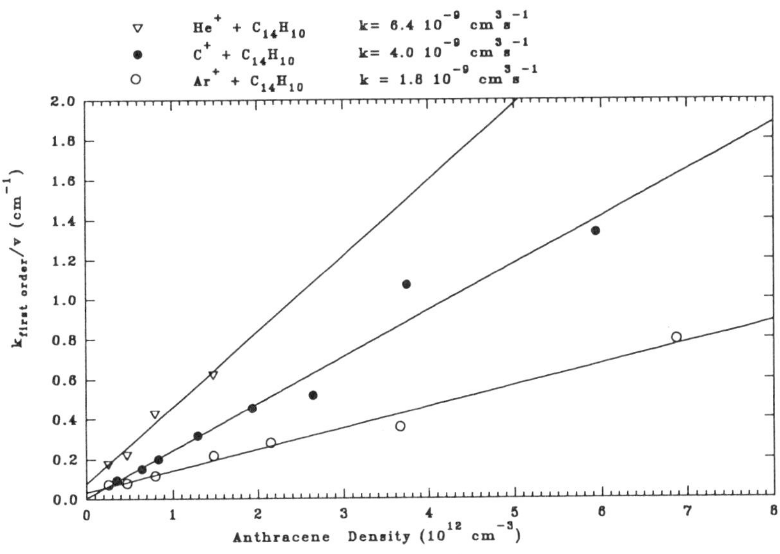

Fig. 6. First order decay constants (divided by flow velocity) plotted versus anthracene density.

These measurements have been carried out with high electron densities (typically 10^{10} cm^{-3}) so that the obtained rate constants are slightly overestimated due to the electron attachment process. It is however interesting to observe that the values for the rate coefficients are very close to the Langevin rate given by:

$$k_L = 2\pi q\sqrt{\frac{\alpha}{\mu}} \quad (7)$$

where μ is the reduced mass of anthracene and the reacting ion, q the ion charge and α the polarizability of anthracene. With $\alpha = 25$ Å^3 we have calculated Langevin rate coefficients of 5.9×10^{-9} cm^{3}s^{-1} 2.0×10^{-9} cm^{3}s^{-1} and 3.5×10^{-9} cm^{3}s^{-1} for He^+, Ar^+ and C^+ respectively.

Due to the high electron density in the flow no attempt was made to obtain the branching ratios of the product ions. We can however qualitatively say that in the case of C^+ two products have been detected, namely: $C_{14}H_{10}^+$ and $C_{15}H_9^+$. The anthracene ion seemed to be slightly dominant. In the case of He^+, though $C_{14}H_{10}^+$ was the dominant ion, it is worth of noting that doubly charged anthracene was also observed and several other fragments in lower quantities.

6. CONCLUSION AND PERSPECTIVES

The present experimental results are of great interest for astrochemistry. It is particularly worth noting that ion-anthracene reactions proceeds at a Langevin rate. Further experiments will be carried out in the future to see if this can be generalised to bigger PAHs. An effort will also be done to try to determine the exact nature of the product ions. The very low electron attachment rate coefficient is also of certain interest as it is usually believed that this process is quite efficient for PAHs (Lepp and Dalgarno 1988). We plan experiments at higher temperatures to see if an activation barrier can be observed in the electron attachment process to anthracene. Other PAHs, such as pyrene or phenanthrene are also planned to be studied. Dissociative recombination studies will also be undertaken for non attaching PAHs.

References

Abouelaziz, H., Gomet, J.C., Pasquerault, D., Rowe, B.R., and Mitchell, J.B.A. 1993, in *J. Chem. Phys.*, **99**, 237.

Bohme, D. 1992, in *Chem. Rev.*, **92**, 1487.

Canosa, A., Parent, D.C., Pasquerault, D., Gomet, J.C., Laubé, S., and Rowe, B.R. 1994, in *Chem. Phys. Letters*, to be published.

Duley, W.W. and Jones, A.P. 1990, *Ap. J.*, **351**, L5.

Léger, A. and Puget, J.L. 1984, in *Astr. Ap.*, **137**, L5.

Lepp, S. and Dalgarno, A. 1988, in *Ap. J.*, **324**, 553.

Pineau des Forêts, G., Flower, D.R., and Dalgarno, A. 1988, in *M. N. R. A. S.*, **235**, 621.

Rowe, B.R., Gomet, J.C., Canosa, A., Rebrion, C., and Mitchell, J.B.A. 1992, in *J. Chem. Phys.*, **97**, 1028.

Smith, D., Adams, N.G., and Alge, E. 1984, in *J. Phys. B*, **17**, 461.

Tobita, S., Meinke, M., Illenberger, E., Christophorou, L., Baumgartel, H., and Leach, S. 1992, in *Chem. Phys.*, **161**, 501.

Van der Zwet, G.P. and Allamandola, L.J. 1985, in *Astr. Ap.*, **146**, 76.

CHAPTER IV

Diffuse Bands and Interstellar Dust

THE DIFFUSE INTERSTELLAR BANDS AND INTERSTELLAR DUST

JOHN S. MATHIS
The University of Wisconsin - Madison

ABSTRACT.

As far as we now know, interstellar dust and the diffuse interstellar bands (DIBs) are not directly related; there is no really good correlation between any DIB and a specific feature of interstellar extinction. I argue that it is not useful to consider the λ2175 "bump", by far the strongest feature in the interstellar extinction law (in terms of equivalent width in frequency units), a *bona fide* DIB. The bump is probably not formed by the same processes that cause the DIBs; it is so strong that it is almost surely produced by carbon in some form (probably graphitic), likely by a different physical process (a plasma resonance rather than a bulk resonance). However, the bump's broadening, which might well be produced by coatings on graphitic grains, suggests that there are rather efficient interactions between grains and the material that produces the coatings. The systematics of the interstellar extinction law among various lines of sight also strongly suggests rapid interactions among grains of various sizes, since apparently the small grains coagulate into large ones in dense-gas environments.

Interstellar depletions, or removal of certain gas-phase ions from the gas phase of the ISM onto grains, also provide strong evidence for interactions of grains and gas atoms. Nonthermal motions of the grains are very likely required to produce the depletion observations. These rapid interactions must also apply to the carriers of the DIBs, so they must be capable of temporarily residing on grain surfaces or else being formed very rapidly in the gas phase. The quantitative determination of depletions is confused by the selection of reference abundances, or total elemental abundances (relative to H) in the ISM. There seem to be incompatible observations regarding whether the solar abundances or those of B stars should be preferred. In any case, along one line of sight (towards ξ Persei) there is substantial gas-phase carbon, so the amount of C used to produce DIBs and visual extinction is limited. It is not at all understood why certain elements are strongly depleted and others are only lightly so.

1. INTRODUCTION

The interstellar extinction law is now rather well known (see e.g. Mathis 1990 for a review) for a large number of lines of sight over wavelengths ranging from the near-infrared (Martin & Whittet 1990) through the optical to about 0.12 μm, the short-wavelength limit of the *International Ultraviolet Explorer* (*IUE*) satellite. There are a few observations down to the Lyman limit at 0.09 μm. One of the most important

A. G. G. M. Tielens and T. P. Snow (eds.), The Diffuse Interstellar Bands, 241–252.

aspects of the extinction law is that it varies very smoothly with wavelength, with very few spectral lines or bands besides the diffuse interstellar bands (DIBs) that are the subject of this conference. For the general diffuse interstellar medium (ISM), outside of molecular clouds, the only examples of broad absorption features (i.e., not sharp lines of gas-phase atoms, ions, or molecules) are the two silicate bands at 9.7 and 18 μm, some weak absorptions near 3.4 μm, probably caused by aliphatic hydrocarbons (Sandford *et al.* 1991; Pendleton et al. 1994), and, by far the strongest in terms of equivalent width in frequency, the λ2175 "bump". This lack of obvious spectral signatures has led to the rejection of many candidates for interstellar grain materials, since most solids produce absorptions not seen in the ISM.

This paper will first discuss whether or not the "bump" should be considered a DIB. Then we will consider some recent observations of depletions in the ISM that have strong bearing on the interactions between the dust grains among themselves and also with the gas atoms, and by implication with the various carriers of the DIBs. Finally I will discuss elemental depletions that are relevant to carbon, a material strongly implicated as being involved in the DIBs, and the difficulty of determining the total chemical composition of the present local ISM, including both gas and dust.

As regards the direct and known connection between extinction laws and DIB strengths, there are none besides a *general* correlation between DIB strengths and extinction, in the sense that both generally increase with the amount of ISM between the background star and the observer. The ultraviolet extinction law can be characterized by six parameters (Fitzpatrick & Massa 1986, 1988) that give an excellent *mathematical* description of the extinction law. Two are the slope and intercept of the "linear background", or the portion of the extinction law varying linearly with x ($\equiv 1/\lambda$); three are the strength, central wavenumber, and FWHM of the bump; the sixth is the strength of the "far-ultraviolet rise" (FUV rise), the coefficient of a certain function of x that fits the wavelength dependence of the excess extinction over the linear background for $x \geq 5.9\ \mu\mathrm{m}^{-1}$. The three bump coefficients probably represent the contribution of a single material, the carrier of the bump (discussed below); the remaining three coefficients probably have little physical significance individually, since there is no substance known that shows the corresponding extinction law of any of them. Correlations between strengths of various DIBs and each of the six coefficients are generally weak and unconvincing (Jenniskens, this conference). Benvenuti & Porceddu (1989) have investigated the relation between DIBs and the bump. In no case has the strength of a specific feature of the extinction law been shown to be directly related to any particular DIB.

2. THE λ2175 BUMP

The λ2175 bump is certainly interstellar; it is certainly diffuse; it is certainly a band. In spite of these "minor" facts, I strongly conclude that the bump is *not* a DIB, at least in the ordinary sense, for two reasons: (1) the strength of the bump is so much larger than any of the DIBs, and (2) the bump is probably produced by a different material by a different physical process.

2.1. THE STRENGTH OF THE BUMP

The ultraviolet extinction law can be expressed by:

$$A(\lambda)/A(V) = c_1 + c_2 x + c_3 D(x, x_0, \gamma) + c_4 f(x) \ ,$$

where the c_i are defined here to refer to the extinction relative to $A(V)$, while Fitzpatrick & Massa (1986, 1988) defined them to refer to $E(\lambda - V)/E(B-V)$, the ratio of color excesses. The bump is contained in the function $D(x, x_0, \gamma)$, the Drude function, where x_0 is the central wavenumber and γ is essentially the FWHM. The area of the bump is $\pi c_3/2\gamma$. The area of the bump per H atom, W_λ, follows from $\langle N(H)/E(B-V)\rangle = 5.8 \times 10^{21}$ H atoms cm^{-1} mag^{-1}, as measured in the diffuse ISM (Bohlin, Savage, & Drake 1978), and $\langle A(V)/E(B-V)\rangle = 3.1$. From

$$[\pi e^2/(m_e c^2)] N_{bump} f = W_\lambda/\lambda^2$$

we obtain $N_{bump} f/N(\mathrm{H}) = 9.3 \times 10^{-6}$, where f is the oscillator strength of the bump. For comparison, $N(\mathrm{C})/N(\mathrm{H}) = 400 \times 10^{-6}$, so carbon can produce the bump with ease, but the abundant refractory elements (Fe, Si, Mg) have abundances of $\approx 30 \times 10^{-6}$ relative to H, and other elements 10 times lower or less. Even with the large value $f \approx 1$ these must have substantial fractions of their abundances in the bump grains, while they are actually needed to produce the visual and much of the ultraviolet extinction. It has been suggested that the OH^- ion, in connection with surface Si ions in a silicate lattice, can produce the bump, but the f of that transition has been measured (Fritz, Lüty, & Anger 1963) to be 0.13, so there is not enough Si to produce the bump in this fashion.

The above considerations are the basis for the almost universal agreement that the bump is produced by carbon in some form. This situation is in dramatic contrast to the DIBs. In the well-known star HD 183143, Herbig (1975) and Herbig & Leka (1991) list many DIBs and their equivalent widths. The sum of these is given by $Nf(\lambda 4428)/N(\mathrm{H}) = 2.6 \times 10^{-9}$; $\sum_{\text{all other DIBs}} Nf(\lambda) = 7.0 \times 10^{-9}$; $\sum_{\text{all DIBs}} Nf(\lambda) = 10^{-8}$. All DIBs together need only 10^{-3} of the strength of the bump.

We see that for the DIBs we can have exotic species of carriers and/or rare elements for their production, but for the bump we almost have to use the most abundant refractory material, carbon.

2.2. THE PHYSICS OF THE BUMP AND DIBS

In this section I discuss the oldest (Hoyle & Wickramasinghe 1962) and what I and many others consider the most likely explanation of the bump: a plasma resonance in small graphite grains (see Draine 1989 for a discussion of the various possibilities). This explanation seems plausible because (1) one only needs about $\mathrm{C/H} = 50 \times 10^{-6}$ in graphite, or about 15% of the available C, to produce the average bump strength in the diffuse ISM, and less in dense clouds; (2) small graphite grains will have their absorption occurring at the same wavenumber, independent of their size, in keeping with the relatively invariant nature of the central wavenumber x_0 of the bump (see below); and (3) small graphite grains can readily be produced by the annealing of small amorphous carbon grains, heated by the absorption of a single ultraviolet photon. In fact, it is not easy to see how small graphite grains can be avoided if larger grains of amorphous carbon are injected into the ISM by carbon stars, as is observed, and these grains are shattered by grain-grain collisions within the ISM. Graphite inclusions have been found in the most primitive meteorites. However, the "graphite" in the bump carriers is surely not the pure product whose optical constants have been measured in the laboratory; rather interstellar graphite must contain impurities of various metallic carbides and sulfides, along with lattice defects from cosmic-ray damage, as well as probably coatings of less refractory materials on

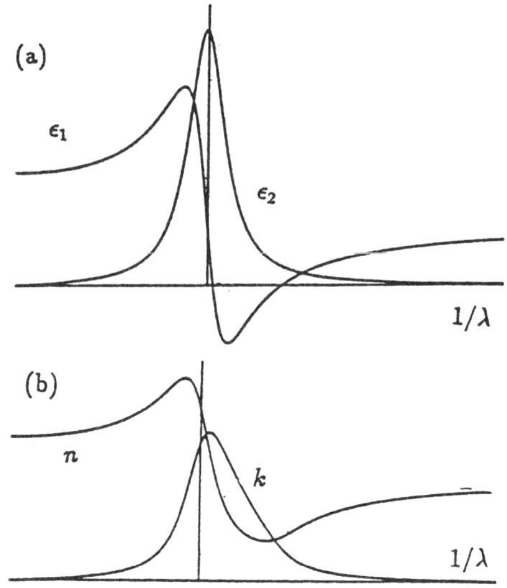

Fig. 1. (a) The real part (ϵ_1) and imaginary part (ϵ_2) of the dielectric constants of a single oscillator as a function of wavenumber, x. (b) The real part, n, and the imaginary part, k, of the index of refraction, m, of the same oscillator, where $m^2 \equiv \epsilon$.

their surfaces. It has been suggested that the bump is caused by "amorphous" carbon, which is simply rather disordered but locally consisting of a graphitic structures with short-range but no long-range order. Quantitative models using the laboratory optical constants of disordered carbon do not show that a feature as strong as the bump can be produced in this fashion.

The physical mechanism by which graphite produces the bump is not the same as that of the DIBs. For relatively narrow transitions like the DIBs there is a sharp change with wavelength in the optical constants of the atom/molecule, similar to that of a damped oscillator resonating at a particular frequency. The optical behavior of such materials is described by the complex dielectric constant, ϵ, shown in Figure 1(a) for a single-oscillator transition (see e.g. Bohren & Huffman 1983). An equivalent description of the optical behavior of the oscillator is by means of the index of refraction, m, related to ϵ by the definition $\epsilon = m^2$, with n and k defined by $m = n + ik$. Note that ϵ_1, the real part of ϵ, can be of either sign, while ϵ_2 is always positive. Light propagating through these oscillators in the z- direction would suffer wavelength-dependent attenuation according to $I = I_0 \exp(-\alpha z)$, where α, the bulk absorption coefficient, is given by $\alpha \propto k(\lambda)/\lambda$. Figure 1($b$) shows the variation of n and k with wavenumber; they vary rather sharply over the same region in which ϵ_2 is large. They are also always positive, which we shall need to recall shortly.

For the DIBs, the range of wavelength represented in Figure 1 is only a few Å, the width of the band. Graphite has optical constants that vary slowly of the whole mid-ultraviolet range, as shown in Figure 2. There we see three values of ϵ_1 and ϵ_2 plotted against wavenumber over an interval of 700 Å.

The cross section C_{ext} of a *small* spherical grain to a passing electromagnetic wave is given by (e.g. Bohren & Huffman 1983):

$$C_{\text{ext}} = (2\pi/\lambda)V\ A\epsilon_2/[(\epsilon_1 + B)^2 + \epsilon_2{}^2]\ , \tag{1}$$

where A and B are *shape-dependent* constants that have the values $A = 9$ and $B = 2$ for spheres. The meaning of "small" here is that $2\pi a/\lambda \ll 1$, where a is the particle radius. One sees that there is a resonance when the denominator of the right-hand term in eqn. (1) is a minimum, or $\epsilon_1 = -B = -2$ for spheres. *This is the condition that produces the bump for small graphite.* Graphite is a tensor material, and the bump is actually produced only by the resonance for waves with the electric field perpendicular to the crystal axis (i.e., parallel to the sheets of the aromatic rings).

Figure 2 shows the values of ϵ for three types of graphite. The dot-dashed curves are the laboratory-measured ones of Borghesi & Guizetti (1991); the dashed curve is the well-known "astronomical graphite" suggested by Draine & Lee (1984); the solid curve shows a suggested modification of the constants (Mathis (1994) that makes the bump narrower than that of Draine & Lee and also at 2175 Å instead of at 2110 Å. (In the Draine & Lee model, based on the MRN power-law distribution, the bump is shifted in wavelength to 2175 Å by the larger grains.) For the particular Mathis constants plotted, the grains are oblate instead of spherical, so the values of A and B in eqn. (1) above are appropriately modified. As one can see from Figure 2, there

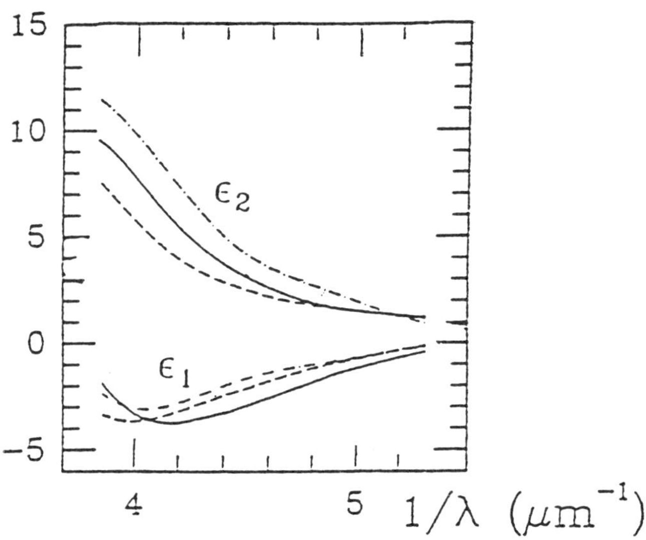

Fig. 2. The real part of the dielectric constant, ϵ_1 (curves < 0) and the imaginary part, ϵ_2, plotted against wavenumber for graphite with the electric field of the wave perpendicular to the axis of the crystal. The upper (dot-dashed) curve of each trio are laboratory measurements on pure graphite (Borghesi & Guizetti 1991); the dashed curves are for the Draine & Lee (1984) "astronomical graphite"; the solid curve is for a suggestion by Mathis (1994).

is no rapid change in optical constants throughout the wavelength region plotted; instead, the resonance occurs because of the minimum of the shape-dependent denominator in eqn. (1). This type of resonance is called a "plasma resonance" or "surface mode", because it so happens that there is a uniform electric field within the grain, with a polarization charge on the grain surface.

Thus, we see that the bump and the DIBs are formed by different processes, and it is not very helpful to consider the bump to be a "super-DIB."

2.3. THE RELEVANCE OF THE BUMP FOR THE DIBS

There is a complete lack of correlation between the properties of the bump (strength and γ) in a given line of sight and the strength of the FUV rise (Greenberg & Chlewicki 1983; Mathis & Cardelli 1992), so the bump and FUV carriers must be distinct and variable in abundance. Perhaps the bump carriers are associated with distinct DIBs, although so far there is no evidence that either property correlates with any DIB. It would be very useful to continue to seek DIBs that seem directly associated with either the bump or the FUV extinction.

It seems likely that there are strong interactions between the bump grains and their surrounding gas. These interactions will likely affect the DIB carriers as well, whether they are in the gas phase or on the surfaces of grains.

Interactions between bump grains and gas are suggested by the bump likely being broadened by coatings on graphitic grains (Mathis 1994), with the coatings accreted from the gas and evaporated by shocks or grain-grain collisions. Bump observations that can be well explained by coatings on graphite, but (I think) not so easily by other mechanisms, are the almost-constancy of the central wavenumber, x_0, among the various lines of sight, along with wide variations in FWHM, γ. The coatings might have optical properties that resemble those of PAHs, but also other substances (e.g., MgO) are also possible. Adding such coatings to small graphite grains initially simply broadens the bump, but if the coating becomes thick enough the central wavenumber is shifted to larger wavenumbers as the bump is broadened. This behavior is exactly as is observed: the broadest bumps are in sight lines in which there is quiescent, cold material, and the central wavenumbers are somewhat larger than the average for diffuse dust (see discussion in Cardelli & Clayton 1991). Within the diffuse ISM, coatings of varying thicknesses occur, serving to broaden the bump in varying degrees. For any environment, it seems reasonable that these coatings accrete on the small grains, which contain most of the grain area.

The most direct evidence of interactions between grains and other grains is the strong linear correlation between $A(\lambda)/A(V)$ with $1/R(V)$ found by Cardelli, Clayton, & Mathis (1989) for $\lambda \leq 0.3$ μm. Here $R(V) \equiv A(V)/E(B-V)$. The implication is that the small grains, responsible for the UV extinction, are relatively lacking along lines of sight that have large $R(V)$. The lack of small grains can be explained either by their being enlarged by the accretion of material from the gas phase to the point that they no longer absorb UV, or else their coagulating onto the larger grains, thereby increasing $R(V)$. (There is also another possibility discussed below.) Depletion studies show that here is not enough condensable material in the gas phase to produce major effects by accretion: both O and N are mostly in the gas phase for both large and small-$R(V)$ lines of sight. The clear implication is that the small grains coagulate onto the large ones.

If coagulation is to occur between grains, there must be non-thermal motions of grains within the clouds in order to achieve the necessary rate of interactions between grains; the small ones must move relative to the larger ones in spite of the drag of the gas. Otherwise, the huge mass of grains relative to individual atoms would imply that they have almost no motions through the gas, or relative to each other. One way of achieving nonthermal motions is through magnetohydrodynamic waves that affect grains of differing sizes (and, therefore, charges) differently.

The conclusion that there is depletion of small grains in regions of large $R(V)$ assumes that there is no change in the *composition* of the grains. This assumption might not be true. Various forms of hydrogenated amorphous carbon (HAC) are strongly suggested by the "Extended Red Emission" (ERE) that is observed in many reflection nebulae (see Witt 1989, 1994 for reviews). HAC has optical constants that depend upon the amount of hydrogenation of the carbon, in the sense that the addition of H makes the material more transparent in the UV. The decrease of the far-UV absorption might be the result of increased hydrogenation in dense regions. However, to my knowledge no quantitative models have been made showing that this process can actually produce the observed effects.

3. INTERSTELLAR DEPLETIONS

The depletions of atoms from the gas phase onto grains have been extensively studied since the launch of the *IUE* in 1978. The Goddard High Resolution Spectrograph (GHRS) aboard the Hubble Space Telescope has recently allowed observations of unprecedented quality. Column densities of various ionic or molecular species arc obtained from the interstellar lines seen against a background stellar spectrum from which the interstellar lines are clearly separable (because of ionization stage, or velocity width) . If the spectral resolution and signal-to-noise (S/N) are sufficient, the velocity profile of each ion almost always shows a spread that exceeds thermal speeds. Often there are clumps of various ions ("clouds") moving with a common speed (although the analogy to terrestrial clouds is not very complete because there probably are less definite outer boundaries to interstellar clouds than to our cumulus).

Within each cloud, the key observations are the abundances of the *dominant* stage of ionization of each element. Usually most of the hydrogen along the line of sight is in the neutral form (with obvious exceptions, such as stars exciting H II regions). For H I regions the dominant stage of ionization is that which has an ionization potential, χ_i, which satisfies the relation $\chi_{i-1} \leq 13.6$ eV $\leq \chi_i$. For the common elements C, Fe, Mg, Si, and many others it is the first stage of ionization (i.e., X^+) that is dominant in H I regions, but for O, N, and, of course, the noble gases, it is the neutral. For calcium it is Ca^{++}, which cannot be observed.

Often the column density of H can be determined by the interstellar Lyman-α line, or perhaps 21-cm radio emission if the line of sight does not contain too much neutral H beyond the star (i.e., the line of sight is fairly well out of the Galactic plane). In this case, the gas-phase abundances of the elements (if their dominant stage of ionization is observed) relative to H can be determined immediately. One then compares the gas-phase abundance to a reference abundance, meaning the total abundance of that element in gas and grains (including complex molecules). The ratio of the gas-phase to the reference abundance is the depletion of the element.

This section will emphasize three points regarding depletions: (1) depletions show considerable regularity with mean gas density along the various lines of sight; (2) the amount of carbon in the gas phase can sometimes be very appreciable, so the carbon budget for grains (and DIB carriers) is tighter than most theories assume; and (3) it is not obvious which is the best reference abundance (e.g., solar or recently-formed stars), and perhaps different lines of sight have different reference abundances, showing inhomogeneities in the present-day ISM.

3.1. SYSTEMATICS OF DEPLETIONS

Depletions of elements are by no means uniform, but (contrary to rumors; see below) form a pattern that is not understood or even characterized by any simple single parameter. Some elements are lightly depleted, typically by less than a factor of 2 (O, N, Zn, and S); others, substantially (roughly a factor of 10 in typical lines of sight), including Fe, Mg, Si, Mn; some, extremely (factor of 100), including Ti, Al, and Ca. Surprisingly, there is a quite tight correlation of some of these depletions with mean H density along the entire line of sight (see e.g., Jenkins 1987 for a review.) To me, one of the most interesting recent results has been the discussion by Crinklaw, Federman, & Joseph (1994) of the ground-based measurements of Ti II and Ca II. Figure 3, adapted from that paper, shows the depletions plotted against $\langle n_{\rm H} \rangle$, the mean H density for various lines of sight. Figure 3(a) shows $[({\rm Ti\ II})/{\rm H}]_{\rm gas}/[({\rm Ti\ II})/{\rm H}]_{\rm solar}$ plotted against $\langle n_{\rm H} \rangle$. Over 3 orders of magnitude in depletion there is only about a factor of 2 for the standard deviation at a given density. Figure 3(b) is the same except for Ca II, even more surprising because Ca^+ is not the dominant stage of ionization. A correction

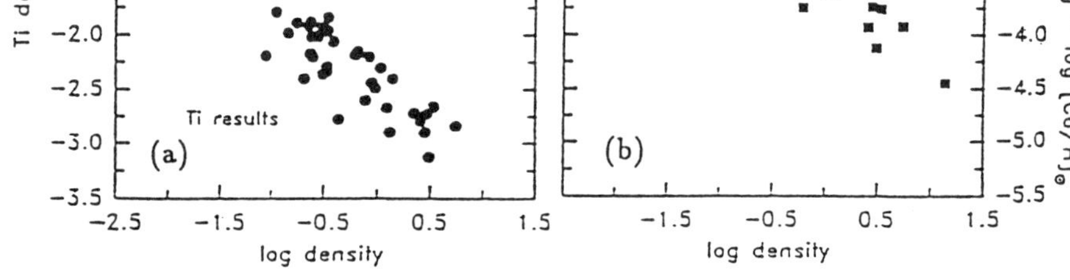

Fig. 3. (a) The interstellar depletions of titanium plotted against the mean H density in the line of sight in which each depletion was measured. The measurements only refer to the H I gas. The figure shows the relatively small spread in the depletion at each mean density. (b) The depletion pattern for Ca as measured from Ca^+, which is *not* the dominant stage of ionization of Ca for H I gas. Even so, there is a rather small spread in the range of depletions at a given density. (Adapted from Crinklaw *et al.* 1994).

factor of about an order of magnitude is needed to convert the Ca^+ abundance to the elemental Ca depletion. This factor is evident in the numerical values of the Ca and Ti depletions, with Ca being lower by about this factor. The Ca^+/Ca ratio depends upon the local electron density and interstellar radiation field for 13.6 eV $\geq h\nu \geq$ 11.8 eV $= \chi(Ca^+)$. In spite of this large and variable correction, the Ca^+ depletions are still very well correlated with $\langle n_H \rangle$.

The implications of the correlations of elemental depletions with $\langle n_H \rangle$, especially those of Ti^+ and Ca^+, are twofold:

1. They suggest that there are strong interactions between grains and the gas atoms, most easily explained by non-thermal motions similar to those that were suggested by the coagulation of the small grains into larger ones in dense regions.

The time required for a typical heavy atom to encounter a grain surface can be estimated as $t_{coll} = 1/(n\sigma v)$, where n is the number density of H nuclei $[= n(H^0) + 2n(H_2)]$ and σ is the cross section of grains per H nucleus, about 3×10^{-21} cm^2 H^{-1}. Let us consider the circulation between the material in the diffuse warm H I gas ($T \approx$ 8000 K; $n_H \approx 0.4$ cm^{-3}) and that in the cool cloud component ($T \approx 80$ K; $n_H \approx 40$ cm^{-3}). The 21-cm observations show that the masses in the two components are about equal. For a Ti^+ ion in the cool cloud with a thermal speed, $t_{coll} \approx 1.4 \times 10^7$ years, comparable to the cloud lifetime of $\approx 10^7$ years. However, Figure 3 shows that the Ti gas-phase abundance must be changed by two orders of magnitude between densities of 0.4 and 40 cm^{-3}, so the observed changes in gas-phase Ti^+ abundance require 4.6 times t_{coll}, or 6×10^7 years. This time seems excessively long to take place within a single cloud, since material probably circulates between the cool clouds and the intercloud medium several times between passages through a spiral arm each 10^8 years.

In the intercloud medium, the encounters of heavy atoms with grains is more difficult than in the clouds, since the density is 100 times lower, while the thermal speeds are only 10 times higher. Thus the encounter times are 10 times longer than in the clouds.

2. The regularity of the depletions with mean gas density, and the very low gas-phase abundances of Ti and Ca, are made more surprising by considering the injection of gas-phase ions into the intercloud ISM by mass loss from hot stars (Jura 1987). Also, the destruction of grains by interstellar shocks injects gas-phase elements into the ISM, and the low depletions might be an observational limitation on the effectiveness of such destructive processes.

I conclude that grains interact very efficiently with other grains and with gas atoms, and by implication with the carriers of the DIBs. Thus, we can easily imagine that grain surface chemistry might be an important of producing (or destroying) the various carriers.

The variations of mean depletions among the various elements represents a *major* unanswered question: what determines the pattern of depletions in the ISM? There have been two suggestions regarding the answer to this question: (1) depletions are related to the "condensation temperature", T_{con}, which is the temperature at which about half of a given element would be in solid form in a cosmic mixture at a pressure comparable to that in a stellar atmosphere; or (2) depletions are related to the first ionization potential of the element. Both ideas seem to fail in the light of recent determinations of depletions of many elements heavier than Zn (up to Pb), in high rows of the periodic table (Cardelli 1994). The elements heavier than Row 4 of the table (i.e, Zn) do not fit well into either suggestion. Since encounters of ions with grains (or molecules) ultimately determine the depletions, I wonder if the pattern might be influenced by interactions of various ions with the hydrocarbon molecules (PAHs?) that are responsible for the unidentified infrared emission bands. Probably most of the area of "grains" is in these molecules. If organometallic complexes are much more readily formed with some ions than others, perhaps these complexes are important in the ISM (Marty *et al.* 1994; Boissel 1994).

3.2. THE REFERENCE ABUNDANCES FOR DEPLETIONS; CARBON ABUNDANCE IN INTERSTELLAR DUST

The question of what is the proper reference abundance, to be used for the determination of the depletion, is not a simple one. At the present time, there seems to be good evidence that neither the solar abundances nor those of recently formed stars (B stars) are completely suitable candidates. The problem is, briefly, that the observed gas-phase carbon in one apparently well-determined sight line (towards ξ Persei) is large, so that a large carbon reference abundance (solar, but not B stars) seems required in order to provide enough carbon for interstellar dust. However, the solar abundance also provides a great deal of oxygen. Then the question is: where is all of the O, if the solar reference abundance is used? B stars provide a low O abundance (along with a low C), so they would seem appropriate as the reference for O (as well having as the obvious advantage of having just been formed out of the ISM).

The element composition of interstellar dust has been discussed by Sofia, Cardelli, & Savage (1994, and refs therein), where the gas-phase abundances from the GHRS are subtracted from two reference abundances to give the dust abundances. The greatly improved resolution over *IUE* and the signal/noise of the GHRS are necessary for reliable determinations of the C/H ratio. Besides the GHRS data, there is only one *Copernicus* measurement of the interstellar C abundance: towards δ Sco (Hobbs, York, & Oegerle 1982). The problem is that the C II line (the dominant stage of ionization of C in H I regions) at λ1334 is always saturated, and without spectral resolution it is impossible to disentangle the effects of overlapping velocity components. However the very weak line λ2325 requires an excellent signal-to-noise ratio for its measurement.

The most interesting interstellar C/H ratio is that for ξ Per (Cardelli *et al.* 1991), for which the GHRS measurements give a gas-phase C/H of $(250 \pm 70) \times 10^{-6}$. This is to be compared with the reference abundances of $(400 \pm 40) \times 10^{-6}$ for solar (Grevasse *et al.* 1991) and $(160 \pm 60) \times 10^{-6}$ for B stars (Gies & Lambert 1992). We see that the B-star reference abundances provide very little gas-phase C if the measurement of ξ Per is correct. We would prefer to use the solar abundance as a reference, although it is very strange that there is a composition difference between B stars (ages of $\approx 10^7 - 10^8$ years) and the present ISM. The lines of sight towards ζ Oph and δ Sco give less gas-phase C (by factors of 0.6). Even with the B-star reference abundance they give no problem in providing enough carbon in the dust to produce the bump and infrared emission bands, if not to help with the visual extinction.

However, adopting the solar reference gets us into trouble with the O/H ratio in the ISM – we cannot account for all of the O that should be present, with solar O/H

$= (850 \pm 50) \times 10^{-6}$ (Anders & Grevesse 1989). The gas-phase O/H in ξ Per and ζ Oph (Savage, Cardelli, & Sofia 1992) are both $(310 \pm 40) \times 10^{-6}$, leaving $(540 \pm 65) \times 10^{-6}$ to be accounted for in the dust if the reference is solar.

Of course, interstellar silicates require oxygen. They can use about 90×10^{-6} (per H atom) if the silicates are pyroxenes ($Fe_xMg_{1-x}SiO_3$), which I like because these silicates would allow Fe, Mg, Al, Cr, etc., to provide oxides for the interstellar extinction in the visual region of the spectrum. These oxides might use an O/H of about 40×10^{-6}. Thus, the total O/H in silicates and oxides is about 130×10^{-6}, a quarter of the "missing" oxygen if the solar reference abundance is used. By contrast, the B stars have O/H $\approx 480 \times 10^{-6}$, so there would be nice agreement between the expected O/H in dust and that missing from the gas phase. The O/H tied up in water or CO in the diffuse ISM are negligible; both have absorptions that are not seen or are weak.

I do not resolve the problem of the reference abundance. Certainly it seems quite unreasonable to use the solar reference abundance for C/H and B-star abundances for O/H: if we adopt the mixed references we would predict C $\approx$ O, incompatible with the ISM (among other objects), where the gas-phase O/H is about 300×10^{-6}, or roughly the total solar C/H and twice the B-star C/H. This observation clearly shows that C:O ≈ 0.5, as both solar and B-star abundances also each require. Most stars show about the same abundance ratio.

Clearly, the C/H towards ξ Per should be remeasured, although the original seems rather secure. Can the ISM have significantly different C/H along various lines of sight? If so, are there such variations in the DIBs? Once they are understood, can they provide clues as to the gas-phase abundance ratios of C/H, or other elements? I leave with the thought that these questions, probably answerable with sufficient GHRS and DIB data, will surely provide excitement for future workers.

ACKNOWLEDGEMENTS. This work is partially supported by NASA grants and contracts GO- 4382.03-92A and NAGW-3833. I have benefited from numerous conversations with colleagues, especially Dr. Jason Cardelli.

REFERENCES

Anders, E. and Grevasse, N. 1989, *Geochim. et Cosmochim. Acta*, **53**, 197.
Benvenuti, P. and Porceddu, I. 1989, *Astr. Ap.*, **223**, 329.
Bohlin, R.C., Savage, B.D., and Drake, J.F. 1978, *Ap. J.*, **224**, 132.
Bohren, C.F. and Huffman, D.R. *Absorption and Scattering of Light by Small Particles*, (New York: John Wiley & Sons).
Boissel, P. 1994, *Astr. Ap.*, **285**, L33.
Borghesi, A. and Guizzetti, G. 1991, in *Handbook of Optical Constants of Solids, II*, ed. E. D. Pallik (New York: John Wiley & Sons), p. 449.
Cardelli, J.A. 1994, *Science*, in press.
Cardelli, J.A. and Clayton, G.C. 1991, *A. J.*, **101**, 1021.
Cardelli, J.A., Clayton, G.C., and Mathis, J.S. 1989, *Ap. J.*, **345**, 245.
Cardelli, J.A., Savage B.D., Bruhweiler, F.C., Smith, A.M., Ebbets, D.C., Sembach, K.R., and Sofia, U.J. 1991, *Ap. J.*, **377**, L57.
Crinklaw, G., Federman, S.R., and Joseph, C.L. 1994, *Ap. J.*, **424**, 748.
Draine, B.T. 1989, in *IAU Symposium 135, Interstellar Dust*, eds. L.J. Allamandola & A.G.G.M. Tielens, (Dordrecht, Kluwer), p. 313.

Draine, B.T. and Lee, H.M. 1984, *Ap. J.*, **285**, 89
Fitzpatrick, E.L. and Massa, D. 1986, *Ap. J.*, **307**, 286.
——. 1988, *Ap. J.*, **28**, 734.
Fritz, B., Lüty, F., and Anger, J. 1963, *Zeit. Physik*, **174**, 240.
Gies, D.R. and Lambert, D.L. 1992, *Ap. J.*, **387**, 673.
Greenberg, J.M. and Chlewicki, G. 1983, *Ap. J.*, **272**, 563.
Grevasse, N., Lambert, D.L., Sauval, A.J., van Dishoeck, E.F., Farmer, C.B., and Norton, R.H. 1991, *Astr. Ap.*, **242**, 488.
Herbig, G.H. 1975, *Ap. J.*, **196**, 129.
Herbig, G.H. and Leka, K.D. 1991, *Ap. J.*, **382**, 193.
Hobbs, L.M., York, D.G., and Oegerle, W. 1982, *Ap. J.*, **252**, L21.
Hoyle, F. and Wickramasinghe, N.C. 1962, *M. N. R. A. S.*, **124**, 417.
Jenkins, E.B. 1987, in *Interstellar Processes*, eds. D.J. Hollenbach and H.A. Thronson (Dordrecht, Reidel), p. 533.
Jura, M. 1987, in *Interstellar Processes*, eds. D.J. Hollenbach and H.A. Thronson (Dordrecht, Reidel), p. 3.
Martin, P.G. and Whittet, D.C.B. 1990, *Ap. J.*, **357**, 113.
Marty, P., Serra, G., Chaudret, B., and Ristorcelli, I. 1994, *Astr. Ap.*, **282**, 916.
Mathis, J.S. 1990, *Ann. Rev. Astr. Ap.*, **28**, 37.
——. 1994, *Ap. J.*, **422**, 176.
Mathis, J.S. and Cardelli, J.A. 1992, *Ap. J.*, **298**, 610.
Pendleton, Y.J., Sanford, S.A., Allamandola, L.J., Tielens, A.G.G.M., and Sellgren, K. 1994, *Ap. J.*, submitted.
Sandford, S.A., Allamandola, L.J., Tielens, A.G.G.M., Sellgren, K., Tapia, M., and Pendleton, Y. 1991, *Ap. J.*, **371**, 607.
Savage, B.D., Cardelli, J.A., and Sofia, U.J. 1992, *Ap. J.*, **401**, 706.
Sofia, U.J., Cardelli, J.A., and Savage, B.D. 1994, *Ap. J.*, in press.
Witt, A.N. 1989, in *IAU Symposium 135, Interstellar Dust*, eds. L.J. Allamandola and A.G.G.M. Tielens, (Dordrecht, Kluwer), p. 87.
——. 1994, this volume.

DUST GRAIN CARRIERS AND THE DIFFUSE BAND PROFILES

PAUL R. SHAPIRO
Department of Astronomy, The University of Texas at Austin, Austin, TX 78712 USA

ABSTRACT. The optical properties of dust grains containing resonant impurities are discussed. Such grains are candidates for explaining the diffuse interstellar bands. The embedded cavity model of Purcell and Shapiro has been used to derive scattering and absorption cross sections, extinction profiles, and anomalous linear and circular polarization for such grains, as well as the effective bulk dielectric constant for the composite material. The results are briefly summarized and conclusions are drawn based upon a comparison of these theoretical extinction profiles with the observed profiles of diffuse interstellar bands.

1. INTRODUCTION

The optical properties of grains with resonant impurities have long been of astronomical interest in connection with the hypothesis that such grains are the source of the diffuse interstellar bands ("DIBs"). In the seven decades since their first discovery, no convincing explanation for these DIBs has yet been found. The DIBs now number over a hundred, at wavelengths between 4400Å and 1.3 μm (cf. Herbig 1993). The band widths range from less than one to tens of angstroms, too wide to be explained as ordinary, thermal-Doppler-broadened atomic or molecular lines under normal interstellar conditions. The bands have no resolved fine structure which might be indicative of molecular band structure. For many years it was thought that the survival of molecules with sufficient abundance to explain the strength and ubiquity of the bands was unlikely under the harsh conditions of the interstellar medium. This led naturally to the dust grain hypothesis, since broad, featureless bands can result easily from extinction by dust grains containing a small concentration of atoms or molecules with resonant absorption lines at the observed band wavelengths. In the meantime, however, the list of fairly complex molecules detected in the interstellar gas has grown in recent years, enough to suggest the possibility that large, stable, gas-phase molecules might indeed be responsible for the DIBs, despite the difficulty in accounting for their formation and survival. Nevertheless, the attempts to distinguish between the gas-phase and dust grain hypotheses by observations of correlations between observed DIB strengths and other properties of the interstellar matter along the same lines of sight have produced ambiguous results (see, for example, Herbig 1993). It is, therefore, worthwhile to compare the observed shapes and widths of individual bands in detail with theoretical predictions for such bands from the dust grain model in the hope of constraining this hypothesis further.

A. G. G. M. Tielens and T. P. Snow (eds.), The Diffuse Interstellar Bands, 253–262.

The focus of this paper is on the extinction profiles which result if interstellar dust grains are the carriers of the resonant absorbers responsible for the DIBs. As such, I will not attempt to identify the species responsible, but will attempt, instead, to constrain the properties of that species and its host grain. Other talks have focused on the search for the identity of the molecule or molecules whose absorption spectrum can match some or all of the observed bands. In that sense, I wish to point out at the outset that the distinction which has been made recently between the origin of the DIBs in some large, gas-phase interstellar molecule and the dust grain origin may be smaller than it appears. Molecules which are observed in emission in the gas-phase may also reside on or within a dust grain. A particularly simple example is provided by the H_2 molecule, which is commonly believed to form on the surface of interstellar grains and subsequently to enter the gas-phase by ejection from the grain. A grain-residence time and/or grain-assisted formation for other, more complicated molecules like polycyclic aromatic hydrocarbons ("PAHs") is also possible, however. Moreover, if the size of the molecules under consideration is large enough, they may have optical properties which begin to approach those of small grains. Indeed, it may not be entirely coincidental that laboratory identifications of bands which resemble some of the DIBs, in the absorption spectra of candidate molecules, often involve the measurement of spectra which result when those molecules are embedded in a solid-state matrix. In this sense, I will address the dust grain origin of the DIBs, not as an *alternative* to the molecular origin, but rather as a possible *site* for the unknown molecule, and ask the following questions: What is the effect on the absorption spectrum of a resonant absorber if it resides as an impurity within a dust grain? What host grain and resonant absorber properties are required to match the observed DIB profiles? This approach is complementary to the direct search for the identity of the resonant absorber by laboratory spectroscopy and the identification of candidate molecules in the gas phase.

Most of what I will report is contained in much greater detail in Purcell and Shapiro (1977; Paper I), Shapiro and Holcomb (1986a; Paper II), and Shapiro and Holcomb (1986b; Paper III). Since those papers appeared, there have been a few misstatements of their results and conclusions in the literature which have, unfortunately, left the false impression that the current observations of DIB profiles are inconsistent with the dust grain hypothesis. For example, Herbig (1993) incorrectly states that "the earlier idea that they might represent a solid state phenomenon caused by impurities within grains has lost favor because present theory (Shapiro and Holcomb 1986) does not explain the observed band asymmetries or the lack of 'emission' wings. . . ." Since the actual conclusion of Shapiro and Holcomb (1986b) was, in fact, just the opposite, it will be useful for us to correct this misimpression in what follows.

2. OPTICS OF GRAINS WITH RESONANT IMPURITIES: THE EMBEDDED CAVITY MODEL

In Paper I, we derived a new model for the optical behavior of grains with resonant impurities. This model, which we called the "embedded cavity model," represented a significant revision of the theory used to predict the extinction cross sections, band profiles, and bulk dielectric constants for such grains. Previous calculations of these properties were based primarily on the approach suggested by van de Hulst (1949, 1957). For a more complete discussion of the differences between the earlier approach and the embedded cavity model, we refer the reader to Paper I. The essential improvement which the new model represented was the explicit treatment of the electrodynamic interaction of the resonant impurity absorber and radiator with the surrounding dielectric host material. The earlier approach had assumed that no interaction between the impurity and the host dielectric occurs which might make the polarizability of the composite grain differ from the sum of the polarizabilities of each, as calculated for each in the absence of the other.

Extinction cross sections and the anomalous linear and circular polarization for spheroidal grains which are small compared with the wavelength were derived analytically for the embedded cavity model. These results also yielded the effective bulk dielectric constant for the composite material which can be used to calculate the optical properties of grains of arbitrary size and shape. The extinction cross sections, band profiles, and bulk dielectric constants calculated in the embedded cavity model differed significantly from the predictions of the old model. For small particles, where $x = 2\pi r/\lambda \ll 1$, r is the particle radius, and λ is the wavelength of incident light, the grain extinction cross section is almost entirely true absorption rather than scattering. In this limit, for spherical grains, van de Hulst's approach yielded a simple symmetric absorption profile centered on the resonant frequency of the embedded impurity absorber in vacuum. The width and shape of the profile in this limit corresponded to that of the Lorentz profile assumed for the molecular polarizability of the resonant impurity, following classical dispersion theory and treating the impurity as an electric dipole represented by a bound harmonic oscillator. As a result, the width of the predicted absorption profile was just the internal self-width of the oscillator, and this width, as well as the shape of the band and the frequency of peak absorption, were completely independent of the value of the background host dielectric constant. As shown in Paper I, on the other hand, the embedded cavity model predicts that the band profiles in this small particle limit have an asymmetric shape which is steeper to the red side of the peak absorption and exhibits an apparent red "emission" wing. (N.B. The term "emission wing" is an unfortunate one since there is not really any emission associated with it. The term means that the extinction to one side of the absorption line is actually less than that which the same host grains would produce at the same wavelength if they contained no resonant impurities.) The interaction between the host dielectric and resonant impurity was found to dominate the self-width of the resonance in determining the profile width and to result in a shift of the peak absorption frequency away from the natural frequency of the oscillators. As a result, the depth, width, and shape of the absorption profile and the frequency of peak absorption were all found to depend on the composition and even the shape of the host grain, the oscillator strength of the resonance, and the concentration of the impurities in the grain, but to be insensitive to the self-width of the oscillators.

Van de Hulst had also discussed the case of spherical particles large compared with the wavelength, in the limit in which the host index of refraction $m(= n - ik)$ is close to unity and the absorption line is weak. Using the old model to solve for the effective index of refraction of the composite material in the vicinity of the line, van de Hulst found a variety of behavior for the extinction profiles of these grains in the course of varying the important dimensionless parameter $x(n-1)$. The symmetric absorption profiles of the small particle limit occurred again, but only for $x(n-1) < 1$. As this parameter was increased, the extinction curve began to resemble the anomalous dispersion shape of the real part n' of the composite index of refraction m', rather than the absorption line shape of the imaginary part k'. On the blue side of the resonant frequency, in fact, a blue "emission" wing appeared. Eventually, as $x(n-1)$ increased further, the profile was found to be entirely one of apparent "emission," with the absorption inside the line everywhere weaker than that by the same grains without any impurities.

In order to determine whether such profile variations are also predicted by the embedded cavity model, Paper II extended the treatment of Paper I to include the case of particles of arbitrary size compared with the wavelength. This work used the embedded cavity model to calculate the bulk dielectric constants of the composite grain material, along with Mie theory solutions (e.g., Debye 1909; van de Hulst 1957) for the scattering and absorption cross sections of homogeneous spherical grains and the series solutions of Güttler (1952) for the cross sections for spherical core-mantle grains. We used these cross sections to investigate the effect on the grain extinction profiles of variations in such parameters as particle size, host grain composition, impurity oscillator strength, impurity concentration, and the confinement of the resonant impurities to a grain mantle or core. As found in Paper I for the small-particle limit, the results of Paper II indicated that the extinction profiles of the embedded cavity model can differ appreciably from the predictions of the old model for grains of any size. For a complete discussion of the embedded cavity model and its predictions, see Papers I and II. For a detailed discussion of the predictions of the *old* model as applied to the diffuse band problem, see Bromage (1972).

A number of results were obtained using these cross sections to calculate the anomalous extinction profiles of the grains, including the following: (1) For homogeneous grains which are neither highly absorbing nor highly transparent, increasing the grain size parameter $x_o = 2\pi r/\lambda_o$ (where c/λ_o = natural frequency of resonant impurity oscillator) from $x_o \ll 1$ to $x_o > 1$ serves to transform the profile shape from the small particle shape found in Paper I (i.e., asymmetric, with absorption which decreases more steeply to the red side of the peak absorption than the blue and with the appearance of a red "emission" wing), to one for $x_o \approx 1$ which reverses this sense of asymmetry but has no "emission" wings, to one for $x_o \gtrsim 1$, which retains this latter asymmetry but also exhibits a pronounced blue "emission" wing, to one for $x_o \gg 1$, which appears to have reversed almost entirely into an apparent "emission" line. (2) This pattern is composition dependent, however. Increasing the real part of the host dielectric constant, for example, decreases the value of x_o at which the blue "emission" wing sets in, in keeping with a rough scaling property according to which the extinction profile for small values of the impurity concentration is dependent on x_o and the host index $m(= n - ik)$ only through the important dimensionless parameter $x_o(n-1)$. For large enough values of the host dielectric constant, such as that of graphite, however, this scaling property breaks down and a greater variety of profile shapes than described above can occur, as x_o is increased. These include a return at moderate values of x_o to the shape of the small particle limit, including a pronounced red "emission" wing! (3) The frequency width and frequency of peak extinction correspond to those of the embedded cavity model bulk index of refraction. As a result, for example, the FWHM of the extinction profiles is roughly proportional to the oscillator strength of the impurities. Increasing the imaginary part of the host dielectric constant while holding this oscillator strength fixed, however, increases the damping of the impurity oscillators by their interaction with the host dielectric and, hence, broadens and lowers the height of the extinction profile. Increasing the real part of the host dielectric, on the other hand, increases the redward shift of the frequency of peak extinction away from the natural frequency of the impurity oscillator. (4) For fixed grain size and host dielectric constant, the anomalous extinction $(Q_{ext} - Q_{ext,o})/Q_{ext,o}$ (where $Q_{ext,o}$ = grain extinction efficiency in the absence of the impurities) scales roughly in proportion to the impurity concentration, as long as the concentration is small. If the grains have a core-mantle structure, however, rather than a homogeneous one, it is the volume-averaged value of the impurity concentration over the entire grain which enters this scaling relation. (5) For relatively large values of the impurity concentration or small values of the imaginary part of the host dielectric constant, or both, large enough departures of the values of the real and imaginary parts of the bulk index of refraction from their background values can occur so as to trigger optical resonance phenomena in the extinction cross sections. The result can be a remarkable distortion of the simple profile shapes described above, including the appearance of dramatic double- and even triple-peaked profile shapes. (6) For core-mantle grains, the shape of the anomalous extinction profiles is, for a given host grain core and mantle composition, determined by the total grain-size parameter, independent of the relative size of the core and mantle and independent of whether the impurities are in the core or the mantle. The host background composition of the part of the grain containing the impurities can influence the value of the size parameter required in order to achieve a particular size-dependent profile shape (e.g., blue "emission" wing), however, just as it does for homogeneous grains. (7) The frequency width and frequency of peak extinction for core-mantle grains correspond to those of the embedded cavity model bulk index of refraction as calculated for the composite material in that part of the grain which contains the impurities. The effects described in (3), above, therefore, characterize the core-mantle grains, as well, in accordance with this statement. (8) Core-mantle grains with impurities in *both* the core and the mantle but different background dielectric constants for the core and mantle exhibit profiles which are a superposition of the resonance in each of the bulk materials. Hence, asymmetric and even double-peaked extinction profiles can occur in this case. (9) *Increasing* the value of the size parameter while holding the host composition and impurity concentration fixed generally *decreases* the anomalous extinction. This is dictated by Babinet's principle in the large x_o limit and the fact that, for $x_o \ll 1$, the anomalous extinction is size independent.

3. CONSTRAINTS ON HOST GRAIN AND RESONANT ABSORBER FROM THE OBSERVED DIB PROFILES

In Paper III, we used the results of the embedded cavity model described above and in Papers I and II to match the high-resolution, high signal-to-noise ratio diffuse band profiles for the narrow bands at 5780 and 6613Å observed in the direction of the star HD187982 by Snell and Vanden Bout (1981) and Herbig and Soderblom (1982), in order to test the dust grain hypothesis for the origin of the diffuse bands. These profiles were mildly asymmetric, with absorption dropping more steeply to the blue side, an asymmetry opposite to that predicted for small grains of a single type by the embedded cavity model. In the latter, the absorption profile shapes ranged from nearly symmetrical profiles for host grains of slightly dirty silicate to profiles pronouncedly steeper to the red side of the line for graphite hosts, including giant red "emission" wings for the latter. Motivated by the knowledge that these new, high resolution, high signal-to-noise DIB profiles did not share the expectations of the embedded cavity model for *small* grains, Paper III set out to determine if the generalization in Paper II to grains of arbitrary size could explain the observed DIB profiles. We succeeded in demonstrating that a subset of the grains already believed to exist in the interstellar medium *can* satisfactorily explain these diffuse bands as the result of the presence of some as yet unidentified resonantly absorbing impurity molecules in the grains. In the process we significantly constrained the properties of the host grains and of the impurities which are capable of explaining the observed bands. Our results may be summarized as follows.

1. We outlined a step-by-step procedure which can be used, along with an observed band profile shape, FWHM, central depth, and equivalent width, as well as the

observed color excess for the same line of sight, to solve for the impurity resonance oscillator strength and natural frequency, the allowed range of impurity concentration values, and the corresponding required range of host grain column densities along the line of sight, for each assumed host grain background dielectric constant and size distribution.

2. We find that small grains, with size parameter $x_o \ll 1$ and with a single value of impurity concentration, cannot be the diffuse band carrier grains for the bands considered here, since they produce a profile shape which does not share the sense of asymmetry of the observed bands.

3. The superposed profiles of small grains, with $x_o \ll 1$ and a broad distribution of values of impurity concentration, *can* match the observed profile shapes, but only if the concentration distribution is finely tuned.

4. Grains composed of "lossy" substances like graphite and magnetite, with fairly large dielectric constants, are not the carrier grains for the bands considered here, since the profile shapes for $x_o \ll 1$ and $x_o \gtrsim 1$ do not match the observed shapes and the band strength per grain is too weak.

5. *Grains composed of dirty silicate or dirty ice with sizes ranging up to* $\sim 0.1\ \mu m$ *can fit the observed band profiles as long as the size distribution is such that grains with* $x_o \lesssim 1$ *dominate the collective extinction profile.* For an MRN power-law size distribution (Mathis, Rumpl, and Nordsieck 1977), for example, this requires that grains with $x_o > 1$ be excluded as band carriers. Otherwise, grains with $x_o > 1$ introduce shape distortions like "blue emission wings" which are not generally seen. The success of dirty silicates and dirty ice in matching the observed profiles is not very sensitive to the detailed choice of background dielectric constant, as long as the imaginary part is not too small. An illustrative match between the theoretical and observed DIB profiles is shown in Figure 1 for spherical dirty silicate host grains ($\epsilon = 3 - 0.5i$) with an MRN size distribution truncated at the upper end at $\sim 0.1\ \mu m$. (N.B. The observed DIB profiles are not very well-defined in the wings, where firmly establishing the stellar continuum level is difficult.)

6. The observed FWHM for the bands at 5780 and 6613 Å requires that the impurity resonance oscillator strengths be in the range $10^{-3} \lesssim f \lesssim 10^{-2}$ for host grain compositions like dirty silicate or dirty ice. (Interestingly, P. Boissel's talk in this meeting indicated that f-values in this range have been measured for electronic transitions in such molecules as naphthalene, anthracene, and coronene.) If the same grain is the carrier of both bands, then $f_{5780}/f_{6613} \approx 0.4$.

7. The concentration values (i.e. fraction of lattice sites occupied by impurities) for the impurities responsible for the $\lambda\lambda$5780 and 6613 bands are generally in the ratio given by $C_{5780}/C_{6613} \approx \frac{2}{3}$. Given the approximate nature of the theoretical fit to the data, this is consistent with a ratio of unity, meaning that a single impurity with two resonant transitions could be responsible for both bands. However, since these bands are thought to be from different "families" of DIBs, this is not required.

8. In order to avoid the shape-distorting effects of optical resonance phenomena on the band profile, carrier grains with astronomically interesting silicate dielectric constants, which are currently estimated to have imaginary parts less than ~ 0.3, must have an impurity concentration less than several percent. This latter value of C_{max} would be lower if the imaginary part of the host dielectric constant were smaller. The value of C_{max} must never fall below C_{min}, however, where C_{min} is the value such

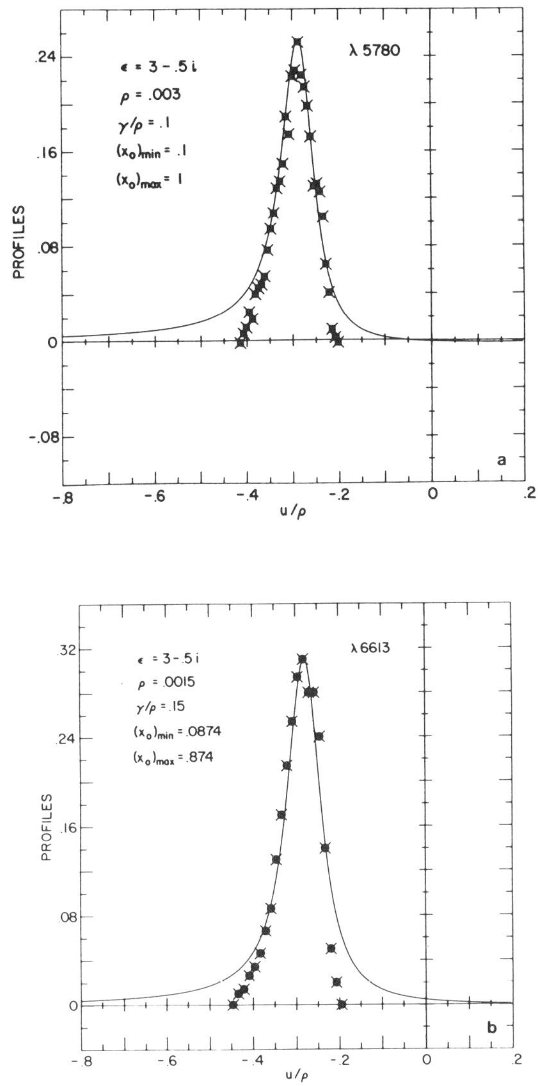

Fig. 1. Theoretical extinction profile (solid line) for spherical host grains of dirty silicate and the observed DIB profile (filled symbols) vs. dimensionless frequency parameter u/ρ [where $u \equiv \lambda_o/\lambda - 1$, $\rho = f\alpha_o/a^3$, f = oscillator strength, $\alpha_o = e^2\lambda_o^2/m_e c^2$, and a = cavity radius = host grain lattice spacing $\cong$ few Å's, and impurity concentration $\cong (1/3)(\gamma/\rho)$], for (a) λ5780, data from Snell and Vanden Bout (1981), and (b) λ6613, data from Herbig and Soderblom (1982).

that carrier grains with a concentration less than C_{min} would exceed the observed $E(B-V)$ if there were enough of them to produce the observed band strength. In principle, therefore, there is a lower limit to the range of acceptable values for the imaginary part of the host grain dielectric constant imposed by the requirement that C_{max} exceed C_{min}. In practice, however, since $C_{max} \gg C_{min}$ for astronomically interesting silicate dielectric constants, we have not explicitly solved for this minimum allowed imaginary part here.

9. The trend reported by Snow and Cohen (1974) for band strength per grain to weaken in ρ Oph with increasing mean grain size for the polarizing grains can be satisfactorily accounted for if the diffuse band carrier grains also vary in size along different lines of sight in proportion to the variations in the observed λ_{max} values (λ_{max} = wavelength of maximum linear polarization of starlight in the optical, known to reflect the mean size of the polarizing grains). However, the size variations of the carrier grains must be the result of the addition of impurity-free mantles of variable thickness, rather than a variation in the total size of the carrier grains due to the addition of more of the core material, containing more of the impurities.

10. An observational test of this latter hypothesis was suggested in which high-resolution, high signal-to-noise ratio profiles of the $\lambda5780$ band are obtained for the stars in ρ Oph. If the band strength variation is the result of grain size variation as described above, then the band profile shapes may vary with λ_{max}. The bands for lines of sight with the largest values of λ_{max}, in fact, may exhibit "blue emission wings." Care must be taken to disentangle the intrinsic profile shape from the shape of the stellar continuum and the shape introduced by multiple velocity components. The latter requires that high resolution atomic absorption line studies of these same lines of sight accompany the DIB observations.

We noted in making this prediction that our assumptions made in explaining the observed variation of band strength in ρ Oph described above if the grain size variations of the polarizing grains are shared by the diffuse band carriers were not unique. They were, rather, merely illustrative of the fact that dust grain DIB carriers could plausibly account for the observed band strength variations. To demonstrate this, for example, we had assumed that the carrier grains were dirty silicates whose particle size for σ Sco, for which $\lambda_{max} = 5550$Å, close to the mean interstellar value, was such that $x_o = 1$ for λ 5780 (i.e. $r \cong 0.1\ \mu m$) while $x_o(\lambda_{max}) = (\lambda_{max}/5550\text{Å})$ for the other lines of sight. We also attributed all of the observed $E(B-V)$ to the carrier grains, in order to be able to solve for the number of carrier grains along each line of sight as we varied the particle size from star to star. Despite the simplicity of these assumptions, we succeeded in explaining the variations in band strength seen with λ_{max}. However, it is also possible that x_o for the carrier grains in σ Sco is less than $x_o = 1$ and that the size of the carrier grains does not vary in direct proportion to the mean size of the polarizing grains. We noted, in fact, that the DIB carrier grains themselves are not likely to be responsible for the polarization of starlight observed in this direction. Attempts to observe the variation of interstellar polarization with wavelength across the bands at 4430, 5780, and 6284Å, expected if the same grains which produce the DIBs also produce most of the polarization, have not detected any such variation in other directions (Martin and Angel 1974, 1975; Fahlman and Walker 1975). Finally, there is no particular reason that all of the observed $E(B-V)$ be attributed to the carrier grains or even that the variation in the total $E(B-V)$ from star to star reflect the variation in the background extinction per star contributed by the carrier grains at wavelengths outside of the DIB profile. Nevertheless, the possibility that carrier grain sizes might vary with λ_{max} in ρ Oph and that this might produce a detectible DIB profile variation, possibly even blue "emission wings," presented an opportunity to confirm the dust grain origin of the DIBs which was worth pursuing.

Recently, Seab and Snow (1995) have followed this suggestion by measuring the profiles for the λ5780 and λ5797 bands towards six stars in the ρ Oph dark cloud, in search of the variation of profile shapes with λ_{max} predicted by Paper III as described above. They did not see any variation with λ_{max} in either the DIB profile shape, the width, or the wavelength λ_c at which the absorption depth is maximum.

The lack of variation of width and of λ_c with λ_{max}, at least, are just as expected from the dust grain hypothesis outlined above. We had already concluded that, if carrier grain sizes vary in proportion to λ_{max} in ρ Oph, only the variation in size caused by the addition of an impurity-free mantle of varying thickness can explain the observed band strength variations with λ_{max}. In that case, point (7) in § 2 above explains that the value of both the band width and the wavelength shift between λ_o and λ_c are properties of the effective bulk dielectric constant for the core material which contains the impurities, invariant with respect to the varying thickness of the impurity-free mantle. However, the observed lack of profile *shape* variation with λ_{max} is not what was expected from the illustrative grain model devised in Paper III to explain the variation of band strength with λ_{max} by assuming the carrier grain size varies with λ_{max}. We can conclude from this absence of profile shape variation that either: (1) The background dielectric constants of the host grains for these DIBs are smaller (i.e. less "lossy") than those of the dirty silicate we assumed ($\epsilon = 3 - i$) in setting $x_o \sim 1$ as the maximum size host grain which can match the observed DIB profile shapes without being too steep on the blue side. If ϵ is decreased, the x_o-value at which blue "emission wings" occur is increased. (2) The real host-grain size parameter for σ Sco is $x_o < 1$, rather than $x_o = 1$ as assumed, so that a variation of x_o with λ_{max} still keeps $x_o \lesssim 1$ for all λ_{max} observed in ρ Oph, thus explaining the lack of blue "emission wings." (3) Host grains in ρ Oph do not vary in size in proportion to λ_{max}, unlike the polarizing grains. (4) Host grains in ρ Oph contribute a portion of the observed $E(B-V)$ which varies with λ_{max}, so that our illustrative model to explain band strength variation with λ_{max} by varying the host grain size with λ_{max} needs to be modified. (5) Dust grains are not the carrier of the DIBs. If some combination of (1)–(4) above are correct, the dust grain hypothesis for the DIBs is still viable. In general, however, the continued failure to detect variations in DIB profiles from one star to the next places a strong constraint on the dust grain model.

ACKNOWLEDGEMENTS. I acknowledge the support of NASA Grants NAGW-2399 and NAG5-2785, and Robert A. Welch Foundation Grant F-1115.

REFERENCES

Bromage, G.E. 1972, *Ap. Space Sci.*, **15**, 426.
Debye, P. 1909, *Ann. Phys.*, **30**, 59.
Draine, B.T. 1995, in *The Infrared Cirrus and Diffuse Interstellar Clouds*, eds. R. Cutri and W. B. Latter (A.S.P. Conference Series), in press.
Fahlman, G.G. and Walker, G.A.H. 1975, *Ap. J.*, **200**, 22.
Güttler, A. 1952, *Ann. Phys.*, **11**, 65.
Herbig, F.H. 1993, *Ap. J.*, **407**, 142.

Herbig, F.H. and Soderblom, D.R. 1982, *Ap. J.*, **252**, 610.
Martin, P.G. and Angel, J.R.P. 1974, *Ap. J.*, **188**, 517.
———. 1975, *Ap. J.*, **195**, 379.
Mathis, J.S., Rumpl, W., and Nordsieck, K.H. 1977, *Ap. J.*, **215**, 425.
Purcell, E.M. and Shapiro, P.R. 1977, *Ap. J.*, **214**, 92 (Paper I).
Seab, C.G. and Snow, T.P., Jr. 1995, *Ap. J.*, in press.
Shapiro, P.R. and Holcomb, K.A. 1986a, *Ap. J.*, **305**, 433 (Paper II).
———. 1986b, *Ap. J.*, **310**, 872 (Paper III).
Snell, R.L. and Vanden Bout, P.A. 1981, *Ap. J.*, **244**, 844.
Snow, T.P. and Cohen, J.G. 1974, *Ap. J.*, **194**, 313.
van de Hulst, H.C. 1949, *Rech. Astr. Obs. Utrecht*, **11**, Part 2.
———. 1957, *Light Scattering by Small Particles* (New York: Wiley).

INTERSTELLAR POLARIZATION

P. G. MARTIN
Canadian Institute for Theoretical Astrophysics

ABSTRACT. Interstellar linear polarization is produced by aligned aspherical grains. Using recent polarimetric observations from infrared to ultraviolet wavelengths it is possible to show that only a subset of all grains causing interstellar extinction actually produce polarization as well. The size and composition dependences are of interest in discussing the origin and evolution of interstellar grains. Possible connections with diffuse interstellar bands are discussed.

1. INTRODUCTION

Interstellar linear polarization (hereafter simply polarization) follows the empirical "modified Serkowski formula" *in the optical and near-infrared* (Whittet *et al.* 1992):

$$p/p_{max} = \exp[-K\ln^2(\lambda_{max}/\lambda)]\,, \quad (1)$$

$$K \simeq 1.66\lambda_{max}\,. \quad (2)$$

The polarization shows a characteristic peak at λ_{max}, which is typically 0.55 μm but ranges over 0.4 – 0.8 μm. Systematic variations such as equation 2 are reviewed in § 2. The basic explanation of the polarization is that at least some of the interstellar grains are both non-spherical and aligned by/with respect to the interstellar magnetic field. The aligned aspherical grains produce differential extinction depending on the orientation of the electric vector with respect to the long axis of the grain profile projected on the plane of the sky (see e.g., Martin 1974). Generally speaking, the grains producing the observed λ_{max} in the optical must have sizes comparable to the wavelength, 0.1 μm, which is also about the size of the grains producing the wavelength-dependent extinction in the optical. For such sizes the electric vector of the transmitted polarized radiation is parallel to the short axis of the projected grain profile, and the projected magnetic field.

What defines the characteristics of this polarizing subset, a matter of some interest, can be probed as follows. The wavelength dependences of polarization and of interstellar extinction are determined by the underlying size distributions of the interstellar particles. A fairly objective study determining the size distribution based on fitting the extinction curve has been presented by Kim, Martin, and Hendry (1994), extending the work of Mathis, Rumpl, and Nordsieck (1977); there is a broad range of sizes present. Less attention has been given to the size distribution and composition of the subset of grains causing polarization, partly because for polarization there are

A. G. G. M. Tielens and T. P. Snow (eds.), The Diffuse Interstellar Bands, 263–270.

more parameters involved, like shape and alignment, and partly because polarization data were not available for a sufficiently wide wavelength range. But since an earlier review (Martin 1989) there has been significant progress both in the infrared (§ 3) and the ultraviolet (§ 4). Therefore it is an opportune time to determine the characteristics of the polarizing grains from the polarization curve. This is a main focus of this review (§ 5), providing some background for polarization studies of diffuse interstellar bands (§ 6).

2. SYSTEMATIC VARIATIONS

The wavelength dependences of both extinction and polarization vary from one line of sight to another, as dust in different galactic environments is probed. However, there are some important systematics to these variations which will be described. At the same time it should be noted that there is real cosmic scatter about these mean trends.

First let us examine extinction. Cardelli, Clayton, and Mathis (1989; see also Mathis 1994) came to the remarkable conclusion that the optical and ultraviolet behavior of diverse extinction curves could be parametrized by a single parameter R_V (the ratio of total to selective extinction A_V/E_{B-V}). This implies that the evolution of grains of all sizes and compositions is tightly interconnected by processes responding efficiently to environmental changes (on much shorter timescales than new grains can be supplied by stars). The sense of the variation is that lines of sight with higher R_V have flatter extinction curves in the ultraviolet. There are correlations with environment too, such as dark (dense molecular) clouds having higher R_V. Martin and Whittet (1990) found that while these major changes in the optical and ultraviolet were occuring little was happening in the infrared. Since the wavelength dependence in any spectral region is produced predominantly by grains comparable in size to the wavelength, the changes must be most dramatic at the small end of the size distribution. This was confirmed and quantified by Kim *et al.* (1994) who interpreted the disappearance of small particles in dense regions as the result of coagulation (obviously shattering would have to occur on dissipation of these regions into more diffuse interstellar atomic clouds).

Now let us turn to polarization. We can already see the foundations for a similar story in equation 1. As demonstrated by Serkowski, the single parameter λ_{max} unifies many seemingly different polarization curves. Serkowski used a fixed value of K, the (inverse) width parameter, but subsequent studies showed another remarkable trend embodied in equation 2. Furthermore, Serkowski and colleagues showed that λ_{max} varies in proportion to R_V, which takes on heightened interest in the above discussion of extinction curves, the environmental relationships, and the evolution of the size distribution. This proportionality is perhaps not surprising if there are such efficient interactions between grains, since the polarizing grains are among those causing extinction. In terms of effective grain size, there must be some increase when λ_{max} increases. But what is really happening across the whole size distribution of polarizing grains? Recall also that all of this is based so far on polarization studies in the optical and near infrared. What about extension of the polarization curve into the infrared or ultraviolet? Will it follow an *extrapolation* of this single-parameter modified Serkowski formula?

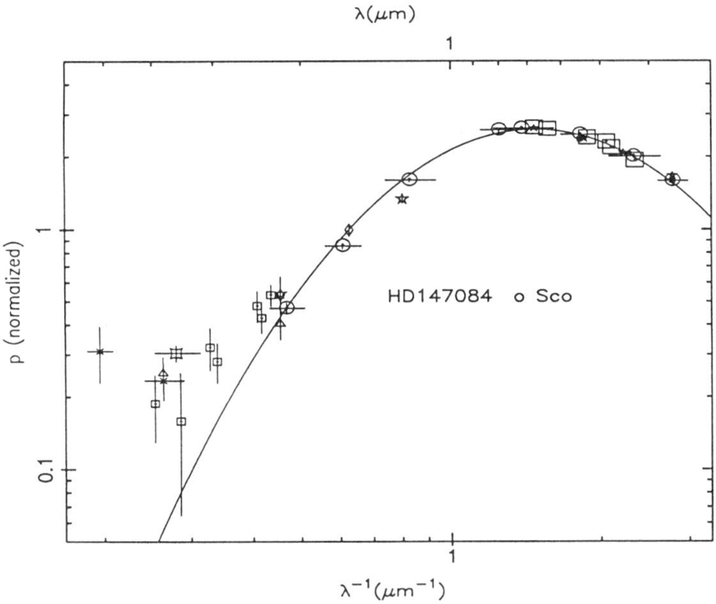

Fig. 1. $\log p$ versus $\log \lambda$ for HD 147084, with different symbols for independent sources of data (Martin *et al.* 1992; data normalized by the polarization at the H passband). The solid curve is a three-parameter fit to the Serkowski formula (parabolic in this plot), treating K as a free parameter, and using optical and near-infrared data at wavelengths shorter than 2.2 μm: $p_{max} = 4.4$ %, $\lambda_{max} = 0.68$ μm, $K = 1.28$.

3. EXTENSION INTO THE INFRARED

Observations have been extended recently to as long as 5 μm for several stars (Nagata 1990; Martin *et al.* 1992). An example is shown in Figure 1. Measurements of the low amounts of infrared polarization are difficult and more are to be encouraged. What appears to be the case is that p lies above a Serkowski extrapolation (Martin *et al.* 1992). Some departure is not surprising since there is no physical basis for the particular Serkowski formula. Statistically the infrared wavelength dependence is more like a common power law with index $\beta \sim 1.6$, *independent of* λ_{max} (and hence independent of the environmental changes that cause the polarization curve to vary in the optical and ultraviolet). This is rather like the common power law behavior (independent of R_V) found for infrared extinction, with similar index (Martin and Whittet 1990).

There has been interesting work on interstellar polarization of the 10-μm silicate feature (§ 5) and on polarized far-infrared and sub-mm emission (electric vector of emission orthogonal to electric vector for transmission) which is beyond the scope of this review. In any case, because of observational constraints these general phenomena have been studied only in selected regions (e.g., dark clouds) rather than the more diffuse interstellar medium being discussed in optical and ultraviolet studies of polarization.

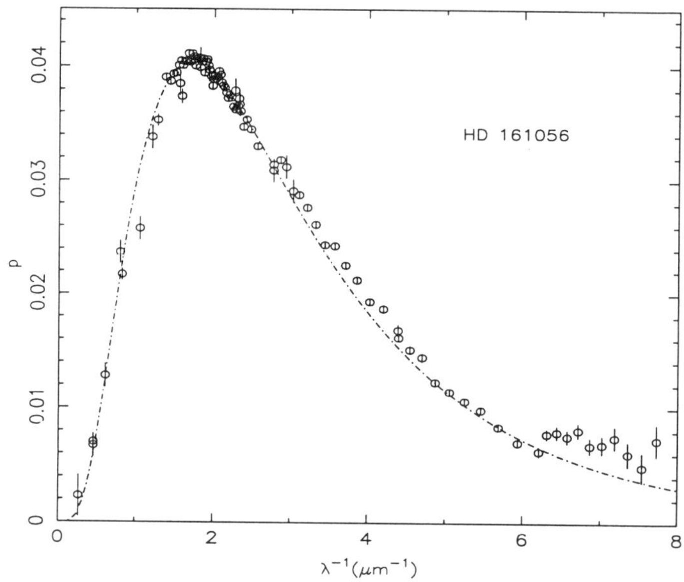

Fig. 2. Interstellar polarization for HD 161056. HST and other data as in Somerville *et al.* (1994a). Dash-dot curve is three-parameter fit of the Serkowski formula to only the ground-based data: $p_{max} = 4.04$ %, $\lambda_{max} = 0.586$ μm, $K = 1.08$.

4. EXTENSION INTO THE ULTRAVIOLET

Gehrels (1974) made balloon-borne polarimetric observations of two stars. There was little spectral resolution but enough to show that the polarization continued to decline from p_{max} in the optical. Ultraviolet spectropolarimetric observations from space have now been reported, making use of the WUPPE experiment on the *Astro-1* mission (Clayton *et al.* 1992; Wolff, Clayton, and Meade 1993) and the Faint Object Spectrograph of the Hubble Space Telescope (Somerville *et al.* 1994a,b; Clayton *et al.* 1994). These have good spectral resolution through the 2175 Å feature out to the continuum beyond 6 μm^{-1}. The *HST* data have somewhat higher spectral resolution and signal-to-noise and most extend to 8 μm^{-1}. A discussion of polarization near the 2175 Å extinction feature is given separately by Martin *et al.* (1994).

An example is given in Figure 2. Here λ_{max} is slightly larger than the average 0.55 μm and the polarization curve is correspondingly narrow. Such narrow widths provide the tightest constraints on the porosity of the aligned grains (Wolff *et al.* 1994). The position angle of the electric vector is constant with wavelength.

In Figure 2 the extrapolated Serkowski formula provides a remarkably close description of the ultraviolet data. However, while this coincidence holds for lines of sight with intermediate to large λ_{max}, it is not a general property. Instead, for smaller λ_{max} (0.50 μm) there tends to be an excess with respect to the extrapolation, despite the fact that the Serkowski curve is intrinsically broader.

5. THE SIZE DISTRIBUTION OF ALIGNED GRAINS

Numerical simulations of interstellar polarization are built up from electromagnetic scattering calculations for single aspherical grains. The simplest computationally is the infinite cylinder which illustrates the main features seen in size distributions for other shapes (e.g., spheroids). An objective way to extract the size distribution from the data (in that no form for the size distribution need be assumed in advance) is the maximum entropy approach. This has been used by Kim *et al.* (1994) for extinction and by Kim and Martin (1994a) for polarization (infinite cylinders). Several compositions were explored with the most satisfactory for polarization being silicate. This is reassuring since the 10-μm silicate feature is polarized (Aitken, Smith, and Roche 1989). A single composition will suffice, in contrast to the case for extinction which requires carbonaceous components in addition to silicates.

Figure 3 shows the size distribution for aligned silicates derived from polarization (λ_{max}= 0.55 μm) for comparison with the size distribution for the silicate component of extinction for diffuse cloud lines of sight ($R_V = 3.1$) shown in Figure 4. Relative to the extinction size distribution, small particles are markedly missing in the aligned grain size distribution (down an order of magnitude by radius $a \sim 0.04$ μm); *small particles are either rounder or harder to align.* This is a general property for all λ_{max} studied (0.49 – 0.68 μm), with interesting systematic behavior (see below).

Mathis (1986) developed an alignment model (shown in Fig. 3) which gives the correct qualitative behavior. The basic premises are that grains are built out of smaller fragments and that alignment, involving magnetic dissipation, depends on a grain containing a superparamagnetic inclusion. Below a certain grain size (which would be 0.1 – 0.2 μm with the normalization adopted here) the probability of containing an inclusion decreases from unity, causing a falloff of the aligned grain size distribution on the small side of the peak.

More recent spheroid-based modeling (Kim and Martin 1994b) has allowed an investigation of the polarization to extinction ratio as a function of grain shape. The observed ratio (< 0.03 at the maximum in the visual) can be reproduced by fairly spherical particles if the alignment is perfect (e.g., oblate spheroids with axial ratio as small as 1.4:1). One gains a factor 2 – 3 leeway using larger axial ratios (2:1 – 4:1). The more realistic spheroid models also provide a better fit to the infrared polarization. The modeling of polarization beyond 6 μm^{-1}, probed by HST, shows a sensitivity to the details of the onset of far ultraviolet electronic absorption in the grain material; the dielectric function must be somewhat different than that of crystalline olivine adopted in "astronomical silicate" (Kim and Martin 1994c).

Systematic variations of the size distribution as λ_{max} increases from 0.55 μm (Fig. 3) are shown in Figure 5. Lines of sight with larger λ_{max} have narrower polarization curves with less ultraviolet polarization (compared to p_{max}) and so the underlying aligned grain size distribution is even more depleted in small grains. Thus the evolution can be interpreted as a narrowing of the size distribution from the small side rather than a major (translational) shift (the effective peak does shift slightly to larger sizes, more or less in proportion to λ_{max}); this is possible because there is so little mass in small grains. Note that when λ_{max} increases so does R_V (on average), and that the extinction size distribution becomes depleted in small particles as well (Kim *et al.* 1994). However, the evolution in the case of aligned grains seems to be more dramatic. Despite the evolution at small sizes, the size distributions (both polariza-

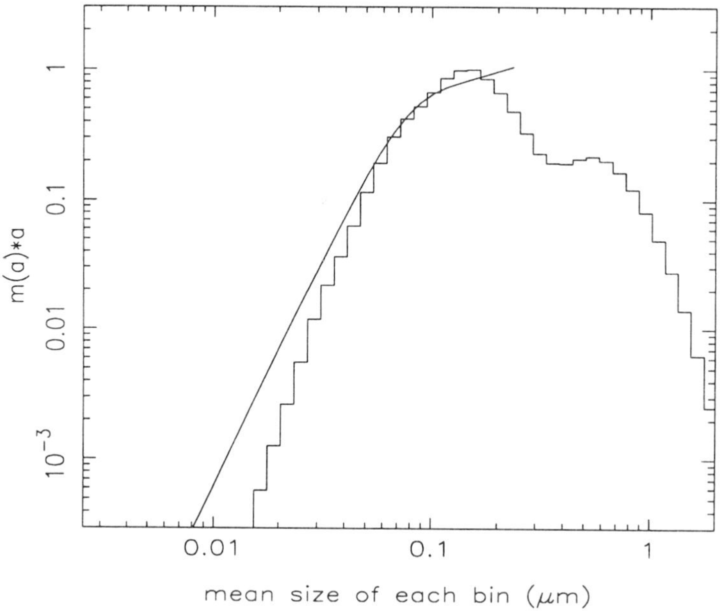

Fig. 3. Maximum entropy silicate mass distribution (based on infinite cylinders; Kim and Martin 1994a) for Galactic average polarization ($\lambda_{max} = 0.55\ \mu$m), normalized to its maximum. The smooth solid line is the model of Mathis (1986), scaled to produce the same polarization near the peak. The slight dip near 0.4 μm is an artefact of using infinite cylinders to try to match the detailed shape of the infrared polarization.

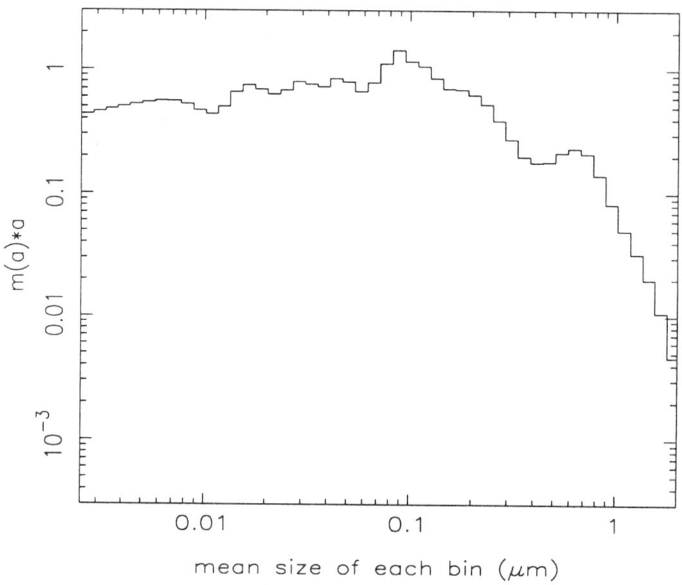

Fig. 4. Mass distribution of silicate component (using infinite cylinders) fitted to the extinction curve with $R_V = 3.1$, normalized so as to coincide with distribution in Fig. 3 for $a > 0.1\ \mu$m.

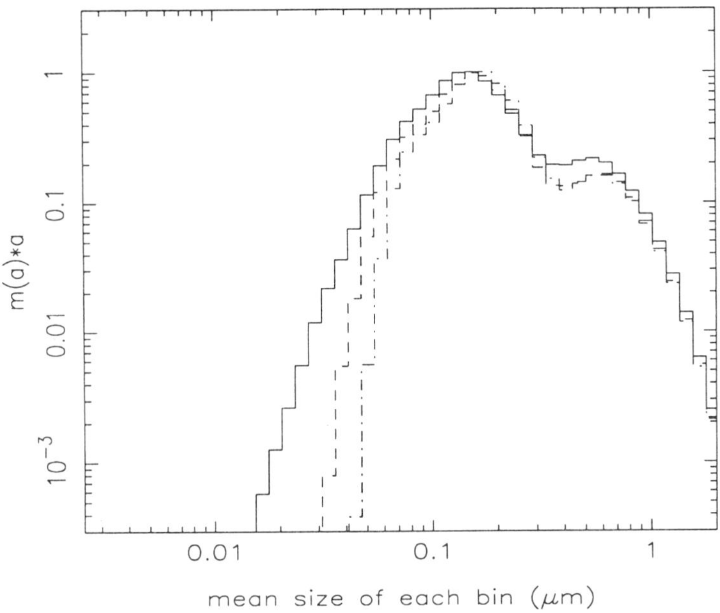

Fig. 5. Aligned grain mass distributions for $\lambda_{max} = 0.55$ (solid curve), 0.615 (dashed) and 0.68 μm (dot-dashed), normalized to their maxima.

tion and extinction) are much less affected for larger sizes (> 0.2 μm), as expected from the relative invariance of the infrared polarization and extinction curves with environment. Evolution by coagulation and shattering seems a worthwhile model to explore.

6. RELATIONSHIP TO STUDIES OF DIBS

Our early polarization studies of diffuse interstellar bands at 4430, 5780, and 6284 Å (Martin and Angel 1974, 1975) were motivated as follows: if DIBs have profiles with asymmetrical structure (even emission wings) then this might be indicative that the carriers are large grains; at least some of the large grains cause optical polarization; therefore, there might be polarization structure associated with the bands. In principle, measurements of structure in the degree of polarization are more straightforward than measurements of the absorption profile, since the underlying stellar spectrum does not have to be known. However, no polarization structure was detected at a level expected of grain carriers. In the meantime the evidence for emission wings has apparently vanished. Observations of similar quality for the 5797 Å feature and other DIBs in the 5600 – 6000 Å region have yielded null results too (Adamson and Whittet 1992, 1994). Thus the DIB carriers cannot be identified with the large "classical-sized" (0.1 μm) grains which cause the continuum polarization described above.

One possibility that might still be consistent with this negative polarimetric evidence is that the DIB carriers are associated with the small grain population. It was found above on the basis of continuum polarization that small silicate particles (especially $< 0.04\mu$m) are either quite round or hard to align. The small carbonaceous (graphite) grains causing the 2175 Å extinction feature are on the whole non-polarizing as well (Martin *et al.* 1994). A second possibility is some independent larger-sized carbonaceous population that does not produce continuum polarization either. However, if these grain models are to be entertained then it is important to consider the implications of grain evolution: for example, if processes like coagulation and shattering are efficient at reshaping the grain size distribution, would one expect the carrier to be associated only with the grains that are small at the present? Furthermore, it is necessary to examine the range of band profiles expected, as Shapiro (1994) has done in detail.

ACKNOWLEDGEMENTS. This research is supported by the Naturual Sciences and Engineering Research Council of Canada. I thank S.-H. Kim for discussions and comments on the manuscript.

REFERENCES

Adamson, A.J. and Whittet, D.C.B. 1992, *Ap. J. (Letters)*, **398**, L69.

Adamson, A.J. and Whittet, D.C.B. 1994, in *The Diffuse Interstellar Bands*, eds. A. G. G. M. Tielens and T.P. Snow, (Dordrecht: Kluwer), this volume.

Aitken, D.K., Smith, C.H., and Roche, P.F. 1989, *M. N. R. A. S.*, **236**, 919.

Cardelli, J.A., Clayton, G.C., and Mathis, J.S. 1989, *Ap. J.*, **345**, 245

Clayton, G.C., *et al.* 1992, *Ap. J. (Letters)*, **385**, L53.

Clayton, G.C., Wolff, M.J., Allen, R.G., and Lupie, O. 1994, in preparation.

Gehrels, T. 1974, *A. J.*, **79**, 590.

Kim, S.-H. and Martin, P.G. 1994a, *Ap. J.*, **431**, 783.

Kim, S.-H. and Martin, P.G. 1994b, *Ap. J.*, in preparation.

Kim, S.-H., and Martin, P.G. 1994c, *Ap. J.*, submitted.

Kim, S.-H., Martin, P.G., and Hendry, P.D. 1994, *Ap. J.*, **422**, 164.

Martin, P.G. 1974, *Ap. J.*, **187**, 461.

Martin, P.G. 1989, in *IAU Symposium 135, Interstellar Dust*, eds. L.J. Allamandola and A.G.G.M. Tielens, (Dordrecht: Kluwer), p. 55.

Martin, P.G., Adamson, A.J., Whittet, D.C.B., Hough, J.H., Bailey, J.A., Kim, S.-H., Sato, S., Tamura, M., and Yamashita, T. 1992, *Ap. J.*, **392**, 691.

Martin, P.G. and Angel, J.R.P. 1974, *Ap. J.*, **188**, 517.

Martin, P.G. and Angel, J.R.P. 1975, *Ap. J.*, **195**, 379.

Martin, P.G., Somerville, W.B., McNally, D., Whittet, D.C.B., Allen, R.G., Walsh, J.R., and Wolff, M.J. 1994, in *The Diffuse Interstellar Bands*, eds. A. G. G. M. Tielens and T.P. Snow, (Dordrecht: Kluwer), this volume.

Martin, P.G. and Whittet, D.C.B. 1990, *Ap. J.*, **357**, 113.

Mathis, J.S. 1986, *Ap. J.*, **308**, 281.

Mathis, J.S. 1994, in *The Diffuse Interstellar Bands*, eds. A. G. G. M. Tielens and T.P. Snow, (Dordrecht: Kluwer), this volume.

Mathis, J.S., Rumpl, W., and Nordsieck, K.H. 1977, *Ap. J.*, **217**, 425.

Nagata, T. 1990, *Ap. J. (Letters)*, **348**, L13.

Somerville, W.B. *et al.* 1994a, *Ap. J. (Letters)*, **427**, L47.

Somerville, W.B. *et al.* 1994b, in preparation.

Whittet, D.C.B., Martin, P.G., Hough, J.H., Rouse, M.F., Bailey, J.A., and Axon, D.J. 1992, *Ap. J.*, **386**, 562.

Wolff, M.J., Clayton, G.C., Martin, P.G., and Schulte-Ladbeck, R.E. 1994, *Ap. J.*, **423**, 412.

Wolff, M.J., Clayton, G.C., and Meade, M.R. 1993, *Ap. J.*, **403**, 722.

POLARIZATION OF THE 2175 Å FEATURE

P. G. MARTIN
Canadian Institute for Theoretical Astrophysics
W. B. SOMERVILLE, D. MCNALLY
University College London
D. C. B. WHITTET
Rensselaer Polytechnic Institute
R. G. ALLEN
University of Arizona
J. R. WALSH
European Southern Observatory
M. J. WOLFF
University of Colorado

ABSTRACT. Although the 2175Å feature in the interstellar extinction curve is often attributed to absorption by small particles of graphite, the identification is not secure. Any further characterization of the carrier would be valuable. One question is whether the feature is polarized. Polarization properties in the 2175Å region put important constraints on the asphericity, optical anisotropy, and alignment of the carrier. Ultraviolet spectropolarimetric observations have now been reported for several bright reddened Galactic stars. These made use of the WUPPE experiment on the Astro-1 mission (Wolff, Clayton, & Meade 1993) and the Faint Object Spectrograph of the *Hubble Space Telescope* (Somerville *et al.* 1994a,b); the latter have somewhat higher spectral resolution and signal-to-noise. A polarization signature accompanying the 2175Å feature appears to be rare. Both the observations and the theoretical interpretation are discussed.

1. INTRODUCTION

There is a prominent feature in the extinction curve at 2175Å (see e.g., Mathis 1994), discovered about three decades ago and immediately attributed to graphite (Stecher and Donn 1965). A graphite component of interstellar dust, of circumstellar origin, had been proposed earlier by Hoyle and Wickramasinghe (1962). Cayrel and Schatzman (1954) discussed interstellar linear polarization produced by aligned flakes of graphite. Subsequent analyses summarized by Wickramasinghe (1967) and Greenberg (1968) showed that whereas sizes of about 0.05 μm would be required for optical extinction, and 0.02 μm or less for the ultraviolet feature, the wavelength dependence of optical interstellar polarization would require somewhat larger grains, about

A. G. G. M. Tielens and T. P. Snow (eds.), The Diffuse Interstellar Bands, 271–278.

0.1 μm. Thus it was clear from the early models that a unified explanation of both extinction and polarization using simply graphite was not straightforward.

Small (0.01 μm) particles of graphite still figure importantly in many current grain models in order to explain the 2175Å extinction feature (Draine 1988, 1989; Mathis 1994). This appears consistent with measurements of the ultraviolet albedo (Witt 1989; Witt *et al.* 1993 and references therein) which show a slight reduction across this feature. Laboratory measurements of extinction by graphite microparticles confirm the existence of the surface plasmon resonance of the π-electrons, although a detailed match of the astronomical feature has not yet been achieved (Huffman 1989; Lorusso *et al.* 1993).

It is possible to explain continuum interstellar linear polarization simply with intermediate to large-sized silicate grains (Kim and Martin 1994). Nevertheless, if the 2175Å feature were due in part to aligned particles (whether graphite or not) there might be an interesting frequency dependence diagnostic of the shape, alignment, and optical anisotropy of the particles (Draine 1988). Gehrels (1974) made preliminary balloon observations and such ultraviolet polarization investigations have been a goal since the planning days of the WUPPE and HST projects (§ 4).

2. THE EXTINCTION FEATURE

Many salient observational constraints were reviewed by Draine (1988). Empirical descriptions of the ultraviolet extinction data often describe this feature by a Drude profile,

$$f(w) = \frac{\gamma^2}{(w - w_o^2/w)^2 + \gamma^2}, \tag{1}$$

(here $w = \lambda^{-1}$) superimposed on some continuuum. A Lorentzian has been used too. The peak position w_o is fairly constant at 4.6 μm^{-1} and there is some variation in the FWHM ($\sim \gamma$) around the mean 1.0 μm^{-1}. By virtue of both this large width and the strength compared to other features discussed at this meeting ($\Delta\tau/E_{B-V}$ typically 3.4; see Table 1 below), this feature might be dubbed the mother of all DIBs.

The longstanding explanation in terms of small particles of graphite can be modeled using the DDA (discrete dipole array) method for electromagnetic scattering. The validity of this method has been checked by performing DDA calculations for spherical particles and comparing them with Mie theory results using the same graphite dielectric function (Draine 1988). A representation of the DDA disk, or short cylinder, used in our calculations is given in Figure 1. The graphitic planes are parallel to the end surfaces of the cylinder. Graphite is optically anisotropic and appropriate dielectric functions have been estimated by Draine (1987). The feature in question arises when the incident electric vector is oriented perpendicular to the c-axis, i.e., parallel to the planar structure of graphite. Given the adopted dielectric functions, Draine found that the particles must be small ($\sim 0.01\mu$m) and not too elongated or flattened in order to reproduce the observed w_o and γ. The disk in Figure 1 has axial ratio 1.55:1, has "equivalent radius" $a_{eq} = 0.01$ μm^{-1}, and consists of $N = 15140$ dipoles. Comparison with supplementary calculations using $N = 960$ indicates that the results are converged with respect to N.

Extinction efficiency factors Q (cross sections normalized by πa_{eq}^2) were calculated

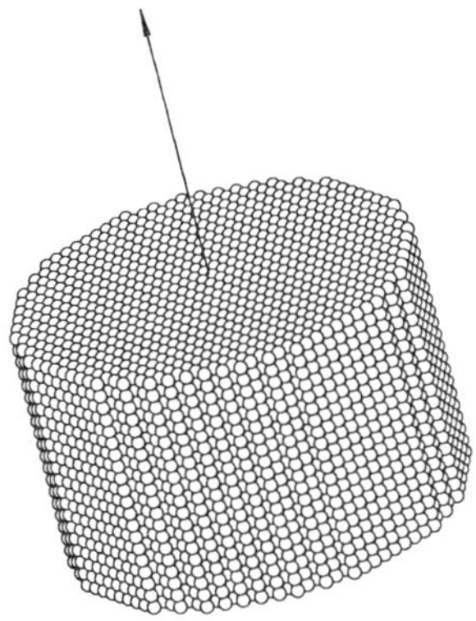

Fig. 1. Representation of DDA graphite grain, showing projection of the 15140 dipoles (not carbon atoms). Aspect ratio is 1.55:1. Axis shown is the c-axis, perpendiclar to the planar structure of the graphite disk. An electric vector orientation perpendicular to the c-axis gives rise to the 2175Å feature in extinction.

first for three special geometries: two for propagation vector perpendicular to the c-axis, the so-called H polarization for electric vector orientation perpendicular to the c-axis, and the E polarization for orientation along the c-axis; and the third for propagation vector along the c-axis, called here the K polarization. We verified that for a particle this small compared to the wavelength, the electric dipole approximation was fairly good, so that $Q_K \simeq Q_H$; therefore the extinction for random orientation is $Q_{random} \simeq (Q_H + Q_E + Q_K)/3 \simeq (2Q_H + Q_E)/3$. This too was verified by direct calculation for an explicitly randomly oriented ensemble. Note that the approximation used by Voshchinnikov (1990), in which the dielectric functions rather than the cross sections are averaged in this $2/3 - 1/3$ manner, is incorrect.

The extinction for random orientation, which would approximate the interstellar situation for poor grain alignment, is shown in Figure 2. We have carried out a 6-parameter fit to the DDA results using the Drude function (eqn. 1) plus a simple quadratic continuum finding $w_o = 4.56\ \mu m^{-1}$ and $\gamma = 1.04\ \mu m^{-1}$ (see Fig. 2), in satisfactory agreement with the observations. This can be compared to Figure 12 of Draine (1988) which should be similar but instead has a lower peak and a somewhat larger width. This discrepancy is puzzling since our polarization calculations (differences of the same efficiency factors – see § 3) agree so well.

3. POLARIZATION BY ALIGNED GRAPHITE PARTICLES

If the grains are aligned, then transmitted light will be linearly polarized. The most likely orientation would have the c-axis of the disk aligned on average along the Galactic magnetic field. Consider perfect alignment and propagation perpendicular to the magnetic field. Extinction is stronger for the in-plane H polarization (Fig. 3), so that the electric vector for transmitted radiation is parallel to the Galactic magnetic field, just as in the continuum polarization. The magnitude of polarization p is proportional to $(Q_H - Q_E)/2$ (Martin 1974). The frequency dependence is again well described (not shown) by a Drude function (plus negligible continuum), with *the same* w_0 *and* γ; the polarization and birefringence results agree well with those in Figure 16a of Draine (1988). The ratio of polarization (differential extinction) to extinction, p/τ, is high (~ 0.8) for this perfect alignment. Imperfect alignment would

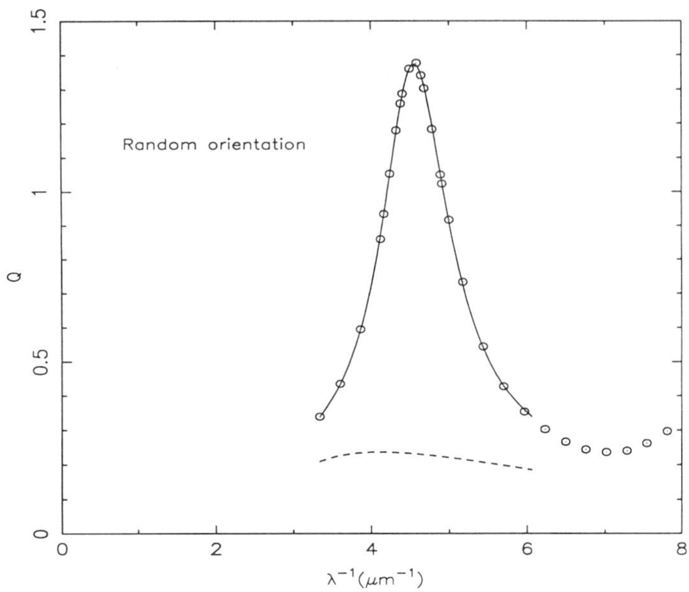

Fig. 2. Frequency dependence of the extinction efficiency factor Q for random orientation of the grain in Fig. 1. Open circles: DDA calculations. Solid line: Drude fit. Dashed line: underlying continuum.

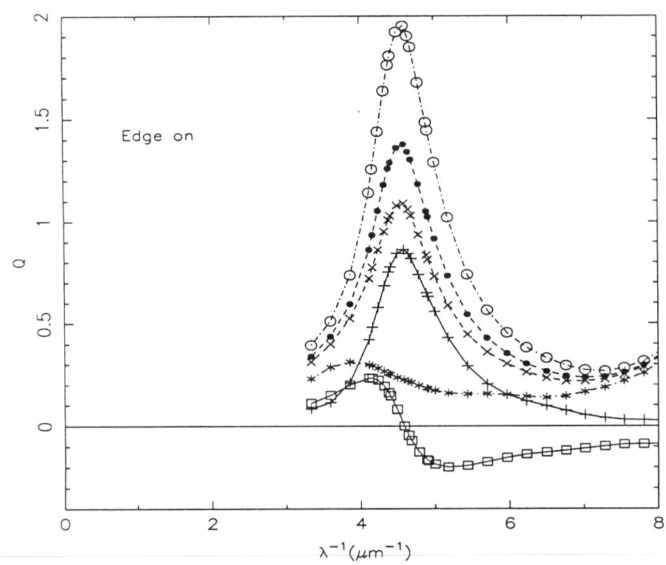

Fig. 3. Various combinations of DDA extinction efficiency factors for edge-on orientation of the grain in Fig. 1, identified by symbol. Open circle: Q_H. Star: Q_E. Filled circle: $(2Q_H + Q_E)/3$ (random orientation approximation). Cross: $(Q_H + Q_E)/2$ (proportional to extinction τ). Plus: $(Q_H - Q_E)/2$ (proportional to polarization p). Square: corresponding function for birefringence (see Martin 1974).

lower p/τ but *the same frequency dependence would hold for* p since these particles are sufficiently small that the electric dipole approximation is fairly good.

Different shapes could be considered, for example a stack of disks making a cylinder with length somewhat greater than diameter. In this case the grains would be spinning about a short axis, with angular momentum vector along the Galactic magnetic field but the projected c-axis now perpendicular to the field. The extinction would still be stronger for the in-plane polarization, so that the electric vector for transmitted radiation would be perpendicular to the Galactic magnetic field, *opposite to the continuum polarization.* Assuming that the size and axial ratio were adjusted (chosen) to match the observed extinction, the frequency dependence of p would still be described by a Drude function with a similar w_o and γ. Some examples are given by Draine (1988).

Even spheres composed of parallel sheets of graphite would produce polarization, if the c-axes of the spheres could be aligned. The polarization comes from the structural order and anisotropic dielectric function of the material. The importance of an aspherical shape is to give a preferred axis for alignment of the angular momentum vector (whence the Galactic magnetic field) with respect to the body axes.

4. OBSERVATIONS

Gehrels (1974) made balloon-borne polarimetric observations of two stars. There was little spectral resolution to distinguish the feature from the continuum, but these data were enough to show that the polarization was less than the maximum value (p_m) in the visual and that there was no dramatically strong feature.

Spectropolarimetric observations from space have been carried out in the 1990's, first by *WUPPE* (Clayton *et al.* 1992; Wolff, Clayton, and Meade 1993) and then by *HST* (Somerville *et al.* 1994a,b). These have good spectral resolution through the 2175Å feature out to the continuum beyond 6 μm^{-1}.

The stars for which the complete analysis of § 5 was possible are listed in Table 1, which includes p_m and the wavelength λ_m (in μm) at which it occurs, the colour excess E_{B-V}, the continuum optical depth τ_c at the 2175Å feature, and the feature strength $\Delta\tau$ (the latter two based on data in Carnochan 1986).

5. DERIVED POLARIZATION TO EXTINCTION RATIOS

Based on the discussion and specific calculations presented in §§ 2 and 3, we fit the observed ultraviolet polarization with a simple local continuum plus a Drude function with $w_o = 4.6\ \mu m^{-1}$ and $\gamma = 1.0\ \mu m^{-1}$; this gives p_c (the underlying continuum at w_o) and the amplitude of the feature Δp. An example is given in Figure 4 and the results are collected in Table 1. Generally Δp is small (the percentage values listed are often limits as noted) and only for HD 197770 does Δp have statistical significance (§ 6).

From the basic data in Table 1 we form the ratios p_m/τ_V, p_c/τ_c, and $\Delta p/\Delta\tau$. Forming the ratios removes the effects of differing column density from star to star. Comparing values of p_m/τ_V measures differences in alignment efficiency (including

TABLE 1

Polarization efficiency in the visual, the ultraviolet, and at the 2175 Å Feature

HD	λ_m	p_m	E_{B-V}	p_m/τ_V	p_c	τ_c	p_c/τ_c	Δp^a	$\Delta\tau$	$\Delta p/\Delta\tau$
7252	0.52	3.5[b]	0.34	3.6–02[c]	2.0[b]	1.9	1.1–02[c]	0.046[d]	1.3	3.4–04[c]
96985	0.58	2.1	0.23	3.2–02	0.75	1.2	6.1–03	0.053[d]	0.7	7.6–04
99872	0.56	3.1	0.36	3.0–02	1.3	1.9	6.8–03	0.048	1.3	3.6–04
161056	0.59	4.0	0.63	2.2–02	1.5	3.5	4.2–03	0.055[d]	2.1	2.6–04
25443	0.49	5.2	0.61	3.0–02	3.6	3.5	1.0–02	0.15[d]	2.0	7.3–04
30614	0.51	1.6	0.30	1.9–02	1.2	1.7	6.8–03	0.060[d]	1.0	6.0–04
37903	0.68	2.0	0.35	2.0–02	0.34	1.8	1.9–03	0.091[d]	1.1	8.1–04
62542	0.58	1.4	0.33	1.5–02	0.74	1.9	3.9–03	0.25[d]	1.1	2.3–03
99264	0.55	2.7	0.30	3.1–02	1.1	1.6	7.0–03	0.097	1.0	9.5–04
197770	0.51	3.8	0.57	2.3–02	2.1	3.2	6.5–03	1.0	1.9	5.4–03

[a] maximum of the absolute value of the amplitude of the polarization feature or its formal error[d] from the parameterized fit, in percent.
[b] polarization in percent, except in ratios.
[c] ratio and exponent.
[d] limit based on 1σ from the parameterized fit, in percent.

field orientation) for different lines of sight. For a given star the continuum ratio depends on frequency (lower at w_o than in the visual) because of changes in the effective grain size compared to the wavelength; also stars with larger λ_m have relatively less ultraviolet polarization. Because of the selection criteria for ultraviolet polarization observations, the continuum ratios are all fairly high.

Except for HD 197770, the ratio $\Delta p/\Delta\tau$ is very small compared to both the theoretical maximum (0.8) and the continuum ratios just discussed, indicating no significant polarization signature (the position angle of the electric vector does not change across the feature either). Evidently the small (graphite) grains causing the extinction feature are poorly aligned. Recall that silicate particles of comparable small size ($\sim 0.01\mu$m) are also poorly aligned (Martin 1994).

6. HD 197770

Of these ten stars (and four others) now observed by *WUPPE* and *HST*, HD 197770 alone shows a significant Δp. Nevertheless, $\Delta p/\Delta\tau$ is still more than two orders of magnitude less than the theoretical maximum. The data and standard fit are shown in Figure 5. Note that Δp is positive, in the sense of aligned disks rather than aligned cylinders (§ 3). The position angle is the same as in the continuum. As can be judged from Figure 5, a significantly better fit is obtained if the parameters of the Drude function are optimized: γ is broadened to 1.11 μm^{-1} and w_o is lowered to 4.43 μm^{-1} (2260Å), formally three sigma away from the standard 4.6 μm^{-1}. Confirmation of these differences with respect to the values for the extinction feature would have interesting implications for the composition, size, and shape of the *aligned* carriers.

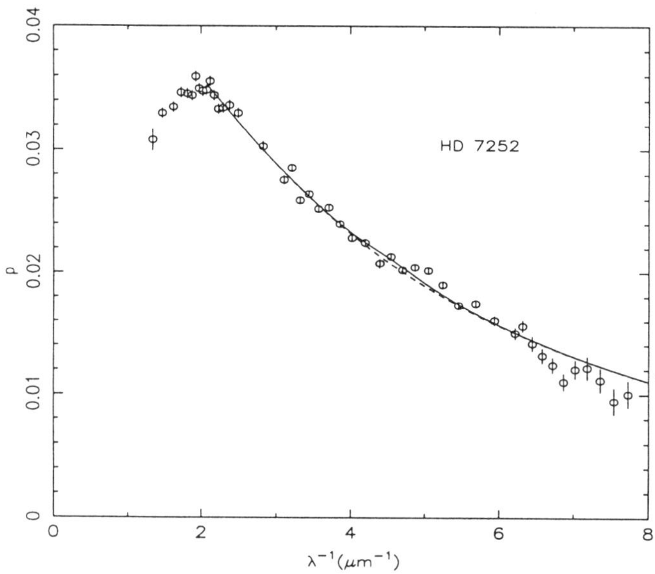

Fig. 4. *HST* ultraviolet linear polarization observations of HD 7252 (Somerville *et al.* 1994a). Dashed line: underlying local continuum, with $p_c = 2.0$ %. Solid line: total fit including continuum and Drude function with amplitude $+0.041 \pm 0.046$ %.

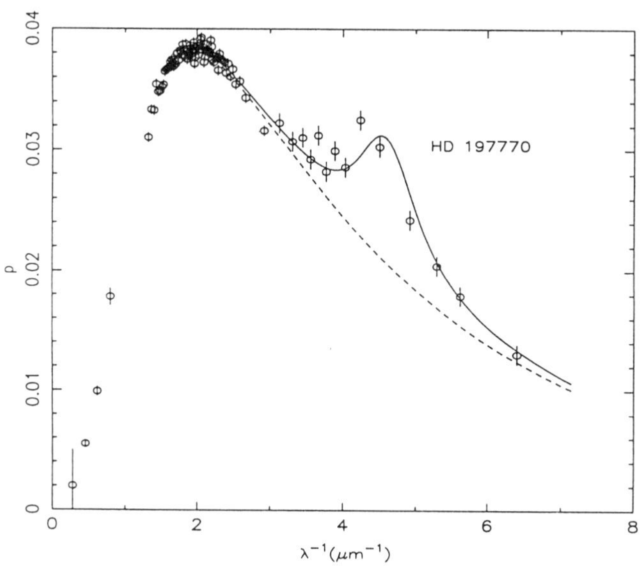

Fig. 5. *WUPPE* ultraviolet linear polarization observations of HD 197770 (Wolff *et al.* 1993). Dashed line: underlying local continuum, with $p_c = 2.07$ % (like Serkowski formula extrapolation from infrared and visual *if* width parameter $K = 1.66\lambda_m$, and so perhaps not super-Serkowski as is typical for small λ_m). Solid line: total fit including continuum and Drude function (fixed $w_o = 4.6\ \mu\text{m}^{-1}$ and $\gamma = 1.0\ \mu\text{m}^{-1}$) with amplitude $+1.04 \pm 0.14$ %.

7. THE FUTURE

Installation of COSTAR during the *HST* refurbishment mission in December 1993 has affected the instrumental polarization of the FOS spectropolarimeter. Prior to COSTAR three gratings were used to cover the full ultraviolet spectral range from 1300 to 3300 Å, but post-COSTAR polarimetry will be attempted in the first instance only in the longest-wavelength grating, G270H, which includes the long-wavelength half of the 2175Å feature. Assuming the calibration is successful, Cycle 4 time awarded to Somerville *et al.* should be sufficient to observe six further stars. Unfortunately, HD 197770 was not observed with *HST* before the installation of COSTAR. However, WUPPE has been scheduled for more observations (*circa* January 1995) which should allow detailed spectropolarimetry of this star and many more.

As emphasized, polarization properties in the region of the 2175Å feature already put important constraints on the asphericity, optical anisotropy, and alignment of the carrier. Planned ultraviolet spectropolarimetry should place us in a better position to assess such interesting questions as how the carrier of the 2175Å (or related) feature can sometimes be aligned with respect to the Galactic magnetic field.

ACKNOWLEDGEMENTS. This paper is based in part on observations with the NASA/ESA Hubble Space Telescope, obtained at the Space Telescope Science Institute, which is operated by the Association of Universities for Research in Astronomy, Inc., under NASA contract NAS5–26555.

REFERENCES

Carnochan, D. J. 1986, *M. N. R. A. S.*, **219**, 903.
Cayrel, R. and Schatzman, E. 1954, *Ann. d'ap.*, **17**, 555.
Clayton, G.C. *et al.* 1992, *Ap. J. (Letters)*, **385**, L53.
Draine, B.T. 1987, *Princeton Observatory Preprint 213*, Princeton University.
Draine, B.T. 1988, *Ap. J.*, **333**, 848.
Draine, B.T. 1989, in *IAU Symposium 135, Interstellar Dust*, eds. L. J. Allamandola and A. G. G. M. Tielens, (Dordrecht: Kluwer), p. 313.
Gehrels, T. 1974, *A. J.*, **79**, 590.
Greenberg, J.M. 1968, in *Nebulae and Interstellar Matter*, eds. B. M. Middlehurst and L. H. Aller, (Chicago: Univ. Chicago Press), p. 221.
Hoyle. F. and Wickramasighe, N.C. 1962, *M. N. R. A. S.*, **124**, 417.
Huffman, D.R. 1989, in *IAU Symposium 135, Interstellar Dust*, eds. L. J. Allamandola and A. G. G. M. Tielens, (Dordrecht: Kluwer), p. 329.
Kim, S.-H. and Martin, P.G. 1994, *Ap. J.*, **431**, 783.
Lorusso, G.F., Capozzi, V., Milani, P., Minafra, A., and Lojacono, D. 1993, *Solid State Comm.*, **85**, 729.
Martin, P.G. 1974, *Ap. J.*, **187**, 461.
Martin, P.G. 1994, in *The Diffuse Interstellar Bands*, eds. A. G. G. M. Tielens and T.P. Snow, (Dordrecht: Kluwer), this volume.
Mathis, J.S. 1994, in *The Diffuse Interstellar Bands*, eds. A. G. G. M. Tielens and T.P. Snow, (Dordrecht: Kluwer), this volume.
Somerville, W.B. *et al.* 1994a, *Ap. J. (Letters)*, **427**, L47.
Somerville, W.B. *et al.* 1994b, in preparation.
Stecher, T.P. and Donn, B.D. 1965, *Ap. J.*, **142**, 1681.
Voshchinnikov, N.V. 1990, *Sov. Astron. Letters*, **16**, 215.
Wickramasinghe, N.C. 1967, *Interstellar Grains*, (London: Chapman and Hall).
Witt, A.N. 1989, in *IAU Symposium 135, Interstellar Dust*, eds. L. J. Allamandola and A. G. G. M. Tielens, (Dordrecht: Kluwer), p. 87.
Witt, A.N., Petersohn, J.K., Holberg, J.B., Murthy, J., Dring, A., and Henry, R.C. 1993, *Ap. J.*, **410**, 714.
Wolff, M.J., Clayton, G.C., and Meade, M.R. 1993, *Ap. J.*, **403**, 722.

CHAPTER V

Theoretical Studies of Molecular Carriers

SPECTROSCOPY AND PHOTOPHYSICS OF POSSIBLE LARGE MOLECULE CARRIERS OF THE DIFFUSE INTERSTELLAR BANDS

S. LEACH*
Observatoire de Paris-Meudon, France

ABSTRACT. A review is presented of spectroscopic and photophysical properties of possible large molecule carriers of the diffuse interstellar bands (DIBs). These include neutral and ionic polycyclic aromatic hydrocarbons (PAHs), fullerenes, and the carbon chain species C_n, polyenes, polyynes and polyynyl radicals. Their spectral properties in the 200 - 1200 nm region are discussed and a comparison is made with those of observed DIBs, reinforcing a recent suggestion that the carriers are polyynyl radicals C_nH. Rotational contours are simulated for vibronic bands of these various potential DIB carriers and are compared with observed DIB profiles. Line broadening mechanisms are discussed and suggestions are made concerning the size of DIB carriers. It is concluded that the observed DIB bandwidths are consistent with molecular carriers containing more than 10 - 15 atoms.

1. INTRODUCTION

The diffuse interstellar bands (DIBs) are absorption lines, whose widths are between 1 and 100 cm^{-1}, seen in the lines-of-sight of a variety of galactic and extragalactic objects. Assignment of the DIBs to specific carriers is an outstanding unresolved problem in astrophysics, although many suggestions have been made (Bromage 1987; McNally 1994; Snow, this volume). In particular, a number of gas phase species have been proposed as generic DIB carriers or as carriers of specific bands. These include polycyclic aromatic hydrocarbons in neutral or ionic form (van der Zwet and Allamandola 1985; Léger and d'Hendecourt 1985), fullerenes (Kroto 1987, 1988, Gerhardt *et al.* 1987, Léger *et al.* 1988), hydrogenated fullerenes (Webster 1992), linear carbon chains C_n, polyynes C_nH and polyenes HC_nH (Douglas 1977; Thaddeus *et al.* 1993; Fulara *et al.* 1993). The present review discusses the spectroscopic and photophysical properties of these species and of their ionic variants in the context of identification of DIB carriers.

*Département Atomes et Molécules en Astrophysique, Observatoire de Paris-Meudon, 92195-Meudon, France

A. G. G. M. Tielens and T. P. Snow (eds.), The Diffuse Interstellar Bands, 281–305.

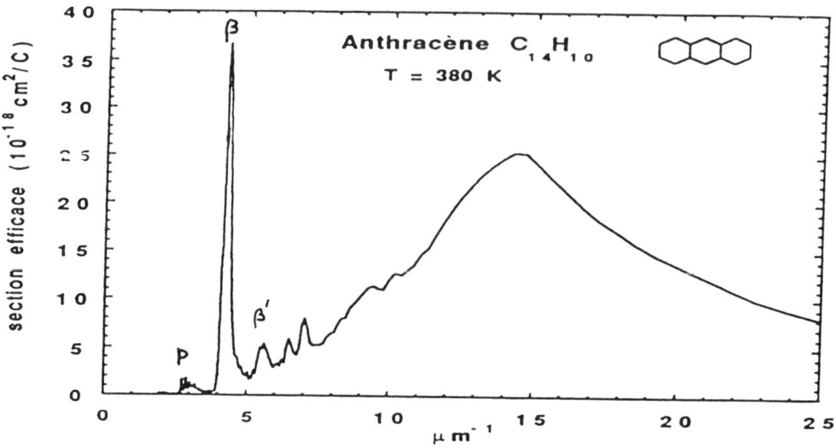

Fig. 1. Absorption spectrum of anthracene (after Joblin, 1992). The absorption cross-section (section efficace) is given in units of 10^{-18}cm^2 per carbon atom.

2. POLYCYCLIC AROMATIC HYDROCARBONS AND THEIR IONS

2.1. *THE SPECTROSCOPY AND IONIZATION YIELDS OF NEUTRAL PAHS*

The lowest energy singlet-singlet transitions of neutral PAHs are usually described by the band notations p, α, β, β' of the empirical model of Clar (1941), or ^1L_a, ^1L_b, ^1B_b, ^1B_a based on the extended free-electron model of Platt (1949), as discussed in detail elsewhere (Leach 1989). These two notations allow one to compare related transitions for PAH molecules which have different nuclear symmetries. The p and β' bands normally have respectively the lowest and highest energies of the four transitions, and are respectively weak and relatively strong. The other two transitions, corresponding to the α and β bands, are degenerate in energy at the single configuration level. Configuration interaction can be sufficiently important so as to cause the α band, usually very weak, to lie at lower energies than the stronger p band, for certain PAHs. The intensities and oscillator strengths of these four PAH transitions can be characterized as follows: α: very weak, $f \approx 0.001$; p: weak, $f \approx 0.01$ - 0.1; β: very strong, $f \approx 1$; β': medium strong, $f \approx 0.01$ - 1. These oscillator strengths are valid for catacondensed PAHs but suffer some exceptions for pericondensed PAHs. Comparative data are listed elsewhere (Clar 1964; Leach 1989).

As an example of the absorption spectra of a catacondensed PAH, Figure 1 shows the spectrum of anthracene. Increasing the energy of photon excitation beyond the absorption threshold leads to the formation of a variety of stable and dissociative valence and Rydberg states. The absorbance of PAHs increases fairly smoothly above about 6 μm^{-1} (7.4 eV), culminating in an absorption maximum at about 14 μm^{-1} (17.4 eV). This, and the similar absorption maximum observed in this spectral region for a large variety of hydrocarbons, can have as components $\pi - \sigma*, \sigma - \pi*, \sigma - \sigma^*$ and Rydberg spectral transitions (Jochims *et al.* 1994a). Coupling can exist between the various valence and Rydberg states formed, giving rise to peculiarities in the intramolecular dynamics. The Rydberg states above the first ionization potential (I.P.) are known as superexcited states and can have an additional relaxation pathway, autoionization, which provides ways of reaching the ionization continuum in addition

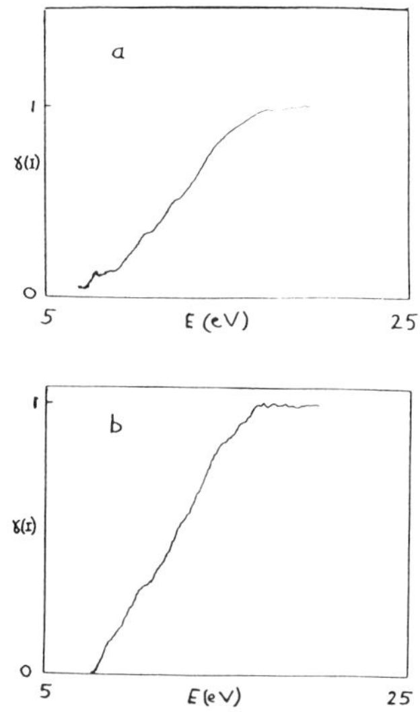

Fig. 2. Ionization yield $\gamma(I)$ as a function of incident photon energy. (a) anthracene; (b) phenanthrene.

to that of direct ionization.

A measure of the importance of nonionic relaxation channels of superexcited states is given by the photoionization efficiency $\gamma(I)$, defined as the number of ions produced per photon absorbed. For molecules, $\gamma(I)$ does not reach unity until several electron volts above the ionization threshold (Table 1). Photoionization yield curves have recently been measured for a number of PAHs (Jochims, Baumgärtel and Leach 1994) and some examples are given in Figure 2. It is found that for the PAHs given in Table 1, which range from C_6H_6 to $C_{24}H_{12}$, the ionization yield $\gamma(I)$ increases approximately linearly from the I.P. threshold up to the maximum value $\gamma(I) = 1$ which occurs at a photon energy E_I . The average value of E_I - I.P. is 9.3 $\pm$ 0.9 eV. The existence of a quasi-linear relation, in the range E_I - I.P., between the excess energy and $\gamma(I)$ makes it possible to propose as "rules of thumb" the following relations for determining the value of $\gamma(I)$ for a PAH at any value E of the photon energy:

$$\gamma(I) \approx 1 \text{ at } E \geq I.P. + 9.3\,\text{eV} \tag{1}$$

$$\gamma(I) \approx (E - I.P.)/9.3 \text{ at } E \leq I.P. + 9.3\,\text{eV} \tag{2}$$

Knowledge of $\gamma(I)$ in the range E_I - I.P. is necessary for modelling PAH ionization rates in circumstellar and interstellar media and also for understanding the energy budget of the diffuse interstellar medium (Lepp and Dalgarno 1988b; Verstraete *et al.* 1990).

2.2. *THE SPECTROSCOPY AND PHOTOPHYSICS OF PAH MONOCATIONS*

The PAH monocations usually considered as being relevant to astrophysics are all open-shell species. The energy separation $\Delta E(D_1 - D_0)$ between the ground electronic state (doublet state D_0) and the first excited state of the same multiplicity

TABLE 1

Energy domain of competitive nonionic decay process of PAH superexcited states

Molecule	I.P (eV)	E_I [γ(I) =1] (eV)	E_I - I.P. (eV)
Benzene C_6H_6	9.24	17.0	7.76
Naphthalene $C_{10}H_8$	8.12	16.65	8.53
Azulene $C_{10}H_8$	7.38	17.2	9.82
Anthracene $C_{14}H_{10}$	7.45	17.2	9.75
Phenanthrene $C_{14}H_{10}$	7.90	16.45	8.55
Benzanthracene $C_{18}H_{12}$	7.43	16.70	9.27
Pyrene $C_{16}H_{10}$	7.45	17.7[a]	10.25
Coronene $C_{24}H_{12}$	7.27	17.7[a]	10.43

a: Verstraete *et al.* (1990)

(doublet state D_1, corresponding to the HOMO-1 -> HOMO molecular orbital promotion) is less than in the nearest related closed-shell species. This is illustrated in a tabulation given elsewhere (Leach 1987) which lists the energy of the first excited state, relative to that of the ground state of the same spin multiplicity, for some open-shell cationic hydrocarbons and their parent closed-shell species. For example, for anthracene the $S_1 - S_0$ transition is at 3.31 eV (375 nm), whereas the monocation $D_1 - D_0$ (forbidden) transition is at 1.1 eV (1126 nm); for coronene the $S_1 - S_0$ transition is at 2.90 eV (427 nm), while the monocation $D_1 - D_0$ transition is at 1.29 eV (961 nm). The lowest non-Koopmans state of PAH monocations, often designated as the A-state (Khan 1983, 1992), results from HOMO -> LUMO π-electron excitation and it invariably lies at a lower energy than that of the p-band transition in the neutral (Khan, Husain, and Haselbach 1993). For anthracene and coronene, their monocation A-bands occur at 2.0 eV (620 nm) and at 2.65 eV (468 nm), respectively. Work on the spectroscopy of PAH monocations, which has revealed a number of spectral transitions in the visible and near-ultraviolet, has been carried out by various techniques, including γ-irradiation of PAHs in organic glasses (Kira, Imamura and Shida 1976) and photoionization of PAHs in rare-gas matrices (Andrews, Friedman and Kelsall 1985; Salama and Allamandola 1993) or by matrix trapped ions formed by electron impact (Vala *et al.* 1994). Theoretical predictions of the electronic energy levels and transitions of PAH monocations have also been made (Du *et al.* 1993; Vala *et al.* 1994).

It should be noted that the dissociation limit energy for breaking a similar C-C or C-H bond will be about the same for monocation and parent neutral, so that the D_1 state in PAH open-shell ions will tend to lie below the first dissociation limit, whereas in the corresponding closed-shell neutral parent, the S_1 state will generally have an energy much closer to that of the lowest dissociation threshold.

Another property of interest in the context of the comparison of the photophysics of neutral and monocationic PAHs concerns interelectronic coupling involving the lowest excited states in these two cases. For the neutral closed-shell species, the S_1 excited state can be coupled to vibronic levels of at least two lower-lying states, namely, the first excited triplet T_1 and the electronic ground state S_0. However, in the open-shell cationic species the vertical excitation energy to reach the lowest-lying quartet Q_1 state from the D_0 ground state will be greater than that of D_1, so that coupling leading to radiationless transitions of D_1 (internal conversion) will occur for this state to vibronic levels of only one lower-lying state, i.e. the electronic ground state D_0 (Leach 1987, 1989).

Following absorption of an ultraviolet photon in the H I region of the interstellar medium, PAHs and their monocations undergo isoenergetic internal conversion to their respective ground electronic states. They would then be in highly excited vibrational states whose infrared radiative rates are considered to be greater or equal to $10^2 s^{-1}$. The only process competitive with infrared emission from these excited vibrational levels is dissociation. Measurements have recently been made on the dissociation rates of a series of PAH monocations, thus providing for the first time experimental data on the effect of molecular size on the photostability of PAHs and their cations (Jochims *et al.* 1994b). Photoion mass spectrometry, with monochromatized synchrotron radiation as excitation source, was used to determine the internal energy E_{int} at which the rate of dissociation of the PAH is $10^4 s^{-1}$. Results on the hydrogen atom loss reaction, fitted to an RRK model calculation, were then scaled to determine the internal energy E_{crit} at the astrophysically critical dissociation rate $10^2 s^{-1}$. Data were also obtained on the H_2 and C_2H_2 loss channels.

Figure 3 shows the effect of size of PAH cations on the internal energy E_{crit} required for H-atom loss. The quasi-linear dependence of E_{crit} on N_A, the total number of atoms, and on N_C, the number of carbon atoms, is demonstrated by these results, and has also been modelled (Jochims *et al.* 1994b). Theoretical values of the dependence of E_{crit} on N_A, obtained on the basis of quantal RRK calculations (Tielens *et al.* 1987; Allamandola, Tielens and Barker 1989), predict E_{crit} values over 1 eV greater than our experimentally based results.

The results of the study of Jochims *et al.* (1994b) indicate that the upper limit of PAH size below which neutral PAHs in H I regions would mainly relax by photodissociation rather than by infrared emission is a species containing 30-40 carbon atoms. Furthermore, PAH monocations of any size formed by single photon ionization of the neutral PAH in H I regions should be stable and relax essentially by emission of infrared photons. Dissociation of PAH monocations would require absorption of a second photon of appropriate energy. In the study reported by Jochims *et al.* (1994b), some evidence was found that the E_{crit} values of substituted or less symmetric condensed PAHs are smaller than those for regular PAHs of similar size. This has recently been substantiated by unpublished results from the same group, and indicates that in the interstellar medium photodissociation processes would tend to eliminate the less regular PAHs. Furthermore, deuterium isotope effects on fragmentation threshold energies (Jochims *et al.* 1993) suggest the possible existence of photolytically induced

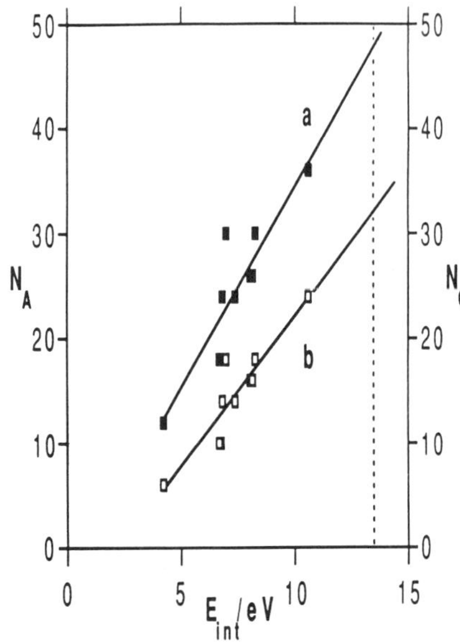

Fig. 3. Effects of size of PAH monocations on the energy required for H-atom loss. Experimental values of $E_{int}(= E_{crit})$, the critical internal energy of the parent cation for a dissociation rate $k_d = 10^2 s^{-1}$, as a function of (a) N_A, the total number of atoms and (b) N_C, the number of carbon atoms. The vertical dotted line at $E_{int} = 13.6$ eV corresponds to the upper limit of photon energy in H I regions of the interstellar medium.

deuterium fractionation processes for PAHs in the interstellar medium. Large deuterium enrichment is known for hydrocarbon phases in carbonaceous meteorites and in interplanetary dust particles (Geiss and Reeves 1981; Allamandola, Sandford and Wopenka 1987).

2.3. THE SPECTROSCOPY AND PHOTOPHYSICS OF PAH ANIONS

The electronic ground state of a PAH anion is characterized by the addition of an electron to the lowest normally unfilled molecular orbital, the LUMO. An electronically excited state of the anion is created when the electron in the LUMO is promoted to one of the higher-lying virtual MOs. In analogy to the one-particle-two-hole (1p-2h) non Koopmans states of cations, electron capture by the neutral PAH can be accompanied by a simultaneous electronic excitation in the molecule, leading to two electrons in the normally unfilled MOs, a situation known as a two-particle-one-hole (2p-1h) state. The latter usually lies above the ionization limit of the anion and is then referred to as a resonance, of limited lifetime.

For alternant hydrocarbons, insofar as the frozen-core approximation is reasonably valid, there is a considerable similarity between certain aspects of the spectroscopy of monocations and anions. This is due to the fact that HOMO–>LUMO transitions are common to both species. Thus, under conditions in which both cationic and anionic PAHs can be produced simultaneously it is not evident, without subsidiary experiments, which spectral features are to be assigned to one or the other species (Shida and Iwata 1973).

The parent anion can be formed in its ground electronic state (M^-), generally with vibrational excitation (M^{-*}), or as a 2p-1h resonance (indicated as M^{*-}), on capture of an electron by a neutral PAH, under single collision conditions. The nascent negative ion can then evolve in the following way:

$$M^{-*}(M^{*-}) --> M + e^- \tag{3}$$

$$M^{-*}(M^{*-}) --> R + X^- \tag{4}$$

Transfer of electronic into vibrational energy can delay autodetachment (channel 3) to an extent which allows direct mass spectrometric detection of the parent ion. The dissociative attachment channel (4) can occur if thermodynamically stable, negatively charged fragments exist and if such channels are energetically accessible at the energy of the temporary negative ion. Experiments have been carried out on negative ion formation from a series of PAHs in which a quasi-monoenergetic electron beam is crossed with an effusive beam containing the PAH (Tobita *et al.* 1992). Some representative results are presented in Figure 4.

Resonances at zero energy correspond to formation of long-lived parent anions with little or no internal excitation. Of the ten PAHs studied, long-lived parent negative ions (autodetachment lifetimes $> 10^{-6}$ s) were observed when the electron affinity (EA) of the PAH exceeds a few tenths of an electron volt. For the larger hydrocarbons, with EA(M) ≥ 0.5 eV, the long-lived parent negative ion was found to form via as many as three negative ion states above 0.0 eV. These resonances at higher energies are considered to be due to electron-excited Feshbach resonances (two-electron negative ion states). They involve concomitant electronic excitation of M (the PAH) by a "fast" electron and capture of the "thermalized" electron by the so-produced electronically excited molecule M^*, i.e.

$$e^-(fast) + M --> M^* + e^-(thermal) --> M^{*-}. \tag{5}$$

The only fragment anions observed for the PAHs studied were $(M-H)^-$ species, in which H-atom loss has occurred. These dissociative attachment processes (channel 4) most probably reflect fast direct dissociation on a steep potential energy surface, since the dissociation energy for H-loss should be of the order of 3 eV whereas the $(M-H)^-$ resonance maximum is in the 7-8 eV region.

Details of the anion peak energies and cross-sections are given elsewhere (Tobita *et al.* 1992). It should be noted that PAH anions are considered to play an important role in the chemistry of interstellar clouds (Lepp and Dalgarno 1988a; Lepp *et al.* 1988; Bakes and Tielens 1994). The type of PAH anion data discussed here can be useful in modelling such processes. It should be noted that hydrogen atom loss dissociative attachment of PAHs could be one of the processes leading to dehydrogenation of interstellar PAHs. The electrons of sufficient energy for this process can be generated by cosmic rays in dense interstellar clouds.

2.4. THE FORMATION AND FRAGMENTATION OF PAH DICATIONS

PAHs have been suggested as the carriers of the so-called unidentified infrared emission bands (UIRs; Léger and Puget 1984, Allamandola, Tielens and Barker 1985).

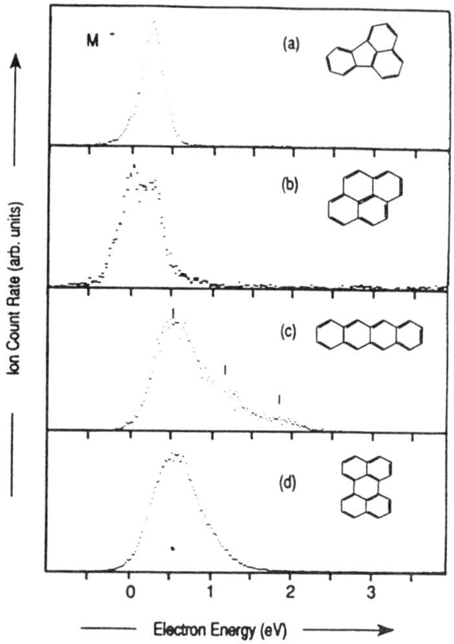

Fig. 4. Relative intensity of long-lived parent PAH negative ions as a function of the incident electron energy.

The ambient physical conditions where these emissions occur are such that the PAHs are mainly in cationic form (Allamandola, Tielens and Barker 1989). It has previously been proposed (Leach 1986, 1987) that these physical conditions permit the formation of PAH dications in the interstellar medium, leading to the partial dehydrogenation of PAHs in H I regions, and to their complete destruction in H II regions. The formation process proceeds by two stages in H I regions:

$$PAH + h\nu_1 - - > PAH^+ + e^- \tag{6}$$

$$PAH^+ + h\nu_2 - - > PAH^{2+} + e^- \tag{7}$$

The minimum energy for the first step (6) is the ionization potential (I.P. $= I^+$) of the PAH. The ionization potentials of PAHs are between 6 and 9 eV (Leach 1986, 1987), so that these species are readily ionized by absorption of photons of less than 13.6 eV, the latter being the upper photon energy limit in H I regions. For the second step the minimum energy is the difference $\Delta I = I^{2+} - I^+$ between the double and single ionization potentials. In these H I regions, the maximum amount of internal energy E_{int} in the nascent PAH^{2+} dication is $E_{int} = 13.6\ eV - \Delta I$.

The probability of fragmentation, and the type of fragmentation process in the parent dication p^{2+} (Leach 1986), i.e. charge separation into two monocations,

$$p^{2+} - - > m_1^+ + m_2^+ (+m_3) \tag{8}$$

or covalent dissociation into a doubly charged and a neutral fragment ,

$$p^{2+} --> m_1^{2+} + m_2 \tag{9}$$

depends on the value of E_{int} (Leach, Eland and Price 1989a,b). Thus it is important to have accurate values of the single and double ionization potentials of PAHs. In the earlier work (Leach 1986), the values used were from the literature, those for I^{2+} being exclusively from electron impact mass spectrometry appearance potential measurements. The latter are therefore vertical ionization potentials from the neutral ground state, whereas, in many cases, adiabatic energies were available for I^+. The values of ΔI, so deduced, were therefore considered as upper limiting values. Furthermore, it later became apparent (Rühl, Price and Leach 1989a) that I^{2+} values obtained from photon impact experiments can be 1 eV or more lower than those from electron impact measurements given in the literature. As discussed below, there is a strong dependence on internal energy of the relative yields of the fragmentation pathways of the dications of PAHs and related molecules (Leach, Eland and Price 1989a,b; Rühl, Price and Leach 1989; Hagan and Eland 1991; Braitbart *et al.* 1992). It thus became imperative to obtain I^{2+} values measured by photon impact and, as far as possible, to obtain adiabatic energies.

This was recently achieved in a photoionization mass spectrometry study in which single and double ionization potentials of 21 polycyclic aromatic hydrocarbons containing up to 24 carbon atoms were measured by photon impact and compared with the values obtained by electron impact (Tobita *et al.* 1994). Differences of up to 4 eV were observed between the values for double ionization measured by the two techniques (Table 2). It was proposed that this could be due to various factors such as the breakdown of the Born approximation for electrons in the 20 eV range, differences in the threshold behaviour for double ionization in the two excitation cases, and differences in dynamic electron correlations resulting from the much longer interaction time of photons with the target molecules as compared with electron-molecule interactions in the same energy range.

The ratio $\alpha(= I^{2+}/I^+)$ of the double to single ionization potentials of PAHs, 2.65 ± 0.06, for photon impact, was found to be significantly lower than for electron impact and for charge stripping measurements (Table 2). The data in Table 2 concerns PAHs containing up to 24 carbon atoms but a larger range of PAH molecules and ions are thought to exist in the interstellar medium (Allamandola, Tielens and Barker 1989). In order to verify whether the sequential two-photon model of double ionization (eq. 6 and 7) in the H I regions of the interstellar medium is valid for the whole range of PAH molecules it is necessary to determine the values of $\Delta I = I^{2+} - I^+$ for such species. The single ionization potentials are known for a large number of PAHs so that the corresponding ΔI values, (ΔI being the significant parameter in the astrophysical model), can be directly determined, using the values of the first ionization potential I^+, from the relation $\Delta I = I^+(\alpha\text{-}1)$. If we take the Table 2 value $\alpha = 2.65$ this leads to values of $\Delta I \leq 13.6$ eV for all PAHs whose ionization potential $I^+ \leq 8.24$ eV. This property corresponds to all of the 200 *polycyclic* aromatic hydrocarbons whose gas phase ionization potentials have been measured (Eilfeld and Schmidt 1981) and which include PAHs containing up to 48 carbon atoms. Other large molecules, in particular the fullerenes, have been suggested to exist in the interstellar medium (Kroto 1987, 1988), so that published ionization data on these pure carbon species has also been considered in Table 2. The results of this investigation support the proposition (Leach

TABLE 2
Single and double ionization potentials of PAHs and fullerenes

Formula	Molecule	Single	Double Ionization Potential			Ratio of Double to Single IP		
		$I^{+\,a}$	$I^{2+\,b}$	I^{2+}	$I^{+}+Q$	I^{2+}/I^{+}	I^{2+}/I^{+}	I^{2+}/I^{+}
	PAHs	Photon Impact (eV)	Photon Impact (eV)	Electron Impact (eV)	Charge stripping (eV)	Photon Impact	Electron Impact	Charge stripping
C_6H_5D	Benzene - d_1	9.22	25.3	26.1	25.4	2.74	2.83	2.75
$C_{10}H_7D$	Naphthalene - d_1	8.15	21.5			2.64		
$C_{10}H_8$	Naphthalene	8.15	21.5	22.7	22.7	2.64	2.79	2.79
$C_{12}H_8$	Biphenylene	7.58	19.7			2.60		
$C_{12}H_8$	Acenaphthylene	7.74	21.4			2.76		
$C_{12}H_{10}$	Biphenyl	8.16	21.3	21.9		2.61	2.68	
$C_{12}H_{10}$	Acenaphthene	7.72	21.7			2.81		
$C_{12}H_{10}$	2-Vinylnaphthalene	7.91	21.3			2.69		
$C_{13}H_{10}$	Fluorene	7.89	21.0			2.66		
$C_{14}D_{10}$	Anthracene - d_{10}	7.46	19.6	21.1		2.63	2.83	
$C_{14}D_{10}$	Phenanthrene - d_{10}	7.90	20.6	21.5	21.7	2.61	2.72	2.75
$C_{14}H_{10}$	Diphenylacetylene	7.99	21.2	21.4		2.65	2.92	
$C_{14}H_9N$	Acridine	7.92	21.1			2.66		
$C_{14}H_{12}$	9,10-Dihydrophenanthrene	7.90	20.6			2.61		
$C_{16}H_{10}$	Pyrene	7.44	19.8	20.5	21.8	2.66	2.76	2.93
$C_{16}H_{10}$	Fluoranthene	7.87	20.3			2.58		
$C_{18}H_{12}$	Tetracene	6.96	18.6	22.1	20.1	2.67	3.18	2.89
$C_{18}D_{12}$	Triphenylene - d_{12}	7.83	20.1	24.1	22.3	2.57	3.08	2.85
$C_{20}H_{12}$	Perylene	6.85	18.0	20.0		2.63	2.92	
$C_{22}H_{14}$	Pentacene	6.61	17.4	19.6	19.4	2.63	2.97	2.94
$C_{24}H_{12}$	Coronene	7.20	18.7	21.0		2.60	2.92	
						2.65±0.06[c]	2.87±0.14[c]	2.80±0.07[c]
	Fullerenes							
C_{60}	Fullerene Bf C_{60}	7.61±0.02	19.00±0.03		19.5	2.50		2.56
C_{70}	Fullerene Bf C_{70}	7.47±0.02	18.84±0.03		19.2	2.52		2.57

a : Single ionization potential. Precision ≤ 0.02 eV; b : Double ionization potential. Precision ≤0.2 eV; c : average value for PAHs.

1986, 1989) that sequential formation of PAH (and fullerene) dications could occur in H I regions of the interstellar medium and that the formation of PAH^{2+} ions by this mechanism could lead to their partial dehydrogenation.

As mentioned above, there are two main fragmentation channels of PAH dications, the charge separation (Coulomb dissociation) process of the parent dication p^{2+} (eq. 8), and covalent dissociation (eq. 9). It is of interest to determine the yields of these two processes and how these yields depend on the amount of internal energy in the parent dication. Also of relevance to astrophysical aspects of PAH dications is the effect of PAH size on these fragmentation pathways and yields. Photoelectron-photoion-photoion triple coincidence (PEPIPICO) techniques have made possible the study of PAH dication fragmentation, using either monochromatized synchrotron radiation or rare-gas lamp high energy photon sources (Leach, Eland and Price 1989a,b; Hagan and Eland 1991; Braitbart *et al.* 1992). The covalent process dominates close to the parent dication threshold, but dissociative charge separation increases more rapidly with increasing internal energy, becoming the major fragmentation process at an internal energy that increases with the size of the PAH. For example, for the benzene dication, charge separation dominates after about 2 eV of maximum internal energy (Hagan and Eland 1991), whereas for the naphthalene dication, it requires about 12 eV of maximum internal energy before this becomes the major fragmentation channel (Leach, Eland and Price 1989a). Studies on the PAH size dependence of the relative importance of the charge separation and covalent dissociation pathways have been carried out at a fixed incident photon energy of 40.8 eV on PAHs containing up to 20 carbon atoms (Hagan and Eland 1991). The results show that dissociative charge separation becomes an increasingly minor process, under these conditions where the maximum internal energy of the parent PAH dication is of the order of 20 - 22 eV, for PAHs whose carbon atom numbers are greater than 10.

The astrophysical implications of these PAH dication fragmentation studies are as follows. The formation of PAH dications by a sequential two-photon process in H I regions of the ISM would lead to a PAH^{2+} containing no more than about 4 eV of internal energy, as can be deduced from Table 2. One can therefore conclude that for the majority of PAHs of astrophysical interest, dication fragmentation will proceed mainly by the covalent channel (eq. 9), giving rise to small neutral products such as $H, H_2, C_2H_2, C_2H_3, C_4H_2$ (Leach, Eland and Price 1989a,b). On the other hand, in H II regions, where the available photons can extend to 54 eV (He II region), PAHs could be destroyed by charge separation processes. It should be noted that $C_2H_2{}^+$ and $C_3H_3{}^+$ are major ionic products of the charge separation reactions of PAH dications (Leach 1989). These small ions are considered to play a role in the formation of certain interstellar molecules, in particular cyclopropenylidene C_3H_2 and carbon chain molecules. Charge separation reactions of PAH dications could therefore contribute to the formation of these interstellar species in appropriate regions of the ISM (Leach, Eland and Price 1989b).

The spectroscopic properties of PAH dications are as yet unknown.

3. THE SPECTROSCOPY AND PHOTOPHYSICS OF FULLERENES

3.1. BUCKMINSTERFULLERENE C_{60}

The C_{60} molecule contains 60 identical carbon atoms arranged in 20 six-membered and 12 five-membered rings; each atom lies at the vertex of one five-membered and two six-membered rings. Thus C_{60} is a truncated icosahedron belonging to the I_h symmetry group. Although the σ and π orbitals of the fullerene are not pure, they may be treated as approximately so because the $\sigma - \pi$ interaction which exists in this nonplanar system is relatively weak.

The totally symmetric ground state 1^1A_g of C_{60} issues from the closed shell "π" electron configuration:

$$...3a_g{}^{(2)}4t_{1u}{}^{(6)}4h_g{}^{(10)}4t_{2u}{}^{(6)}4g_u{}^{(8)}4g_g{}^{(8)}7h_g{}^{(10)}4h_u^{(10)} \qquad ......1^1A_g.$$

For this non-alternant species, the first three excited configurations, and corresponding electronic state symmetries are :

$$...7h_g{}^{(10)}4h_u{}^{(9)}5t_{1u}{}^{(1)} \qquad ......{}^{1,3}T_{1g},{}^{1,3}T_{2g},{}^{1,3}G_g,{}^{1,3}H_g$$

$$...7h_g{}^{(9)}4h_u{}^{(10)}5t_{1u}{}^{(1)} \qquad ......{}^{1,3}T_{1u},{}^{1,3}T_{2u},{}^{1,3}G_u,{}^{1,3}H_u$$

$$...7h_g{}^{(10)}4h_u{}^{(9)}2t_{1g}{}^{(1)} \qquad ......{}^{1,3}T_{1u},{}^{1,3}T_{2u},{}^{1,3}G_u,{}^{1,3}H_u$$

Transitions from the ground state electron configuration to these excited configurations correspond to HOMO–> LUMO, HOMO-1–> LUMO and HOMO–> LUMO+1, respectively. Electric dipole transitions from the 1^1A_g ground state are only allowed to $^1T_{1u}$ states. All other singlet-singlet transitions from 1^1A_g are forbidden but can occur weakly via Herzberg-Teller vibronic coupling. Triplet <–singlet transitions are spin forbidden but can appear extremely weak due to spin-orbit coupling effects.

Absorption spectra of C_{60} in n-hexane solution have been taken at room temperature with a Cary 210 spectrophotometer in the 190 - 700 nm region at a resolution of 1 nm (Leach *et al.* 1992). An overview of the spectrum is shown in Figure 5. Over 40 vibronic bands were observed and have been assigned to 14 different electronic transitions. The measured wavelengths, frequencies and assignments of the observed features are given in detail elsewhere (Leach *et al.* 1992). It is sufficient to describe here the gross spectral features as follows:

(1) A strong band region between 190 and $\approx$ 350 nm which contains three intense broad peaks at 211, 256 and 328 nm; there are shoulders or inflexions at $\approx$ 195, 227 and 295 nm.
(2) A region of much weaker bands between 350 and 430 nm some of which are sharply structured and are due to electronic transitions exhibiting some vibrational structure.
(3) A broad weak continuum between 430 and 640 nm, whose maximum is at about 540 nm, and on which are superimposed several clearly evident peaks and some weaker shoulders or inflexions. These features also appear to be due to vibronic transitions.
(4) Four extremely weak bands between 627 and 690 nm.

Absorption spectra taken at 77 K show only minor changes as compared to the 300 K spectra, mainly in reduced bandwidths and enhanced band resolution. The peak wavelengths remain essentially the same but the narrowest bands have half-widths of about 120 cm^{-1} at room temperature, reduced to about 70 cm^{-1} at 77 K and have widths comparable to the latter in an argon matrix at 5 K (Gasyna *et al.* 1991).

The principal bands in the 190-410 nm region have been assigned to nine orbitally allowed singlet-singlet $^1T_{1u} - 1^1A_g$ transitions. The absorption bands in the 410 -

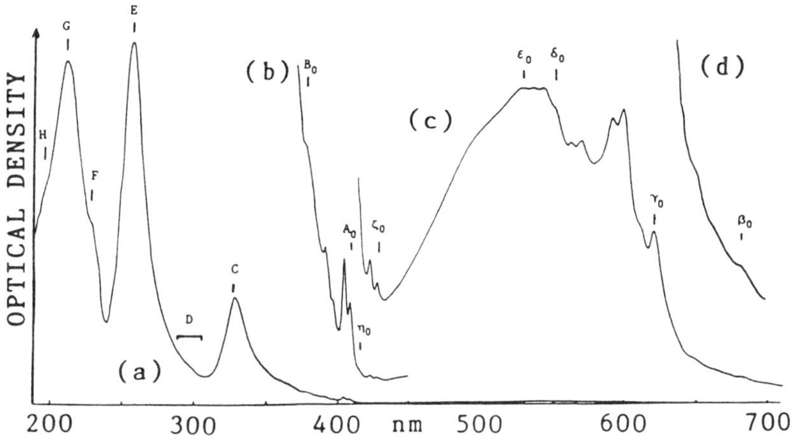

Fig. 5. Absorption spectra of C_{60}-*n*-hexane solutions at 300 K. (a) 190 - 700 nm. (b) 365 - 450 nm; *a* x 20. (c) 410 - 700 nm; a x 200. (d) 630 - 700 nm; *a* x 1000. Spectral resolution = 1 nm.

700 nm region are due to a set of singlet-singlet transitions and triplet-singlet transitions (Leach *et al.* 1992). The laboratory spectra have been compared directly with interstellar extinction curve observations (Leach 1993). Figure 6 compares the absorption spectrum of C_{60} (*a*) with normalized profiles for the broadest (*b*) and the narrowest (*c*) extinction humps in the 220 nm (4.6 μm^{-1}) region observed in a study of 45 lines-of-sight towards reddened Milky Way stars (Fitzpatrick and Massa 1986). The set of features in Figure 6(*a*) at 328.4 nm (3.05 μm^{-1}), 256.6 nm (3.90 μm^{-1}) and 211 nm (4.74 μm^{-1}) correspond to the three most intense $^1T_{1u} - 1^1A_g$ transitions of C_{60}. For these strong transitions, and given the high density of electronic states of C_{60}, the corresponding gas-phase absorption bands should also be broad, due to inter-electronic state coupling. This set of bands is obviously not present in the 220 nm region of the interstellar extinction curves of our galaxy (Figure 6(*a*), (*b*)).

Determination of an upper limit of C_{60} abundance from the interstellar extinction curves requires more precise information about the widths of the C_{60} ultraviolet absorption bands in the gas phase. A gas phase absorption spectrum of C_{60} has been measured over a wide spectral range (mainly 200 - 400 nm) for C_{60} (Qihuang Gong *et al.* 1994; Smith 1994), but at temperatures of the order of 600 - 950 K so that comparison with spectra at 300 K is not relevant in this case. A recent electron energy loss (EEL) spectrum of gas phase C_{60} at about 650 K (Bulliard, Allan and Leach 1993) exhibits spectral features very similar in energy to those observed in n-hexane solution at 300 K. However, the energy resolution in the gas phase EEL experiment was about 60 meV (about 500 cm^{-1}), which is insufficient for very detailed comparison of vibronic structure. More appropriate is a resonant two-photon ionization (R2PI) measurement, made at a spectral resolution of 3 to 6 cm^{-1}, of the supersonic jet gas phase spectrum of C_{60} (Haufler *et al.* 1991) which, however, has been restricted to two small spectral regions. One is around 400 nm, (375 - 415 nm), where the features observed can clearly be correlated with absorption bands found in n-hexane solutions at 77 and 300 K. In the other spectral region, 595 - 635 nm, the features observed do not correspond at first sight as evidently to solution spectra of C_{60} at 77 or 300 K but several of them can indeed be correlated with bands observed

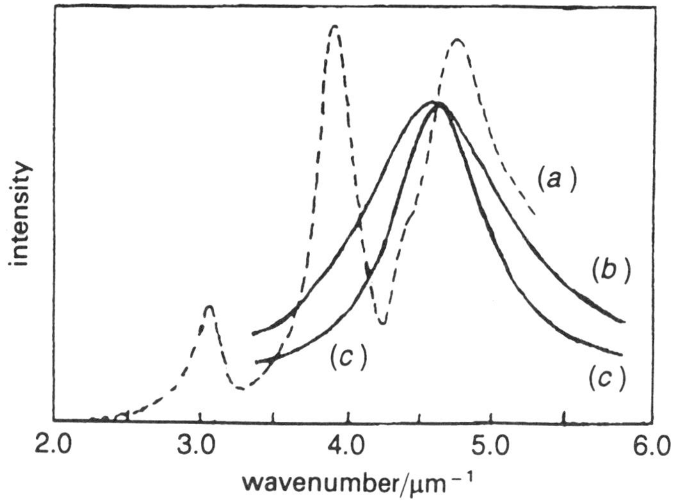

Fig. 6. (a) Part of absorption spectrum of C_{60} in *n*-hexane solution at 300 K. (b) and (c) are respectively the broadest and narrowest 220 nm humps in the extinction curves in 45 lines-of-sight in the Milky Way.

in the latter (Leach 1994b).

Although the gas phase spectrum of C_{60} in the 595 - 635 nm region has many features that are very close to some of the observed diffuse interstellar bands, one can eliminate this fullerene as a carrier of the DIBs since the far more intense allowed transitions of C_{60} are not observed, as discussed above.

The photophysical properties of the lowest lying excited singlet and triplet states of C_{60} have been studied, mainly in solution, using flash photolysis and pulse radiolysis techniques. Lifetimes of excited states, their transient absorption spectra and extinction coefficients, and the quantum efficiencies of their formation have been measured. Details and appropriate references are given by Bensasson *et al.* (1993a).

Electronic absorption transitions of C_{60} at energies above 6 eV have not been observed directly by optical spectroscopy but gas phase EELS experiments have been carried out up to 30 eV (Keller and Coplan 1992; Burose *et al.* 1993) which have revealed several broad structures in the 7 - 13 eV region and a peak at about 18 eV assigned to a $\pi - \sigma$ plasmon by Burose *et al.* (1993). Because of its large number of internal degrees of freedom, C_{60} requires a considerable amount of internal energy, much greater than the lowest bond energy, before dissociation becomes an important detectable relaxation process. This is also seen in the large kinetic shift for the ion dissociation process $C_{60}^+ \rightarrow C_{58}^+ + C_2$ (Yoo *et al.* 1992; Lifshitz *et al.* 1992). Analogous physical reasons can be invoked for the observation of "slow thermionic emission" in the photoionization of C_{60} (Campbell *et al.* 1991; Wurz and Lykke 1992). Indeed, photodissociation and photoionization ("thermionic emission") can be competitive (Foltin *et al.* 1993).

3.2. THE BUCKMINSTERFULLERENE MONO- AND DI-CATIONS C_{60}^{+} AND C_{60}^{2+}

Absorption spectra of the monocation C_{60}^{+} have been observed in pulse radiolysis experiments in solution at room temperature (Bensasson *et al.* 1993a), in γ-irradiated solutions at 77 K (Kato *et al.* 1991) and in rare gas matrices at 5 K (Fulara, Jacobi and Maier 1993b). The spectral region explored was 450 - 1300 nm at 77 K, and 800 - 1150 nm at 5 K. The single band observed at 980 nm at 77 K is resolved in the 5 K spectrum into two distinct features, separated by 67 cm^{-1}, at 9583 and 9645Å, respectively. Several weaker features occur at lower wavelengths. Two diffuse interstellar bands have recently been observed at 9577Å and 9632Å and have been assigned to C_{60}^{+} by Foing and Ehrenfreund (1994). This would correspond to a gas to neon matrix red shift of 7 cm^{-1} for the 9577Å band and 14 cm^{-1} for the 9632Å band. These displacements are smaller than those observed for neutral C_{60} and C_{70} (Table 3) in the 595 - 635 nm region and which are of the order of 45 cm^{-1} and 60 cm^{-1}, respectively. Verification of the C_{60}^{+} assignment should therefore be considered as provisional until confirmed by observation of the laboratory spectrum in the gas phase.

The ionization potentials for forming the mono- and di-cations of C_{60} are given in Table 2. It is clear that C_{60}^{2+} could be formed in H I regions of the interstellar medium by the sequential absorption mechanism discussed earlier (eqs. 6 and 7). The possible role of C_{60}^{2+} in cosmochemical processes has been discussed by Millar (1992) and by Petrie *et al.* (1993).

3.3. THE FULLERENE C_{70}

The atomic arrangement in C_{70} has D_{5h} symmetry and there ensues 210 occupied molecular orbitals. The ground and lowest mono-excited electron configurations and the electronic states that arise after electron-electron and spin-orbit interactions are taken into account are as follows (Bensasson *et al.* 1993b):

.........$15\, e_1'^{(4)} 13e_2''^{(4)} 4a_2'^{(2)} 13e_1''^{(4)} 10a_2''^{(2)}$ $\quad 1^1A_1{}'$

..........$15\, e_1'^{(4)} 13e_2''^{(4)} 4a_2'^{(2)} 13e_1''^{(4)} 10a_2''^{(1)} 14e_1''^{(1)}$ $\quad {}^{1,3}E_1{}'$

......$15\, e_1'^{(4)} 13e_2''^{(4)} 4a_2'^{(2)} 13e_1''^{(3)} 10a_2''^{(2)} 14e_1''^{(1)}$ $\quad {}^{1,3}A_1{}', {}^{1,3}A_2{}', {}^{1,3}E_2{}'$

..........$15\, e_1'^{(4)} 13e_2''^{(4)} 4a_2'^{(2)} 13e_1''^{(4)} 10a_2''^{(1)} 4a_1''^{(1)}$ $\quad {}^{1,3}A_2{}'$

..........$15\, e_1'^{(4)} 13e_2''^{(4)} 4a_2'^{(2)} 13e_1''^{(3)} 10a_2''^{(2)} 4a_1''^{(1)}$ $\quad {}^{1,3}E_1{}'$

Optical electric dipole transitions are allowed when $\Gamma'.\Gamma''$, the product of representations of the upper and lower electronic states, contains the representation E_1' or A_2", and would be polarized respectively in the (x,y) plane and the (z) direction.

It is not clear which of the two molecular orbitals $14e_1$" and $4a_1$" is the LUMO of C_{70}, since they are predicted to be close in energy. Consequently, calculations of C_{70} electronic state energies involving very extensive configuration interactions, and using realistic geometries and bond lengths are required in order to establish the true symmetry assignments of the lowest singlet and triplet states. The analyses of C_{70} spectra given below were carried out using currently available calculation results.

TABLE 3
C_{70} absorption features in the 595-640 nm region.
Gas phase and neon matrix data and comparison with observed DIBs

DIB[a] ν/cm^{-1}	C_{70} Gas phase[b] ν/cm^{-1}	C_{70}/Ne 5 K[c] ν/cm^{-1}	Gas - matrix Δν/cm^{-1}
	d	15690	d
15895	15891	15841	+ 50
15923	15923	15869	+ 54
	15979	15929	+ 50
16030	16024	15962	+ 62
16093	16090	16036	+ 54
16104	16102	16046	+ 56
	16146	16089	+ 57
16184	16186		
	16213	16149	+ 63
	16279	16215	+ 54
	16332	16265	+ 67
16344	16345		
16367	16372	16305	+ 67
16416	16409		
	16432		
16537	16535	16464	+ 71
	16599	16532	+ 67
	16664	16601	+ 63
	16734	16668	+ 66

a : Jenniskens and Désert F.-X. (1994). DIB frequency values in vacuum .
b : Data based on measurements of Haufler *et al.* (1991).
c : Fulara, Jakobi and Maier (1993b) .
d : not measured.

Band assignments of the absorption spectrum of C_{70} in solution in the spectral range 200 - 700 nm have been made (Leach 1994b), based on correlations with the results of the INDO/CI calculations of Feng *et al.* (1991). In the ultraviolet region, the observed strong bands differ considerably from the interstellar extinction curve in the 220 region, thus eliminating C_{70} as the carrier of the 220 nm interstellar peak. The first, forbidden, electronic transition in the visible is predicted to be $1^1A_2' - 1^1A_1'$ at 684 nm. A series of weak bands are observed in the spectral region 570 - 700 nm which at first sight could be considered to be associated with this forbidden transition, appearing through Herzberg-Teller vibronic interactions. Peak absorption coefficients ($dm^3mol^{-1}cm^{-1}$) in representative "allowed" and "forbidden" transition spectral regions can be compared for the two fullerenes C_{70} and C_{60} (Leach *et al.* 1992; Bensasson *et al.* 1993b): a) 330 - 360 nm: 21500 (C_{70}); 52000 (C_{60}); b) 620-640 nm: 1300 (C_{70}); 400 (C_{60}).

Comparison of room temperature solution absorption spectra of C_{70} in the visible region has been made (Leach 1994b) with a recent absorption spectrum of C_{70} isolated in a 5 K neon matrix obtained by Fulara, Jakobi and Maier (1993a), and with the gas phase spectrum of C_{70} obtained by the R2PI technique (Haufler *et al.* 1991). A number of sharp features were observed in the gas phase study in the 595 - 635 nm range but none in the other region studied, between 375 and 415 nm. Similarly, in the matrix study sharp band features were observed only in the 570 - 640 nm region.

Our room temperature absorption in this range is essentially the envelope of the gas phase or neon matrix absorption displaced by redshifts of the order of 300 - 360 cm^{-1} with respect to the gas phase, i.e. of the same order of magnitude as for the bands of C_{60} in the same spectral region. A careful frequency and intensity comparison between features in the gas phase and in the matrix spectra (Leach 1994b) shows that most of the gas phase features are to be found in the matrix phase, indicating that few hot bands are present in the former, consistent with the low temperature expected for the R2PI supersonic jet experiment. Data on features in the matrix spectrum extend 800 cm^{-1} to higher frequencies than that reported for the R2PI spectrum. The gas-matrix frequency shift is 50 - 70 cm^{-1} (Table 3), which is of the order of magnitude found for weak electronic transitions in aromatic species (Leach 1987).

Similar to the case of C_{60}, the gas phase spectrum of C_{70} in the 595 - 635 nm region has many features that are very close to some of the observed diffuse interstellar bands (Table 3). However, one can also eliminate this fullerene as a carrier of the DIBs since the far more intense transitions of C_{70} are not observed in the 210 - 250 nm region of interstellar extinction.

The photophysical properties of the lowest lying excited singlet and triplet states of C_{70} have been studied, in solution, in similar fashion to those of C_{60}, using flash photolysis and pulse radiolysis techniques. The results are discussed by Bensasson *et al.* (1993b). Furthermore, the photostability and delayed ionization effects observed for C_{70} are very similar to those observed for C_{60}.

3.4. *THE FULLERENE MONO- AND DI-CATIONS* C_{70}^{+} *AND* $C_{70}{}^{2+}$

The absorption spectrum of C_{70}^{+} has been observed in matrices at low temperature, in particular in a neon matrix at 5 K (Fulara, Jakobi and Maier 1993a). It consists in a series of well structured bands between 700 and 800 nm. The origin band of the transition assigned as $^2E_1' <---{}^2E_1''$ is at 12533 cm^{-1} (798nm).

The ionization potentials for forming the mono- and di-cations of C_{70} are given in Table 2. The values indicate that the dication could be formed in H I regions of the interstellar medium by the sequential absorption mechanism of eqs. 6 and 7. Petrie *et al.* (1993) have discussed ion-molecule reactions involving $C_{70}{}^{2+}$ and how they could play a role in the chemistry of the interstellar medium.

3.5. *SPECTROSCOPY OF THE HIGHER FULLERENES*

There are few studies of electronic spectra of fullerenes C_n for $n > 70$. Diederich and Whetten (1992) have reported the room temperature absorption spectra between 200 and 900 nm of C_{76} (D_2 symmetry) and of two isomers of C_{78} (C_{2v} and D_3 symmetries) in dichloromethane. The room temperature absorption spectra between 500 and 1100 nm, in benzene solution, have been recorded by Kikichi *et al.* (1992) for $C_{76}, C_{78}, C_{82}, C_{84}, C_{90}$ and C_{96}. Agreement is good between the Diederich and Whetten and the Kikuchi *et al.* spectra for the common spectral region in the case of C_{76} but there are noticeable differences in the case of C_{78}. Five isomers that obey the isolated pentagon rule (Smalley 1992) are predicted for C_{78} (Diederich and Whetten

1992); it is therefore possible that Kikuchi *et al.* have isolated an isomer that is other than the C_{2v} or D_3 forms of C_{78}.

C_{76} and C_{78} exhibit strong absorption at wavelengths below 330 nm, as is the case for C_{60} and C_{70}. Absorbance coefficients are of the order of 100 000 $dm^3mol^{-1}cm^{-1}$ in the region below 300 nm. In the 500 - 1100 nm region, absorption is relatively weak with respect to the ultraviolet region, e.g. for C_{76} the peak at 715 nm has an absorbance of 3080 $dm^3mol^{-1}cm^{-1}$ (Diederich and Whetten 1992) and there is an increased blurring of band features as the number of carbon atoms increases in the series $C_{76} - C_{96}$ (Kikuchi *et al.* 1992). At the same time the apparent long wavelength limit of absorption appears generally to shift to the red with increasing size of the fullerene. Detailed analysis of the spectra of these higher fullerenes is premature at the present time. Gas phase spectra have not been measured for these species. Their general spectral characteristics in solution do not conform with interstellar extinction curve features or DIB observations. Nothing is known concerning the spectra of the cations of these higher fullerenes.

4. THE SPECTROSCOPY OF POLYENES, POLYYNES AND THEIR IONS

4.1. *POLYENES*

Conjugated linear polyenes containing n double bonds have the generic formula $H(CH = CH)_nH$. Visible-UV absorption spectra of these species have been reported up to at least $n = 10$, mainly in solution. Gas phase absorption spectra are known for some of these, notably 1,3-butadiene ($n = 2$) and 1,3,5-hexatriene ($n = 3$) (Gavin, Risemberg and Rice 1973; Gavin and Rice 1974). The first allowed singlet-singlet transition ($1^1B_u - 1^1A_g$ in the idealized C_{2h} symmetry) of the polyenes exhibits 4 main band groups separated by a spectral interval of the order of 1630 cm^{-1} which corresponds to the totally symmetric C=C double bond stretching motion in the excited state. Figure 7 shows that the frequency of the most intense band of conjugated polyenes ($n = 2$ - 10) in paraffinic solutions decreases with increasing length of the conjugated chain (Jaffé and Orchin 1962). Simple free-electron theory predicts that the frequency of the first allowed absorption transition of linear polyenes will decrease indefinitely, eventually to zero, with increasing n. However, experimentally it is found that with increasing n the peak frequency approaches a lower bound of the order of 18000 cm^{-1}, indicating the existence of a finite band gap between the HOMO and LUMO for an infinitely long linear polyene. This behaviour can be rationalized by taking into account of the alternation of bond lengths in an infinite linear polyene, by introducing a sinusoidal potential at the bottom of the box-like potential well in the free-electron model, or by other, more sophisticated molecular orbital treatments (Salem 1966). As n increases it is found that the polyene peak intensity increases, its molar absorption coefficient going from 21000 dm^3mol^{-1} for $n = 2$, to 138000 dm^3mol^{-1} for $n = 6$. Indeed the increase in intensity as a function of the number of double bonds is approximately linear in the case of the α,ω-diphenylpolyenes (Jaffé and Orchin 1962). The general behaviour of the first allowed transition of the linear polyenes can be followed up to $n = 19$ in the case of the carotenoids. The main peak frequencies of the first allowed transition of the polyenes is not very sensitive to the particular form of their *cis* and *trans* isomers, but the peak absorption intensities lose about half their value in going from the all-*trans* to the all-*cis* isomer (Bensasson, Land and Truscott 1993).

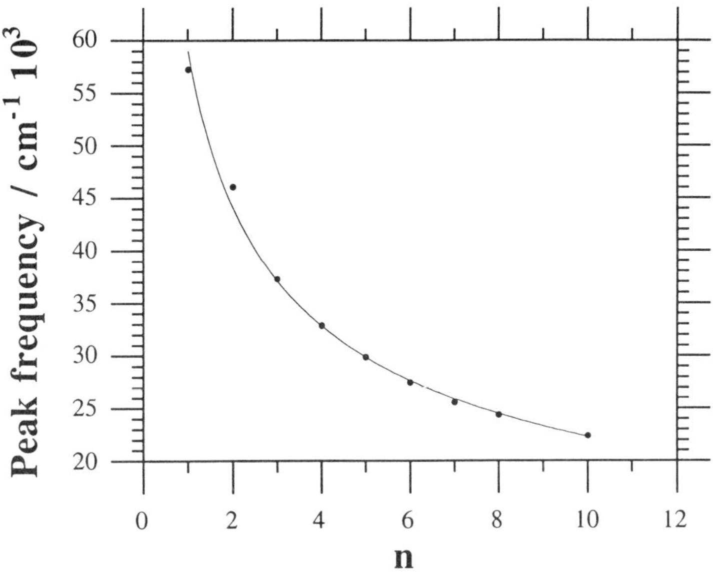

Fig. 7. Frequency of the most intense absorption band of conjugated polyenes as a function of n, the number of C=C bonds. The plot includes data for n = 1.

The first singlet-singlet transition of linear polyenes is a forbidden transition (^1A_g - ^1A_g in C_{2h} symmetry) which generally appears at 0.2 - 0.8 eV below the first allowed transition (Hudson and Kohler 1974; Kohler 1993). Its intensity is about 20 to 30 times less than the latter. The transitions of conjugated linear polyenes at higher energies than the first allowed transition have been relatively little studied. For the case of gas phase *trans*-hexatriene, Gavin and Rice (1974) have reported the $2^1B_u - 1^1A_g$ transition between 210 nm and 170 nm to have similar vibrational progressions as that of the $1^1B_u - 1^1A_g$ transition but to be about 10 times weaker in intensity than the latter. Further bands, down to about 150 nm, were also observed, some of which are Rydberg bands leading to an ionization limit at 8.27 eV. Tentative assignments of *cis*-hexatriene absorption bands observed in the 210 - 150 nm region were also made.

Studies on the electronic structure and spectra of a few polyene monocations have been made by photoelectron and optical spectroscopy. The He I photoelectron spectra of *trans*- and *cis*-1,3,5, hexatriene (n = 3) and the gas phase emission spectra of their monocations have been measured (Allan and Maier 1976; Allan, Dannacher and Maier 1980), as have the corresponding spectra for *trans*-1,3,5,7-octatetraene (n = 4) (Jones and Maier 1979). They exhibit a number of emission bands in the 615 - 740 nm region (n = 3) and in the 740 - 870 nm region for n = 4.

4.2. POLYYNES AND POLYYNYL RADICALS

The polyynes are unsubstituted polyacetylenes, of generic formula $H(C\ C)_nH$. Their instability has made spectroscopic study difficult and rare. Nevertheless, absorption spectra have been investigated for polyynes from n = 1 (acetylene) to n = 12 (dodeca-acetylene). Absorption spectra in solution have been measured at room temperature by Eastmond, Johnson and Walton (1972) for the polyynes n = 4 - 10 and n = 12 between 200 and 500 nm. Gas phase absorption spectra have been determined in the spectral range 110 - 400 nm for the polyyne species n = 1, 2, 3 and 4 (Kloster-Jensen, Haink and Christen 1974), who also measured solution spectra for the species n = 2 - 5 at room temperature and at 120 K or at 150 K. There are three groups of bands in polyyne absorption. The first (A) corresponds to two forbidden transitions $^1\Sigma_u^-$ - $^1\Sigma_g^+$ and $^1\Delta_u$ - $^1\Sigma_g^+$, while the second (B) and third (C) are due respectively to the allowed transitions $^1\Sigma_u^+$ - $^1\Sigma_g^+$ and $^1\Pi_u$ - $^1\Sigma_g^+$. The frequencies of the various band groups shift to the red with increasing values of n, the number of carbon-carbon triple bonds.

The polyyne monocations $H(C\ C)_nH^+$ (n = 2 - 4) have been studied by photoelectron and emission spectroscopies (Allan, Kloster-Jensen and Maier 1976). The first electronic transition of the cation, observed by emission spectroscopy is allowed, $A^2\Pi_u - X^2\Pi_g$ for n = even and $A^2\Pi_g - X^2\Pi_u$ n = odd. The origin band of this transition occurs in the gas phase at 19727, 16670 and 14160 cm^{-1} for n = 2, 3 and 4, respectively. The second electronic transition, $B^2\Pi_g - X^2\Pi_g$ for n = even and $B^2\Pi_u - X^2\Pi_u$ for n = odd, is forbidden, but could give rise to a weak transition via Herzberg-Teller interactions. The energy of this transition can be estimated from photoelectron spectra as 6.44 eV (193 nm), 3.39 eV (366 nm) and 3.05 eV (406 nm) for n = 2, 3 and 4, respectively. The third transition is allowed, $C^2\Pi_u - X^2\Pi_g$ for n = even and $C^2\Pi_g - X^2\Pi_u$ for n = odd; its energy is 3.88 eV (320 nm) for n = 4, from photoelectron spectroscopy (Allan *et al.* 1976).

Polyynyl radical chains, C_nH, n = 2 - 6, have been observed by radiofrequency spectroscopy both in the laboratory and in the interstellar medium (Irvine *et al.* 1991). Recently, Fulara *et al.* (1993) have obtained absorption spectra which are very probably due to polyynyl radical chains C_nH (n = 6 - 12) in rare gas matrices. The spectra have bands in the visible and near-infrared, but little absorption in the near ultra-violet. Comparison with DIB observations has been made (Fulara *et al.* 1993).

5. THE SPECTROSCOPY OF CARBON CUMULENE CHAINS C_N

The carbon cumulene chains C_n have geometrical and electronic structures that are a function of size and also of n being even or odd. Theoretical calculations for n = 2 - 18 have predicted that when n is less than ten, the species should be linear chains in their ground states, whereas the larger carbon clusters should form monocyclic rings (Pitzer and Clementi 1959; Hutter, Lüthi and Diederich 1994). The linear even-numbered chains were predicted to have $^3\Sigma$ ground electronic states whilst the odd-numbered linear chains to have $^1\Sigma$ ground electronic states and to be more stable than the even-numbered species. Infrared spectroscopic studies have borne out these predictions up to n = 9 (Heath and Saykally 1993; Saykally 1993). Earlier work on the structure and spectroscopy of the C_n species has been reviewed by Weltner and Van Zee (1989). Visible-UV absorption spectra of these species in the visible and near UV regions have been reported in matrices and have been compared with the diffuse interstellar bands (Krätschmer 1986, 1993).

6. DIFFUSE INTERSTELLAR BAND PROFILES AND LINE BROADENING MECHANISMS

The three principal criteria of consistency in comparing the spectra of possible carriers of the diffuse interstellar bands with astrophysical observations are band peak wavelengths, band profiles, and band correlations. Attention is paid here to the second of these criteria. It is useful to recall the remark of Herbig (1975) that many DIBs fall into two general classes, class A, in which the profiles have somewhat steep wings and a well- defined peak, and class B, where the bands are broad, with no easily discernable peak or fine structure. Furthermore, most bands have symmetric or slightly asymmetric profiles, but a few bands are clearly asymmetric. Following the recent work of Jenniskens and Désert (1994), who carried out a DIB survey at a resolution of 0.3 Å, there appear at present to be of the order of 210 DIBs, although a certain number of these still require confirmation (Krelowski and Sneden 1993). Examination of the reported 210 diffuse lines indicates that 78% have FWHM of less than 8 cm^{-1}, whereas only relatively few have a width approaching that of the line at 4430Å (100 cm^{-1}) (Table 4). Most DIB observations have been carried out at spectral resolutions between 0.15 cm^{-1} and 2 cm^{-1} (Cossart-Magos and Leach 1990).

Vibronic band profiles have been simulated for a number of species that have been suggested as carriers of the diffuse interstellar bands. These include cumulene chains C_n, polyynyl radicals C_nH and polyacetylene (polyyne) HC_nH linear chains, for n = 4 - 30 (Leach 1994a), as well as PAHs (Cossart-Magos and Leach 1990) and fullerene (Edwards and Leach 1993; Leach and Edwards 1994) species. The rotational band contours were calculated mainly at a resolution of 0.15 cm^{-1}, which is that of the "intrinsic profiles" measured for five DIBs by Westerlund and Krelowski (1988a,b), and for various temperature ranges, 3 - 100 K for the linear chains, 3 - 500 K for the fullerene C_{60}, and 3 - 1000 K for the PAHs studied. The latter include the symmetric rotors benzene C_6H_6 and coronene $C_{24}H_{12}$ and the asymmetric rotors pyrene $C_{16}H_{10}$ and ovalene $C_{32}H_{14}$. Only for large cumulene chains and for C_{60} could profiles be obtained, at $T \geq 100$ K, which resembled those of some, asymmetric, interstellar diffuse lines. Symmetric or near-symmetric DIB profiles of observed FWHM were not reproduced. Furthermore, the simulations showed no substructures resembling the 3 - 10 cm^{-1} interval features observed by Jenniskens and Désert (1993) for the 5280 Å and 6284Å DIBs, the band pairs observed by Herbig and Leka (1991), or the 35 cm^{-1} spaced features observed by Herbig (1988) for a series of 30 weak bands in the 6767 - 6862Å region.

Further calculations on the linear chains and on the fullerenes have taken into account possible intramolecular line broadening mechanisms, such as vibronic state congestion, vibrational redistribution and interelectronic state coupling (Leach 1994a; Edwards and Leach 1994; Leach and Edwards 1994). While the first mechanism is unlikely to play a major role, the other two, which are variants of the line-broadening mechanism proposed by Douglas (1977), were found to give adequate broadening to match the profiles of some symmetric DIBs observed by Westerlund and Krelowski (1988a,b). This behavior was found to be generally characteristic of large molecules. Using the coronene monocation ($C_{24}H_{12}{}^+$) as an example, and from a knowledge of its electronic level energies and symmetries, reasonable values of the intramolecular level density and coupling parameters required to give adequate band broadening for an interelectronic state coupling mechanism have been determined (Leach and Edwards 1994). It has thus been possible to show (Leach 1994a; Leach and Edawards 1994) that with appropriate broadening mechanisms, a molecule the size of C_{60} can give a good match to the diffuse lines at 5780Å (at 100 K), 5797Å (at 80 K) and 6196Å (at 5 K) though a fit to the line at 6284Å (at 200 K) was less satisfactory. From these DIB profile simulation studies one can conclude that the carriers of most of the DIBs could be large molecules containing more than 10 to 15 atoms, whose upper states are intramolecularly coupled with a dense isoenergetic manifold of vibronic states arising from the lower lying electronic levels. Table 4 gives the number of DIBs in various categories of bandwidth (FWHM), derived from the catalogue of diffuse interstellar bands given by Jenniskens and Désert (1994), and the intramolecular nonradiative (isoenergetic) relaxation rate k_{nr} consistent with representative bandwidths in these categories. Further details of these contour studies are given elsewhere (Leach and Edwards 1994).

TABLE 4
Linewidths (FWHM) of diffuse interstellar bands and corresponding nonradiative intramolecular coupling rates k_{nr}

FWHM (cm^{-1})	N° of DIBs	FWHM (cm^{-1})	k_{nr} (s^{-1})
< 2	62	1	1.9×10^{11}
2-4	71	3	5.7×10^{11}
4-8	29	6	1.1×10^{12}
8-16	13	12	2.3×10^{12}
16-32	12	24	4.5×10^{12}
32-64	14	48	9.1×10^{12}
> 64	9	96	1.8×10^{13}

7. CONCLUSION

This review has presented the spectroscopic and photophysical properties of species which have been proposed on various occasions as large molecule carriers of the diffuse interstellar bands. The species discussed are neutral, cationic and anionic polycyclic aromatic hydrocarbons, the fullerenes C_{60} and C_{70} as well as higher fullerenes up to C_{96}, and the carbon chain species C_n polyenes, polyynes and polyynyl radicals.

It is well known that the diffuse interstellar bands are observed in the 400 - 1200 nm region, but that no bands in the 200 - 400 nm ultraviolet region have been detected that can be clearly designated as DIBs. In Table 5 is summarized the spectral properties of the species previously discussed, as well as of the hydrogenated fullerenes $C_{60}H_2$ (Henderson and Cahill 1993), $C_{60}H_4$, $C_{60}H_{18}$ and $C_{60}H_{36}$, (Leach 1994c). This table shows that the only molecules for which it has been clearly demonstrated that absorption occurs in the visible but not in the ultraviolet are the polyynl radicals C_nH, thus reinforcing their possible assignment as the carriers of the DIBs in preference to the other species listed.

Rotational contours have been simulated for vibronic bands of these various potential DIB carriers and have been compared with observed DIB profiles. The results indicate that intramolecular relaxation in the excited electronic states of these carriers must occur in order to account for the observed DIB bandwidths. This has led to the conclusion that the DIB carriers are molecules having at least 10 to 15 atoms.

TABLE 5
Spectral properties of proposed molecular carriers of the diffuse interstellar bands

Species	Electronic transitions in 400 - 1200 nm region	Known or predicted strong(er) transitions in 200 - 400 nm region
PAH	$C_n \geq 10$	All PAHs
PAH^+	All PAH^+	All PAH^+
PAH^{2+}	N/A	N/A
<u>Fullerenes</u>		
C_{60} - C_{96}	C_{60} - C_{96} ; [$C_n > 96$ N/A]	C_{60} - C_{96} ; [$C_n > 96$ N/A]
<u>Fullerene monocations</u>		
C_{60}^+, C_{70}^+	C_{60}^+, C_{70}^+	C_{60}^+, C_{70}^+
<u>Hydrogenated fullerenes</u>		
$C_{60}H_2$, $C_{60}H_4$	$C_{60}H_2$, $C_{60}H_4$	$C_{60}H_2$, $C_{60}H_4$
$C_{60}H_{18}$, $C_{60}H_{36}$	$C_{60}H_{18}$, $C_{60}H_{36}$	$C_{60}H_{18}$, $C_{60}H_{36}$
others unknown		
C_n (linear)	n = 3 - 9	N/A
C_nH	n = 2 - 12	absent or very weak
H $(CH = CH)_n$ H	$n \geq 8$	$n \geq 1$
H $(CH=CH)_n$ H^+	n = 3,4	n = 3,4
H $(C\equiv C)_n$ H	$n \geq 6$	$n \geq 1$
H $(C\equiv C)_n$ H^+	n = 2,3,4	n = 2,3,4

REFERENCES

Allamandola, L.J., Sandford, S.A., and Wopenka, B. 1987, *Science*, **237**, 56.
Allamandola, L.J., Tielens, A.G.G.M., and Barker, J.R. 1985, *Ap. J.*, **290**, L25.
Allamandola, L.J., Tielens, A.G.G.M., and Barker, J.R. 1989, *Ap. J. Suppl.*, **71**, 733.
Allan, M. and Maier, J.P. 1976, *Chem. Phys. Lett.*, **43**, 94.
Allan, M., Kloster-Jensen, E., and Maier, J.P. 1976, *Chem. Phys.*, **17**, 11.
Allan, M., Dannacher, J., and Maier, J.P. 1980, *J. Chem. Phys.*, **73**, 3114.
Allan, M, Heilbronner, E., Kloster-Jensen, E., and Maier, J.P. 1976, *Chem. Phys. Lett.*, **41**, 228.
Andrews, L., Friedman, R.S., and Kelsall, B.J. 1985, *J. Phys. Chem.*, **89**, 4016.
Bakes, E.L.O. and Tielens, A.G.G.M. 1994, *Ap. J.*, **427**, 822.
Bensasson, R.V., Land, E.J., and Truscott, T.G. 1993, Excited States and Free Radicals in Biology and Medicine: Contributions from Flash Photolysis and Pulse Radiolysis, (Oxford, Oxford University Press).
Bensasson, R.V., Hill, T., Lambert, C., Land, E.J., Leach, S., and Truscott, T.G. 1993a, *Chem. Phys. Lett.*, **201**, 326.

Bensasson, R.V., Hill, T., Lambert, C., Land, E.J., Leach, S., and Truscott, T.G. 1993a, *Chem. Phys. Lett.*, **206**, 197.
Braitbart, O., Tobita, S., Leach, S., Roy, P., and Nenner, I. 1992, Synchrotron Radiation and Dynamic Phenomena, AIP Conf. Ser. **258**, 42.
Bromage, G.E. 1987, *Q. Jl. R. Astron. Soc.*, **28**, 294.
Bulliard, C., Allan, M., and Leach, S. 1993, *Chem. Phys. Lett.*, **209**, 434.
Burose, A.W., Dresch, T., and Ding, A.M.G. 1993, *Z.Phys.D - Atoms, Molecules and Clusters*, **26**, S 294.
Campbell, E.E.B., Ulmer, G., and Hertel, I.V. 1992, *Z.Phys.D - Atoms, Molecules and Clusters*, **24**, 81.
Clar, E. 1941, Aromatische Kohlenwasserstoffe, (Berlin: Springer).
Clar, E. 1964, Polycyclic Hydrocarbons, (New York: Academic).
Cossart-Magos,C. and Leach, S. 1990, *Astr. Ap.*, **233**, 559.
Diederich, F. and Whetten, R.L. 1992, *Acc. Chem. Res.*, **25**, 119.
Douglas, A.E. 1977, *Nature*, **296**, 130.
Du, P., Salama, F., and Loew, G.H. 1993, *Chem. Phys.*, **173**, 119.
Eastmond, R., Johnson, T.R., and Walton, D.R.M. 1972, *Tetrahedron*, **28**, 4601.
Edwards, S.A. and Leach, S. 1993, *Astr. Ap.*, **272**, 533.
Edwards, S.A. and Leach, S. 1994, in *Physical Chemistry of Molecules and Grains in Space*, AIP Conf. Ser., in press.
Eilfeld, P. and Schmidt, W. 1981, *J. Electron. Spectr. Rel. Phen.*, **24**, 101.
Feng, J., Li, J., Li, Z., and Zerner, M.C. 1991, *Intern. J. Quantum Chem.*, **39**, 331.
Fitzpatrick, E.I. and Massa, D. 1986, *Ap. J.*, **307**, 286.
Foing, B.H. and Ehrenfreund, P. 1994, *Nature*, **369**, 296.
Foltin, M., Lezius, M., Scheier, P., and Märk, T.D. 1993, *J. Chem. Phys.*, **98**, 9624.
Fulara, J., Jakobi, M., and Maier, J.P. 1993a, *Chem. Phys. Lett.*, **206**, 203.
Fulara, J., Jakobi, M., and Maier, J.P. 1993b, *Chem. Phys. Lett.*, **211**, 227.
Fulara, J., Lessen, D., Freivogel, P., and Maier, J.P. 1993, *Nature*, **366**, 439.
Gasyna, Z., Schatz, P.N., Hare, J.P., Dennis, T.J., Kroto, H.W., Taylor, R., and Walton, D.R.M. 1991, *Chem. Phys. Lett.*, **183**, 283.
Gavin, R.M. Jr. and Rice, S.A. 1974, *J. Chem. Phys.*, **60**, 3231.
Gavin, R.M. Jr., Risemberg, S., and Rice, S.A. 1973, *J. Chem. Phys.*, **58**, 3160.
Geiss, J. and Reeves, H. 1981, *Astr. Ap.*, **93**, 189.
Gerhardt, P., Loffler, S., and Homann, K.H. 1987, *Chem. Phys. Lett.*, **137**, 306.
Hagan, D.A. and Eland, J.H.D. 1991, *Rapid Comm. Mass Spectrom.*, **5**, 512.
Haufler, R.E., Chai, Y., Chibante, L.P.F., Fraelich, M.R., Weisman, R.B., Curl, R.F., and Smalley, R.E. 1991, *J. Chem. Phys.*, **95**, 2197.
Heath, J.R. and Saykally, R.J. 1993, in *On Clusters and Clustering, From Atoms to Fractals*, ed. P.J. Reynolds, (Amsterdam: Elsevier), p. 7.
Henderson, C.C. and Cahill, P.A. 1993, *Science*, **259**, 1885.
Herbig, G.H. 1975, *Ap. J.*, **196**, 129.
Herbig, G.H. 1988, *Ap. J.*, **331**, 999.
Herbig, G.H. and Leka, K.D. 1991, *Ap. J.*, **382**, 193.
Hudson, B. and Kohler, B. 1974, *Annu. Rev. Phys. Chem.*, **25**, 437.
Hutter, J., Lüthi, H.P., and Diederich, F. 1994, *J. Am. Chem. Soc.*, **116**, 750.
Irvine, W.M., Ohishi, M., and Kaifu, N. 1991, *Icarus*, **91**, 2, and references therein.
Jaffé, H.H. and Orchin, M. 1962, The Theory and Applications of Ultraviolet Spectroscopy, (New York:Wiley).
Jenniskens, P. and Désert, F.X. 1993, *Astr. Ap.*, **274**, 465.
Jenniskens, P. and Désert, F.X. 1994, *Astr. Ap. Suppl.*, **106**, 39.
Joblin, C. 1992, doctoral thesis, Université Paris 7.
Jochims, H.W., Rasekh, H., Rühl, E., Baumgärtel, H., and Leach, S. 1993, *J. Phys. Chem.*, **97**, 1312.
Jochims, H.W., Baumgärtel, H., and Leach, S. 1994, manuscript in preparation.
Jochims, H.W., Rühl, E., Baumgärtel, H., Tobita, S., and Leach, S. 1994a, manuscript in preparation.
Jochims, H.W., Rühl, E., Baumgärtel, H., Tobita, S., and Leach, S. 1994b, *Ap. J.*, **420**, 307.
Jones, T.B. and Maier, J.P. 1979, *Int. J. Mass Spectrom. Ion Phys.*, **31**, 287.
Kato, T., Kodama, T., Shida, T., Nakagawa, T., Matsui, Y., Suzuki, S., Shiromaru, H., Yamauchi, K., and Achiba, Y. 1991, *Chem. Phys. Lett.*, **180**, 446.
Keller, J.W. and Coplan, M.A. 1992, *Chem. Phys. Lett.*, **193**, 89.
Khan, Z.H. 1983, *Can. J. Spectrosc.*, **28**, 57.
Khan, Z.H. 1992, *Acta Phys. Polon.*, **A82**, 937.
Khan, Z.H., Husain, M.M., and Haselbach, E. 1993, *Appl. Spectrosc.*, **47**, 2140.
Kikuchi, K., Nakahara, N., Wakabayashi, T., Honda, M., Matsumiya, H., Moriwaki, T., Suzuki, S., Shiromaru, H., Saito, K., Yamauchi, K., Ikemoto, I., and Achiba, Y. 1992, *Chem. Phys. Lett.*, **188**, 177.
Kira, A., Imamura, M., and Shida, T. 1976, *J. Phys. Chem.*, **80**, 1445.

Kloster-Jensen, E., Haink, H.J., and Christen, H. 1974, *Helv. Chim. Acta*, **57**, 1731.
Kohler, B. 1993, *Chem. Rev.*, **93**, 41.
Krätschmer, W. 1986, *Astrophys. Space Sci.*, **128**, 93.
Krätschmer, W. 1993, *J. Chem. Soc. Faraday Trans.*, **89**, 2285.
Krelowski, J. and Sneden, C. 1993, *Publ. Astron. Soc. Pacific*, **105**, 1141.
Kroto, H.W. 1987, *Nature*, **329**, 529.
Kroto, H.W. 1988, *Science*, **242**, 1139.
Leach, S. 1986, *J. Electron. Spectr. Rel. Phen.*, **41**, 427.
Leach, S. 1987, in *Polycyclic Aromatic Hydrocarbons and Astrophysics*, eds. A. Léger, L. d'Hendecourt and N. Boccara, (Dordrecht: Reidel), p.99.
Leach, S. 1989, in *Interstellar Dust*, IAU Symposium 135, eds. L.J. Allamandola and A.G.G.M. Tielens, (Dordrecht: Kluwer), p.155.
Leach, S. 1993, *J. Chem. Soc. Faraday Trans.*, **89**, 2305.
Leach, S. 1994a, in *Summary of RAS Discussion Meeting on The Diffuse Interstellar Absorption Lines*, McNally, D. 1994, The Observatory, **114**, 97.
Leach, S. 1994b, in *The Physics and Chemistry of Fullerenes*, ed. K.Prassides, (Dordrecht: Klewer), in press.
Leach, S. 1994c, data presented at this meeting, from work in progress.
Leach, S. and Edwards, S.A. 1994, manuscript in preparation.
Leach, S., Eland, J.H.D., and Price, S.D. 1989a, *J. Phys. Chem.*, **93**, 7575.
Leach, S., Eland, J.H.D., and Price, S.D. 1989b, *J. Phys. Chem.*, **93**, 7583.
Leach, S., Vervloet, M., Després, A., Bréheret, E., Hare, J.P., Dennis, T.J., Kroto, H.W., Taylor, R., and Walton, D.R.M. 1992, *Chem. Phys.*, **160**, 451.
Léger, A. and Puget, J.L. 1984, *Astr. Ap.*, **137**, L5.
Léger, A. and d'Hendecourt, L. 1985, *Astr. Ap.*, **146**, 81.
Léger, A., d'Hendecourt, L., Verstraete, L., and Schmidt, W. 1988, *Astr. Ap.*, **203**, 145.
Lepp, S. and Dalgarno, A. 1988a, *Ap. J.*, **324**, 553.
Lepp, S. and Dalgarno, A. 1988b, *Ap. J.*, **335**, 769.
Lepp, S., Dalgarno, A., van Dishoek, E.F., and Black, J.H. 1988, *Ap. J.*, **329**, 418.
Lifshitz, C., Gotkis, I., Sandler, P., and Laskin, J. 1992, *Chem. Phys. Lett.*, **200**, 406.
McNally, D. 1994, Summary of RAS Discussion Meeting on The Diffuse Interstellar Absorption Lines, *The Observatory*, **114**, 97.
Millar, T.J. 1992, *Mon. Not. R. Astron. Soc.*, **259**, 35P.
Petrie, S., Javahery, G., and Bohme, D.K. 1992, *Astr. Ap.*, **271**, 662.
Pitzer, K.S. and Clementi, E. 1959, *J. Am. Chem. Soc.*, **81**, 4477.
Platt, J.R. 1949, *J. Chem. Phys.*, **17**, 484.
Qihuang Gong, Yuxing Sun, Ziwen Huang, Xihuang Zhu, Z.N.Gu, and Di Qiang, 1994, *J. Phys. B: At. Mol. Opt. Phys.*, **27**, L199.
Rühl, E., Price, S.D., and Leach, S., 1989, *J. Phys. Chem.*, **93**, 6312.
Salama, F. and Allamandola, L.J. 1993, *J. Chem. Soc. Faraday Trans.*, **89**, 2277.
Salem, L. 1966, The Molecular Orbital Theory of Conjugated Systems, (New York: Benjamin).
Saykally, R. 1993, in *Astronomical Infrared Spectroscopy: Future Observational Directions*, ed. Sun Kwok, ASP Conference Series, **41**, 233.
Shida, T. and Iwata, S. 1973, *J. Am. Chem. Soc.*, **95**, 3721.
Smalley, R.E. 1992, *Acc. Chem. Res.*, **25**, 98.
Smith, A.L. 1994, in *The Physics and Chemistry of Fullerenes*, ed. K.Prassides, (Dordrecht: Klewer), in press.
Thaddeus, P., Gottlieb, C.A., Mollaaghababa, R., and Vrtilek, J.M. 1993, *J. Chem. Soc. Faraday Trans.*, **89**, 2125.
Tielens, A.G.G.M., Allamandola, L.J., Barker, J.R., and Cohen M. 1987, in *Polycyclic Aromatic Hydrocarbons and Astrophysics*, eds. A. Léger, L. d'Hendecourt and N. Boccara, (Dordrecht: Reidel), p.273.
Tobita, S., Meinke, M., Illenberger, E., Christophorou, L.G., Baumgärtel, H., and Leach, S. 1992, *Chem. Phys.*, **161**, 501.
Tobita, S., Leach S., Jochims, H.W., Rühl, E., Illenberger, E., and Baumgärtel, H. 1994, *Can. J. Phys.*, in press.
Vala, M., Szczepanski, J, Pauzat, F., Parisel, O., Talbi, D., and Ellinger, Y. 1994, *J. Phys. Chem.*, **98**, 9187, and references therein.
van der Swet, G.P. and Allamandola, L.J. 1985, *Astr. Ap.*, **146**, 76.
Verstraete, L., Léger, A., d'Hendecourt, L., Dutuit, O., and Défourneau, D. 1990, *Astr. Ap.*, **237**, 436.
Webster, A.S. 1992, *Astr. Ap.*, 257, 750.
Weltner, W., Jr. and Van Zee, R.J. 1989, *Chem. Rev.*, **89**, 1713.
Westerlund, B.E. and Krelowski, J. 1988a, *Astr. Ap.*, **189**, 221.
Westerlund, B.E. and Krelowski, J. 1988b, *Astr. Ap.*, **203**, 134.
Wurz, P. and Lykke, K.R. 1992, *J. Phys. Chem.*, **96**, 10129.
Yoo, R.K., Ruscic, B., and Berkowitz, J. 1992, *J. Chem. Phys.*, **96**, 911.

LARGE MOLECULES IN THE INTERSTELLAR MEDIUM

ERIC HERBST
The Ohio State University, Columbus, OH 43210, USA

ABSTRACT. The synthesis of large molecules in dense interstellar clouds and carbon-rich circumstellar envelopes is discussed. Gas phase reactions can produce observed abundances of molecules with as many as 11 atoms in dense clouds at so-called "early times" of $10^5 - 10^6$ yr despite the oxygen-rich nature of the medium. Extensions of current models to incorporate much larger molecules can only show that efficient synthesis of such species is possible, given the large uncertainties in relevant rate coefficients. Alternately, the molecules produced in dense clouds may serve as seeds for the gas phase production of larger species in diffuse clouds. The synthesis of large molecules on the surfaces of dust particles in interstellar clouds is possible if hydrogenation of reactive radicals on grain surfaces to saturated stable forms does not dominate. In carbon-rich circumstellar envelopes, the condensation of carbonaceous grains occurs near the stellar photosphere, the production of PAH's proceeds somewhat farther out at temperatures near 1000 K, and the synthesis of carbon clusters and unsaturated hydrocarbons occurs in the outer envelope.

1. INTRODUCTION

Following the early work of Johnson (1972), Douglas (1977) and others, a variety of authors have attributed the diffuse interstellar bands to organic molecules sufficiently large in size that they are either relatively immune to photodestruction (Jochims *et al.* 1994) or chemical destruction into smaller species. Specific types of molecules suggested include carbon clusters, polycyclic aromatic hydrocarbons (PAH's), fullerenes, fulleranes, porphyrins, and others (see chapters by Maier, Thaddeus, Léger, Webster, and Johnson in this volume). The production of large molecules in diffuse clouds, however, presents grave difficulties because the small molecules from which they are synthesized are not immune to radiation and thus may possess very small abundances. Two partially distinct solutions to this problem can be imagined. In one, the difficulty of producing large molecules in the harsh environment of the diffuse interstellar medium is reduced by providing sufficiently large abundances of "seed" molecules, perhaps 10-20 atoms in size, from a different environment. The truly large species can subsequently develop from the seeds. In the second solution, the large molecules are synthesized almost completely in a more benign region. In both cases, this region then evolves into the diffuse material seen along the lines of sight towards stars. In this chapter, we discuss the production of organic molecules in two regions which appear to be better suited to organic molecule development than the diffuse interstellar medium - dense interstellar clouds and carbon-rich circumstellar envelopes.

A. G. G. M. Tielens and T. P. Snow (eds.), The Diffuse Interstellar Bands, 307–314.

2. GAS/GRAIN MODELS OF DENSE INTERSTELLAR CLOUDS

Our current gas phase and gas/grain models of dense interstellar clouds contain up to 4000 gas phase reactions involving nearly 400 species, composed of 12 elements. For homogeneous, quiescent sources, we assume the clouds to have gas densities in the vicinity of 10^4 cm^{-3} and temperatures of 10 K. The material is protected from the interstellar radiation field to some extent both by dust particles and the dominant gas phase species H_2 and CO. Cosmic rays penetrate all but the largest dense clouds and cause a fractional ionization. The models utilize as initial conditions either atoms or a combination of atoms and diatomic molecules. The relative gas phase elemental abundances with respect to hydrogen are assumed to show modest to severe depletions vis-a-vis standard cosmic abundances. Most importantly, the oxygen-to-carbon elemental abundance ratio is greater than unity, reflecting the fact that most matter is oxygen-rich. Molecular hydrogen is produced from its constituent atoms on grain surfaces (Hollenbach and Salpeter 1970), after which the H_2 molecule leaves the grain surface via evaporation or during the process of formation. Once in the gas, hydrogen molecules are occasionally ionized by cosmic ray particles, leading to a rich and complex ion-molecule chemistry (Herbst and Leung 1989). Ion-molecule and ion electron recombination reactions have long been expected to dominate low-temperature interstellar chemistry because these processes are known to occur at the collision-limited rate without short-range potential ("activation") energy barriers in most instances (Anicich and Huntress 1986; Anicich 1993), whereas neutral-neutral reactions often possess activation energy barriers (Baulch *et al.* 1992). Recent research, however, indicates that the role of certain classes of neutral-neutral reactions, typically involving at least one radical, or reactive neutral, has been overlooked (Sims *et al.* 1994; Bettens and Brown 1992; Herbst *et al.* 1994). Even with all of the gas phase reactions included in dense cloud models, the largest neutral molecule in the models - HC_9N - contains only 11 atoms. The inclusion of still larger molecules at the current detailed level of treatment will not be facile.

Two features of the results of our purely gas phase models are quite relevant here. First, the abundances of organic molecules are all predicted to peak at a so-called "early time" of $10^5 - 10^6$ yr before reaching sharply lower steady-state values at significantly later times. The calculated abundances within a period of time around the peak calculated values are in reasonable agreement with observed gas phase abundances in dark clouds (Herbst and Leung 1989, 1990). The diminution in abundance after early-time values derives from the oxygen-to-carbon elemental abundance ratio chosen; in carbon-rich models, large abundances of organic molecules are calculated to remain at steady state. This difference may not be critical, because gas/grain models show that at times after early time, much of the original gaseous matter condenses onto the cores of dust particles to form rich ice-dominated mantles (Hasegawa *et al.* 1992). Although some scientists think that non-thermal desorption processes can return a good percentage of the adsorbed mantles back into the gas phase (Willacy and Williams 1993), our calculations indicate that desorption does not occur readily for organic species (Hasegawa and Herbst 1993a,b). The second feature of our models is that despite the dominant nature of the element hydrogen, most of the abundant organic molecules produced are strongly unsaturated in nature; viz., carbon clusters and hydrocarbons with only one or two hydrogens seem preferred because of the difficulty of finding exothermic ion-H_2 reactions leading to further hydrogenation (Herbst *et al.* 1983; McElvany *et al.* 1987).

2.1. SYNTHETIC AND DESTRUCTIVE REACTIONS

The three types of ion-molecule reactions contributing to the synthesis of hydrocarbon ions, in order of decreasing importance, are labelled carbon insertion, condensation, and radiative association. Their roles are discussed by Smith (1992). In carbon insertion, a carbon ion or atom inserts itself into a hydrocarbon molecule to produce a larger species while one or more hydrogen atoms is ejected. In condensation, a hydrocarbon ion and a hydrocarbon neutral react to form a larger hydrocarbon ion plus one or more hydrogen atoms. Finally, in radiative association, the ion and the neutral collide and manage to stick together by emitting excess energy in the form of a photon. Some examples of these classes of reactions are:

$$C^+ + C_2H_2 \longrightarrow C_3H^+ + H \tag{1}$$

$$C_2H_2^+ + C_2H_2 \longrightarrow C_4H_3^+ + H \tag{2}$$

$$CH_3^+ + H_2 \longrightarrow CH_5^+ + h\nu \tag{3}$$

$$C_4H_2^+ + C_2H_2 \longrightarrow c - C_6H_4^+ + h\nu. \tag{4}$$

Although carbon insertion and condensation reactions are well studied processes in the laboratory (Anicich and Huntress 1986; Anicich 1993), radiative association has been mainly the province of theoreticians, who have shown these reactions to be generally inefficient for small reactants but to improve in efficiency as the size of the reactants increases, especially at low temperatures and for large bond energies (Bates and Herbst 1988a). Recent experiments in low pressure ion traps seem to agree moderately well with theoretical results (Gerlich and Horning 1992; Dunbar 1994). Two types of radiative association reactions are included in gas phase models of dense interstellar clouds: reactions between ions and H_2 and reactions between ions and heavy neutrals. The former need not be nearly as efficient to be competitive given the large abundance of H_2. For hydrocarbon ions containing more than a few carbon atoms, however, association reactions involving H_2 often do not occur, presumably because of activation energy barriers (Herbst *et al.* 1983).

In addition to radiative association reactions between ions and H_2, hydrogenation can occur via H-atom transfer reactions; e. g.,

$$C_3H^+ + H_2 \longrightarrow C_3H_2^+ + H \tag{5}$$

although such reactions are not known to put more than a few hydrogen atoms onto carbon skeletons for ions with more than several carbon atoms because of a combination of potential barriers, as in the association case, and endothermicities (Maluendes *et al.* 1994). It is not even clear that one hydrogen atom can be added to C_n^+ clusters for n > 9 by either association or H-atom transfer, presumably because the most stable form becomes a non-reactive ring (Weltner and van Zee 1989; McElvany *et al.* 1987). A scheme for the growth of carbon clusters without any hydrogenation is advanced by Thaddeus in this volume. Basing his arguments on the earlier work of Suzuki (1983) and Freed *et al.* (1982), he suggests that radiative association reactions between C^+ and carbon clusters:

$$C^+ + C_n \longrightarrow C_{n+1}^+ + h\nu, \tag{6}$$

occur with unit efficiency as long as n is greater than a few.

Analogous radical-neutral examples for the classes of reactions discussed here are now also thought to play a role in synthesis, although the laboratory evidence is nowhere near as complete and, indeed, few products have been measured. Perhaps the condensation reaction

$$C_2H + C_2H_2 \longrightarrow C_4H_2 + H_2 \tag{7}$$

has received the most attention because it is rapid at room temperature and slightly below (Pedersen *et al.* 1993) and the reaction products have been studied. A large number of reactions have been inserted into models based on this one studied case. Carbon insertion reactions may occur as well; for example,

$$C + C_2H_2 \longrightarrow C_3H + H \tag{8}$$

although only the rate of reaction at room temperature is known (Clary et al. 1994). Hydrogenation reactions all appear to involve activation energy barriers, although tunneling under them may occur at low temperatures (Herbst 1994). In any case, hydrogenation reactions will not lead to saturated species but to hydrocarbons with at most a few hydrogen atoms.

Ion-molecule and neutral-neutral reactions both destroy molecular complexity by converting complex molecules into more simple ones. In oxygen-rich material, reactions involving oxygen atoms appear to play a key role. Oxygen atoms are known to react with neutral radicals rapidly at room temperature (Baulch *et al.* 1992); e.g.,

$$O + C_2H \longrightarrow CH + CO, \tag{9}$$

and to react rapidly with hydrocarbon ions (Anicich and Huntress 1986); e.g.,

$$O + C_2H_2^+ \longrightarrow HCO^+ + CH; HC_2O^+ + H, \tag{10}$$

although there are channels in which the degree of molecular complexity is not changed. Rather, oxygenated organic species are formed.

Even in the absence of the above destruction reactions, recombination reactions with electrons destroy molecular ions to produce the observed neutral molecules. For small ions, these processes are known to be dissociative, although both the actual product ratios and the mechanisms by which they are formed are still poorly understood for most systems. Recent experimental work (Larsson 1994; Herd *et al.* 1990) on the dissociative recombination of two small ions - H_3^+ and H_3O^+ - suggests that the dominant channels show considerable fragmentation:

$$H_3^+ + e^- \longrightarrow H + H + H(80\%); H_2 + H(20\%) \tag{11}$$

$$H_3O^+ + e^- \longrightarrow OH + ...(65\%). \tag{12}$$

Presumably as the size of the polyatomic ion becomes sufficiently large, a sizeable fraction of electron collisions should result in radiative stabilization, probably proceeding via Rydberg intermediates. Detailed theoretical estimates of the minimum ionic size necessary for such radiative capture of electrons have not been made, although Thaddeus (this volume) has suggested 12-14 atoms for bare carbon ions. Sticking of electrons to neutral species to form negatively charged ions is likely to be important as well for neutrals with moderate to large electron affinities (Herbst 1981; Bakes and

Tielens, this volume). If the negatively charged ions reach significant abundances (Lepp and Dalgarno 1988), they may compete with electrons in reacting with positive molecular ions. Mutual neutralization reactions between positive and negative ions presumably lead to larger neutral fragments than do dissociative recombination reactions (Bates and Herbst 1988b).

Especially in more diffuse regions, the photodestruction of molecules must be considered as a destruction mechanism. Recent work does appear to confirm the statistical argument that large molecules are relatively immune to photodissociation if not photoionization because decomposition following excitation can be slower than radiative relaxation (Jochims et al. 1994). Of course, for larger molecules there are more molecular bonds that can break so it is not apparent that the immunity will work in all instances. We are presently calculating these competing effects in some detail for bare carbon clusters exposed to the interstellar radiation field.

The synthesis of organic molecules on the surfaces of grains may also occur in interstellar clouds (Allen and Robinson 1977). It is normally assumed that atoms and molecules are only weakly adsorbed (physisorbed) onto grains and that the lightest of the adsorbed species (atoms and H_2) can diffuse appreciably via classical migration or tunneling from site to site on the grain surfaces (Tielens and Hagen 1982). Association reactions between species on grains can be efficient for colliding adsorbates since the excess energy need not be radiated away as in the gas but can be imparted to the grain phonon modes. To avoid activation energy barriers once they collide with one another, the reactants should be reactive atoms or radicals. Reactions involving molecular hydrogen are an exception since tunneling under activation energy barriers is a distinct possibility (Tielens and Hagen 1982; Hasegawa and Herbst 1993a). One mechanism for molecular growth on grain surfaces is the successive association of rapidly moving carbon atoms with bare reactive carbon clusters or hydrocarbon radicals. This process must compete with hydrogenation reactions involving either atomic or molecular hydrogen, which produce unreactive saturated species. In our most recent models, hydrogenation appears to have the upper hand, so that molecule growth is not competitive with that occurring in the gas (Hasegawa and Herbst 1993a,b). On the other hand, the possibility of an active photochemistry on grain surfaces which can regenerate reactive radicals must also be taken into account (Shalabiea and Greenberg 1994).

2.2. A MODEL OF COMPLEX MOLECULE SYNTHESIS

We have previously considered an extension of our dense cloud gas phase models to include molecules containing up to thirty carbon atoms (Herbst 1991). To accomplish this goal, we considered hydrocarbons only and divided the hydrocarbons into groups according to their charge (neutral or positive) and the number of carbon atoms they possess. No attempt was made to follow the degree of saturation. Generic rate coefficients were utilized for chain growth from seed molecules (obtained in our normal models) via association/condensation/insertion reactions as well as for chain neutralization, fragmentation, and reaction. Photodestruction was ignored since the chemistry was assumed to occur in dark clouds. A wide variety of results were obtained depending on the values of the generic rate coefficients. In one "optimistic" scenario, a total fractional abundance for hydrocarbon neutrals containing 20-30 carbon atoms of 10^{-7} with respect to the gas density was obtained within a million years. The model is similar to (although less optimistic than) that of Thaddeus (this volume) except that he considers bare carbon clusters growing from seeds in a diffuse environment, which necessitates the inclusion of photodestructive processes. From both of these approaches it can be concluded that the growth of large molecules in the gas phase of interstellar clouds is possible but that much more information about the processes utilized in the models is required.

3. MOLECULAR SYNTHESIS IN C-RICH CIRCUMSTELLAR ENVELOPES

Once they leave their stable main sequence existence, stars of low to intermediate mass go through two red giant stages. The second, or asymptotic giant branch (AGB) stage, produces an extensive envelope of outward flowing gas and dust which is often carbon-rich ($C > O$), in which case it is known to be a fecund source of organic molecules. The best studied object of this class is the infrared source IRC+10216 (Omont 1993). Near the star, the envelope is hot and dense, but with increasing radial distance, both the temperature and density decrease. In the hot and dense inner envelope, thermodynamic equilibrium is thought to prevail and carbonaceous dust is thought to form, possibly via induced nucleation (Cadwell *et al.* 1994). The abundant gas phase molecules coming out of this regime include C_2H_2. The chemistries occurring in cooler regions make use of the high abundance of acetylene, which is a well-known precursor of more complex species in carbon-rich flames.

3.1. PAH FORMATION

At somewhat cooler but still warm temperatures ($\approx$ 1100 K), PAH's can form from the "seed" acetylene, especially since the gas is carbon-rich. According to the model of Frenklach and Feigelson (1989), which is based on soot formation in the laboratory, after initial neutral rings are formed via ring closure reactions, growth of PAH's occurs via successive condensation reactions with acetylene molecules at reasonably high densities. As opposed to interstellar schemes, in which synthesis is thought to be dominated by carbon insertion rather than condensation reactions, this mode of synthesis also effectively adds hydrogen atoms to the aromatic structures so that bare clusters do not dominate. For the condensation reactions to occur rapidly, the aromatic species must be reactive radicals. Such radicals form from stable aromatic species via H-abstraction reactions with atomic H to form H_2, which tend to have activation energy barriers and be close to thermoneutral. Generation of a sufficient abundance of aromatic radicals requires the high temperatures. A second look at the model of Frenklach and Feigelson (1989) has been undertaken by Cherchneff *et al.* (1992) who show that under the customarily assumed physical conditions and flow rates, PAH formation is not very efficient. They stress, however, the uncertainties present even in the study of soot formation under laboratory conditions.

3.2. OUTER ENVELOPE CHEMISTRY

It is the outer portions of carbon-rich circumstellar envelopes that have been looked at in most detail. Found to be rich in a variety of unsaturated, chain-dominated organic molecules by radio observers, these sources have a gas phase chemistry that is known to resemble its interstellar cousin without the difficulty of oxygen-rich conditions (Glassgold *et al.* 1986; Howe and Millar 1990; Cherchneff *et al.* 1993; Cherchneff and Glassgold 1993; Millar and Herbst 1994). Molecules in the outer envelope are subjected to dissociating and ionizing radiation in small doses from *external* sources. Both UV photons and cosmic rays produce radicals and a small fractional abundance of ions; radical-neutral and ion-molecule chemistries then synthesize more complex molecules via all of the processes described in Section 2. Condensation reactions are important, especially in the neutral-neutral chemistry, and have been used to produce carbon clusters and very unsaturated hydrocarbons, based on the well-studied reaction between C_2H and C_2H_2 (Pedersen *et al.* 1993). In the most recent model (Millar and Herbst 1994), reasonable agreement with observation for both the abundance and cloud extent of individual carbon clusters, hydrocarbon radicals, and other species such as cyanopolyynes ($HC_{2n+1}N$) is achieved. The model is severely truncated in two aspects, however: (1) the size of the hydrocarbons is followed only up to ≈ 10 atoms, and (2) the production of cyclic species other than the simple c-C_3H and c-C_3H_2 is not considered (Maluendes *et al.* 1993).

The dichotomy between the zones of PAH chemistry and unsaturated chain chemistry may or may not be artificial. Certainly, if models of the outer envelope are extended to unsaturated hydrocarbons with more than nine carbon atoms, it will be difficult to avoid cyclization of carbon clusters since these become more stable than linear chains. Without hydrogenation, however, the end result of cluster synthesis should be fullerenes rather than PAH's. On the other hand, ion-molecule reactions such as reaction (4) are known to produce aromatic species (Anicich *et al.* 1990; Bohme 1992). It seems possible therefore that a variety of types of complex species can be synthesized.

The role of grains in the cool outer envelope has also not been explored. Even if grain chemistry does not occur on a short enough time scale to be of crucial importance, grains must be considered as a template on which organic molecules synthesized in the gas can subsequently adsorb.

4. CONCLUSION

Gas phase organic molecules of more than ten atoms in size have been observed via high resolution spectroscopy in two types of sources: dense interstellar clouds and the outer portions of carbon-rich circumstellar envelopes. Both of these objects possess similar low temperatures and densities and both are subjected to small amounts of ionizing radiation. If the carriers of the diffuse interstellar bands are complex molecules, then it is quite possible that these types of sources play some role in the synthesis of these complex molecules. One likely role is as a site for the production of seed molecules large enough to withstand the interstellar radiation field and chemical destruction in more diffuse material. The gas phase chemistry of both dense clouds and outer circumstellar envelopes has been studied extensively, and detailed models are available with abundance predictions for molecules up to approximately 10 atoms in size. Extensions of these model predictions to include larger molecules are highly tentative at this point owing to the lack of laboratory data. Still, it is possible that molecules large enough to be stable in a diffuse environment are made in reasonable abundances in both types of sources and that these molecules are related to the carriers of the diffuse interstellar bands. The difficulty of hydrogenating species via both ion-molecule and neutral-neutral reactions has led investigators to develop synthetic routes towards very unsaturated species including bare clusters. It is not clear, however, that the role of ion-molecule and radical-neutral reactions in forming cyclic and even aromatic species under low density, low temperature conditions is fully appreciated. Current models for PAH formation in carbon-rich envelopes posit significantly higher temperatures and densities, as are likely to occur in the inner rather than the outer envelope. These models are similar to models of soot formation in flames, which themselves are somewhat uncertain. The formation of cyclic molecules under low density, low temperature conditions should be investigated more fully.

ACKNOWLEDGEMENTS. I would like to acknowledge the support of the National Science Foundation (USA).

References

Allen, M. and Robinson, G. W. 1977, *Ap. J.*, **212**, 396.
Anicich, V.G. 1993, *Ap. J. Suppl.*, **84**, 215.
Anicich, V.G. and Huntress, Jr., W.T. 1986, *Ap. J. Suppl.*, **62**, 553.
Anicich, V.G., Sen, A.D., Huntress, Jr., W.T., and McEwan, M.J. 1990, *J. Chem. Phys.*, **93**, 7163.
Baulch, D.L., Cobos, C.J., Cox, R.A., Esser, C., Frank, P., Just, Th., Kerr, J.A., Pilling, M.J., Troe, J., Walker, R.W., and Warnatz, J. 1992, *J. Phys. Chem. Ref. Data*, **21**, 411.
Bates, D.R. and Herbst, E. 1988a, in *Rate Coefficients in Astrochemistry*, eds. T. J. Millar and D. A. Williams, (Dordrecht: Kluwer), p. 17.
Bates, D.R. and Herbst, E. 1988b, in *Rate Coefficients in Astrochemistry*, eds. T. J. Millar and D. A. Williams, (Dordrecht: Kluwer), p. 41.
Bettens, R.P.A. and Brown, R.D. 1992, *M. N. R. A. S.*, **258**, 347.
Bohme, D.K. 1992, *Chem. Rev.*, **92**, 1487.
Cadwell, B.J., Wang, H., Feigelson, E.D., and Frenklach, M. 1994, *Ap. J.*, July 1.
Cherchneff, I., Barker, J.R., and Tielens, A.G.G.M. 1992, *Ap. J.*, **401**, 269.
Cherchneff, I. and Glassgold, A.E. 1993, *Ap. J. (Letters)*, **419**, L41.
Cherchneff, I., Glassgold, A.E., and Mamon, G.A. 1993, *Ap. J.*, **410**, 188.
Clary, D.C., Haider, N., Husain, D., and Kabir, M. 1994, *Ap. J.*, **422**, 416.
Douglas, A.E. 1977, *Nature*, **269**, 130.
Dunbar, R.C. 1994, in *Unimolecular and Bimolecular Ion-Molecule Reactions*, eds. C. Y. Ng and T. Baer, (New York: Wiley), in press.
Freed, K.F., Oka, T., and Suzuki, H. 1982, *Ap. J.*, **263**, 718.
Frenklach, M. and Feigelson, E.D. 1989, *Ap. J.*, **341**, 372.
Gerlich, D. and Horning, S. 1992, *Chem. Rev.*, **92**, 1509.
Glassgold, A.E., Lucas, R., and Omont, A. 1986, *Astr. Ap.*, **157**, 35.
Hasegawa, T.I. and Herbst, E. 1993a, *M. N. R. A. S.*, **261**, 83.
Hasegawa, T.I. and Herbst, E. 1993b, *M. N. R. A. S.*, **263**, 589.
Hasegawa, T.I., Herbst, E., and Leung, C.M. 1992, *Ap. J. Suppl.*, **82**, 167.
Herbst, E. 1981, *Nature*, **289**, 656.
Herbst, E. 1991, *Ap. J.*, **366**, 133.
Herbst, E. 1994, *Chem. Phys. Letters*, **222**, 297.
Herbst, E., Adams, N.G., and Smith, D. 1983, *Ap. J.*, **269**, 329.
Herbst, E., Lee, H.-H., Howe, D.A., and Millar, T.J. 1994, *M. N. R. A. S.*, **268**, 335.
Herbst, E. and Leung, C.M. 1989, *Ap. J. Suppl.*, **69**, 271.
Herbst, E. and Leung, C.M. 1990, *Astr. Ap.*, **233**, 177.
Herd, C.R., Adams, N.G., and Smith, D. 1990, *Ap. J.*, **349**, 388.
Hollenbach, D. and Salpeter, E.E. 1970, *J. Chem. Phys.*, **53**, 79.
Howe, D.A. and Millar, T.J. 1990, *M. N. R. A. S.*, **244**, 444.
Jochims, H.W., Rühl, E., Baumgärtel, H., Tobita, S., and Leach, S. 1994, *Ap. J.*, **420**, 307.
Johnson, F.M. 1972, *Mem. Soc. Roy. des Sc. Liege*, **III**, 6th series, 391.
Larsson, M. 1994, private communication. See also 1993, *J. Chem. Soc. Faraday Trans.*, **89**, 2216.
Lepp, S. and Dalgarno, A. 1988, *Ap. J.*, **324**, 553.
Maluendes, S.A., McLean, A.D., and Herbst, E. 1993, *Ap. J.*, **417**, 181.
Maluendes, S.A., McLean, A.D., and Herbst, E. 1994, *Chem. Phys. Letters*, **217**, 571.
McElvany, S.W., Dunlap, B.I., and O'Keefe, A. 1987, *J. Chem. Phys.*, **86**, 715.
Millar, T.J. and Herbst, E. 1994, *Astr. Ap.*, in press.
Omont, A. 1993, *J. Chem. Soc. Faraday Trans.*, **89**, 2137.
Pedersen, J.O.P., Opansky, B.J., and Leone, S.R. 1993, *J. Phys. Chem.*, **97**, 6822.
Shalabiea, O.M. and Greenberg, J.M. 1994, *Astr. Ap.*, in press.
Sims, I.R., Queffelec, J.-L., Defrance, A., Rebrion-Rowe, C., Travers, D., Bocherie, P., Rowe, B.R., and Smith, I.W.M. 1994, *J. Chem. Phys.*, **100**, 4229.
Smith, D. 1992, *Chem. Rev.*, **92**, 1473.
Suzuki, H. 1983, *Ap. J.*, **272**, 579.
Tielens, A.G.G.M. and Hagen, W. 1982, *Astr. Ap.*, **114**, 245.
Weltner, W., Jr. and Van Zee, R.J. 1989, *Chem. Rev.*, **89**, 1713.
Willacy, K. and Williams, D.A. 1993, *M. N. R. A. S.*, **260**, 635.

CHARGE VARIATION OF POLYCYCLIC AROMATIC HYDROCARBONS IN THE DIFFUSE INTERSTELLAR MEDIUM

E. L. O. Bakes
Princeton University, USA
A. G. G. M. Tielens
NASA Ames Research Center, USA

ABSTRACT.

We have modelled the charge balance for several polycyclic aromatic hydrocarbon (PAH) molecules, the fullerene C_{60} and small graphitic grains in the diffuse interstellar medium (ISM). For the smaller PAH molecules, we used pyrene ($C_{16}H_{10}$), pentacene ($C_{22}H_{14}$) and coronene ($C_{24}H_{12}$). We then calculated the charge distribution for each particle as a function of electron density, radiation field and depth at which it was situated in a diffuse cloud. For typical conditions in the diffuse ISM, the charge distribution is determined to a large extent by PAH electron affinity, which determines whether an electron will stick to it. For the PAH molecules, we compare the charge distribution calculated using experimentally measured cation-electron recombination rates with theoretically determined rates. We conclude that while a PAH mixture may exist in a variety of singley charged and neutral states, larger disk shaped and spherical grains will be predominantly neutral.

1. INTRODUCTION

The single carrier hypothesis for the DIBs requires a stable, complex, polyatomic molecule which is an efficient absorber in the UV (Herbig 1975). The carrier must not be optically polarizable and must be correlated with interstellar reddening. Scarrott, Watkin, Miles & Sarre (1992) found that the wavelengths of the four most prominent extended red emission bands in the Red Rectangle correspond with one of the families of DIBs, implying a common carrier composed of gas phase molecules. Analysis of infrared spectra of many interstellar sources has revealed that PAH molecules containing 20–50 carbon atoms are an ubiquitous and abundant component of the ISM (Allamandola *et al.* 1989). Hence, PAHs fulfil all of the above conditions. Small neutral PAHs do not absorb in the visible wavelength region. However, PAHs are expected to be charged in the ISM and positive or negatively charged PAHs do show visible absorption bands (Crawford *et al.* 1985, Léger & d'Hendecourt 1985). Therefore, considering PAH charge in detail is important. Two near–infrared DIBs have been attributed to the C_{60} cation and in this paper, we determine its charge state for the diffuse ISM. The effects of PAH charging on the diffuse interstellar bands is considered in depth in the forthcoming paper of Salama *et al.*

A. G. G. M. Tielens and T. P. Snow (eds.), The Diffuse Interstellar Bands, 315–321.

2. THEORY OF CHARGE BALANCE

Interstellar grains can charge by collisions with electrons or ions and by the photoelectric ejection of electrons. We have extended the study of collisional charging by Draine & Sutin (1987) to include grain charging via the photoelectric effect. The probability of finding a grain at charge Ze is defined as f(Z). By detailed balance

$$f(Z)[J_{pe}(Z) + J_{ion}(Z)] = f(Z+1)J_e(Z+1) \quad (1)$$

where J_{pe} is the rate of photoelectron emission, J_{ion} is the accretion rate of ions and J_e is the accretion rate of electrons.

Unlike classical grains (radius $> 0.1\ \mu$m), a molecular sized dust grain is polarized by an approaching electron or ion. The induced image charge produces an attractive electrostatic force. The smaller the grain, the greater is the attractive electrostatic force. In a cool gas, the approaching charged particle will move relatively slowly and will have more chance of being captured by the grain. The diffuse ISM has a low temperature, and so PAH molecules will have a high probability of capturing electrons or ions in the gas.

3. METHOD

We have modelled the charge balance using the formalism of Bakes & Tielens (1994) for several PAH molecules, C_{60} and small graphitic grains in the diffuse ISM. For the smaller PAH molecules, we used pyrene ($C_{16}H_{10}$), pentacene ($C_{22}H_{14}$) and coronene ($C_{24}H_{12}$). The ionization cross section for these molecules was taken from Verstraete *et al.* (1990) and we used experimentally measured electron affinities and ionization potentials. For C_{60}, we have calculated the ionization cross section using measurements from Leach *et al.* (1992) and taken ionization potentials and electron affinities from Chang *et al.* (1991). For the small graphitic grains, we considered a disk shaped grain composed of 100 carbon atoms and a spherical grain of 1000 carbon atoms. The particles investigated are shown in Figure 1. The ionization potentials for these grains were calculated using Smith (1961). Their physical properties are summarized in Table 1. We then calculated the charge distribution for PAHs and their agglomerates as a function of electron density, radiation field and depth at which it was situated in the cloud in front of ζ Per using Model F of van Dishoeck & Black (1986).

4. RESULTS

The typical conditions for diffuse interstellar clouds are taken as : electron density $n_e = 7.5 \times 10^{-3}$ cm^{-3}, gas temperature T=100K and the UV flux in Habings as G_o=1. This produces an interesting charge mixture in the diffuse cloud, which depends sensitively on the cation-electron recombination rate and the UV radiation field. The results are sensitive to the adapted electron recombination rate. In our initial study, we extrapolated the electron recombination rate derived for classical grains (Draine & Sutin 1987) to the molecular domain. With this relatively high recombination rate (10^{-7} s^{-1} for positively charged PAHs), we obtain pyrene and coronene providing mainly neutral particles and a very small fraction of positively charged particles ($\sim$ 2%). Pentacene (which has $s_e = 1$ for the neutral PAH) produces an approximately equal number of neutral and negatively charged particles with the larger PAHs providing

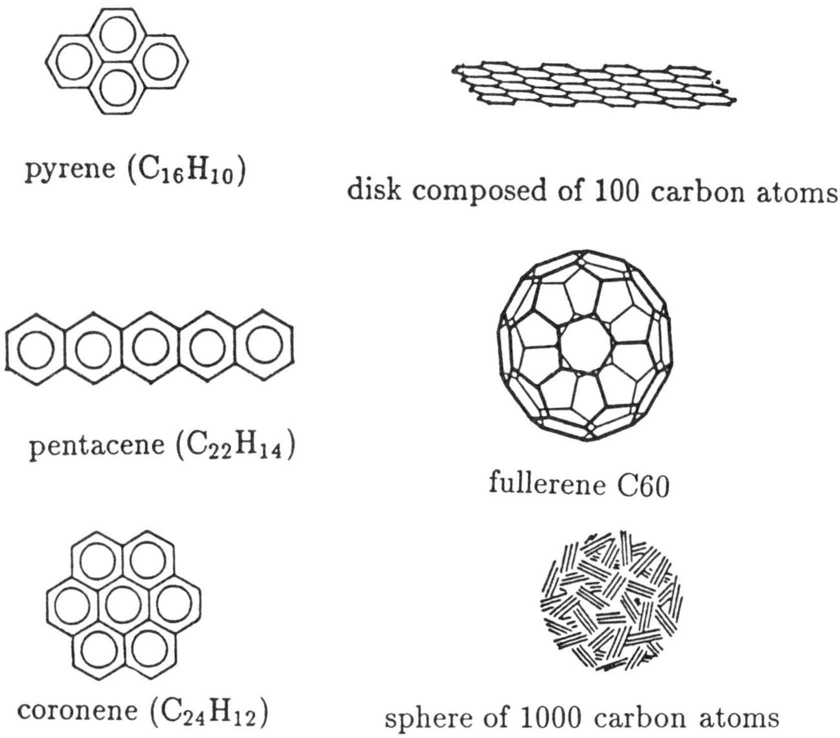

Fig. 1. This shows the symmetric PAHs, C_{60} and the different geometries of carbonaceous particles investigated in order to ascertain typical charging within the diffuse ISM.

TABLE 1
Physical properties of polycylic aromatic hydrocarbons, C_{60} and small graphitic grains. We list the particle and its first electron affinity, first ionization potential and the electron sticking coefficient.

PAH Molecule	EA_1 (eV)	IP_1 (eV)	$s_e(Z=0)$
Pyrene $C_{16}H_{10}$	0.55	7.41	10^{-3}
Pentacene $C_{22}H_{14}$	1.1	6.74	1
Coronene $C_{24}H_{12}$	0.64	7.36	10^{-3}
C_{60}	2.7	7.6	1
N_c=100	3.14	5.66	1
N_c=1000	3.84	4.96	1

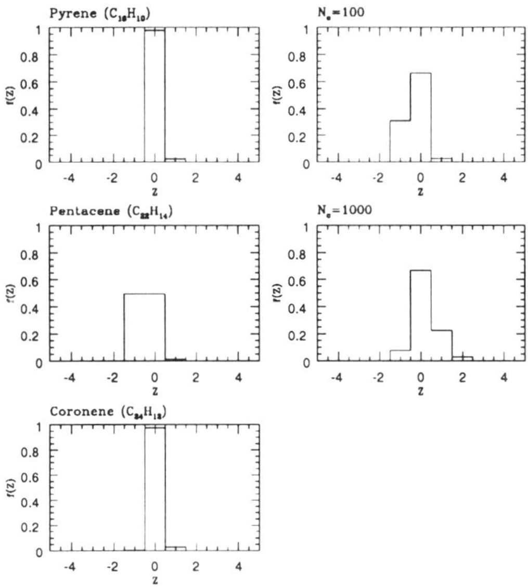

Fig. 2. These charge distributions were evaluated for typical conditions in the diffuse ISM corresponding to electron density $n_e = 7.5 \times 10^{-3}$ cm^{-3}, gas temperature T=100K and G_o=1.

a sizeable neutral population. This is illustrated in Figure 2. Recently, the electron recombination rates for the small PAHs, benzene and napthalene, were measured to be an order of magnitude less than that obtained on the basis of the extrapolation from the classical grains (Abouelaziz *et al.* 1993). Adopting this value, we find a much increased proportion of positively charged PAHs in the diffuse ISM (see Figure 3).

The results of grain charging for Model F of van Dishoeck and Black (1986) using the electron-cation recombination rate of Abouelaziz *et al.* (1993) are shown in Figure 4. The neutral fraction for pentacene remains almost constant. Pentacene becomes more negative further into the cloud because the grain-electron recombination rate increases with hydrogen nucleus density. Pentacene has a relatively high electron affinity and so electrons can stick to it more easily than coronene. The fraction of positively charged pentacene molecules decreases with depth into the cloud as the UV radiation and hence, the photoelectric ejection rate is exponentially attenuated.

As cloud depth increases in the same model, coronene still remains mostly neutral whilst the positive fraction decreases. Coronene's lower electron affinity means it has a significant probability of being neutral. Its neutral fraction increases with the depth of the cloud as expected and conversely, the positive fraction decreases. For the 100 carbon atom particle, the grain charge has an equal probability of being neutral or negative. As with pentacene, the negative fraction increases with depth into the cloud and the neutral fraction decreases.

The results for the fullerene C_{60} use the grain-electron recombination rates of Draine & Sutin (1987) because of the lack of availability of experimentally measured values. Lezius *et al.* (1993) note that electrons attach to neutral C_{60} efficiently with a cross section approaching 10^{-12} cm^2 (compared to the value calculated using Draine

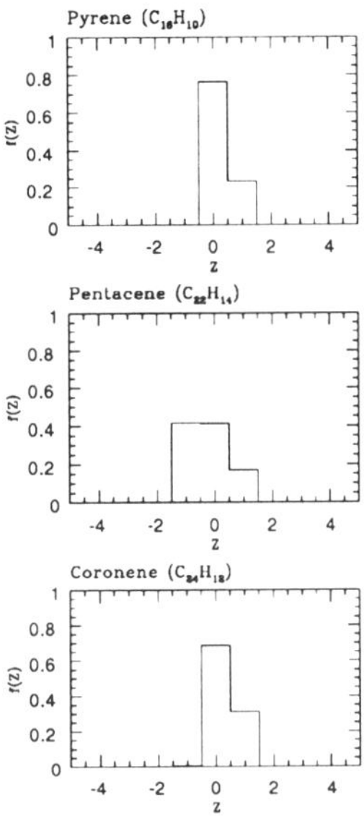

Fig. 3. As Figure 2, substituting the cation-electron recombination rate of Draine & Sutin (1987) with Abouelaziz *et al.* (1993). This produces considerably more positively charged PAHs.

and Sutin's formalism of 1.1× 10^{-13} cm^2). They also note that no doubly charged anions are seen. We have used electron attachment cross sections of Lezius *et al.* for neutral C_{60}. The photoionization yield is calculated using the formalism of Bakes & Tielens (1994). The results for its charge distribution are shown in Figure 5 and illustrate that C_{60} can have negative, neutral and positive charge states. This is similar to pentacene because both have large electron affinities, making them highly efficient electron scavengers.

5. CONCLUSIONS

For typical conditions in the diffuse ISM, small PAHs with electron affinities less than 0.5 eV (i.e. small electron sticking coefficients) will be mainly neutral with a significant fraction positively charged. A PAH like pentacene with an electron affinity of 1.1 eV has an almost equal chance of being either negatively charged or neutral and also exhibits significant positive charging. The fullerene C_{60} will be mainly neutral with a substantial fraction of singly negative and positive states. Following the experimental results of Lezius *et al.* (1993), no doubly charged C_{60} anions are found in this study.

Larger disk shaped and spherical grains will be predominantly neutral. If PAHs are the carriers of the DIBs, they will be a mixture of neutral, positively and negatively charged molecules at the surface of a diffuse cloud. Further into the diffuse cloud, the PAH charge will be predominantly neutral or negative. Hence DIBs due to PAH cations are localised to cloud surfaces. In contrast, DIBs associated with anions will become stronger with depth A_v into a cloud. Since observations suggest that at least some DIBs are a cloud surface phenomena, cationic carriers are more likely, since neutrals do not absorb in the visible (Jenniskens *et al.* 1994). Nevertheless, in

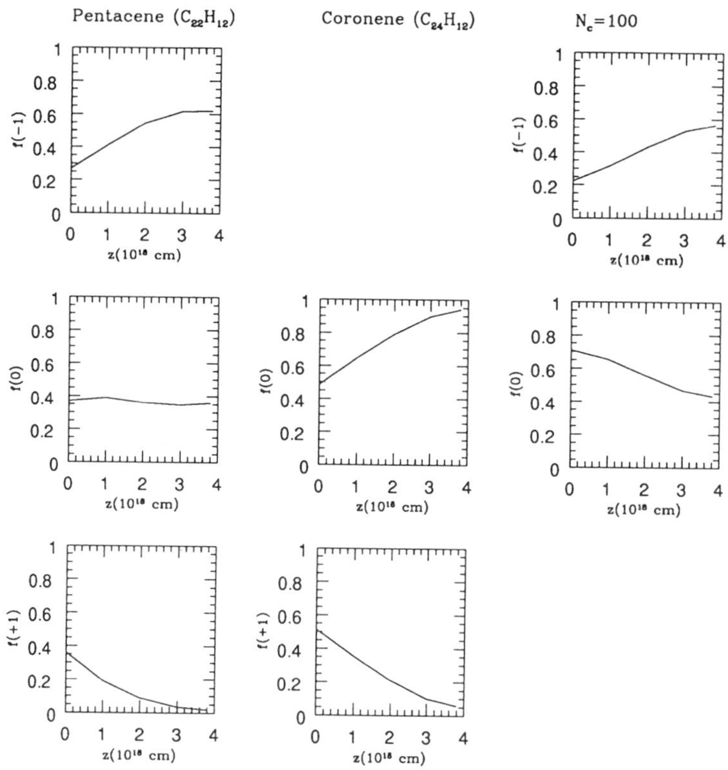

Fig. 4. The charge variation f(Z) for PAHs with depth z into the diffuse cloud model F from van Dishoeck and Black (1986). As electron density increases and FUV flux decreases, these particles become increasingly neutral or negatively charged.

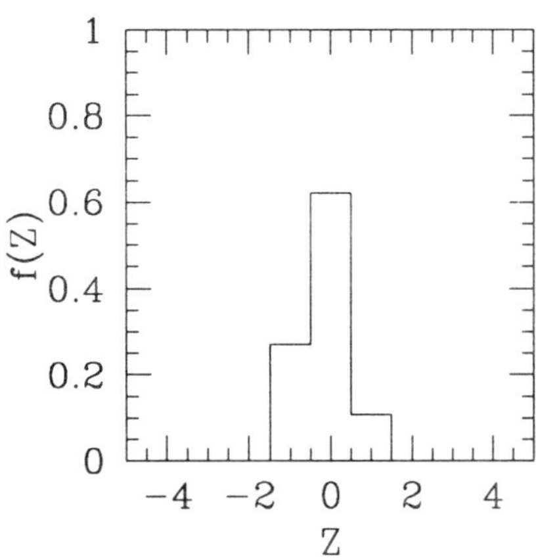

Fig. 5. The fullerene C_{60} will be mainly neutral in typical diffuse ISM, but will exhibit positive and negative charge states as well.

laboratory studies of small PAHs, as well as C_{60}, it may pay to consider the anionic, as well as cationic form. This is a very general conclusion. Any molecular species which stabilises itself against UV photodissociation by internal conversion and has an electron affinity larger than 0.5 eV will have an appreciable fraction of anions in the ISM. Besides pentacene and C_{60}, this includes small hydrocarbon chain radicals (C_nH).

REFERENCES

Abouelaziz, H., Gomet, J.C., Pasquerault, D., and Rowe, B.R. 1993, *J. Chem. Phys.*, **99**, 237.
Allamandola, L.J., Tielens, A.G.G.M., and Barker, J.R. 1989, *Ap. J. Suppl.*, **71**, 733.
Bakes, E.L.O and Tielens, A.G.G.M. 1994, *Ap. J.*, **427**, 822.
Chang, A.H.H., Ermler, W.C., and Pitzer, R.M. 1991, *J. Phys. Chem*, **95**, 9288.
Crawford, M.K., Tielens, A.G.G.M., and Allamandola, L.J. 1985, *Ap. J.*, **293**, L45.
Draine, B.T. and Sutin, B. 1987, *Ap. J.*, **320**, 803.
Herbig, G.H. 1975, *Ap. J.*, **196**, 129.
Jenniskens, P., Ehrenfreund, P., and Foing, B. 1994, *Astr. Ap.*, **281**, 517.
Krelowski, J. 1989, in *Interstellar Dust*, IAU Symposium No. 135, p. 67, eds. L.J. Allamandola and A.G.G.M. Tielens (Kluwer, Dordrecht).
Leach, S. 1991 in *Molecular Clouds*, eds. R.A. James and T.J. Millar (Cambridge University Press).
Leach *et al.* (1992), *Chem. Phys.*, **160**, 451.
Lezius, M., Scheier, P., and Märk, T.D. 1993, *Chem. Phys. Letters*, **203**, 232.
Salama, F., Bakes, E.L.O., Allamandola, L.J., and Tielens, A.G.G.M. 1994, *Ap. J.*, submitted.
Scarrott, S.M., Watkin, S., Miles, J.R., and Sarre, P.J. 1992, *M. N. R. A. S.*, **255**, 11P.
Smith, F.T. 1961, *J. Chem. Phys.*, **34**, 793.
van Dishoeck, E.F. snf Black, J.H. 1986, *Ap. J. Suppl.*, **62**, 109.
Verstraete, L., Léger, A., d'Hendecourt, L., Dutuit, O., and Défourneau, D. 1990, *Astr. Ap.*, **237**, 436.

CHAPTER VI

Proposed Carriers

OBSERVATIONAL CONSTRAINTS ON THE CARRIERS OF THE DIFFUSE BANDS

THEODORE P. SNOW
Center for Astrophysics and Space Astronomy, University of Colorado, Boulder, CO

ABSTRACT. This paper presents a summary of observational factors which must be considered and satisfied by any successful hypothesis for the origin of the diffuse interstellar bands. Included are a brief summary of some of the proposed carriers, a lengthy itemization of observational "facts" or trends regarding the behavior of the features, and a proposed checklist that could be used as a screen for proposed carriers. We conclude with a list of cautions for observers and others who may wish to propose new mechanisms for diffuse band formation.

1. INTRODUCTION

The diffuse interstellar bands have been staring astronomers in the face for over 70 years now, and a few astronomers (and more recently, chemists) have taken on the challenge of attempting to explain them. After an initial surge of interest following the explicit recognition of the features as both unidentified and interstellar (Merrill 1934), there followed a lengthy period of relative inactivity during the 1940s and 1950s, and then a steady build-up of interest starting in the mid-1960s. There was a relative peak in activity in the early 1970s, stimulated largely by three novel mechanisms that were proposed at that time (see Duley 1968; Rudkjøbing 1969a; Huffman 1970; Manning 1970). Starting in the 1980s, two developments stimulated another boom in diffuse band research: (1) the widespread availability of CCDs and other linear detectors, allowing for unprecedented data quality in observing the diffuse bands; and (2) the proposal that polycyclic aromatic hydrocarbons (PAHs), and by inference other large molecules, might be common in the diffuse interstellar medium. This renaissance in diffuse band research has been sustained and reinvigorated in the 1990s by laboratory programs that have lent support to the notion that the diffuse bands may be formed by large molecules or molecular ions (e.g. several papers, this volume).

Today not only astronomers, but also chemists, are becoming more and more active in the quest for a carrier. It was in this environment of growing interest in the problem and proliferation of data on the diffuse bands that the Boulder conference was conceived and conducted. Because of the growth in interest in the diffuse bands, and because the problem of their origin is now attracting scientists whose backgrounds are not in traditional astronomy, it is important to review and explain the observational constraints, derived from observations, that must be addressed by any successful hypothesis for the diffuse bands.

A. G. G. M. Tielens and T. P. Snow (eds.), The Diffuse Interstellar Bands, 325–340.

2. A SUMMARY OF PROPOSED CARRIERS

It might be helpful, for the purposes of illustration and discussion, to provide a list of proposed mechanisms for the diffuse bands. The following is not complete, but it does provide examples of most of the classes of carriers that have been put forth. A bibliography of diffuse band publications, dating back to the first papers known to refer to the features, has been published elsewhere (Snow, Barnes, and Heitzmann 1994); for an updated and current copy of this compilation, contact the author.

In the following brief paragraphs, we summarize a number of the diffuse band mechanisms that have been proposed over the years. The hypotheses are presented approximately in chronological order, according to their dates of introduction into the astronomical literature.

Molecular carriers in general: The basic idea that the diffuse bands should have molecular origins was mentioned, more or less in passing, by Russell (1935), in a lengthy treatise on astronomical spectroscopy. This is probably of historical interest only, since no specific molecule was proposed. Swings and Ohman (1939) followed up with some general comments on types of molecular transitions, but again no specific model was provided. In the meantime, Eyster (1937) criticized molecular carriers in general.

Starting in the 1950s, Herzberg (1955, 1966, 1967, 1971) published the suggestion that pre-dissociative transitions in molecules would have the appropriate properties to produce the diffuse bands. Much more recently, Danks and Lambert (1976) reviewed these and other mechanisms for creating diffuse features in large molecules, and helped set the stage for specific molecular proposals that were yet to come (these are described later).

Solid oxygen: In the mid-1950s, McKellar, Welsh, and Stephenson (1955) made the first explicit proposal for a solid-state carrier of the diffuse bands, by noting that solid oxygen should have a transition that closely matched one of the observed bands.

Metastable states in H_2: Herbig (1963) entered the arena of diffuse band research with his paper suggesting that the 4430Å band might be produced by H_2 molecules in metastable excited states, in solid dust grains (this paper also was the first to raise the issue of matrix shifts, although in the opposite sense of most modern papers: in Herbig's proposal, it was necessary to invoke a shift from the gas phase transitions that were measured in the lab to solid-state transitions that were suggested to take place within interstellar dust grains). Malville (1964) later published a paper debunking the metastable H_2 idea, on the grounds that insufficient column densities would be produced under reasonable interstellar conditions.

Defects in dust grains: From the mid-1960s through the late 1970s, most suggestions for diffuse band origins invoked optical processes in solid particles, in part because there was skepticism regarding the possibility of forming gas-phase molecules in sufficient quantities, and in part because a number of researchers with expertise in solid-state physics became interested in the problem. Among the first of the solid-state suggestions were the ones by Wickramasinghe *et al.* (1968) and Wolstencroft *et al.* (1969), which pointed out that lattice defects in crystals could give rise to absorption centers that would produce broad, strong absorption bands. These defects were originally envisioned as sites where bound electrons would give rise to electronic transitions; the defects themselves were thought to be caused by cosmic rays and energetic photons in circumstellar environments around O and B stars.

Impurity centers within interstellar grains: A number of authors, many of them named Duley, have suggested that the diffuse bands are created by impurity atoms or ions embedded within solid grains. In a series of papers starting in the late 1960s, Duley and co-workers focused on metallic ions in oxide grains (MgO was a particular favorite; Duley 1968; Duley and Graham 1969; Duley 1975, 1976, 1977, 1979, 1981, 1982, 1983). Simultaneously, Duley and McCullough (1977) also suggested molecular ions embedded on grain surfaces as the band carriers, and more recently Duley (1995) has proposed aromatic molecules embedded within hydrogenated amorphous hydrocarbon (HAC) grains as the carriers. Duley probably deserves credit (or blame!) for having proposed a wider variety of mechanisms (some of them in competition with each other) for the diffuse bands than any other author. Strong criticisms of some of the particular proposals of Duley and co-workers were published by Nuth and Donn (1983, 1984) and by Herbig and Leka (1991).

In a similar vein, Huffman (1970) and Manning (1970a,b,c) suggested a specific ion (Fe^{+++}) in grains of various compositions as the carrier of some of the bands, while other authors proposed calcium atoms in graphite (see McIntyre and Williams 1970, who debunked the idea, which they blamed on unspecified, but multiple, papers in the literature).

There are substantial issues related to the optical properties of impurity absorbers, and these were explored in a generic sense in a series of papers by Shapiro and co-workers (Purcell and Shapiro 1977; Shapiro and Holcomb 1986a,b; Shapiro, this volume). These papers demonstrate that it is possible to reproduce most of the observed properties of the diffuse bands with an impurity absorber model, although just barely (see the summary paper on the panel discussion; Snow, this volume).

Negative ions: Some suggested diffuse band origins have a single champion, who perseveres despite all indications that he or she is pursuing a dead end, and the idea of interstellar negative ions as the band carriers had such a champion. In a series of papers starting in the late 1960s, Rudkjøbing proposed first H^- and later O^- as the culprits, on the theoretical grounds that these species should have autodetachment transitions in the optical (Rudkjøbing 1969a,b, 1970, 1973, 1978; Ingemann-Hilberg and Rudkjøbing 1970; Gammelgaard and Rudkjøbing 1980; Gammelgaard 1984). Unfortunately for this idea, very strong astronomical arguments could be and were made against it, by Snow (1973a) and by Herbig and Leka (1991), and spectroscopic problems were raised by Herbst *et al.* (1974) and by Spence and Inokuti (1974).

Interstellar porphyrins: Another idea championed by a single determined individual was the proposal that tetrabenzporphyrin molecules embedded in paraffin matrices was responsible for the diffuse bands, in this case for *all* of them (Johnson 1967, 1972a,b, 1991; Johnson, Bailey, and Wegner 1975; Johnson, this volume). This proposal was criticized by Donn and Khanna (1980) on the basis of laboratory data, and was similarly discussed at the Boulder conference (see Snow, this volume). The hypothesis is that these complex molecules which, interestingly, contain aromatic ring structures, have rich optical spectra, and provide laboratory matches in wavelength with all of the (now nearly 200) known diffuse bands. The molecules are embedded in special matrices (Shpol'ski sites) which are claimed to preserve invariant wavelengths and narrow band profiles.

Carbon chain molecules: A harbinger of things to come was published by Douglas (1977), who suggested that linear carbon chain molecules could produce optical spectra that resemble the diffuse bands. This was based also on the fact that radio detections of carbon chains were starting to occur, lending credibility to the idea. In the review of Smith, Snow, and York (1977), specific types of transitions in carbon chains that could produce the diffuse bands were discussed; similar discussions can be found in the earlier paper of Danks and Lambert (1976). Mitchell and Huntress (1979) made general consistency arguments favoring long carbon chains as the diffuse band carriers, and Thaddeus (c. 1980) took up the idea, arguing that there should be an interstellar ecology of carbon chains, with small ones building up and then being cut apart by photons, resulting in a steady-state propulation of moderate- sized chains that would have optical properties consistent with the diffuse bands.

The carbon chain hypothesis has been strongly boosted recently by the solid-state laboratory measurements of the Basel group (Fulara *et al.* 1993; Freivogel, Fulara, and Maier 1994; Maier it et al., this volume). In these studies several near-coincidences in wavelength were found between observed diffuse bands and transitions measured in the matrix for molecules of the form C_nH_m, where m is small. In this volume, Maier *et al.* argue that more coincidences were found than would be expected statistically. Thaddeus has maintained his advocacy of the carbon-chain idea (Thaddeus 1994, 1995). Krätschmer (1986) conducted laboratory studies of carbon chains which showed promising optical properties. In all, the carbon chain hypothesis has withstood most of the challenges presented to it (see Snow, later in this volume), and remains viable.

Carbon "gunk": Another carbonaceous carrier for the diffuse bands has been proposed, this time on the basis of laboratory data, rather than speculations about the potential of some class of molecules for having optical absorption bands with the right properties. Unfortunately, the experiments in question do not reveal the identify of the material that is producing the possible diffuse band analogs. Wdowiak (1980) and Wdowiak *et al.* (this volume) have produced an organic substance in the lab by subjecting a mixture of methane and argon to an electrical discharge. The spectrum of the resulting amorphous compound reveals a number of features that coincide fairly closely with diffuse bands, including a very broad, strong feature near 4500Å, which plausibly could be the 4430Å diffuse band, shifted by matrix effects. It will be helpful when Wdowiak and colleagues are able to isolate and identify the species in their experiments that produce this and the other absorption bands.

Ionized PAHs: In 1984 a revolutionary new picture of the diffuse interstellar medium was introduced by compelling evidence for the widespread presence of PAHs (Léger and Puget 1984; Allamandola, Tielens, and Barker 1985). This evidence was based on a series of infrared emission features which appear in many kinds of environments where there is a source of ultraviolet radiation to excite the emission. While neutral PAHs (except for very large ones) do not have rich optical spectra and therefore are themselves unlikely candidates as the diffuse band carriers, positively ionized PAHs do and are. Consequently, several authors proposed PAH^+ ions as the diffuse band carriers, nearly simultaneously (Léger and d'Hendecourt 1985; Crawford, Tielens, and Allamandola 1985; van der Zwet and Allamandola 1985).

Until recently, the PAH^+ hypothesis was interesting but untestable, given the total lack of laboratory data. But now some spectroscopic information is available, based on matrix measurements, and tantalyzing near-coincidences in wavelength between interstellar and laboratory features have been found (Salama and Allamandola 1992a,b; Léger, d'Hendecourt, and Défourneau 1994). In each case, unknown matrix shifts must be allowed for, and the identifications must be regarded as only possibilities until gas-phase spectra can be obtained.

Some have argued against the PAH^+ proposal, on the basis of spectroscopic considerations (Leach 1987; Donn, Allen, and Khanna 1989; Cossart-Magos and Leach 1990; Lee and Wdowiak 1993). In a recent analysis, Snow *et al.* (1995) have raised questions about whether the weakness of the diffuse bands in reflection nebulae is consistent with PAH cations as the carriers, but show that for reasonable conditions the PAHs ought to be neutral in these regions, so there may not be a serious problem here.

Buckminsterfullerenes and their relatives: Almost at the same time that the PAH bandwagon began to roll, another one hit the streets: the giant cage-like molecules known as buckminsterfullerenes, or "buckyballs". These are cage-like molecules of the general form C_{60}, C_{70}, etc., and their existence was announced originally by Kroto *et al.* (1985). Even in that seminal paper, the possibility that these molecules could produce the diffuse bands was mentioned. The "ball" was then carried, so to speak, by Léger *et al.* 1988, who proposed ionized C_{60} as the carrier, but this was soon criticized by Snow and Seab (1989). Much more recently, ionized C_{60} has reappeared in the literature, with the claim by Foing and Ehrenfreund (1994) that two newly-discovered near-infrared diffuse bands are coincident (within uncertainties introduced by the dearth of laboratory data) with expected features of $C_{60}{}^+$. This is regarded by many as a probable identification of two diffuse bands, though certainly questions remain (e.g. the Snow and Seab result cited above).

The interest in fullerenes in general continued unabated. Kroto (1989); Krätschmer *et al.* (1990); Ballester, Antoniewicz, and Smoluchowski (1990); Braga, Rosen, and Larson (1991); Braga *et al.* (1991); Hare and Kroto (1992); and Kroto and Jura (1992) have continued to suggest C_{60}, though not necessarily in the ionized form, as a band carrier. Edwards and Leach (1993) calculated simulated rotational contours for C_{60} bands, and found that profiles similar to those of observed diffuse bands could be constructed for reasonable conditions.

Partially hydrogenated fullerenes (i.e. fulleranes) have entered the picture as well (Webster 1992; McIntosh and Webster 1992, 1993; Webster 1993a,b,c). This is another difficult proposal to test, not so much because it is hard to do laboratory work (although it is), but more because of the near-infinite variety of possible candidates.

Two-photon absorption by vibrationally excited H_2: At the Boulder conference, Glownia and Sorokin (1994) introduced a novel new idea for producing diffuse bands. This hypothesis envisions absorption from excited states of molecular hydrogen ions near hot stars by a two-step process: absorption of a Lyman-α photon, leaving the molecular ion in an electronically excited state; followed rapidly by absorption from that state at optical wavelengths. Little discussion of this idea has occurred to date, although it does appear to face serious astronomical hazards.

3. OBSERVATIONAL CONSTRAINTS ON THE DIFFUSE BAND CARRIERS

Having reviewed the candidate diffuse band carriers that have been proposed to date, we now take a look at the spectroscopic and astronomical "facts" about diffuse bands that must be confronted by any successful hypothesis. These are listed in no particular order in the following paragraphs.

The large number of diffuse bands: The number of bands has grown (though certainly not linearly) at an average rate of about three per year since the very first reports (Heger 1922; Merrill 1934), and seems to be limited only by technology and the level of scrutiny that is applied. Heger noted that the 5780 and 5797Å features that showed up in stellar spectra were "stationary", implying interstellar, and Merrill noted the presence of four broad interstellar features of unknown origin. By the time of Herbig's (1975) landmark paper, 39 bands were reported; Sanner, Snell, and Vanden Bout (1978) and later Herbig and Leka (1992) added to the list, with the result that a decade ago, there were 104 known diffuse bands. The red optical region, and the near-infrared as well, have proven to be rich hunting grounds (e.g. Herbig and Leka), and today it seems every time someone looks closely at high-quality spectra of reddened stars, new features are found (e.g. Krełowski and Sneden 1993). In a recent broad (in spectral coverage) survey by Jenniskens and Désert (1993, 1994) many new bands were found, bringing the total close to 200. It now seems untenable that a single carrier could account for all of these bands.

Invariance of diffuse band wavelengths and profiles: The question of whether or not the diffuse bands ever vary in wavelength or profile shape is important, because there are classes of models (those invoking impurity absorbers in solid grains) in which variations would be expected under reasonable interstellar conditions. Early observations, which focused most often on the very broad 4430Å feature, discussed possible profile variations, but were hopelessly limited by poor data quality in photographic spectra. Thus the question of band profile invariance was not properly addressed until bands other than λ4430 were studied, and especially until the advent of electronic detectors. To date no convincing evidence of any variations (except those introduced by line-of-sight velocity component structure) has been found (e.g. Savage 1976; Danks and Lambert 1976; Welter and Savage 1977; Westerlund and Krełowski 1988a,b,c; Herbig and Soderblom 1982; Krełowski and Sneden (1993). Snow and Seab (1991) reported unusually narrow profiles of the 5780 and 5797Å bands in the Taurus cloud, and suggest that this is due to unusually low rotational excitation of the carrier, which is therefore presumed to be a molecule. A very similar conclusion was reached by Chlewicki *et al.* (1986), based on the fact that the band profiles are not only narrow, but also display different forms of asymmetry.

In two explicit searches for band profile variations in regions known to have variations in mean dust grain size, the diffuse band profiles and central wavelengths were found to be stubbornly stable (Snow, Timothy, and Saar 1982; Seab and Snow 1995). It seems justified to conclude that in general, the diffuse band wavelengths are invariant.

The lack of emission wings: Some models of diffuse band formation, particularly those invoking solid-state processes such as impurity absorbers, predict the presence of emission wings (along with shifted absorption wavelengths), depending on such parameters as concentration of absorbers within the grains, or grain size. There are reports in the literature of emission wings having been detected to the violet of some of the broad diffuse bands. Brück and Nandy (1968) claim to have found such an emission wing for the 4430Å band, while Walker and Yang (1979) make a similar claim for λ6177. Oddly enough, just prior to the Brück and Nandy report, one of their frequent collaborators, Seddon (1967) reported failure to see emission associated with λ4430. Inspection of the spectra claimed to show emission is not convincing; it appears that continuum placement problems can account for the claimed detections. There has been little follow-up on this issue, as far as we know, largely because during the recent observational renaissance of activity in diffuse band research, most of the attention has been devoted to the relatively narrow features. It would be important to utilize modern detector systems to obtain high-quality data on the broad diffuse bands as well, and lay to rest the remaining uncertainties about the possible existence of emission wings, as well as to study the profiles of these bands in general.

For now it appears to be generally safe to conclude that the profiles of the diffuse interstellar bands lack emission wings. This is quite distinct from the reports (discussed below) that diffuse bands *purely in emission* have been seen, at least in one very unusual nebula.

The narrow widths of some diffuse bands: Certain hypotheses (e.g. absorbers embedded in solid grains) have grave difficulties in reproducing diffuse bands as narrow as some of those that are seen; this is discussed by Smith, Snow, and York (1977) as a general criticism of the entire solid-state class of carriers. Some of the diffuse bands are less than 1Å in width (e.g. Herbig 1975; Smith *et al.* 1981). The only apparent mechanism for producing such narrow bands in solid-state absorbers is a special kind of matrix arrangment called a Shpol'ski site; Johnson (this volume) has proposed such matrices as the hosts for the complex organic molecules that he claims to be the carriers of the diffuse bands.

The lack of polarization structure within the diffuse bands: Yet another prediction of the solid-state models for the diffuse band carriers is that there should be a variation of linear polarization across the band profiles. This is a consequence of scattering theory, and must be present *if* the grains that produce the observed general linear polarization are also the carriers of the diffuse bands. In several searches for polarization structure within diffuse bands, none has been found (A'Hearn 1972; Martin and Angel 1974, 1975; Fahlman and Walker 1975; Adamson and Whittet 1992). This is generally regarded as proof that the diffuse bands are not carried by the same population of grains that are responsible for linear polarization and extinction at visible wavelengths. This constraint does not rule out the possiblity that the diffuse bands could be formed by the much smaller grains thought to be the agents of far-ultraviolet extinction.

General correlation with extinction and other interstellar properties: Once significant databases of diffuse band strengths began to accumulate, it became possible to analyze the degree of correlation with other interstellar quantities. Most commonly it was extinction, more specifically the color excess E_{B-V}, that was considered (e.g. Duke 1951; Walker 1963; Gammelgaard 1968; York 1971; Wu 1972; Murdin 1972; Snow 1973b; to name a few of the more extensive ones). All of these surveys show a general correlation, helping to verify that the diffuse bands are indeed interstellar.

Other parameters have been correlated with the diffuse bands., These include the 2175Å extinction bump (Dorschner, Friedmann, and Gürtler 1977; Danks 1980; Benvenuti and Porceddu 1989); ultraviolet extinction in general (Wu, York, and Snow 1981; Witt, Bohlin, and Stecher 1983; Seab and Snow 1984; Krełowski *et al.* 1987; Sneden, Woszczyk, and Krełowski 1991); infrared extinction (Sneden *et al.* 1968); depletions of interstellar gas-phase elements (Millar and Duley 1979; Zimmermann 1982; Niedzielski and Krełowski 1989; Herbig 1993); and polarization (Krełowski *et al.* 1987). All of these studies showed some degree of correlation between diffuse bands and the particular interstellar parameters being considered, and some conclusions (though usually ambiguous) could be drawn regarding the significance of these correlations. The key point is that *nothing* was found to correlate perfectly with the diffuse bands; the entire upshot of all this work was that no clear-cut identification or association of the diffuse band carriers with any single interstellar quantity was found.

Perhaps the last (or at least the latest) word on correlation analyses of the diffuse band comes from Herbig (1993), who carefully compared the strengths of the 5780 and 5797Å diffuse bands against a large variety of parameters, including H I, H_2, Na I, C I, K I, and Ti II column densities, as well as ultraviolet extinction, the ratio of total to selective extinction in the optical, and the presence or absence of interstellar shocks. Herbig found that the best correlation is with H I (and that there is no significant relationship with H_2 or carbon-bearing diatomic molecules); that there is no significant correlation with the depletion of titanium; no dependence on the presence of shocked gas; and that there is no relationship to either ultraviolet or optical extinction. The best extinction correlation was with E_{B-V}. Herbig's conclusion from all of this is that the diffuse band carriers behave more like neutral gas-phase species than anything else, and that their behavior relative to neutral species such as C I, Na I, and K I is consistent with a carrier whose ionization potential is slightly higher than 5 eV.

Correlations among the diffuse bands: It has been clear from early on that the diffuse bands do not always correlate well with each other; in almost every study of correlations, a few outstanding anomalies are pointed out. Sometimes these are due to observational error, but it can hardly be doubted that anomalous diffuse band behavior does occur. Eventually these disparities led to general analyses of intercorrelations among the bands, with the result that some observers have reported finding groupings that seem to correlate especially well internally, and therefore may constitute "families" of diffuse bands formed by a common carrier (Krełowski and Walker 1987; Josafatsson and Snow 1987; Westerlund and Krełowski 1989). Unfortunately, different authors using different samples of stars find different groupings of the diffuse bands, and in some cases further investigation fails to confirm groupings previously assigned (e.g. Scarrott *et al.* 1992; Fossey and Sommerville, this volume). In this kind of work, it is imperative that the data acquisition and data quality, as well as the interpretation and measurement of the data, be internally consistent, and this makes it difficult to enlarge the database by combining measurements from different observers. Therefore, until someone undertakes a really massive observational program, controlling as many factors as possible, a proper analysis of the "families" of diffuse bands will be lacking.

There do seem to be significant trends in the *ratios* of certain diffuse bands and interstellar parameters, but as in the case of general correlation studies (discussed above), there are uncertainties caused by the small sizes of the samples that have been examined. The most striking of the possible correlations is the finding that the relative strengths of the 5780 and 5797Å bands seem to reverse, depending on the nature of the ultraviolet extinction curve (Sneden, Woszczyk, and Krełowski (1991), with the λ5780 being stronger where the far-ultraviolet extinction is low, and the 5797Å band being stronger where the far-UV extinction is high. These conditions, in turn, seem to correlate with low and high molecular abundances, respectively, and with high and low UV radiation field intensities (Snow 1992). This suggests very strongly that the ratio of these two diffuse bands is very sensitive to some local physical condition (possibly the radiation field intensity, which is linked in turn to the grain size distribution, hence the shape of the extinction curve).

It has also been reported that the relative strengths of different diffuse bands appear to vary systematically with depth inside of interstellar clouds; this is discussed in a later paragraph, below.

Perhaps the most important message from these studies is that the bands certainly do *not* correlate perfectly among themselves, and that therefore there must be multiple carriers, or a variety of physical states having different spectroscopic properties.

Scarcity of diffuse bands below 4430Å: It has been very striking, right from the time of the earliest systematic studies of the diffuse bands, that they do not seem to extend to wavelengths below the very strong band at 4430Å. As noted by Smith (this volume), this surely must say something important about the physical properties of the carriers.

Hibbens *et al.* (1994a) have conducted a new, sensitive search for diffuse bands between 3900 and 4200Å, and report finding five possible bands, all of them quite broad and shallow. They conclude that there are definitely no sharp diffuse bands in this spectral region.

There have been searches for ultraviolet diffuse bands, but to date they have been limited in sensitivity by two factors: (1) the high level of ultraviolet extinction, which limits the optical depth to which one can probe; and (2) the fact that hot stars have very crowded photospheric spectra in the ultraviolet, presenting grave difficulties due to confusion. Both *Copernicus* (Snow, York, and Resnick 1977) and the *IUE* (Seab and Snow 1985) were employed in unsuccessful searches for ultraviolet diffuse bands. More recently, a possible ultraviolet diffuse band has been reported by Tripp, Cardelli, and Savage (1994), on the basis of a serendipitous discovery using the Goddard High Resolution Spectrograph on the *Hubble Space Telescope*.

The *HST* can certainly provide sensitive data on the possible presence of ultraviolet diffuse bands, particularly if the Faint Object Spectrograph could be brought to bear on the problem, but it has so far proven very difficult to persuade the time allocation panels of the importance of doing so. This is especially frustrating in view of the fact that large molecules such as PAHs, now believed to be of possible importance in the diffuse interstellar medium, have strong ultraviolet transitions, meaning that the use of the *HST* for this task could, in principle, settle unambiguously the question of whether these molecules are present, and in what quantities.

Consistency with cosmic abundances: Any proposed carrier of the diffuse bands certainly must obey the constraint that there has to be a sufficient supply of the raw materials available to form the carriers in adequate quantities. Many authors have undertaken the exercise of estimating the abundances needed, usually by assuming some reasonable value for the oscillator strength and then estimating the column density needed to produce the observed band strengths (e.g. Smith, Snow, and York 1977). Generally it is found that column densities toward typical diffuse band stars must be at least in the range of 10^{13} to 10^{14} cm^{-2}, or about 10^{-8} to 10^{-7} of the total hydrogen column density. These values are high enough to rule out any rare elements as the carriers.

Most current models invoke cosmically abundant species, however, and thus do not face serious problems on this score. Carbon in particular is of interest, since so many of the large molecules that have been proposed are carbonaceous. Among these, the fullerane hypothesis (Webster 1992; McIntosh and Webster 1992, 1993; Webster 1993a,b,c; 1995) probably requires the greatest carbon abundance, but even this model fits with what is currently known of the interstellar depletion of carbon from the gas phase, though just barely (see remarks in Snow, this volume). It is noteworthy also that currently there is significant uncertainty about carbon depletions, because of discrepancies in the "cosmic" carbon abundance derived from solar system measures and from young stars (Sofia, Cardelli, and Savage 1994; Mathis, this volume).

Formation and stability of the carrier under interstellar conditions: The diffuse interstellar medium is not a hospitable place, and is especially hard on species that cannot withstand a strong dose of hard ultraviolet radiation. It is clear that the diffuse bands are formed in environments subjected to such radiation, which implies that they must have sufficiently high thresholds for photoionization or photodissociation to maintain a significant population in the presence of UV photons having energies up to 13.6 eV. The carriers may also be required to withstand interstellar shocks, although the observational data on the potential relationship of the bands to shocked regions in as yet ill-defined (see Herbig 1993).

Some of the current hypotheses for diffuse band formation appear to be flirting with trouble on this score. For example, at the Boulder conference we heard that neutral carbon chain molecules appear to have appropriate spectroscopic properties (Maier, this volume), but also that these species may be photodissociated at uncomfortably low energies (see discussion summarized by Snow, this volume). Laboratory data bearing on this question are simply not available for many of the proposed diffuse band carriers, so it would be appropriate for those studying potential carriers in the lab to include measurements of physical properties along with the spectroscopy.

Very few proponents of specific diffuse band carriers have offered any explanation of how their favorite species might form in the interstellar medium. This problem must eventually be faced, but for the present it seems the emphasis lies in finding candidates that pass the observational tests outlined here, with any effort to explain the origins of the carriers being deferred until we know which carriers we have to explain. This is probably a reasonable approach, but it certainly begs some interesting questions, many of which we face anyway, in view of the independent evidence that large molecules such as PAHs may be common.

Weakness of the diffuse bands in dark clouds: It was first noted by Wampler (1966), and later confirmed by Bromage and Nandy (1973), by Snow and Cohen (1974), and more recently by Adamson, Whittet, and Duley (1991) and by Jenniskens, Ehrenfreund, and Foing (1994) that the diffuse bands tend to be deficient in strength (with respect to color excess) in relatively dense clouds. As a rule, in lines of sight dominated by a single dense cloud rather than an assortment of more diffuse clouds, the diffuse bands are systematically weaker with respect to extinction. This tendency is enhanced if normalization relative to total extinction A_V is used instead of E_{B-V}.

The tendency of the diffuse bands to be weak in dense clouds is perhaps one of the more perplexing observational results, particularly for those advocating a molecular origin, and it contributes to the conclusion that the diffuse bands do not correlate well with commonly observed molecules (such as the diatomics CN and CH), nor with H_2 (e.g. Herbig 1993). Apparently if the diffuse bands are formed by molecules, the carriers either require some ultraviolet radiation for their formation, or are destroyed chemically in dense clouds.

Absence of diffuse bands in circumstellar material: Many searches have been made for diffuse bands formed in circumstellar material, but no unambiguous detection has yet been made (see Seab, this volume). Among the earliest searches were those by Snow and Wallerstein (1972) and Snow (1973c), who probed a number of different types of circumstellar regions. More recently, Le Bertre (1990) and Le Bertre and Lequeux (1993) have reported detection of circumstellar diffuse bands in the spectrum of the carbon star CS 776, but there are doubts as to whether the bands are actually circumstellar. In any case of a detection of diffuse features in the spectrum of a source having a circumstellar envelope, it is difficult to unambiguously distinguish circumstellar from interstellar contributions, and this is true not only of the reported detections for CS 776, but also those reported for the planetary nebula nucleus and *IRAS* infrared source IRAS 21282+5050 (Cohen and Jones 1987) and for the planetary nebula central star NGC 7027 (Pritchet and Grillmair 1984), the latter case being further compounded by a complex emission-line spectrum which makes identification of the continuum level difficult.

Perhaps more notable than the putative detections of circumstellar diffuse bands cited above are the non-detections in systems where one might expect them, based on current models for the carriers. One such is the proto-planetary nebulae central star HR 4049, which has an extremely carbon-rich envelope yet no trace of diffuse bands (Waters *et al.* 1989). Similarly, the circumstellar dust which forms periodically around the young variable star HD 5999 is also devoid of diffuse band absorption (Anderson, Gahm, and Krełowski 1982). Other cases are described by Seab (this volume).

It has also been found that diffuse bands are deficient in the lines of sight to T Tauri stars (Herbig 1966; Meyer and Ulrich 1984), which may be another manifestation of the general weakness of the bands in all kinds of circumstellar environments. Similarly, Be stars tend to show weak diffuse bands (e.g. Wu 1972), which probably means that they are absent in the circumstellar shells around these stars, but it is not clear how significant this is, since these shells are usually hot and ionized, and are quite unlike typical diffuse interstellar medium material.

The general lack, or at least weakness, of circumstellar diffuse bands is perplexing, and may eventually have a lot to tell us about the formation and subsequence alteration of circumstellar material as it makes its way into the general interstellar medium.

Weakness of diffuse bands in reflection nebulae: A few studies have incorporated reflection nebula in the samples of diffuse band lines of sight, finding either a deficiency in band strength or, at best, "normal" band strengths. Morgan (1944) originally commented that the 4430Å band is deficient in the Orion Trapezium region. Later, A'Hearn (1971) found no systematic weakness of the 4430Å band in other reflection nebulae, although the data quality and the extinctions probed were probably insufficient to allow detections of this band with enough accuracy to see subtle effects. Certainly the enhanced $\lambda 4430$ strengths predicted the next year by Bromage (1972) were not seen.

More recently, Josafatsson and Snow (1987) undertook a survey of the 5780 and 5797Å diffuse bands in reflection nebulae, and found them either normal or deficient, relative to E_{B-V}. Interestingly, this study included some lines of sight now known to have strong infrared emission features associated with PAHs. This may seem to argue against ionized PAHs as the carriers, but the analysis of Bakes and Tielens (this volume) shows that neutral PAHs may dominate, even in these regions of high ultraviolet flux, due to the very high rate of recombination for $PAH^+ + e^-$ (see also Snow *et al.* 1995).

Diffuse band emission in the Red Rectangle: The bizarre nature of the nebula known as the Red Rectangle has puzzled spectroscopists for some time, as it displays emission lines that appear very much to follow molecular series, yet no clear cut identifications were forthcoming, except for CH^+ (Balm and Jura 1992; Hall *et al.* (1992), and C_2 (Miles, Sarre, and Scarrott, this volume). Recently it was shown, quite convincingly, by Sarre (1991), Fossey (1991), and by Scarrott *et al.* (1992) that some of these features match interstellar diffuse bands, with offsets in wavelength corresponding to different molecular excitations. This demonstrates that at least the particular diffuse bands in question ($\lambda\lambda$5797, 5849, 6376/9, and 6614) *must* be molecular in origin.

The presentation of diffuse band emission by the Red Rectangle is one of the most significant of all the observational constraints to be found in the observational lore of the diffuse bands, for it virtually proves that gas-phase molecular carriers must be responsible, at least for some of the more narrow of the bands.

Fine structure in the 6614Å band: As this review was being finalized, word came in from Hibbens *et al.* 1994b) that consistent fine structure, resembling P, Q, and R branches of a molecular band, have now been detected in the 6614Å diffuse band in a few stars. A very high spectral resolving power ($\lambda/\Delta\lambda$ = 500,000) was used for this observation, and it confirms tentatively-identified structure reported at the Boulder conference (See Sarre, this volume). This finding adds further weight to the conclusion that at least some of the diffuse bands are carried by molecules.

4. THE ESSENTIAL CONSTRAINTS: A CHECK LIST

We can summarize the foregoing scattershot list of observational "facts" about the diffuse bands by organizing the constraints into a short list of general criteria:

A serious candidate as the carrier of any or all of the diffuse bands *must*:

- Produce invariant wavelengths and profiles;
- Be able to form and survive under conditions appropriate for the diffuse interstellar medium;
- Be consistent with "cosmic" abundances;
- Explain the lack of polarization or emission wings associated with the diffuse bands;
- Be spectroscopically consistent with the observed diffuse bands (i.e. must not predict diffuse bands that are not seen; must match the observed strength ratios for bands presumed to be formed by the same carrier; and must not have consequences in other wavelength bands such as the infrared or ultraviolet that are not seen);
- Be consistent with observed environmental dependencies of the diffuse bands (e.g. must match the observed deficiencies of diffuse bands in dark clouds, in reflection nebulae, and in circumstellar environments; the variations in diffuse band strength ratios; the correlations that are observed with other interstellar quantities).

5. ADMONITIONS TO OBSERVERS

My review of the observational literature has shown that there are many possible pitfalls for observers. It may be useful to conclude this review by itemizing some of the more common among these, as an aide to those who may wish to continue the pursuit of this generally vexing problem.

Negative results can be as important as positive ones: Observers may have a tendency to not publish the results of searches or attempted confirmations of models which do not produce positive results. This tendency to not publish is probably most pronounced when someone's own hypothesis is being tested. It is, however, extremely important and useful that the non-detections and non-confirmations be published, for they can contribute as much as positive results in some cases. Even data that seem inconclusive at the time, one way or the other, may end up having significance. So the advice from here is to publish all observational results, even if their implication is not clear, and even if they do not support your favorite views on the origin of the diffuse bands.

Beware of stellar lines: The diffuse band literature is marked by many examples where confusion with stellar photospheric lines was not recognized, and erroneous conclusions about the diffuse bands were reached.

Beware of small statistical samples: It seems the diffuse band literature, more so than other fields of astronomical research, is rife with papers containing statistical conclusions that lack sufficient data to support those conclusions. It simply is not possible or justified to claim a correlation or a trend on the basis of a handful of stars in a sample. Be patient; collect enough data to be sure, and in the process be careful to include cases that represent enough range in properties (i.e. enough controls) to allow the real trends to be isolated.

Avoid steamrollers and bandwagons: Another common phenomenon seems to be ready acceptance of an idea, once it is implanted in the literature, and therefore sometimes the hard questions are not asked. It is important to think critically about the underlying premises of observational tests, in order to properly interpret their significance.

There is no such thing as too much S/N: The more closely you scrutinize any observational phenomenon, the more you will learn from the details. Many observers tend to collect large samples at the expense of data quality, and then are forced to rely on inferior spectra as they struggle to reach conclusions about correlations or details of profiles. It is almost always better to get superior data on just a few objects than to collect inferior data on many objects.

6. ACKNOWLEDGEMENTS

I am very grateful to all those who participated in the Boulder conference, for their comments and feedback when I first presented this review orally.

REFERENCES

Adamson, A.J. and Whittet, D.C.B. 1992, *Ap. J. (Letters)*, **398**, L69.
Adamson, A.J., Whittet, D.C.B., and Duley, W.W. 1991, *M. N. R. A. S.*, **252**, 234.
A'Hearn, M.F. 1971, *A.J.*, **76**, 264.
A'Hearn, M.F. 1972, *A.J.* **77**, 302.
Allamndola, L.J., Tielens, A.G.G.M., and Barker, J.R. 1985, *Ap. J. (Letters)*, **290**, L25.
Andersen, J., Gahm, G.F., and Krełowski, J. 1982, *Astr. Ap.*, **113**, 176.

Ballester, J.L., Antoniewicz, P.R., and Smoluchowski, R. 1990, *Ap. J.*, **356**, 507.
Balm, S.P. and Jura, M. 1992, *Astr. Ap.*, **261**, L25.
Benvenuti, P. and Porceddu, I. 1989, *Astr. Ap.*, **223**, 329.
Braga, M., Rosen, A., and Larsson, S. 1991, *Z. Phys. D. Atomic and Molecular Clusters*, **19**, 435.
Bromage, G.E. 1972, *Ap. Spa. Sci.*, **15**, 426.
Bromage, G.E. and Nandy, K. 1973, *Astr. Ap.*, **26**, 17.
Brück, M.T. and Nandy, K. 1968, *Nature*, **220**, 46.
Chlewicki, G., van der Zwet, G.P., van Ijzendoorn, L.J., Greenberg, J.M., and Alvarez, P.P. 1986, *Ap. J.*, **305**, 455.
Cohen, M. and Jones, B.F. 1987, *Ap. J. (Letters)*, **321**, L151.
Cossart-Magos, C. and Leach, S. 1990, *Astr. Ap.*, **233**, 559.
Crawford, M.K., Tielens, A.G.G.M., and Allamandola, L.J. 1985, *Ap. J. (Letters)*, **193**, L45.
Danks, A.C. 1980, *Pub. A. S. P.*, **92**, 52.
Danks, A.C. and Lambert, L. 1976, *M. N. R. A. S.*, **174**, 571.
Donn, B. and Khanna, R.K. 1980, *Ap. Spa. Sci.*, **68**, 19.
Dorschner, J., Friedemann, C., and Gürtler, J. 1977, , *Ap. Spa. Sci.*, **58**, 201.
Douglas, A.E. 1977,*Nature*, **269**, 130.
Duke, D. 1951, *Ap. J.*, **113**, 100.
Duley, W.W. 1968, *Nature*, **218**, 153.
Duley, W.W. 1975, *Ap. Spa. Sci.*, **36**, 345.
Duley, W.W. 1976, *Ap. Spa. Sci.*, **45**, 253.
Duley, W.W. 1977, *Ap. Spa. Sci.*, **47**, 185.
Duley, W.W. 1979, *Ap. J.*, **227**, 824.
Duley, W.W. 1981, *Ap. Spa. Sci.*, **78**, 145.
Duley, W.W. 1982, *Ap. Spa. Sci.*, **88**, 501.
Duley, W.W. 1983, *Ap. Spa. Sci.*, **95**, 213.
Duley, W.W. 1995, in *The Diffuse Interstellar Bands*, eds. A. G. G. M. Tielens and T. P. Snow [Dordrecht:Kluwer], *this volume*.
Duley, W.W. and McCullough, J.D. 1977, *Ap. J. (Letters)*, **211**, L145.
Edwards, S.A. and Leach, S. 1993, *Astr. Ap.*, **272**, 533.
Eyster, E.H. 1937, *Ap. J.*, **86**, 486.
Fahlman, G.G. and Walker, G.A.H. 1975, *Ap. J.*, **200**, 22.
Foing, B.H. and Ehrenfreund, P. 1994, *Nature*, **369**, 296.
Fossey, S.J. 1991, *Nature*, **353**, 393.
Freivogel, P., Fulara, J., and Maier, J. 1994, *Ap. J. (Letters)*, **431**, L151.
Fulara, J., Lessen, D., Freivogel, P., and Maier, J. 1993, *Nature*, **366**, 439.
Gammelgaard, P. 1984, *Ap. Spa. Sci.*, **135**, 77.
Gammelgaard, P. 1968, *J. d'Observatoire*, **51**, 297.
Gammelgaard, P. and Rudkjøbing, M. 1980, *Ap. Spa. Sci.*, **72**, 319.
Glownia, J.H. and Sorokin, P.P. 1994, in *NASA Conf. Pub. 10144*, ed. A.G.G.M. Tielens [NASA:Ames Research Center], p. 99.
Hall, D.I., Miles, J.R., Sarre, P.J., and Fossey, S.J. 1992, *Nature*, **358**, 629.
Hare, J.P. and Kroto, H.W. 1992, *Acc. Chem Res.*, **25**, 106.
Heger, M.L. 1922, *Lick Obs. Bull.*, **10**, No. 337, 146.
Herbig, G.H. 1963, *Ap. J.*, **137**, 200.
Herbig, G.H. 1966, *Z. f. Ap.*, **64**, 512.
Herbig, G.H. 1975, *Ap. J.*, **196**, 129.
Herbig, G.H. 1993, *Ap. J.*, **407**, 142.
Herbig, G.H. and Leka, K.D. 1991, *Ap. J.*, **382**, 193.
Herbig, G.H. and Soderblom, D.R. 1982, *Ap. J.*, **252**, 610.
Herbst, E., Patterson, T.A., Norcross, D., and Lineberger, W.C. 1974, *Ap. J. (Letters)*, **191**, L43.
Herzberg, G. 1955, *Mém. Roy. Soc. Liége 4*, **15**, 291.
Herzberg, G. 1966, *Mol. Spec. and Mol. Struct. III: Electronic Spectra and Electronic Structure of Polyatomic Molecules* [New York:Van Nostrand].
Herzberg, G. 1967, *IAU Symp. 31*, ed. H.P. van Woerden [London and New York:Academic Press], p. 91.
Herzberg, G. 1971, in *Vistas in Astronomy*, ed. A. Beer, **2**, 415.
Hibbens, R.E., Miles, J.R., Sarre, P.J., and Herbig, G.H. 1994a, in *NASA Conf. Pub. 10144*, ed. A.G.G.M. Tielens [NASA:Ames Research Center], p. 31.
Hibbens, R.E., Miles, J.R., Kerr, T.H., and Sarre, P.J. 1994b, *Newsletter of the Anglo-Australian Obs., Oct.*.
Huffman, D.R. 1970, *Nature*, **225**, 833.
Ingemann-Hilberg, C. and Rudkjøbing, M. 1970, *Ap. Spa. Sci.*, **6**, 101.
Jenniskens, P. and Désert, F.X. 1993, *Astr. Ap.*, **274**, 465.
Jenniskens, P. and Désert, F.X. 1994, *Astr. Ap. Suppl. Ser.*, **106**, 39.
Jenniskens, P., Ehrenfreund, P., and Foing, B. 1994, *Astr. Ap.*, **281**, 517.
Johnson, F.M. 1967, in *Interstellar Grains; NASA Spec Pub. SP-140*, eds. J.M. Greenberg and T.P. Roarke [NASA:Washington].

Johnson, F.M. 1972a, *Ann. Bull. N.Y. Acad. Sci*, **187**, 186.
Johnson, F.M. 1972b, *Mém. Roy. Soc. Liége*, **3**, 391.
Johnson, F.M. 1991, *Bull. AAS*, **23**, 933.
Johnson, F.M., Bailey, D.T., and Wegner, P.A. 1975, in *Interstellar Dust and Related Topics*, IAU Symp. 52, eds. J.M. Greenberg and H.C. van de Hulst [Dordrecht:Reidel], p. 317.
Josafatsson, K. and Snow, T.P. 1987, *Ap. J.*, **319**, 436.
Krätschmer, W. 1986, *Ap. Spa. Sci.*, **128**, 93.
Krätschmer, W., Lamb, L.D., Fostiropoulos, K., and Huffman, D.R., *Nature*, **347**, 354.
Krełowski, J. and Sneden, C. 1993, *Pub. A. S. P.*, **105**, 1141.
Krełowski, J. and Walker, G.A.H. 1987, *Ap. J.*, **312**, 860.
Krełowski, J., Walker, G.A.H., Grieve, G.R., and Hill, G.M. 1987, *Ap. J.*, **316**, 449.
Kroto, H.W. 1989, *Ann. Phys.*, **14**, 169.
Kroto, H.W., Heath, J.R., O'Brien, S.C., Curl, R.F., and Smalley, R.E. 1985, *Nature*, **318**, 162.
Kroto, H.W. and Jura, M. 1992, *Astr. Ap.*, **263**, 275.
Leach, S. 1987, in *Polycyclic Aromatic Hydrocarbons and Astrophysics*, eds. A. Leger, L.B. d'Hendcourt, & N. Boccara (Dordrecht:Reidel), 99.
Le Bertre, T. 1990, *Astr. Ap.*, **236**, 472.
Le Bertre, T. and Lequeux, J. *Astr. Ap.*, **274**, 909.
Lee, W. and Wdowiak, T.J. 1993, *Ap. J. (Letters)*, **410**, L127.
Léger, A. and d'Hendecourt, L. 1985, *Astr. Ap.*, **146**, 81.
Léger, A., D'Hendecourt, L., and Défourneau, D. 1995, *Astr. Ap.*, in press.
Léger, A., d'Hendecourt, L., Verstraete, L., and Schmidt, W. 1988, *Astr. Ap.*, **203**, 145.
Léger, A. and Puget, J. 1984, *Astr. Ap.*, **137**, L5.
Malvill, J.M. 1964, *Ap. J.*, **139**, 198.
Manning, P.G. 1970a, *Nature*, **226**, 829.
Manning, P.G. 1970b, *Nature*, **227**, 1121.
Manning, P.G. 1970c, *Nature*, **228**, 844.
Martin, P.G. and Angel, J.R.P. 1974, *Ap. J.*, **188**, 517.
Martin, P.G. and Angel, J.R.P. 1975, *Ap. J.*, **195**, 379.
McKellar, A., Welsh, H.L., and Stephenson, F.C. 1955, *A.J.*, **60**, 170.
McIntosh, A. and Webster, A. 1992, *M. N. R. A. S.*, **255**, 37p.
McIntosh, A. and Webster, A. 1993, *M. N. R. A. S.*, **261**, L13.
McIntyre, H.A.J. and Williams, D.A. 1970, *M. N. R. A. S.*, **148**, 53.
Merrill, P.W. 1934, *Pub. A. S. P.*, **46**, 206.
Meyer, D.M. and Ulrich, R.K. 1984, *Ap. J.*, **283**, 98.
Mitchell, G.F. and Huntress, W.T. 1979, *Nature*, **278**, 722.
Morgan, W.W. 1944, *A.J.*, **51**, 21.
Murdin, P. 1972, *M. N. R. A. S.*, **157**, 461.
Nuth, J.A. and Donn, B. 1983, *Ap. Spa. Sci.*, **95**, 209.
Nuth, J.A. and Donn, B. 1984, *Ap. Spa. Sci.*, **103**, 353.
Purcell, E.M. and Shapiro, P.R. 1977, *Ap. J.*, **214**, 92.
Pritchet, C.J. and Grillmair, C.J. 1984, *Pub. A. S. P.*, **96**, 349.
Rudkjøbing, M. 1969a, *Ap. Spa. Sci.*, **3**, 102.
Rudkjøbing, M. 1969b, *Ap. Spa. Sci.*, **5**, 68.
Rudkjøbing, M. 1970, *Ap. Spa. Sci.*, **6**, 157.
Russell, H.N. 1935, *M. N. R. A. S.*, **95**, 610.
Salama, F. and Allamandola, L.J. 1992a, *Nature*, **358**, 42.
Salama, F. and Allamandola, L.J. 1992b, *Ap. J.*, **395**, 401.
Sanner, F., Snell, R., and Vanden Bout, P. 1978, *Ap. J.*, **226**, 460.
Savage, B.D. 1976, *Ap. J.*, **205**, 122.
Sarre, P.J. 1991, *Nature*, **351**, 356.
Scarrott, S.M., Watkin, S., Miles, J.R., and Sarre, P.J. 1992, *M. N. R. A. S.*, **255**, 11p.
Seab, C.G. and Snow, T.P. 1984, *Ap. J.*, **277**, 200.
Seab, C.G. and Snow, T.P. 1985, *Ap. J.*, **295**, 485.
Seddon, H. 1967, *Nature*, **214**, 257.
Shapiro, P.R. and Holcomb, K.A. 1986a, *Ap. J.*, **305**, 443.
Shapiro, P.R. and Holcomb, K.A. 1986b, *Ap. J.*, **310**, 872.
Smith, W.H., Snow, T.P., Jura, M., and Cochran, W.D. 1981, *Ap. J.*, **248**, 128.
Smith, W.H., Snow, T.P., and York, D.G. 1977, *Ap. J.*, **218**, 124.
Sneden, C., Woszczyk, A., and Krełowski, J. 1991, *Pub. A. S. P.*, **103**, 1005.
Snow, T.P. 1973a, *Ap. J.*, **184**, 135.
Snow, T.P. 1973b, *A. J.*, **78**, 913.
Snow, T.P. 1973c, *Pub. A. S. P.*, **85**, 590.
Snow, T.P. 1992, *Aust. J. Phys*, **45**, 543.
Snow, T.P., Bakes, E.L.O., Buss, R., and Seab, C.G. 1995, *Astr. Ap.*, in press.
Snow, T.P. and Seab, C.G. 1989, *Astr. Ap.*, **213**, 291.
Snow, T.P. and Cohen, J.G. 1974, *Ap. J.*, **194**, 313.
Snow, T.P. and Seab, C.G. 1991, *Ap. J.*, **382**, 189.
Snow, T.P., Timothy, J.G., and Saar, S. 1982, *Ap. J.*, **262**, 611.

Snow, T.P. and Wallerstein, G. 1972, *Pub. A. S. P.*, **84**, 492.
Snow, T.P., York, D.G., and Resnick, M. 1977, *Pub. A. S. P.*, **89**, 758.
Sofia, U.J., Cardelli, J.A., and Savage, B.D. 1994, *Ap. J.*, **430**, 650.
Spence, D. and Inokuti, M. 1974, *J. Quant. Spec. Rad. Trans.*, **15**, 823.
Swings, P. and Ohman, Y. 1939, *Observatory*, **62**, 150.
Thaddeus, P. 1980, *various unpublished talks.*
Thaddeus, P. 1994, in *Molecules and Grains in Space*, ed. I. Nenner (Amer. Inst. of Phys. Conf. Proc. **312**) [New York:AIP Press], p. 711.
Tripp, T.M., Cardelli, J.A., and Savage, B.D. 1994, *A. J.*, **107**, 645.
van der Zwet, G.P. and Allamandola, L.J. 1985, *Astr. Ap.*, **146**, 76.
Walker, G.A.H. 1963, *Pub. A. S. P.*, **75**, 418.
Walker, G.A.H. and Yang, S. 1979, *J. Roy. Astr. Soc. Can.*, **73**, 304.
Wampler, E.J. 1966, *Ap. J.*, **144**, 921.
Waters, L.B.F.M., Mathlener, E., Trams, N.R., van Hoof, P.A.M., Lamers, H.J.G.L.M., Snow, T.P., Waelkens, C., Seab, C.G., and Stanga, R. 1989, *Astr. Ap.*, **211**, 208.
Wdowiak, T.J. 1980, *Ap. J. (Letters)*, **241**, L55.
Webster, A. 1992, *M. N. R. A. S.*, **255**, 41p.
Webster, A. 1993a, *M. N. R. A. S.*, **262**, 831.
Webster, A. 1993b, *M. N. R. A. S.*, **263**, 385.
Webster, A. 1993c, *M. N. R. A. S.*, **265**, 421.
Welter, G.L. and Savage, B.D. 1977, *Ap. J.*, **215**, 788.
Westerlund, B.E. and Krełowski, J. 1988a, *Astr. Ap.*, **189**, 221.
Westerlund, B.E. and Krełowski, J. 1988b, *Astr. Ap.*, **203**, 134.
Westerlund, B.E. and Krełowski, J. 1989, *Astr. Ap.*, **218**, 216.
Wickramasinghe, N.C., Ireland, J.G., Nandy, K., Seddon, H., and Wolstencroft, R.D. 1968, *Nature*, **217**, 412.
Witt, A.N., Bohlin, R.C., and Stecher, T.P. 1983, *Ap. J. (Letters)*, **267**, L47.
Wolstencroft, R.D., Ireland, J.G., Nandy, K, and Seddon, H. 1969, *M. N. R. A. S.*, **144**, 245.
Wu, C.-C. 1972, *Ap. J.*, **178**, 681.
Wu, C.-C., York, D.G., and Snow, T.P. 1981, *A. J.*, **86**, 755.
York, D.G. 1971, *Ap. J.*, **166**, 65.

PORPHYRINS WITHIN INTERSTELLAR GRAINS

F. M. Johnson
Physics Department, California State University, Fullerton, California, USA.

ABSTRACT. It is postulated that two main chromophores, the tetrabenzoporphyrins MgTBP and H_2TBP are situated within interstellar paraffin grain matrices, plus a small amount of pyridine. This hypothesis is supported by an impressive array of lab spectroscopic data, including a total, so far, of 160 lab and DIB coincidences to +/- 1Å. The lab data requires no additional matrix shifts or any other theoretical manipulations for fitting the DIB data. In addition, data by Platenkamp and Canters (1981) of MgTBP in a single crystal of n-octane establishes an additional 27 lab and DIB coincidences (to +/- 1Å) for the special case of HD29647 and HD21389 respectively. This increases the number of matching lab/DIB data to a total of 187. The outstanding properties of these tetrabenzoporphyrins are (a) thermodynamic stability: more stable than virtually all other microwave interstellar mini-molecules; (b) overall strong absorption features: *f* values of order unity. Furthermore, they are well shielded within the crystal grains.

1. INTRODUCTION

The purpose of this conference was to identify the chromophores responsible for the DIBs. The proposed solution to this six decade old DIB enigma utilizes the extensive accumulated observations by Herbig (1966, 1967, 1975, 1982, 1988, 1991, & 1993), recent survey data by Jenniskens and Désert (1994) and the Red Rectangle fluorescence emissions by Scarott *et al.* (1992) and Miles (1994) plus all the lab data (Johnson and Rose 1994).

The totality of DIB data, their wavelengths, line widths, and intensities, plus 25 years of accumulated lab data form the basis of the proposed identification. The two postulated DIB chromophores are within two special classes of organic grain compounds composed of the solvent pyridine and mixtures of paraffins. More recently discovered DIBs by Krełowski and Sneden (1993) and by Jenniskens and Désert (1993) seen as substructures, are consistent with the proposed solution.

2. LAB SPECTRA VS. DIBS: PROPOSED SOLUTION

The following series of postulates are based on an extensive array of lab data on the Tetrabenzoporphyrins (MgTBP and H_2TBP). Preliminary results have been given previously (Johnson, 1972 a & b, 1973, 1976, 1991, and 1993).

A. G. G. M. Tielens and T. P. Snow (eds.), The Diffuse Interstellar Bands, 341–347.

There are three reasons why the chromophores have to be within the grains. One, line width consideration: in supersonically cooled molecules (Evan *et al.* 1982), vibronic lines have widths of the order 1 cm^{-1}, which is inconsistent with the observed DIB line widths (hence free molecules in interstellar space cannot produce the wide range of observed DIB band widths). Two, in special Shpols'ki type matrices, narrow lines are produced comparable to those observed in the DIB spectra (Shpols'ki, 1960 and 1962). Wider DIB lines arise from overlapping vibronic transitions or inhomogeneously broadened lines. Three, DIB wavelengths and line widths are indeed reproduced in the lab data. A total of **187** DIBs have so far been matched with the lab data to ± 1Å. The primary assumption is that the main chromophores H_2TBP and MgTBP are situated within a set of grain types where the A type produces the quasilines (via Shpol'ski matrices) and the B type produces the inhomogeneously broadened features.

The lab data implies that on a purely random basis, a large variety of possibilities can arise: differing pyridine/paraffin ratio, different sets of paraffins, crystal structures, chromophore configurations, monomers, dimers, and with/without pyridine ligands attached to the central Mg atom. These factors in turn will affect transition probabilities and electronic origins. A large variety of these possibilities are expected to produce a complete jumble (i.e. broad features), and that is indeed the inhomogeneous case. However, over 120 repetitive lab experiments showed that on a random basis, five distinct "sites" or configurations (Site VI being the paraffin only case) are favored and occur most frequently. Consequently, the accumulated absorption and fluorescence lab data (Johnson and Rose 1994) can be confronted with the DIB data. In addition, knowledge of the complete set of vibrational frequencies, and the various identified electronic (0 - 0) origins of the S_1 and S_2 states, allow for decoding and identifying **all** of the reported DIBs. MgTBP is the primary molecule. The free base derivative H_2TBP (DIBs at 6597Å, 6591Å, 6177Å, & others) and a non-descript derivative Φ (DIB 4882Å, lab 4890Å), all confirmed with lab spectra are separated out from the DIB list, the remaining DIBs all apparently belong to MgTBP. The set of possible vibronic transitions in MgTBP: S_0 –> S_1(v), S_0 –> S_2(v), produce a fraction of the observed DIBs. The DIBs in the red region originate from transitions S_0(v) –> S_1(0). The strongest lab (Soret) band at 4428Å is also the strongest DIB. Its huge (lab) molecular extinction coefficient (1.5 x 10^6) allows for interstellar abundance determination. e. g. for HD183143, there are 2 x 10^{14} MgTBP molecules per cm^2 in this line of sight.

The DIB spectra does show matrix effects on the chromophores. For example, the DIB spectra of HD29647 (Snow and Seab 1991; Seab 1994) in the 5700-5900Å region is consistent in wavelength (± 1Å) with the lab spectra (Platenkamp and Canters, 1981). This lab spectra was taken of MgTBP at 4K in a single crystal of n-octane. The unusual new features of the DIB spectra of HD29647 are 19 sharp emission lines whose wavelengths (but not intensities) coincide with the lab spectra within experimental errors. Another example is HD21389, (Jenniskens and Désert, 1994) where 8 new DIBs, including the Soret band coincide with the lab data of Platenkamp and Canters (1981) in the DIB blue region. These two special cases are example MgTBP in interstellar paraffin grain matrices devoid of pyridine. The n-octane crystal grains in the line of sight of HD29647 are apparently of high quality. Its MgTBP components are periodically left in long lived excited states (via intersystem crossings to and from the triplet state) facilitating the possibility of induced emissions and cascade de-excitation. The observed sharp emission features have all the earmarks of laser lines. Their pulse periodic nature is hereby predicted. These regions of interstellar space could be considered potential interstellar amplifiers for use in ultra long distance communication (Johnson 1963).

At room temperature (Ehrenberg and Johnson 1990) in a pyridine solution, the dominant spectral features are the Soret band at 4422Å, the Q band at 6344Å, the weaker Q(0,1) broad structure at 5900Å and a vibronic structure at 4161Å. These features are far better resolved at temperatures of 77K and below, with slight temperature shifts. The peak of the 4422Å Soret shifts to 4428Å (and is weakly dependent on the pyridine/paraffin ratio); the 4161Å band becomes 4175Å, recently confirmed in the DIB data (Hibbins, Miles, Sarre, and Herbig, 1994). The Q (0,1) and the Q bands show well resolved structures, both in the lab and DIB data. The lab integrated intensities of each of the above four spectral regions were confronted with the corresponding integrated intensities of the sum of the DIB equivalent widths (EW), for each of the band regions (Jenniskens and Désert, 1994). Indeed, a completely consistent result is obtained. There is a linear one to one correspondence of lab and DIB intensities. These plots were repeated for data of the four stars in the J & D catalog. E(B-V) values for each of the four stars were found to be directly proportional to the y axis intercepts of each of the intensity plots. Thus, an intimate relationship exists between interstellar extinction and the energy absorbed by the interstellar porphyrins, presumably via inter-system crossing to the triplet state, rather than the previously assumed scattering phenomenon.

Since a large number of DIBs consist of overlapping vibronic transitions, the intensity (I) of each DIB resulting from such a combination is given by

$$I = \sum_i n_i \cdot m_i \cdot p_i \cdot q_i \cdot r_i$$

n= number of crystal sites involved in the line of sight
m= transition probability for each vibronic transition
p= probability that a certain site is occupied
q= probability that a certain configuration (chromophore and lattice) is involved
r= for the case of the red region only, probability that a particular molecule is in a given excited state.

An experimental measure of the p factor is obtained by the number of times a particular transition is observed in the lab spectra, and is indicated as such in the lab data of Johnson and Rose (1994). All the major DIBS, ranked in decreasing order of intensity, are seen in the lab data. Furthermore, the Red Rectangle emission bands corroborate 35 out of a possible 57 RE bands with the lab data. Thus, a total, so far, of over 187 DIBs (including a large number of new ones and the 27 Platenkamp and Canters, 1981) coincidences are corroborated in the lab data within ± 1Å.

Next consider internal self-consistency: over 200 vibrations seen in the DIB data are confirmed by independently measured lab data. Furthermore, there are a very large number of DIB vibronic transitions involving identical vibrational frequencies. The latter implies that the nuclear potential surfaces in the S_0 and S_1 states are essentially identical, which is independently confirmed by various lab measurements on the tetrabenzoporphyrins. Table 1 confronts the lab spectra of MgTBP with the DIBs. All DIBs designated as "probable" were measured from the scans from Jenniskens and Désert (1994); H refers to DIBs from Herbig. Figures 1, 2, and 3, are representative examples of individual spectral scans of MgTBP in Shpols'ki matrices at 77 K.

TABLE 1
Lab spectra, quasiline bands vs. DIBs

Probable DIBs	Lab Wide	Notes	n	Lab Å
4131			6	4130.0
4169.9			5	4170.5
4175		H	4	4175.0
4186		b	5	4184.9
4187.5		b	2	4187.7
4268			2	4268.0
4277			6	4276.5
4283		vs	5	4282.9
4286			1	4287.5
4301			2	4301.5
4305			2	4304.5
4308			2	4307.8
4314			4	4315.1
4320		b	3	4320.4
4326			6	4325.5
4329			3	4329.5
4333			1	4334.1
4350			1	4351
4357			1	4357.5
4426			1	4424.8
4428		H / Soret	11	4428.2
4437.5			2	4437.1
4440			1	4438.8
4444			4	4444.5
4447.5			2	4447.2
4453			3	4454.1
4512.8			1	4512.3
4522.5			1	4521.6
4528 br			3	4527
4550			1	4551.2
4581 br			1	4579.8
4614			2	4613.9
4621 br			1	4622.6
4631			4	4631.5
4635.8			1	4636.2
4642.8			1	4642.6
?4682.5			1	4681.3
?4710.5			1	4710.4
5264.8			2	5265.5
5711.5			1	5712.4
			3	5755.8
5762.5			3	5761.8
5768			1	5767.5
5797 v. s.		H	5	5798
5813.5			4	5813.8
5844		H	1	5843.4
			1	5863.7
5872			3	5872.1
5899			8	5898
6016			1	6017.1
6030			4	6032.4
6038			3	6039
6071			1	6069.8
6078			2	6078
6247.5				6248
				6254
			1	6258.8
				6264
6270		H	4	6269.4
6272			1	6273.5
6279			7	6278.2
6281			5	6282.0
6284		H	4	6284.0
6288			4	6288.5
6290		absorption	6	6291
			5	6294.1
			10	6297.3
			4	6301.6
6307		H	2	6306.7

Probable DIBs	Lab Wide	Notes	n	Lab Å
6314		H	18	6314.6
6318		H	6	6318.3
			7	6322.9
			10	6328.4
6331				6331.1
6335				6333
6345	6345		6	6345.7
6354	6354	H	1	
em6357 ?			7	6355.9
6362	6361	H	1	
6365	6365		5	6364.7
6367		H	4	6367.3
	6370		2	6373.7
6376		H	1	6377
6379		H	1	6379
? 6383				6381
6397		H	1	6397
6400			1	6398.8
6410			1	6409.5
6432.8			1	6431.7
6439.5		H	2	6441.2
6445.3		H	3	6445.8
6449.3		H	1	6451.0
6456				6455.5
? 6466.8			5	6465.7
6474			9	S6476.9
6482			8	6481.6
6489			5	6488.7
6492			1	6491.5
wide 6495			10	S6494.9
6502			2	6502.8
6507	6508		2	6508.8
6509			9	6511.5
			12	6513
	6515		9	6514.5
			2	6530.5
			5	6533.0
	6538		5	6536.7
			2	6552.4
			4	6562.6
			1	6568.7
			3	6574.0
				6576
6578.5			4	s6578.2
6583			8	6582.0
			1	6587.0
				6590
			2	6591.5
6597			5	6595.5
6600				6600
w 6603			8	6602.9
	6604		2	6605.4
?6609	6608		2	6609.2
6611			2	6611.3
6614			2	6614.6
			2	6618.0
			9	6620.7
6622.5	6621		2	6623.5
6627	6627		4	6627.9
6633			1	6632.6
6635.3			3	6635.1
w 6638			2	6638.5
6642			1	6640.0
6649	wide		4	6647.6
			2	6650.6
6654			3	6654.4
6657			3	6656.3
			4	6658.4
6660.6		H	1	6662.6

Probable DIBs	Lab Wide	Notes	n	Lab Å
6672			2	6672.7
			1	6677
			1	6680.2
6689.4			3	6690.2
6694.5			4	6694.5
6704.2			1	6705
6740.8		H	1	6740
6743.5				6744
6746			2	s6747.3
6750.5			1	6749
6752.7			1	6752.6
6756			2	s6755.3
6758			1	6758
6775			1	6774.9
6792.4		H	1	6790.7
6795.2		H	1	6794.6
6809.3		H	3	6808.6
6811.1	wide	H	3	6811.0
6823.3		H	1	6824.2
6832.8		H	1	6832.2
6847.6		H	4	s6848.3
6852.5		H	3	s6853.3
			1	6865.1
			1	6882.2
			3	6889.5
			3	s6892.6
	6906		2	6908.2
6944.5	6931	H	2	6944.6
			1	6970.9
6993		H	2	6993.1
7023.3			1	7022.9
7027.4			2	7027.5
7037	wide		1	7038
7050.2	wide		1	7050.2
He			3	7065.8
7068.2			2	7068.0
7069.5			1	7070.3
7075			1	7075.4
7088			1	7088
7282.3			1	7283.4

n = number of measurements
H = Herbig designated DIBs
br = broad
v. s. = very sharp
w = weak

Probable DIBs measured from Survey paper by Jenniskens & Desert

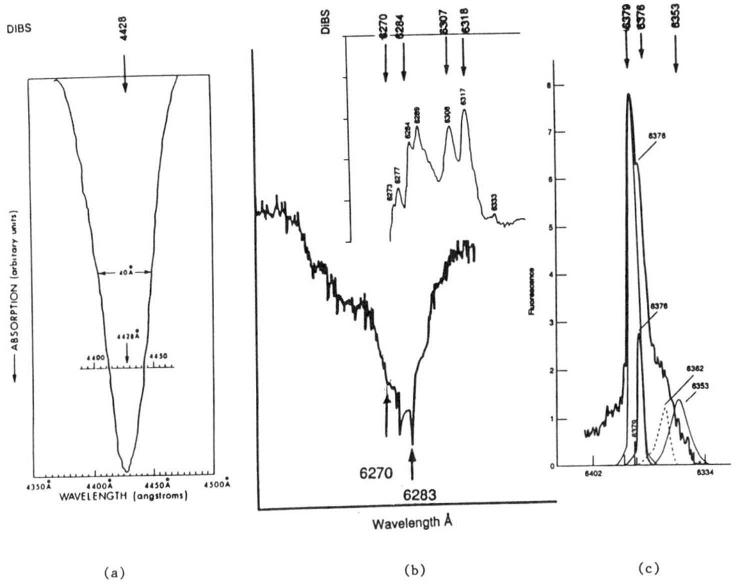

Fig. 1. Composite absorption and emission spectra of MgTBP in Shpols'ki matrices at 77K: (a) The Soret band at 4428Å.
(b) Q-Band region absorption and emission spectra.
(c) Fluorescence spectra of MgTBP showing well resolved 6376 and 6379 DIBs.

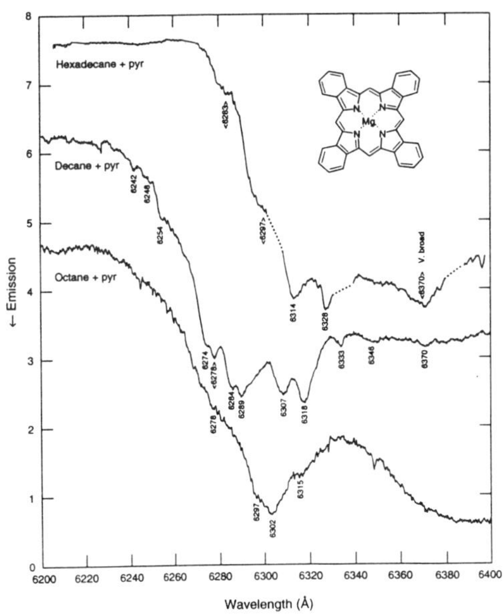

Fig. 2. Fluorescence spectra of MgTBP in Shpols'ki matrices at 77K: Q-Band region.

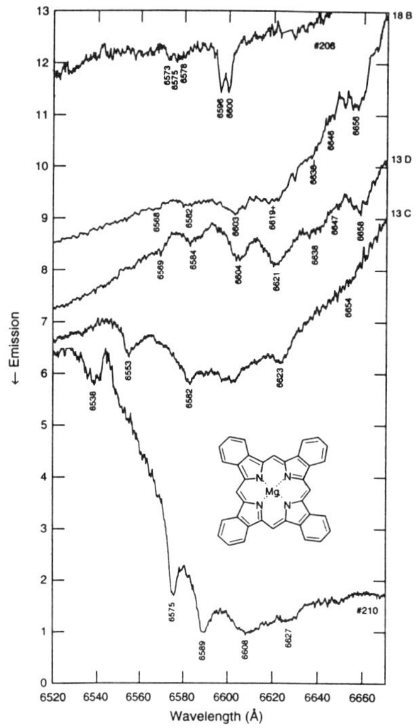

Fig. 3. Fluorescence spectra of MgTBP in Shpols'ki matrices at 77K: 6500Å to 6670Å region.

3. CONCLUSION

A large amount of accumulated experimental evidence provides proof for the presence of tetrabenzoporphyrins in the interstellar medium. Interstellar optically pumped MgTBP molecules can exhibit periodically a population inversion of their excited states, thus facilitating the possible emission of pulsed coherent lasing signals. This latter process should occur in certain specified direction in the interstellar medium, viz. HD29647 and the Red Rectangle.

ACKNOWLEDGEMENTS. I wish to thank Mr. Perry Rose for carefully remeasuring and plotting all of the Shpols'ki data. Dr. John Kerns was among more than a dozen physics majors who participated in the project. I am indebted to Dr C. G. Seab for sending me unpublished data of HD29647, to Dr. P. J. Sarre, Dr. J.R. Miles, and to Dr. P. Jenniskens for preprints of their papers.

REFERENCES

Ehrenberg, B. and Johnson, F.M. 1990, *Spectrochimica Acta*, 46A, **10**, 1521.
Even, U., Magen, J., and Jortner, J. 1982, *J. Chem. Phys.*, **77**, 4384.
Herbig, G.H. 1966, *Z.f.Astrophys.*, **64**, 512.

——. 1967, in *Radio Astronomy and the Galactic System*, IAU Symp. 31, H. van Woerden ed., 85.
——. 1975, *Ap.J.*, **196**, 129.
——. 1988, *Ap.J.*, **331**, 999.
——. 1993, *Ap.J.*, **407**, 142.
Herbig, G.H. and Sonderblom, D.R. 1982, *Ap.J.*, **252**, 610.
Herbig, G.H. and Leka, K.D. 1991, *Ap.J.*, **382**, 193.
Hibbins, R.E., Miles, J.R., Sarre, P.J., and Herbig, G.H. 1994, proceedings of this DIB Symposium.
Jenniskens, P. and Désert, F.-X. 1993, *Astron. Astrophys.*, **274**, 465.
Jenniskens, P. and Désert, F-X. 1994, *Astron. Astrophys*, in press.
Johnson, F.M. 1963, *Bull.Am.Phys.Soc.*, **8**, 625.
Johnson, F.M. 1972a, *Mem.Soc. de Sci.Liege*, **6**, III, 391.
——. 1972b, *Ann.N.Y.Acad.Sci.*, **187**, 186.
Johnson, F.M. 1976, *Voyage into Astronomy*, (Kendall Hunt Pub.).
Johnson, F.M., Bailey, D.T., and Wegner, P.A. 1973, IAU Symp. 52, 317.
Johnson, F.M. 1991, *BAAS*, **23**, 2, 933.
——. 1993, *BAAS*, **25**, 2, 820.
Johnson, F.M. and Rose, P. 1994, submitted for publication.
Krełowski, J. and Sneden, C. 1993, *P.A.S.P.*, **105**, 1141.
Miles, J.R. 1994, PhD Thesis.
Platenkamp, R.J. and Canters, G.W. 1981, *J. Phys. Chem.*, **85**, 56.
Scarott, S.M., Watkin, S., Miles, J.R., and Sarre, P.J. 1991, *M.N.R.A.S.*, **252**, 229.
Seab, C.G. 1994, private communication.
Shpols'ki, E.V. 1960, *Usp.Fiz.Nauk.*, **71**, 215.
——. 1960, *Soviet Phys. Usp.*, **3**, 372.
——. 1962, *Usp. Fiz Nauk.*, **77**, 321.
——. 1962, *Soviet Phys. Usp.*, **5**, 522.
Snow, T.P. and Seab, C.G. 1991, *Ap.J.*, **382**, 189.

FULLERANES AND THE DIFFUSE INTERSTELLAR BANDS

ADRIAN WEBSTER
Royal Observatory,
Blackford Hill,
Edinburgh EH9 3HJ,
Scotland

ABSTRACT. It is discussed whether the fulleranes, the hydrides of C_{60}, might have properties appropriate to the carriers of the diffuse interstellar bands. The average number of hydrogen atoms on an interstellar population of these molecules is expected to vary in a systematic way with the interstellar conditions, and it is shown how this variation might explain not only the existence of the families of bands but also the vanishing of the carrier in dense gas. The carriers of the broad bands might be allyl radicals embedded in the surface of the molecules, whilst those of the narrow bands might be highly-symmetrical isomers of light hydrides of C_{60}. A brief summary is given of how the fulleranes might help account for other interstellar phenomena.

1. INTRODUCTION

There are many poorly-understood phenomena in the interstellar medium whose carriers are believed to be rich in carbon, so the discovery of a radically new form of that element, such as the 'soccerball' molecule C_{60} (Kroto *et al.* 1985), is automatically of great interest. C_{60} is of interest in astronomy for more specific reasons, because it is created at very high temperatures and is highly resistant to destruction, whether by ultraviolet photolysis or by mechanical impact. It could be that C_{60} forms abundantly in hot, carbon-rich stellar atmospheres, perhaps those of carbon stars and the progenitors of planetary nebulae, and then survives for long periods in the interstellar medium in conditions which are hostile to most other species of molecule.

The carriers of the puzzling interstellar phenomena may be rich in carbon, but it does not follow that they consist purely of carbon. Indeed, there is clear evidence in two cases that they are hydrocarbons, because the unidentified infrared emission (Gillett, Forrest & Merrill 1973) and the infrared absorption spectrum of the interstellar medium, as typified by the line of sight to the galactic centre (Soifer, Russell & Merrill 1976), both have features near $3000\,\mathrm{cm}^{-1}$, the characteristic frequency of stretching vibrations of the CH group. This evidence of hydrocarbons is one of the reasons we have been studying the possibility that the 'fulleranes' $C_{60}H_m$, the hydrides of C_{60}, might be of greater importance in astrophysics than C_{60} itself. Each carbon atom in C_{60} is bonded to three others but the valency is four, so one hydrogen atom may also be bonded. If it is present, such a hydrogen atom is expected to take

A. G. G. M. Tielens and T. P. Snow (eds.), The Diffuse Interstellar Bands, 349–358.

up station radially outward from the carbon atom. At most one hydrogen atom may be bound to each carbon atom, from which it follows that the number of hydrogen atoms on any fullerane $C_{60}H_m$ is expected to lie in the range $m = 1$ to 60.

Two great difficulties have to be faced when trying to establish whether molecules of this family could be important interstellar species. The first is the lack of laboratory spectra. Although there is a great deal of work on C_{60} and its relatives, with about a thousand papers published annually, the hydrides receive much less attention, with perhaps ten papers annually. As a result, few ultraviolet-visible spectra of the fulleranes are available for comparison with astronomical spectra, and indirect ways of attacking the problem have to be found. The second difficulty is the number of hydrides of C_{60}, which is so large that they cannot all be studied individually. There are sixty different chemical formulae $C_{60}H_m$ as m takes all values from 1 to 60, which itself is no small number, but there are many structural isomers with any given formula, each having a different geometrical arrangement of the hydrogen atoms on the carbon framework. Thus, there are 23 different dihydrides, 303 trihydrides and so on, and the total number of different isomers of all the hydrides of C_{60} is 9 607 689 885 269 311, all with different spectra (Webster 1993b).

2. THE HYDROGENATION-DEHYDROGENATION EQUILIBRIUM

There is nevertheless a way of proceeding. It is astrochemical in nature because it exploits the likely dependence of $\overline{m}$, the average number of hydrogen atoms per fullerane, on spatial variations of the environmental conditions in the interstellar medium. The hydrogen in $C_{60}H_m$ is not bound as tightly as the carbon, and it is to be expected that ultraviolet starlight will frequently cleave the CH bonds, resulting in the loss of hydrogen atoms. The opposite process is also expected to be a frequent event because hydrogen atoms are known to attach to C_{60} in a facile chemical reaction. As the ultraviolet flux and the gas density vary from place to place in the interstellar medium $\overline{m}$ will vary in response, the value being determined by a dynamical equilibrium between these competing processes. $\overline{m}$ will as a result show a systematic dependence on the environment, taking high values in the dense gas of molecular clouds but low values in the tenuous gas of the diffuse interstellar medium. This characteristic trend might be detectable as a corresponding systematic behaviour of any phenomenon which happens to be carried by the fulleranes (Webster 1993a).

It may be possible to be a little more specific about the systematic behaviour of the visible and near ultraviolet absorption spectra of an interstellar population of fulleranes. The transitions of the fulleranes in this region are expected to be excitations of the π-electrons in the molecules, and the number of these electrons varies inversely with the number of covalently-bonded hydrogen atoms. C_{60} has sixty of them, $C_{60}H_{60}$ has none and an arbitrary fullerane $C_{60}H_m$ has $(60 - m)$. There may as a result be two effects in the interstellar medium. The first, a gross effect, is simply that ultraviolet and visible features carried by the molecules should be strong in the diffuse interstellar medium, where the molecules bear many π-electrons, but weak or absent in dense gas, where they bear few. The second, and more subtle, is that there should be a systematic variation in the relative intensities of features carried by fulleranes with different values of m in response to changes in the average value $\overline{m}$. The spectra of fulleranes bearing few hydrogen atoms will be relatively strong on lines of sight along which $\overline{m}$ is small, whereas the spectra of those with more hydrogen atoms will be strong where $\overline{m}$ is larger, and this dependence on $\overline{m}$ will give rise to systematic differences in the preferred habitats of the corresponding features that may be discernible observationally.

3. THE HABITATS OF THE DIFFUSE INTERSTELLAR BANDS

As it happens, the carriers of the diffuse interstellar bands appear to show such spatial effects. It was established some time ago that the bands are absent from dense gas (Snow & Cohen 1974), in accord with the predicted gross effect, and more recently it was discovered that the ratios of the intensities of different bands can vary from star to star, giving rise to the phenomenon known as 'families' of bands (Chlewicki *et al.* 1987, Josafatsson & Snow 1987, Krełowski & Walker 1987). This variation may be identified with the subtle effect if the carriers of the different families are taken to be different fulleranes, distinguished from each other by bearing different numbers of hydrogen atoms. The underlying cause of the observed variations in the intensity ratios defining the families is then the shifting balance of the hydrogenation equilibrium.

The variation in the relative strength of any two families should not be random if the theory is correct but should be correlated with the hydrogenation conditions. A search for this relationship has been made in the light of a sample of stars by seeing whether the ratios of pairs of bands from different families show a correlation with the height of the stars above and below the plane of the Milky Way, because the density decreases systematically with height (Fig. 1), and evidence has been found for the effect. The results of this study and others seem consistent with the idea that the number of hydrogen atoms on the carriers increases in going from family to family in the order $\lambda5797$, $\lambda5780$ and $\lambda4430$.

4. THE π-ELECTRON SYSTEMS

Whilst the existence of the families can be discussed in the terms of the fullerane theory without assuming anything more about the carriers than that they bear different numbers of hydrogen atoms, much more has to be specified to make the theory complete. Which hydrides would carry which bands, which isomers would be involved, what are their electronic structures, and what kind of transition would give rise to any particular band? Given the enormous number of isomers it is a daunting task to try to pin down the specific carriers, but it may be possible to make progress by classifying the π-electron systems in a simple way.

An unhydrogenated carbon atom on a fullerane has one π-electron, whereas a hydrogenated atom has none. The orbitals of the π-electrons on adjacent unhydrogenated atoms join to form molecular orbitals, and it is these orbitals that determine the visible spectrum. We classify these molecular orbitals into 'global' and 'local' π-electron systems, depending on whether they contain most of the carbon atoms in the molecule or only a few. C_{60} has a global system that encompasses all the atoms in the molecule, and lightly-hydrogenated fulleranes also have global systems, albeit punctuated here and there by hydrogenated atoms. As more hydrogen atoms are attached there comes a point at which the global system breaks up and several local systems are to be found instead. Each local system consists of a handful of neighbouring unhydrogenated atoms surrounded by hydrogenated atoms isolating it from the other π-electron systems on the molecule.

In an interstellar population of fulleranes this behaviour could give rise to a new systematic effect dependent on $\overline{m}$. It would take the form of a progressive change

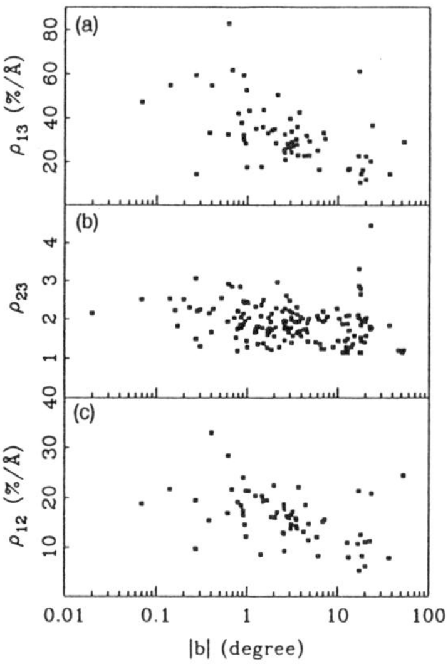

Fig. 1. The ratios of the strengths of three diffuse interstellar bands plotted against the modulus of the galactic latitude. a) 4430/5797; b) 5780/5797; c) 4430/5780. If the carriers of these bands have different habitats, with some preferring denser gas than others, then there should be a dependence of their relative abundance on height above the galactic plane because the gas density falls off with height. This dependence on height might then be observable as a correlation with the modulus of the galactic latitude. The interstellar medium is inhomogeneous so a scatter diagram is to be expected, and that is what is found, but for all three band ratios there is a correlation of modest statistical significance ($\sim 3-5\sigma$). The sense of the correlation is that the carrier of λ5797 prefers the most tenuous gas, that of λ4430 prefers the densest, and that of λ5780 prefers intermediate conditions. In the terms of the fullerane theory this systematic dependence on gas density may be attributed to the workings of the hydrogenation-dehydrogenation equilibrium, which would lead one to expect molecules in the low-density gas away from the galactic plane to bear fewer hydrogen atoms than those lower down in the denser gas near the plane. It follows that the carrier of λ5797 bears fewer hydrogen atoms than that of λ5780, which itself bears fewer than the carrier of λ4430. From McIntosh & Webster (1992).

from a condition in which the π-electron systems are predominantly global to one in which they are predominantly local as $\overline{m}$ increases. It would be possible to look for the transition if some property of the diffuse interstellar bands could be identified which depends on m, the number of hydrogen atoms on the carrier.

On the basis of the available evidence there may be such a property, namely the width of the band profiles. Of the twelve bands classified into families by Krełowski & Walker (1987) ten are narrow, with widths of order a few Å, and two are broad, with widths of order a few tens of Å. The two broad bands (λ4430 and λ6177) form a family by themselves, and all the narrow bands fall into two other families whose archetypes are λ5797 and λ5780. Evidence was mentioned above that the carrier of λ4430 bears more hydrogen than those of λ5797 and λ5780, so the tentative hypothesis may be formulated that the broad bands are carried by local systems of π-electrons and the narrow bands by global systems (Webster 1993b).

In the next two sections we see whether this hypothesis is at all helpful in the attempt to find specific candidates for the carriers of the bands.

5. THE BROAD BANDS

Most of the local systems have known analogues in small molecules. The following are simple examples: two adjacent unhydrogenated carbon atoms surrounded by hydrogenated atoms form a double bond between them, and may be regarded as a substituted or embedded molecule of ethylene; a longer system of four atoms is embedded butadiene; and an unhydrogenated hexagon surrounded by hydrogenated atoms is embedded benzene. These examples all have even numbers of π-electrons and have closed electronic shells, making them no more likely to have transitions in the visible band than the small molecules to which they are analogous. Systems with odd numbers of carbon atoms are different, however, because they have odd numbers of π-electrons, which cannot all pair off to give a closed shell. They are radicals, some of which are quite likely to have transitions in the visible band.

Odd-alternant radicals may be the most likely. Each radical of this type has a weak transition in the visible, which is interesting with regard to the diffuse interstellar bands, and a much stronger one at shorter wavelength, which is not. On the contrary, the latter feature is usually an embarrassment because it tends to fall near 3000Å where no corresponding interstellar absorption feature is known, ruling out many of these radicals as band-carriers. There is one splendid exception, however, which happens to be the smallest of the odd-alternant radicals. It is the allyl radical, $\cdot C_3H_5$, the embedded version of which has three unhydrogenated carbon atoms. The ultraviolet band falls near 2200Å (Fig. 2) and is therefore hidden beneath the much stronger interstellar feature at that wavelength. The visible band of $\cdot C_3H_5$ falls at 4083Å, close enough to 4430Å to be interesting. It is not known whether the differences between $\cdot C_3H_5$ and its embedded equivalents are appropriate to shift the wavelength to correspond to any of the diffuse interstellar bands, and a laboratory measurement would be of great value here. These embedded allyl radicals are the best candidates which have been found so far for carriers of broad bands in the terms of the new theory (Webster 1993e).

The allyl radical constitutes only a small part of the fullerane molecule in which it is embedded, and it may be that the spectrum varies a little depending on how many hydrogen atoms are to be found on the rest of the molecule and on where they are sited. If so, this dispersion could be the reason the corresponding interstellar bands have the broadest profiles.

If the hypothesis is adopted that λ4430 is carried by an embedded allyl radical, it is possible to estimate the amount of interstellar carbon required. The oscillator strength of the visible band of $\cdot C_3H_5$ is small, of order 10^{-3}, and with an average of one allyl radical per fullerane a few tenths of the interstellar carbon would be needed in these molecules. This is a high value, as is discussed below in section 7.

6. THE NARROW BANDS

The bands of the other families are sufficiently narrow that it seems unlikely that they are made up of blends of lines from different isomers, so each of them may be the transition of just one isomer. If all the isomers of any given hydride are equally abundant in the interstellar population, no one of them would ever be noticed because its fractional abundance would be so small, and it seems necessary to suppose that the carriers are 'magic' isomers, meaning that they are much more abundant than their fellows. Another way in which the spectrum of a particular isomer may be prominent

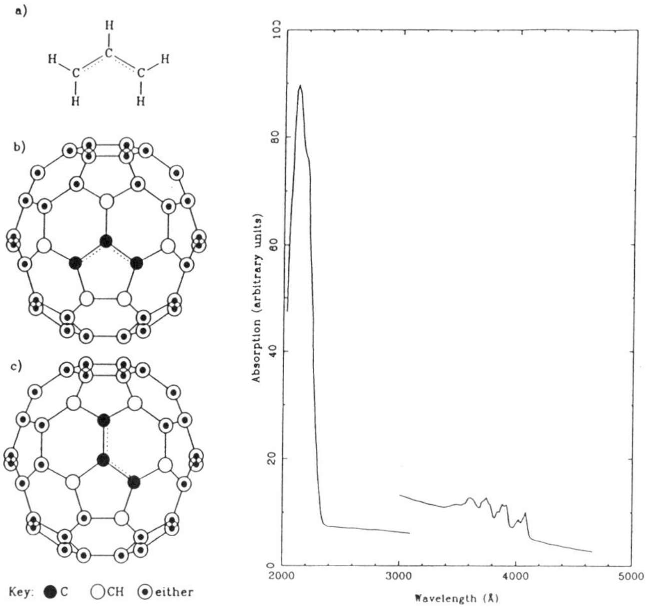

Fig. 2. a) The allyl radical $\cdot C_3H_5$. The ultraviolet-visible absorption spectrum of this species in an argon matrix is shown at the right. The band near 4000Å is much weaker than the ultraviolet band and is shown on an expanded vertical scale. b) and c) Embedded allyl radicals in the surface of a fullerane of the C_{60} family, the first on a pentagon, the second on a hexagon of the structure. Note that the embedded radical is neither a group attached to the outside of the fullerane nor a caged species trapped inside it; the atoms making up the radical are three of the sixty making up the framework of the fullerane itself. From Webster (1993e).

is if it consists of only a few, strong transitions; a different isomer with the same total equivalent width spread out over many weak transitions would be less likely to be noticed in the interstellar spectrum.

These considerations may help restrict the otherwise enormous number of possible carriers because they both suggest that the carriers have high symmetry. Symmetry often bestows stability, as in benzene or in C_{60} itself, and stable species are automatically good candidates for magic species. Furthermore, symmetric species invariably have fewer and stronger transitions than asymmetric ones. Although C_{60} has the highest symmetry possible for any molecule, that of the point group I_h, the icosahedral group with inversion, the same is not true of its hydrides. Adding hydrogen usually reduces the symmetry or eliminates it entirely, and only the saturated $C_{60}H_{60}$ has the full symmetry I_h. Unsaturated hydrides have the symmetry of one or other of the subgroups of I_h, the largest of which (corresponding to the highest symmetry) are the tetrahedral groups T_h and T, followed by the dihedral groups D_{5d} and D_{3d}. All the tetrahedral fulleranes with $m < 30$ are shown in Fig. 3, where it can be seen that symmetry considerations are potentially very useful in reducing the number of isomers to consider.

While symmetry may bring stability, stability is not guaranteed by symmetry, and some of the isomers in the figure do not seem to be plausible magic species, e.g. g) and i) with their isolated radical sites. Others have extensive conjugated systems

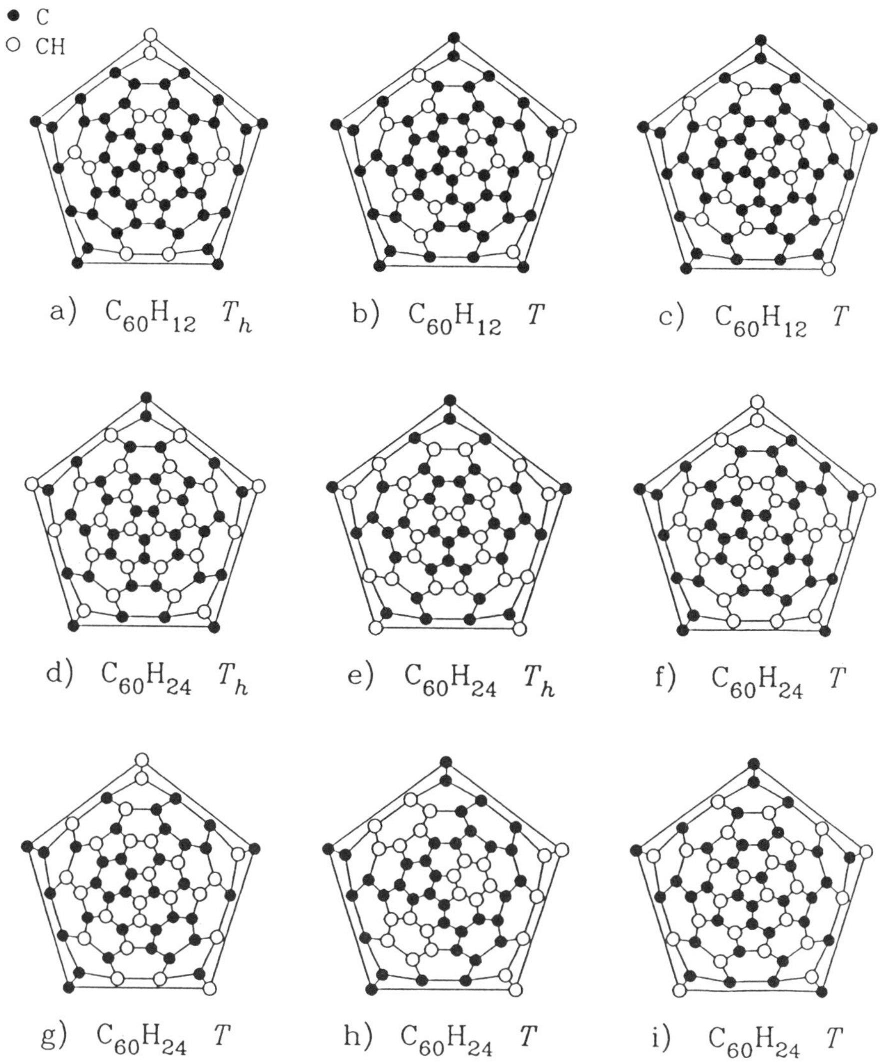

Fig. 3. The tetrahedral isomers of the $C_{60}H_m$ fulleranes with $m < 30$. Each diagram is a plane projection of the C_{60} molecule, in which the circles represent carbon atoms and the lines C-C bonds. The presence or absence of a hydrogen atom bonded to any carbon atom is indicated by the filling of the circle according to the key at top left. Each isomer is labelled by the chemical formula and the point group, either T for the tetrahedral group or T_h for the tetrahedral group with inversion. Isomers belonging to T are chiral, and have mirror images to which they are not equivalent but which are not shown. The tetrahedral isomers with $m > 30$ may be obtained by reversing the interpretation of the key and replacing m with its 60's complement; as an example, h) becomes an interesting isomer of $C_{60}H_{36}$, containing as it does four embedded benzene rings. From Webster (1993b).

which might be expected to be very stable, e.g. a), b), g) and h), and it would be interesting to see the spectra of such isomers and their ions (Webster 1993b).

The unidentified infrared emission was the first astronomical phenomenon to be investigated in the context of the fulleranes (Webster 1991). It is a fluorescence with a rich spectrum of lines and bands (Gillett, Forrest & Merrill 1973), which are presumably the characteristic vibrational transitions of large interstellar molecules. The vibrations of a force-field model of $C_{60}H_{60}$ show some resemblance to this spectrum, the best being a strong CH rocking feature near $\lambda 7.7\,\mu$m, the wavelength of the strongest of the narrow features in the astronomical emission (Webster 1993d).

The extended red emission is a visible-light fluorescence of interstellar and circumstellar matter (Greenstein & Oke 1977) which is almost always found with the unidentified infrared emission. A recent laboratory determination of the spectrum of the fluorescence of C_{60} makes an interesting comparison with this astronomical feature (Fig. 4).

A surprising finding is that several phenomena usually attributed to dust also show the spatial variation expected as a consequence of the shifting hydrogenation equilibrium of the fulleranes. A prime example is the rising component of the far-ultraviolet extinction, which is absent from dense gas but strong in the diffuse medium (Webster 1992a). By comparing the strength of this component of the extinction with the measured ultraviolet opacity of a film of C_{60} it has been estimated that something like a quarter of the interstellar carbon is needed in the fulleranes. This tallies with the estimate obtained by supposing that $\lambda 4430$ is the visible band of an allyl radical, and points to a general property of the new theory: most or all of the depleted carbon has to be in the form of the fulleranes if they are to stand a chance of explaining anything (Webster 1992b).

In the visible band the very broad structure, a low-amplitude ripple on the extinction curve (Whiteoak 1966), varies with the far-ultraviolet extinction in the manner expected of a fullerane feature (Webster 1992b).

Another great surprise is the strong and broad absorption feature at 2200 Å (Stecher 1965), which is known to be weak in dense gas and which therefore bears the stigmata. The strength increases in the diffuse medium but decreases again where the far-ultraviolet extinction is highest, much as $\lambda 4430$ does, and so it may be another diffuse interstellar band. Like $\lambda 4430$ it would be carried by a local system of π-electrons, and the great intensity of the interstellar feature would require one whose product of abundance and oscillator strength is very high, such as embedded benzene or ethylene (Webster 1993e). There is an isomer of $C_{60}H_{36}$ which has four benzene rings embedded (Fig. 3), and the strong feature in the ultraviolet part of the absorption spectrum of this isomer presented by Dr. Leach (these proceedings) is of interest in the context of $\lambda 2200$.

The discovery by Drs. Foing and Ehrenfreund (these proceedings) of near-infrared astronomical features close to those of C_{60}^{+} is clearly of great interest. The low abundance required, about 0.3 % of the interstellar carbon, is consistent with a high abundance of fulleranes if this species with $m = 0$ is regarded as the tail of a distribution of hydrides, with the total carbon amounting to a hundred times that contained in the one particular species.

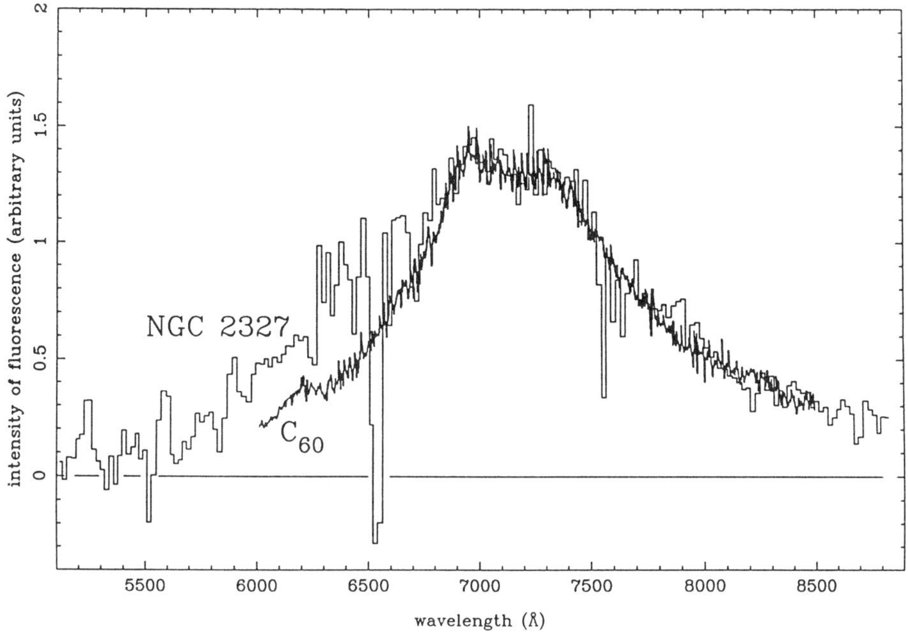

Fig. 4. Laboratory and interstellar fluorescences. A spectrum of the fluorescence of a solution of C_{60} in toluene at room temperature, excited by 5250-Å radiation, is shown together with a spectrum of the extended red emission from NGC 2327. From Webster (1993c).

7. OTHER PHENOMENA

Over all, the fullerane theory has for many years shown features which correspond to known properties of a surprisingly large number of astronomical phenomena, but the absence of relevant laboratory measurements has made it impossible to establish firm connections. In the last year or so this situation has begun to change and several relevant comparisons have emerged (two of them at this conference!), none of which has so far proved to be discouraging.

REFERENCES

Chlewicki, G., De Groot, M.S., van der Zwet, G.P., Greenberg, J.M., Alvarez, P.P., and Mampaso, A. 1987, *Astr. Ap.*, **173**, 131.
Gillett, F.C., Forrest, W.J., and Merrill, K.M. 1973, *Ap. J.*, **183**, 87.
Greenstein, J.L. and Oke, J.B. 1977, *Pub. A. S. P.*, **89**, 131.
Josafatsson, K. and Snow, T.P. 1987, *Ap. J.*, **319**, 436.
Krełowski, J. and Walker, G.A.H. 1987, *Ap. J.*, **312**, 860.
Kroto, H.W., Heath, J.R., O'Brien S.C., Curl, R.F., and Smalley R.E. 1985, *Nature*, **318**, 162.
McIntosh, A. and Webster, A.S. 1992, *M. N. R. A. S.*, **255**, 37p.
Snow, T.P. and Cohen, J.G. 1974, *Ap. J.*, **194**, 313.
Soifer, B.T., Russell, R.W., and Merrill, K.M. 1976, *Ap. J.*, **207**, L83.
Stecher, T.P. 1965, *Ap. J.*, **142**, 1683.
Webster, A.S. 1991, *Nature*, **352**, 412.
Webster, A.S. 1992a, *M. N. R. A. S.*, **255**, 41p.

Webster, A.S. 1992b, *Astr. Ap.*, **257**, 750.
Webster, A.S. 1993a, *M. N. R. A. S.*, **262**, 831.
Webster, A.S. 1993b, *M. N. R. A. S.*, **263**, 385.
Webster, A.S. 1993c, *M. N. R. A. S.*, **264**, L1.
Webster, A.S. 1993d, *M. N. R. A. S.*, **264**, 121.
Webster, A.S. 1993e, *M. N. R. A. S.*, **265**, 421.
Whiteoak, J.B. 1966, *Ap. J.*, **144**, 305.

DIB CARRIERS ON GRAIN SURFACES

W. W. DULEY
Physicspartment
University of Waterloo
Waterloo, Ontario, Canada N2L 3G1

ABSTRACT. The hypothesis that the DIB carriers may arise as surface species on small silicate or carbonaceous grains is discussed. It is concluded that hydrogenated amorphous carbon grains contain PAH and hydrocarbon species that could exhibit the DIB's when liberated from these materials in weak shocks. The condition for Shpol'ski sharpening of spectral lines from PAH molecules may also occur naturally on the surface of HAC grains.

1. INTRODUCTION

One of the fundamental questions concerning the diffuse interstellar bands (DIB's) is whether or not the carriers of these features exist as gaseous species or are attached to or are in dust grains. If they are related to dust grains then the carriers must be associated with small (~20 nm) rather than large (~200 nm) grains. They are also not associated with those grains which produce polarization in the visible and near UV (Martin and Angel 1974).

If the DIB carriers arise in small grains, then the following questions must be addressed:

1) Is the dust/carrier combination abundant enough to account for the strength of the observed DIB's?

2) Are the potential carriers in carbon or silicate grains?

3) Can the carrier yield spectral lines with widths in the 1-10^2 cm^{-1} range as required by observation?

These are fundamental questions, with question 3 being the primary constraint on the identification of the DIB carriers with interstellar grains.

2. ABUNDANCE

The abundance of absorber, β_H, relative to hydrogen, required to yield a DIB with an equivalent width W(Å) (Duley 1979).

$$\beta_H = 1.6x10^{-9}\frac{W}{f}(\text{Å})\frac{(3x10^{21})}{N_{H+2H_2}}\left[\frac{5000}{\lambda(\text{Å})}\right]^2 \quad (1)$$

A. G. G. M. Tielens and T. P. Snow (eds.), The Diffuse Interstellar Bands, 359–362.

where f is the oscillator strength and the normalization is to a hydrogen column density $N_{H+2H_2} = 3 \times 10^{21}$ cm^{-2}. The required abundance relative to that of carbon, where $n_c/n_H = 3.7 \times 10^{-4}$ is

$$\beta_c = \frac{4.3\text{x}10^{-6}}{f} W(\text{Å}) \frac{(3\text{x}10^{21})}{N_{H+2H_2}} \left[\frac{5000}{\lambda(\text{Å})}\right]^2 \tag{2}$$

with $N_{H+2H_2} = 1.9 \times 10^{21} A_v$, this becomes

$$\beta_c = \frac{6.8\text{x}10^{-6}}{A_v} \frac{W(\text{Å})}{f} \left[\frac{5000}{\lambda(\text{Å})}\right]^2 \tag{3}$$

Thus a DIB with $\lambda = 5000$Å having W = 0.1 Å observed along a sightline where $A_v = 1.6^m$, and arising from a transition with $f = 10^{-3}$, gives $\beta_c = 4.3 \times 10^{-4}$. If we assume this absorber to be a PAH molecule with m carbon atoms, then the carbon abundance required is 4.3×10^{-4} m or 4.3×10^{-2} m % of available carbon. On this basis, only PAH molecules with m > 2325 would be excluded by abundance considerations, as the source of this DIB.

These requirements become much more stringent for the stronger DIB's. For example, the 4428Å DIB in HD183143 ($A_v = 4.1^m$) has W = 3.4 Å (Herbig 1994). Then $\beta_c = 7.2 \times 10^{-6}/f$ implying that $f \gtrsim 10^{-2}$ if abundance problems are to be avoided in this case. With $f = 10^{-2}$, the 4428Å carrier would have $\beta_c = 7.2 \times 10^{-4}$ m. A carbon molecule with m = 10 would then require 0.72% of available carbon in HD 183143 to yield a 4428Å band of the observed strength if $f = 10^{-2}$ for the 4428 Å transition. This tells us that the carriers of the strongest DIB's (4428, 6177, 6284, 6533Å) are either very abundant interstellar species (containing ≈ 1% of available carbon) or that the transitions giving rise to these features are intense ($f \gtrsim 10^{-2}$); i.e. are fully allowed electric dipole transitions. If the carbon abundance in any *single* DIB carrier is limited to 1% of that available, and f = 1 for the appropriate transition giving rise to this DIB, the number of C atoms contained in the carrier is m = 208. Such a molecule would have the attributes of a small grain.

3. CARBON OR SILICATE GRAINS?

The case for the assignment of selected DIB's to absorption by impurities in silicate or oxide solids has been extensively discussed elsewhere (see Duley 1979, Manning 1976, Huffman 1977) and will not be repeated here. Sharp lines ($\Delta\nu \lesssim 50$ cm^{-1}) are obtained only for zero phonon transitions associated with lattice defects or with field independent transitions of transition metal ions such as Cr^{3+} and Mn^{4+}. Phonon sidebands add subsidiary structure in the form of broad bands which are often wider than 50 cm^{-1}. The oscillator strength of these transitions is usually small $f \sim 10^{-6} - 10^{-5}$ but can be increased when lattice defects reduce the symmetry at impurity sites or when pairs of ions are involved in a transition. In these cases, f may increase to 10^{-3}. With $f = 10^{-5} - 10^{-6}$, transitions in ions such as Fe^{2+} and Fe^{3+} would be strong enough to account for DIB's with equivalent widths W $\lesssim$200 mÅ, while the DIB at 4430 Å could be produced by Fe^{3+} if $f \sim 10^{-3} - 10^{-4}$. Less abundant transition metals such as Ni (in the form Ni^{2+}) would be abundant enough to yield many of the weaker DIB's. There is therefore no a priori reason why transition metal ions in oxides or silicates should be excluded by abundance constraints as the source of at least some of the DIB's.

The spectroscopic case for an association of DIB's with spectral lines of transition metal ions in silicate or oxide solids, however, provides a stronger constraint. It can be concluded, that despite the availability of considerable laboratory data on the spectra of metal ions in a variety of silicate and oxide hosts, no firm matches appear between these data and the DIB's. If transition metal ions are present in silicate grains in interstellar clouds, at concentrations approaching the cosmic abundances of these species, then d-d transitions should be observable. Since the wavelengths of these lines will be sensitive to the composition and structure of the silicate matrix, detection of new DIB's due to ions such as Cr^{3+} and Mn^{4+} may yield important information on the properties of interstellar silicates.

The possibility that DIB carriers might exist as surface molecules adsorbed on silicate grains was discussed by Duley and McCullough (1977). Ions and radicals of polycyclic aromatic hydrocarbon (PAH) molecules were suggested as likely candidates, as these species have strong absorption lines in the visible region of the spectrum. It was shown that the benzene cation, $C_6H_6^+$, adsorbed on protosilicates yields an absorption feature near 4430 Å. The possibility that other PAH ions could be the source of the DIBs was also discussed at that time (Duley and McCullough 1977) and elaborated at a later date by Crawford et al. (1985). This hypothesis is yet unproven, but remains as a viable possibility in the absence of contradictory data. However, to date, there have been no accurate simulations of a DIB with either gas phase or matrix spectra.

It is now thought that small carbon grains can assume a variety of compositions depending on their thermal and radiative histories and location in the interstellar medium (Jones et al. 1990, Duley 1993). Infrared absorption spectra of galactic centre sources (Sandford et al. 1991) show the presence of a low temperature form of hydrogenated amorphous carbon (Duley 1994). Materials of this sort have mixed aromatic and diamond-like character (Robertson and O'Reilly 1987) and contain a variety of C_6 aromatic rings bonded together by other non-aromatic hydrocarbon species. It has been suggested that the discrete IR emission features at 3.3 μm and other wavelengths arise when HAC is dissociated in interstellar shocks (Duley and Jones 1990).

It is interesting that a structure in which aromatic molecules are dispersed in a matrix of hydrocarbon molecules as observed in HAC, is similar to the condition for production of the Shpol'ski effect (Shpol'ski 1960). Under these conditions, narrow absorption lines can be produced when the dimension of an aromatic molecule is approximately matched by that of neighbouring alkane molecules. Line widths observed under these conditions can be $\simeq 1$ cm^{-1} (Pace and Maple 1985) or comparable to the width of the narrowest DIB's. Similar effects are also observed when aromatic derivatives are placed in a matrix of a similar aromatic (Pace and Maple 1985).

Recent structural information obtained from scanning tunnelling studies of HAC and aC (Cho et al. 1992, Vandentop et al. 1991) show that these materials contain organized regions which mimic the structure of simple C_6 aromatic molecules eg. pyrene, anthracene and coronene. This suggests an interesting possibility, namely that such regions may act as templates which accrete or grow other PAH molecules. If such sites are oriented in the same way on individual grains then the conditions for

Shpol'ski line narrowing might be expected to occur. The requirement for narrowing would involve site organization only on individual grains with no requirement for the relative orientation of individual grains. The possibility that Shpol'ski spectra may be obtained under certain conditions in HAC has not yet been investigated but is worthy of experimental study.

4. DIB ABSORBERS FROM SPUTTERED HAC

The dissolution of HAC in weak shocks has been discussed in some depth elsewhere (Duley and Williams 1988, Duley 1993, Duley and Jones 1990). The effect can be to yield an appreciable column density of alkane, hydrocarbon and PAH fragments, derivatives and radicals. Optical spectra of many of the species likely to derive from HAC under these conditions have not yet been obtained. Difficulty may lie in obtaining a sufficient spatial density under laboratory conditions to permit absorption measurements. Alternatively, laser induced fluorescence of material ablated from HAC could be a means of obtaining spectra. Quenching via supersonic expansion would be necessary to ensure that excited state populations were minimized.

ACKNOWLEDGEMENTS. This research has been supported by grants from the NSERC of Canada.

REFERENCES

Cho, N.H., Veirs, D.K., Ager, J.W., Rubin, M.D., Hopper, C.B., and Bogy, D.B. 1992, *J. Appl. Phys.*, **71**, 2243.
Crawford, M.K., Tielens, A.G.G.M. and Allamandola, L.J. 1985, *Ap. J.*, **293**, L45.
Duley, W.W. 1979, *Ap. J.*, **227**, 824.
Duley, W.W. 1986, *MNRAS*, **219**, 859.
Duley, W.W. 1993, in *Dust and Chemistry in Astronomy*, ed. T.J.Millar and D.A. Williams (Bristol IOP Publ.), 71.
Duley, W.W. 1994, *Ap. J.*, **430**, L133.
Duley, W.W. and Jones, A.P. 1990, *Ap. J.*, **351**, L49.
Duley, W.W. and McCullough, J.D. 1977, *Ap. J.*, **211**, L145.
Duley, W.W. and Williams, D.A. 1984, *MNRAS*, **211**, 97.
Herbig, G. 1994, (preprint).
Huffman, D.R. 1977, *Adr. Phys.*, **26**, 129.
Jones, A.P., Duley, W.W., and Williams, D.A. 1990, *QJRAS*, **31**, 567.
Manning, P.G. 1976, *Chem. Soc. Rev.*, **5**, 233.
Martin, P.G. and Angel, J.R.P. 1974, *Ap. J.*, **188**, 517.
Pace, C.P. and Maple, J.R. 1985, *J. Opt. Soc. Am.*, **B2**, 1582.
Robertson, J.R. and O'Reilly, E.P. 1987, *Phys. Rev.*, **B35**, 2946.
Sandford, S.A., Allamandola, L.J., Tielens, A.G.G.M., Sellgren, K., Tapia, M., and Pendleton, Y. 1991, *Ap. J.*, **371**, 607.
Shpol'ski, E.V. 1960, *Sov. Phys. Uspekhi*, **3**, 372.
Vandentop, G.J., Nascente, P.A.P., Kawasaki, M., Ogletree, D.F., Somarjai, G.A., and Salmeron, M. 1991, *J. Vac. Sci. Technol.*, **A9**, 2273.

PAH CATIONS AS CANDIDATES TO THE DIB CARRIERS

ALAIN LÉGER *Institut d' Astronomie Spatiale, CNRS, bat 121, University of Paris Sud, F 91405 ORSAY, Fr*

ABSTRACT. The discussion of the advantages/disadvantages of grain versus molecular carriers of the DIBs points in favour of the latter. The equivalent widths of the strongest DIBs require that they are carried by abundant species . If a carbon bearing molecule and provided reasonable assumptions on molecular properties are made, the mere carrier of the 4430Å band has to contain over 0.08% of the cosmic C.

PAH cations are specially attractive candidates because: (i) we have independent indications of their abundant presence in the ISM, (ii) they have transitions with appropriate energies, intensities and linewidths. However, no precise fit has been obtained up to now. We conclude that only gas phase experiments will permit unambigous identifications.

Meanwhile, matrix experiments are useful to select candidate species. We report the intringing case of a PAH species, related to CH_3-pyrene which is a candidate carrier for both the 4430Å and the 7565Å DIBs. This proposal has a falsifiable implication: there should be a total correlation in the observed equivalent widths of these two bands in any line of sight.

1. MOLECULAR ORIGIN

The invariant and rather narrow widths of the DIBs in absorption (0.5 to 30Å), the absence of red scattering wings and polarisation across the bands argue against middle to large grains ($a > 100$Å) and favours a molecular origin of the DIBs. In the case of the bands which are observed in emission in the Red Rectangle, *the molecular origin has been established* by Scarrott *et al.* (1992) and Miles and Sarre (1993). Their observations nicely show that the band shapes are rotational structures of electronic transitions with rotational temperatures that decrease with increasing distance to the illuminating star.

2. MOLECULAR PHYSICS CONSIDERATIONS

There are several points of importance for the DIB identification problem in the features of electronic, vibrational and rotational levels of large molecular species. They are illustrated by the spectra of PAH neutrals and cations (Leach 1987; Ehrenfreund *et al.* 1992; Salama and Allamandola 1992a and 1992b; Joblin 1992; Parisel *et al.* 1992).

A. G. G. M. Tielens and T. P. Snow (eds.), The Diffuse Interstellar Bands, 363–368.

2.1. OSCILLATOR STRENGTHS

The very first electronic transitions have f values of the order of $10^{-4} - 10^{-3}$ for closed shell species, e.g. neutrals, and $10^{-2} - 10^{-1}$ for open shell species, e.g. ions or radicals. The next ones are stronger ($f \sim 1$) and occur in the UV. The sum of the f values for all the transitions from the ground state is equal to the electron number:

$$\Sigma_i f_i = N_{el} \tag{1}$$

but most of the contributions come from the far UV.

2.2. WIDTHS

The widths of the first electronic transitions are small ($\Delta\lambda = 2 \times 10^{-4} - 5 \times 10^{-3}$) whereas those of the upper ones are much larger ($10^{-2} - 10^{-1}$).

2.3. IR EMISSION

The absorption of visible - UV photons by free molecules is followed by visible fluorescence and vibrational IR fluorescence (Léger *et al.* 1989; Allamandola *et al.* 1989).

These three points constrain the possible candidates for the origins of the observed DIBs.

3. BASIC ARGUMENT : THE ABUNDANCE REQUIREMENT

As soon as large aromatic molecules have been recognised as possibly abundant in the ISM (Léger and Puget 1984) several groups have pointed out that their cations were good candidates for the DIB's carriers (Van der Zwet and Allamandola 1985; Léger and d'Hendecourt 1985; Crawford *et al.* 1985). A basic argument is the abundance requirement. The equivalent width ΔW_λ of an observed absorption is related to the oscillator strength f_λ of the corresponding transition and the column density N_x of the carrier as (Spitzer 1978)

$$\Delta W_\lambda = a\lambda^2 f_\lambda N_x \tag{2}$$

where a is π times the classical radius of the electron, $a = 8.86\,10^{-13}$cm.

For the 4430Å band, the observations (Jenniskens and Désert 1994) indicate, as an average,

$$\Delta W_{4430} = 720\text{mÅ} \times \text{Av} \tag{3}$$

where Av is the visual extinction.

This implies a large abundance of the 4430Å carrier as shown below.

For a given absorption band, the higher the f value of the transition the lower the abundance of the carrier (eq. 2). A minimum abundance of the 4430Å carrier will be obtained for a maximum f value estimate. However, the modest width of the 4430Å band, $\Delta\lambda/\lambda = 2.8 \times 10^{-3}$, excludes that the band is due to a transition towards a high number electronic level with a very large f value (> 1) but too large a bandwidth, $\Delta\lambda/\lambda > 10^{-2}$, (section 2.2). The next highest f value does not imply a broad band and is provided if the DIB comes from one of the first transitions of an ion or a radical (section 2.1 and 2.2). A maximum estimate of the f value is then $f = 0.1$. If it is a carbon bearing molecule with, say over 15 carbon atoms (one of the smallest PAH considered as resistant in the harsh ISRF conditions is pyrene, $C_{16}H_{10}$), with $N_H/Av = 10^{21} cm^{-2}$ and $N_C/N_H = 4 \times 10^{-4}$, the minimum C amount involved in the 4430Å carrier is

$$[C]_{4430}/[C]_{cosmic} > 0.08\% \tag{4}$$

When the same considerations are applied to the whole set of known DIBs (Jenniskens and Désert 1994), the required amount of C becomes:

$$[C]_{DIBs}/[C]_{cosmic} > 0.5\% \tag{5}$$

These numbers tell us that *the DIB carriers are quite abundant species indeed.* They qualify the DIB problem as an important one for the physics and chemistry of the ISM. They also imply that the carriers must be:

- stable against photo dissociation;
- strong IR emitters (section 2.3), which is a valuable indication for their search.

4. PAH CATIONS AS CANDIDATES

The PAH cations have the advantage to fulfil the preceding requirements:

- when large enough (more than 15-20 C atoms per molecule) they are expected stable against photolysis by the ISRF (Joachims *et al.* 1994);

- they have first electronic transitions in the visible with f values that can reach 0.1 and band widths $\Delta\lambda/\lambda = 2 \times 10^{-4} - 5 \times 10^{-3}$, compared to the observed values $1 \times 10^{-4} - 6 \times 10^{-3}$ (Jenniskens and Désert). It should be noticed that this is *not* true for neutral PAHs (section 2.1)

- they are abundant species in the ISM as indicated by data in a different λ range. They are growing evidence that these species are responsible for the set of IR emission bands from the ISM at 3.3 - 5.2 - 6.2 - 7.7 - 8.6 - 11.3 μm (Léger and Puget 1984; Allamandola *et al.* 1989; Roche *et al.* 1991) and involve 10-15% of the cosmic C. So a given member of that family could satisfy the abundance requirement of section 3. However, whether these IR bands originate from ionised, neutral or radical PAHs is *a serious and unsolved problem* (De Frees *et al.* 1993; Vala *et al.* 1994; d'Hendecourt and Léger, in preparation).

5. HOW TO GO FURTHER?

The preceding arguments plead in favour of PAH^+ but they do not identify any specific DIB carrier. A convincing identification would require a good spectroscopic match between laboratory data and observations. As it has been pointed out several times during this Conference, it could be done *only with gas phase experiments* because of the wavelength shifts generated by other experimental technics: insertion in boric acid glasses, rare gas matrices...

It could be achieved by ion traps if the problem of getting cold species can be solved, or by supersonic jet expansion of a carrier gas where the seeded molecules are cold and isolated. Today, the latter seems a promising approach although delicate for ions.

Meanwhile, we can look for (un)compatibilities between the different proposals and the observation facts as it is done in section 6. We can also use simpler technics to select candidates (section 7).

6. LINEWIDTHS

If DIBs are due to electronic transitions of molecules, with ro-vibrational structures, their rotational widths have the following contributions that we compare to the observed values:

6.1. PQR BRANCH SPREADING

The rotational transitions involve the selection rule: Δ J = -1,0,+1. Starting from the thermally populated rotational ladder of the ground electronic state (with temperature T_{rot}) leads to the well known PQR branch structure that expends roughly on

$$\Delta E_1 = B^{1/2}(kT_{rot})^{1/2} \tag{6}$$

where B is a rotational constant of the molecule.

The observed absence of PQR structure plead against small molecules that would exhibit broad and distinguishable such structures. It is compatible with bigger molecules (smaller B) provided that the other broadening processes wash out their narrow PQR structure.

6.2. CHANGE IN B

When the excited electronic state has a different geometry and therefore rotational constant (B' = B + ΔB) than the ground state (B), the band has a head and a tail expanding on

$$\Delta E_2 = (\Delta B/B)\, kT_{rot} \tag{7}$$

This structure has been nicely demonstrated by the observations of Scarrott *et al.* (1992) and Miles and Sarre (1993) for the DIBs seen in emission in the Red Rectangle. They have also shown that kT_{rot} was a function of the distance to the star. Unfortunately, only the product $(\Delta B/B)\, kT_{rot}$ is measured and assumptions are needed to estimate each factor individually which would give us direct information on the carrier molecule structure and rotational state.

6.3. INTERNAL COUPLING

The coupling between an electronic level and an iso-energetic lower one with vibrational excitation leads to broadening of the level that increases with the number of the electronic level (sect. 2.2). The observed limited widths imply that transitions are, if ionic species, $D_1 \leftarrow D_0$ or $D_2 \leftarrow D_0$ but not $D_5 \leftarrow D_0$ or upper ones.

As a *conclusion*, the observed DIB bandwidths are compatible with the first electronic transitions of rather large PAH cations or radicals.

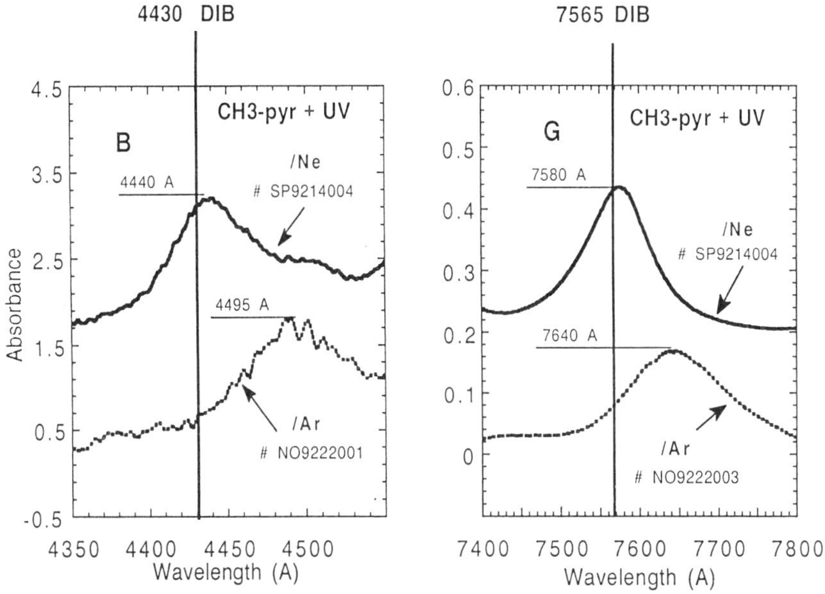

Fig. 1. Visible spectra of CH3 - pyrene in Ne and Ar matrices. The samples are irradiated by a UV H_2 lamp that provides photons in the 1215 - 1600Å (10.2 - 7.8 eV) range. The positions of the 4430 and 7565Å DIBs are indicated. The laboratory data are compatible with exact fits of the DIBs with the bands for the gaseous species because one expects the bands to be redshifted in matrices from their positions in the gas phase and the shifts are significantly larger for Ar than for Ne.

7. SEARCH FOR POSITION MATCHES

Experiments on gaseous molecular ions being difficult and long to carry, easier technics, as rare gas matrix isolation, are valuable tools to select carriers with approximate wavelength matches that could be studied in gas phase experiments as a second step. Several attempts have been done in that direction (Salama and Allamandola 1992 a and b; Ehrenfreund *et al.* 1992; Fulara *et al.* 1993). Up to now, whenever an explicit species has been proposed, it has met internal conflicts, e.g. predictions that were not observed. Specifically, the identification of a molecular band with a DIB often implies the presence of other DIBs that should be observed with the predicted intensities.

We mention thereafter a preliminary result on a candidate that has not yet suffered such internal contradictions but on the other hand which has not yet been explicitly identified (d'Hendecourt *et al.* at this Conference; Léger *et al.* 1994). It is an aromatic species, likely a cation, obtained by UV irradiation of CH3-pyrene in Ar and Ne matrices. Its main bands in the visible have positions in these matrices which are compatible with, but do not prove, an exact match of the gas phase species bands with the 4430Å and 7565Å DIBs (Fig. 1).

8. CONCLUSION

The presently available laboratory data on DIB candidates allow speculations but no firm conclusions. We have shown that *PAH cations are good candidates* because they fulfil the requirements inferred from the DIB observations in the different wavelength ranges (visible absorption/emission and mid-IR emission), including the abundance requirement and that no internal contradiction has yet been pointed out but we have made no actual identification.

Only gas phase experiments will be conclusive when a precise match of the measured and observed λi and $\Delta\lambda i$ will be obtained. To prepare such tests, it is important to select candidates with easier spectroscopical technics, as rare gas matrix isolation. An intriguing example is given of a PAH radical related to CH3-pyrene that appears to be a candidate to the (common) carrier of the 4430Å and 7565Å DIBs. If this interpretation is correct, it has a falsifiable implication: the equivalent widths of these two DIBs should be totally correlated in all the lines of sight of the ISM.

REFERENCES

Allamandola, L.J., Tielens, A.G., and Barker, J.R. 1989, *Ap. J. S.*, **71**, 733.
Allamandola, L., Bregman, J.D., Standford, S.A., Tielens, A.G., Witteborn, F.C., and Wooden, D.H. 1989, *Ap. J. Lett.*, **345**, L59.
Bondibey, V.E. and Miller, T.A. 1983, in *Molecular ions: Spectroscopy, Structure and Chemistry*, Bondibey, V.E. and Miller, T.A. (eds), North Holland, Amsterdam, p. 138.
Boschi, R., Clar, E., and Schmidt, W. 1974, *J. Chem. Phys.*, **60**, 4406.
Crawford, M.K., Tielens, A.G., and Allamandola, L.J. 1985, *Ap. J.*, **293**, L45.
De Frees, D.J., Miller, M.D., Talbi, D., Pauzat, F., and Ellinger, 1993, *Ap. J.*, **408**, 530.
Ehrenfreund, P., d'Hendecourt, L., Verstraete, L., Léger, A., Schmidt, W., and Défourneau, D. 1992, *A & A*, **259**, 257.
Jenniskens, P. and Désert, F.X. 1994, *A & A Suppl.*, **106**, 39.
Joachims, H.W., Rühl, E., Baumgartel, H., Tobita, S., and Leach, S. 1994, *Ap. J.*, **420**, 307.
Joblin, C. 1992, "Spectroscopie de molécules aromatiques - Test du modele PAH en Astrophysique", These 'Université Paris 7, p. 139-143.
Leach, S. 1987 in *Polycyclic Aromatic Hydrocarbon and Astrophysics*, eds. A. Léger, L. d'Hendecourt, and N. Boccara, Reidel.
Léger, A. and d'Hendecourt, L. 1985, *A & A*, **146**, 81.
Léger, A. and Puget, J.L. 1984, *A & A Lett.*, **137**, L5.
Léger, A., d'Hendecourt, L., and Défourneau, D. 1989, *A & A*, **216**, 148.
Léger, A., d'Hendecourt, L., and Défourneau, D. 1994, *A & A Lett.*, in press.
Miles, J.R. and Sarre, P.J. 1993, *J. Chem. Soc. Faraday transactions* 89, 2269 and 2307.
Parisel, O. 1994, Communication at this Conference.
Parisel, O., Berthier, G., and Ellinger, Y. 1992, *A & A Lett.*, **266**, L1.
Roche, P.F., Aitken, D.K., and Smith, C.H. 1991, *MNRAS*, **252**, 282.
Salama, F. and Allamandola, L.J. 1992, *Nature*, **358**, 42.
Scarrott, S.M., Watkin, S., Miles, J.R., and Sarre, P.J. 1992, *MNRAS*, **255**, 11.
Shida, T. and Iwata, S. 1973, *JACS*, **95**, 3473.
Spitzer, L. 1978, in *Physical Processes in the Interstellar Medium*, Wiley, p. 52.
Vala, M., Szczepanski, J., Pauzat, F., Parisel, O., Talbi, D., and Ellinger, E. 1994, *J. Phys. Chem.*, in press.
Van der Zwet, G.P. and Allamandola, L.J. 1985, *A & A*, **146**, 76.

CARBON CHAINS AND THE DIFFUSE INTERSTELLAR BANDS

P. THADDEUS
Center for Astrophysics and Division of Applied Sciences, Harvard University

ABSTRACT. It is shown that a stable population of long carbon chains and large planar rings is maintained by C^+ insertion, photocleavage, and spontaneous isomerization in the diffuse interstellar gas. Chains with ~20-30 atoms are synthesized at about the right abundance to account for the diffuse bands, and it is suggested that these and derivative acetylenic rings comparable in size or slightly larger are an important, possibly dominant, component of large molecules in the diffuse gas. It is further suggested that in a slower subordinate cycle spontaneous isomerization further converts some of the larger planar rings to fullerenes, which then add carbon and possibly other atoms to form even larger molecules and amorphous grains which are recycled only very slowly via grain-grain collisions and supernova shocks.

1. INTRODUCTION

There is now a good deal of evidence, discussed elsewhere in this meeting, that the carriers of the diffuse bands are molecules in the gas phase, and not a component of the interstellar dust grains (e.g., color centers, transition metal complexes, or other embedded chromophores). One of the most compelling reasons for believing in a molecular carrier is the remarkable sharpness of some of the diffuse bands–about 1Å, corresponding to a fractional width of only about 2 x 10^{-4}. It is possible to obtain solid state transitions that sharp in the laboratory, but only under special conditions of preparation, annealing, and purity, and it is very difficult to see how such conditions could be achieved in a cosmic grain assembled under a variety of conditions, a fleck of matter probably containing a mixture of elements, and so small that edge effects should make a significant contribution to line broadening. Even if edge effects are unimportant–in a sufficiently large grain for example–a localized chromophore would probably have to be surrounded by the same configuration of atoms out to a distance of 5-10Å to produce absorption lines as narrow as 1Å and that amounts to a fairly close description of a large molecule. For this reason, and on the basis of the evidence and arguments discussed elsewhere here, I will assume that the carriers of the diffuse bands are molecules in the gas phase, and show why carbon chains, the dominant structural theme in molecular clouds, are interesting candidates.

As the first carbon chains were being discovered by radio astronomers in the 1970's, A. E. Douglas (1977) suggested that these were attractive candidates to explain the DIBs. He pointed out that such chains are conjugated systems similar to organic dyes, and that like such dyes their strongest electronic transition with increasing chain length would tend to progress from the blue toward the red into the spectral

A. G. G. M. Tielens and T. P. Snow (eds.), The Diffuse Interstellar Bands, 369–378.

range where the diffuse bands are observed, i.e., $\lambda > 4430$Å; Douglas also noted that fast radiationless transitions could plausibly account for the wide, featureless shapes of the bands. Today as in 1977 the optical spectra of even short carbon chains like those in Figure 1 are very poorly known, so the chain hypothesis explains one of the most enigmatic aspects of the diffuse bands: the total lack of laboratory identification.

This lack of laboratory identification is a serious objection to the hypothesis that PAHs are the carriers of the diffuse bands, because PAHs are readily synthesized stable molecules, whose optical spectra have been exhaustively studied for many years. The idea that the diffuse bands are produced by ionized PAHs is not a serious mitigation of this objection, because under ionization equilibrium in the diffuse gas of the order of one-half of PAHs (or similar large molecules) are neutral (Lepp and Dalgarno 1988; Bakes and Tielens, this meeting). Why *not one* member of this neutral moiety of PAHs has been identified on the basis of existing laboratory data is an obvious question quite unanswered by the PAH hypothesis.

2. CHAINS TO RINGS TO FULLERENES TO GRAINS

Is there a way to proceed from the kinds of molecules we know to exist in the interstellar gas to the larger systems with 20-100 atoms which might produce the unidentified infrared bands and the diffuse interstellar bands, while avoiding the limitations of the PAH hypothesis? Here I will suggest a scheme grounded in recent laboratory work on carbon clusters, but with roots that go back a number of years. It is well at the start to emphasize the ubiquitous presence of carbon chains in the table of known interstellar molecules. About one-third of the 106 molecules which have been identified in molecular clouds–or circumstellar shells like that of the evolved carbon star IRC+10216–can be classified as chains, including an actual majority of the polyatomic molecules. The largest interstellar chain is $HC_{11}N$, with a molecular weight of 147.

If one considers the fate of a long carbon chain in the diffuse interstellar gas, one notices an important difference between the evolution of a chain and that of a large two or three dimensional molecule of high structural stability such as a PAH or–an extreme case–the slippery closed cage C_{60}. The two and three dimensional systems are subject to rapid accretion, tending to add carbon at an unacceptably high rate, and so becoming larger and larger (Thaddeus 1994). A long chain would probably do the same, rapidly growing longer and longer, because there is little reason to think that it is less susceptible to C^+ interaction and insertion, *except that it is much more liable to photocleavage owing to its linear structure: it is only necessary to break a single bond for fission to occur.* It is paradoxically this very fragility that makes long chains interesting candidates for the large molecules in the diffuse interstellar gas, because fragmentation solves the problem of returning carbon to the C^+ reservoir and thereby avoids indefinite accretion on an unacceptably short time scale.

Let us consider in detail the very simple steps required to accomplish chain growth and the recycling of carbon. Assume for simplicity that all chains of length X_n insert C^+ at the Langevin rate–or more realistically at a Langevin rate which increases linearly with the length of the chain when that is greater than the ion-molecule impact parameter of 10-20 Å. The subsequent behavior of the resulting ionized chain X^+_{n+1} depends critically on length: short chains will dissociatively recombine like other small molecules; long chains, like solids (a graphite electrode is a familiar example) will recombine without fragmentation. The transition between these short, fragile chains,

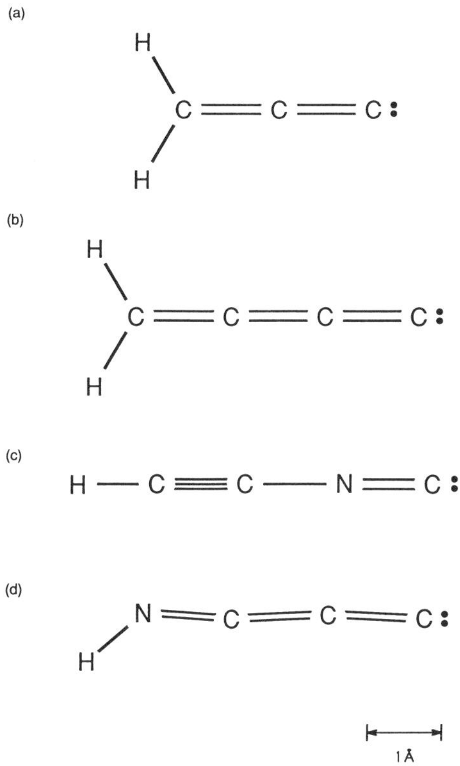

Fig. 1. Four highly polar carbon chains recently detected by radio astronomers.

which I will call "secondary," and the long robust "primary" chains which survive recombination is undoubtedly not sharp, but can be estimated from an RRKM calculation (e.g. Forst 1973) to lie in the vicinity of 12 atoms. By interstellar standards this is not a particularly large number, since it is comparable as we have seen to the number of atoms in the longest chains actually observed in molecular clouds. For both short and long chains it will be assumed that subsequent electron recombination is rapid compared to the Langevin ion-molecule reaction rate (because the recombination and Langevin cross sections are comparable and $n_e \sim n_{C+}$, but the electron velocity is large: $v_e \approx 150\, v_{C+}$).

As an idealization, let us suppose that the transition will occur precisely between n=11 and 12 and assume on the basis of the foregoing reasoning that a new link in the chain–C^+ insertion plus rapid recombination–will occur at the Langevin rate, but that growth is opposed by photocleavage in the interstellar radiation field, some of whose UV photons are able to break the strong localized sigma bonds holding the chain together. Just how and at what rate cleavage will occur is not easy to calculate and there is little relevant laboratory data. Let us therefore assume that it occurs randomly at a constant rate irrespective of the chain length or the location of the bond in the chain–fully aware that this is an oversimplification since the triple bonds in acetylenic chains are stronger than the single bonds. The photocleavage rate is the one free parameter in the model; it will eventually be fixed by fitting the observed amount of carbon depletion in the diffuse gas.

When a chain is cut into two fragments, each with 11 atoms or more, each fragment resumes the process of growth by C^+ insertion followed by non-dissociative electronic recombination, and the chain has essentially reproduced itself. If, however one of the fragments has less than 11 atoms, it continues to react with C^+ at the Langevin rate, and shatters on subsequent recombination. The fragments in turn shatter on further reaction with C^+ or UV photons until the chain dissolves and C^+ is returned to the interstellar gas (or C atoms which in any case will rapidly ionize). If both initial fragments are smaller than 11 atoms, the breakup of the original chain is complete. In the present model I have also allowed random photocleavage of small chains at a rate of $3 \times 10^{-11} \mathrm{sec}^{-1}$, a typical value for small molecules in the interstellar radiation field (Watson 1976).

The coupled differential equations which then specify the time derivative of the number density $N_n(t)$ of chains with n atoms are readily written down:

dN_n/dt = gain from collisions - loss from same
+ gain from photocleavage - loss from same

$$= n_{C^+}[N_{n-1}r(n-1) - N_n r(n)] + S[2 \sum_{j=n+1}^{\infty} N_j - (n-1)N_n], \qquad (1)$$

when $n \geq n_c = 12$, and

$$= n_{C^+}[2 \sum_{j=n+1}^{n_c-2} N_j r(j)/(j-1) - N_n r(n)] + S[2 \sum_{j=n_c}^{\infty} N_j] - S' N_n, \qquad (2)$$

when $n < n_c$, where S is the (unknown) probability per unit time of bond cleavage in the radiation field, $S' = 3 \times 10^{-11} \mathrm{sec}^{-1}$ as assumed for small chains, and provision is made for the Langevin rate r to increase with chain length n. $n_{C^+} = N_1$ is readily determined from conservation of the total number of carbon atoms.

On solving these time-dependent coupled differential equations numerically, one finds, as shown in Figure 2, that if the diffuse gas is initially seeded or nucleated with a small quantity of carbon chains of length $n = 11$ or greater, the population of chains quickly spreads in number and amplifies in total mass, reaching a steady state distribution within a few x 10^5 years. As expected, the equilibrium distribution is found to be quite independent of the amount or distribution with length of the initial nuclei, provided that some are at least 11 atoms in length. The carbon for this rapid growth is supplied by the reservoir of C^+, and when steady state equilibrium is attained, carbon is depleted by an amount which at a given total gas density depends only on the photocleavage rate S. Carbon depletion of ~50%, about that observed in low extinction clouds (Jenkins 1987), fixes S at $\sim 1 \times 10^{-13}$ sec^{-1}, which for a chain with about 20 bonds implies a total photodestruction rate of about 2×10^{-12} sec^{-1}, or about 15 times less than the value we adopted for the small secondary chains. One hesitates to state that this is a reasonable value in the absence of experimental or theoretical data, but it qualitatively agrees with the general finding that large polymers and dyes are photostable relative to small molecules.

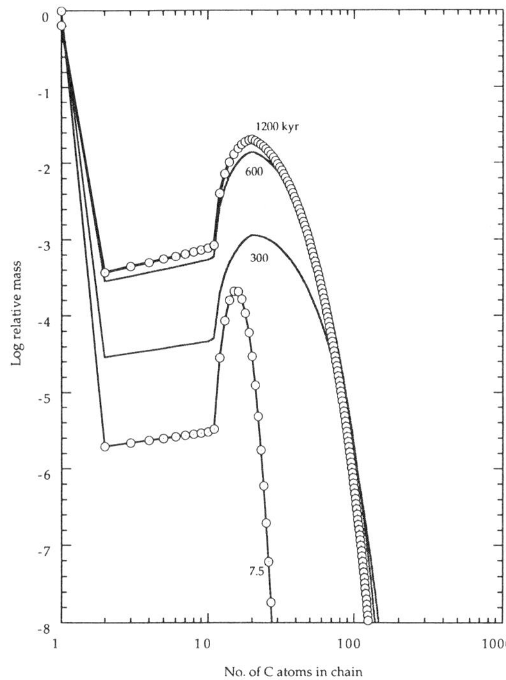

Fig. 2. Distribution with time of carbon chains produced by C^+ insertion and destroyed by photocleavage a numerical solution of the model discussed in the text, where the relative mass of each chain nN_n/N_1 is the quantity plotted. At $t = 0$ it is assumed that all carbon is in the form of C^+ (i.e., $n = 1$), except for a small amount ($N_{12} = 10^{-4}N_1$) in the form of the smallest primary chain, that with 12 atoms. Within a few x 10^5 years the initial chain population has spread, amplified, and converged on an equilibrium population of long primary chains ($n \geq 12$) able to reproduce, and short secondary chains in the process of dissolving to C^+ in the interstellar radiation field but continually replenished by photocleavage of the primary chains. Convergence to the indicated equilibrium distribution is independent of the size and density of the initial seed chains. The distributions shown here are for $A_V = 0$, for which the equilibrium carbon depletion [i.e., $1 - N_1(\text{equil})/N_1$ (t=0)] is 0.36 (see Figure 3).

Because we have neglected the fraction of the time that a given chain is positively ionized and C^+ insertion inhibited, the time required to attain a steady state will be somewhat longer than exhibited in Figure 2. It is however unlikely to be much longer. The ionization equilibrium in the diffuse gas of chains with 20-40 atoms has not to my knowledge been calculated, but it is arguably not greatly different from that of the large molecules considered by Lepp and Dalgarno (1988), who find that such molecules are singly positively charged roughly one-half of the time. The implication is that the equilibration time scale may be roughly twice as long as the few x 10^5 years calculated here, but not much more than that.

An important explicit prediction of the model is that carbon depletion is found to increase with decreasing S, in qualitative accord with the observation that carbon depletion increases in regions of high density and extinction (Jenkins 1987). The change of the depletion [i.e., $1 - N_1(\text{equil})/N_1$ (t=0)] with visual extinction for the

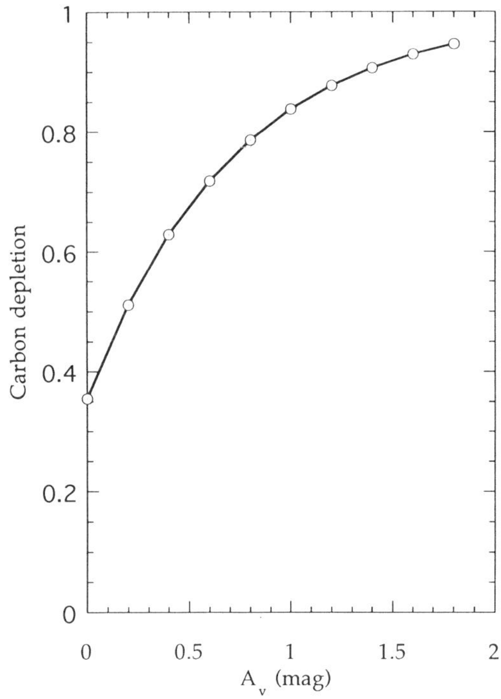

Fig. 3. Carbon depletion as a function of visual extinction A_V through the diffuse cloud where the chains are located. It is assumed that the chains are near the center of the cloud where the extinction of the interstellar radiation field is $\sim A_V/2$, and that the electronic transitions responsible for photocleavage occur in the far ultraviolet where $A_{uv} \sim 3A_V$.

specific model discussed here is shown in Figure 3. The rate of increase of carbon depletion with UV extinction is not very well determined from the available observational data, but it appears to be about what the present model predicts, i.e., that in moderately low extinction regions the depletion increases by about a factor of two when the visual extinction of the cloud increases by 1 mag.

The sharp break in slope in the distribution of small and large chains at n = 12 in Figure 2 is clearly an artifact of the model, and would undoubtedly smooth out under more realistic assumptions. Similarly, the slope in the distribution of the small secondary chains is somewhat sensitive to the assumptions made, and therefore fairly arbitrary. The hump in the distribution of large chains at $n \approx 20$ and the precipitous decline in the distribution when n is greater than about 30 is however a robust[1] feature of the model and its most interesting prediction.

For in the primary chains we have not only synthesized a stable population of large molecules with favorable spectroscopic properties to account for the diffuse bands, we have made them in about the required amount as well. The chains in Figure 2 near the peak of the distribution at $n \approx 20$ are the ones of major interest and the ones on

[1]One indication of the stability of the solutions is that when fairly large random chain-to-chain variation (20% rms) are imposed on both the Langevin and photocleavage rates the distributions remain essentially unchanged.

which to focus attention. All the reasons Douglas (1977) gave for thinking that fairly short chains like those found in molecular clouds might be the carriers of the diffuse bands apply to these longer ones, probably *a fortiori*, and the optical spectra of these longer chains are even less familiar in the laboratory, readily explaining why none of the diffuse bands has yet been identified.

The strength of a diffuse band produced by one of the primary chains is readily calculated. The equivalent width W of an electronic transition in absorption is

$$W = \pi e^2 \lambda^2 f N_{20} L / mc^2, \quad (3)$$

$$= 0.2 - 20 \text{Å}, \quad (4)$$

where in (4) we assume a range of oscillator strengths $f = 0.01$-1, $\lambda \approx 5000$Å, $N_{20} \approx 3\text{x}10^{-6}\text{cm}^{-3}$ from Figure 2, and take the initial number density of C^+ to be 3 x 10^{-3}cm^{-3} and the cloud thickness L along the line of sight to be 10 pc. According to Herbig (1975), W for the strongest diffuse bands lies in the range 0.4-4 Å, nicely within the range just calculated. There is even some intensity to spare, accommodating the dilution that might result from chain variation or embellishment: various end groups, insertion of hetero atoms such as N, etc.–and the possibility that some of the carbon is contained in other types of molecules (perhaps even PAHs).

In a theory of nucleation it is generally unnecessary in the first instance to account for the origin of the nuclei. Fragments of classical dust grains, shattered by grain-grain collisions, supernova shocks, cosmic rays, or other mechanisms, are a possible source in the present model, but just how and at what rate nuclei might be generated from the grains remains speculative. It bears emphasis that the equilibrium population of chains in the diffuse gas calculated here is entirely insensitive to such matters.

Since the primary chains according to our model are a major reservoir of interstellar carbon, it would be peculiar if their only observable manifestation were the diffuse interstellar bands in the visual and infrared. Radio detection is an extremely interesting possibility, *and indeed radio rotational lines at some intensity level are a consequence of any molecular interpretation of the diffuse bands* (unless one supposes that all the carriers are non-polar by reason of symmetry). Polar variants of the chains with n ≈ 20–e.g., chains with a terminal CN group or highly polar carbene chains like those in Figure 1–might be detectable in the cm band in absorption against a strong distant continuum source such as Cas A. Optical depths in the strongest radio lines depend on the unknown dipole moment and degree of rotational excitation, but can be estimated to be possibly as large as 10^{-4}.

To summarize, we have (i) produced a population of large carbon chains in the diffuse gas via explicit processes of formation and destruction, which (ii) exist in stable equilibrium with the reservoir of carbon, (iii) possess favorable spectroscopic properties to account for the DIBs, (iv) have about the abundances required to account for the strength of these bands, and (v) are connected explicitly to the depletion of carbon and the UV extinction–very general processes in the diffuse gas. It is further suggested (vi) that these chains may have new observational consequences, specifically, detectable radio lines. The scheme is apparently hermetic, with no connection to classical grains or large two- and three-dimensional molecules, but recent experiments on the origin of C_{60} and other fullerenes suggest that this may not be so, and that the large chains proposed here may lead naturally in the interstellar gas to the synthesis of specific two and three dimensional larger molecules.

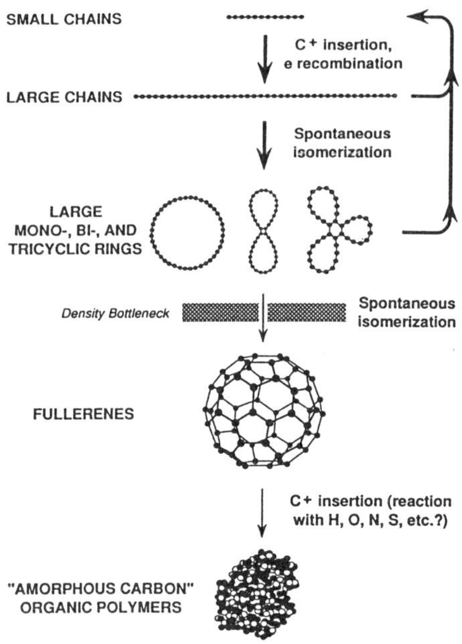

Fig. 4. Schematic illustration of the proposed growth of interstellar carbon clusters via C^+ insertion and isomerization from chains to large monocyclic, bicyclic, and tricyclic rings to fullerenes to amorphous carbon grains.

A major mystery in the fullerene story is the astonishing efficiency with which C_{60} and similar closed carbon cages can now be produced in a simple carbon arc; the yield of C_{60} under the right conditions can reach 50%. Two recent experiments (Hunter *et al.* 1993; von Helden *et al.* 1993) and previous work by some of the same investigators provide a plausible explanation. By a technique of gas phase ion chromatography it has been shown that as the number of atoms increases carbon clusters in the gas phase develop from linear chains to planar polyyne ring systems of the kind shown in Figure 4 to fullerenes. What is new and unexpected is that the transition apparently occurs spontaneously and with high efficiency once a threshold in size is crossed and enough heat is applied to carry the cluster over the activation barrier to isomerization. These remarkable processes are of astrophysical interest because the heat required to isomerize the clusters, of the order of 4 eV for the large monocyclic rings, is comparable to that sporadically supplied in the diffuse gas by the absorption of a UV photon–the very process thought to be responsible for the UIR bands, the ubiquitous 3-12 μ emission with a line-like character which has been attributed to CH and CC bonds. The size threshold to form rings is also significant: it is apparently as low as 20 atoms which means that the chains near the peak of the primary distribution in Figure 2 may be susceptible to spontaneous isomerization.

The model which I propose for the dominant large molecules in space therefore consists of chains with a distribution similar to that in Figure 2, complemented by a significant but indeterminate population of planar rings like those in Figure 4. For the rings the competition between C^+ reaction, photocleavage, and isomerization is undoubtedly complex, as rings are broken, reformed, fragmented, etc., and loops are added or subtracted, and it is not likely that the simple size distribution of chains in Figure 2 applies to the rings. There is probably a cutoff in the ring distribution at some point above n = 30, but where this occurs, and how sharply, is not at all clear at this time.

The threshold for the spontaneous isomerization of planar rings to fullerenes is according to the laboratory experiments in the vicinity of 40 atoms. Most of the carbon which follows this route is lost to the chain-ring cycle, as illustrated schematically in Figure 4, since the fullerenes and their descendants are highly stable structures not very susceptible to photodissociation–they are likely instead to accrete carbon indefinitely until much slower processes discussed in the literature–grain evaporation and grain-grain collisions for example (e.g. Draine 1990; McKee 1989)–are able to return carbon to the interstellar gas; the time scale for this cycle is of order 10^7 years or even longer. A bottleneck to the loss of carbon in the ring-chain cycles in Figure 4 is presumably the precipitous decline in the chain mass spectrum above $\sim$30 atoms, but how effectively this bottleneck operates is difficult to calculate.

Notice that we have proceeded from one to two to three dimensional carbon–i.e., from small to long chains to large planar rings to fullerenes to large amorphous carbon clusters–simply with ion-molecule chemistry and photodissociation, reactions which are universal in the diffuse interstellar gas, plus an appeal to spontaneous isomerization suggested by recent laboratory experiments–without encountering PAHs. Actually, the present scheme *excludes* PAHs, because isomerization of large planar rings to fullerenes avoids graphitic plates owing to their many dangling bonds. Subsequent synthesis of PAHs by hydrogenation is choked off at the source.

There are a number of ways in which hetero atoms might be incorporated into the present model, and for several reasons it is very desirable that this be attempted. Nitrogen atoms for example may be incorporated into the cycle from the small chains on; it is perhaps worth noting in this context that two of the carbon chains recently identified by radio astronomers in molecular clouds contain interior nitrogen atoms (Figure 1). Reaction of many different hetero atoms with the fullerenes and amorphous grains that follow isomerization is another possibility, but one very difficult to work out in much detail. At any rate, in elaborating the present model further, there is no reason whatever to confine attention solely to the chemistry of carbon, particularly in view of the evidence for the depletion of many of the hetero atoms in the interstellar gas. Such depletion is one pressing reason why as a working proposition it should be assumed that the large molecules in the interstellar gas draw on the full architecture of organic chemistry, the foundation of the most intricate structures in nature.

Finally, a comment on the crucial question of laboratory confirmation of the unidentified interstellar features. Any molecular interpretation of the diffuse bands must be regarded simply as a working hypothesis towards *exact* laboratory identification *in the gas phase*. Until that is accomplished, all explanations are entirely provisional. Although the PAH hypothesis has so far failed to explain the diffuse bands, it does provide a fairly plausible explanation for the UIR bands–specifically, emission spectra calculated from laboratory absorption data which are "not dissimilar" to the astronomical bands. Can the present chain or chain-ring scheme give as good a description of the UIR bands–or a better one?

That is presently a difficult question to answer, because of the total lack of infrared spectroscopy on long chains and large planar rings. But a partial answer may be possible. C_{60} was first identified spectroscopically by Kratschmer *et al.* (1990) on the basis of group theoretical analyses of its infrared active vibrational transitions and theoretical calculations of their frequencies (Wu *et al.* 1987; Cyvin *et al.* 1988;

Stanton and Newton 1988; Weeks and Harter 1989), and it might be possible by similar methods to decide whether a reasonable population of large chains and planar rings has a vibrational spectrum which is sufficiently "not dissimilar" to the UIR bands to be viable, or on the contrary is so dissimilar as to be ruled out. Also, laboratory investigations of carbon clusters are now being pursued so vigorously, and modern laser and mass spectrometric techniques are becoming so sensitive, that relevant laboratory spectra in the next ten years are not beyond hope.

An earlier, somewhat longer version of this paper was given in Sept. 1993 at the Mont Sainte-Odile conference *Molecules and Grains in Space*, and appears in the conference proceedings (Thaddeus 1994).

REFERENCES

Cyvin, S.J., Brendsdal, E., Cyvin, B.N., and Brunvoll, J. 1988, *Chem. Phys. Lett.*, **143**, 377.
Douglas, A.E. 1977, *Nature*, **269**, 130.
Draine, B.T. 1990, in *The Evolution of the Interstellar Medium*, Astron. Soc. Pacific Conf. Series, **12**, 193.
Forst, W. 1973, *Theory of Unimolecular Reactions*, (New York: Academic Press).
Herbig, G.H. 1975, *Ap. J.*, **196**, 129.
Hunter, J., Fye, J., and Jarrold, M.F. 1993, *Science*, **260**, 784.
Jenkins, E.B. 1987, in *Interstellar Processes*, eds. D.J. Hollenback and H.A. Thronson, Jr. (Dordrecht: Reidel), p. 533.
Kratschmer, W., Fostiropoulos, K., and Huffman, D.R. 1990, in *Dusty Objects in the Universe*, eds. E. Bussoletti and A.A. Vittone (Dordrecht: Kluwer), p. 89.
Lepp, S. and Dalgarno, A. 1988, *Ap. J.*, **335**, 769.
McKee, C.F. 1989, in *Interstellar Dust*, eds. L.J. Allamandola and A.G.G.M. Tielens (Dordrecht: Kluwer), p. 431.
Stanton, R.E. and Newton, M.D. 1988, *J. Phys. Chem.*, **92**, 2141.
Thaddeus, P. 1994, in *Molecules and Grains in Space*, ed. I. Nenner (Amer. Inst. of Phys. Conference Proceedings 312, New York: AIP Press, 1994), p. 711.
von Helden, G., Gotts, N.G., and Bowers, M.T. 1993, *Nature*, **363**, 60.
Watson, W.D. 1976, *Rev. Mod. Phys.*, **48**, 513.
Weeks, D.E. and Harter, W.G. 1989, *J. Chem. Phys.*, **90**, 4744.
Wu, Z.C., Jelski, D.A., and George, T.F. 1987, *Chem. Phys. Lett.*, **137**, 291.

A CRITIQUE OF SUGGESTED DIFFUSE BAND CARRIERS

THEODORE P. SNOW
Center for Astrophysics and Space Astronomy
University of Colorado
Boulder, CO

ABSTRACT.

We summarize the discussions that took place on the last afternoon of the conference, on the various diffuse band carriers that were advocated by speakers during the week. This discussion was presented in the form of a "shoot-out", in which the speakers were challenged to compare their models with a list of observational criteria that had been presented earlier in the conference. The summary given here is an attempt to reproduce the presentations that were made during the shoot-out, complete with comments on how well the proposed carriers did or did not satisfy the observational criteria. It is hoped that this summary will serve dual purposes: (1) to record the comments and ideas that were presented at the conference, since these remarks were not submitted in writing in any other format; and (2) to provide stimulation, if not inspiration, to future authors of diffuse band carrier models, who may want to see where previous suggestions have either succeeded or failed.

1. INTRODUCTION

During the final afternoon of the conference, a general discussion was held, with the principal goal of comparing the various proposed models for the diffuse band carriers. As a means of structuring the discussion, each model was compared with a list of observational criteria that had been presented earlier in the day (see Snow, this volume). Here are the major factors in this list of criteria:

(1) *Does the model produce invariant DIB wavelengths and profiles?* The diffuse bands are known to be quite invariant in both wavelength and intrinsic profile shape, which would rule out models in which variations in either quantity are expected.

(2) *Can the proposed carrier form and survive under interstellar conditions?* The carriers of the diffuse bands must be hardy enough to withstand the interstellar ultraviolet radiation field, for example.

(3) *Is the composition of the proposed carrier consistent with "cosmic" abundances?* The diffuse band carriers must form from material that is not seen in the atomic gas, so the required abundance of the carriers must be consistent with the elemental abundances and depletions in the observed gas.

A. G. G. M. Tielens and T. P. Snow (eds.), The Diffuse Interstellar Bands, 379–393.

(4) *Does the proposed model explain the lack of polarization within the DIBs, and the lack of emission wings?* Neither polarization nor emission wings of the diffuse bands has been detected, so the proposed carrier must not require either to be present.

(5) *Is the spectrum of the proposed carrier spectroscopically consistent with the observed DIBs?* The proposed diffuse band carrier must not produce features in the laboratory that are not seen in the interstellar medium, and it must not have consequences for other spectral regions (e.g. the infrared or ultraviolet) that are inconsistent with observations. If a single carrier is proposed to explain more than one diffuse band, then the expected band strength ratios must be consistent with observations.

(6) *Is the proposed model consistent with observed environmental dependencies of the DIBs?* Examples would include the known weakness of the diffuse bands in dark clouds; the fact that the ratios of diffuse band strengths vary from region to region; the absence of diffuse bands in circumstellar environments; and the fact that the diffuse bands correlate better with neutral atomic gas than with molecular abundances.

With these criteria in mind, we proceeded to discuss several proposed diffuse band models. In the following sections, these models and the comments made about them are presented in the same order as they were discussed at the conference. The comments attributed to various individuals are not quoted directly, but they are reproduced here on the basis of tape recordings made at the time, so it is hoped that the meanings are reproduced accurately.

2. CARBON CHAINS AS DIFFUSE BAND CARRIERS

This hypothesis, presented jointly by John Maier and Pat Thaddeus, invokes both laboratory data and plausibility arguments to suggest that the diffuse bands are formed by carbon chain molecules. These are linear molecules of carbon atoms, possibly with end groups of other species. One class that has been studied in the lab by Maier have the form C_nH_m, where m is generally very small. The laboratory data, based on molecules isolated in matrices, show several coincidences in wavelength with observed diffuse bands (taking into account reasonable matrix shifts; see references cited by Maier, this conference). The plausibility argument proposed by Thaddeus is that we know carbon chains are abundant in the interstellar medium (by far the majority of large molecules observed are carbon chains), and in addition laboratory data and theoretical considerations suggest that as carbonaceous molecules form, grow (through insertion of C^+ ions), and fragment via photodissociation, an equilibrium will be established in which there is a stable population of small to moderate-sized carbon chains.

Does the model produce invariant wavelengths and profiles? There was general agreement that this model easily meets this criterion; gas-phase molecular transitions do in general.

Is the proposed carrier able to form and survive under interstellar conditions? Alain Léger commented that carbon chains may be very easily destroyed by photodissociation, but this is not known until appropriate laboratory measurements can be made. Here, Sydney Leach commented that photodissociation depends on both the size and the shape of the molecule, and that to date only the size has been considered explicitly in models (of PAHs specifically; it was not clear that any models have been done for chains). For PAHs, within certain groups there is a simple dependence of photodissociation probability on size, but this is not the case when PAHs with different structures are considered. Hence shape must also play an important role. Leach commented that this should be true of chains as well. Lou Allamandola then raised what appeared to be a possibly serious problem: Maier had stated earlier that the particular carbon chains his group was working with in the lab photodissociates under UV radiation at wavelengths below 2500Å. Maier agreed that this appeared to be so.

Allamandola then raised another point: lab data show that as carbon chains grow, they begin to form cyclic species (i.e. PAHs) once the carbon number becomes as large as 12 to 15. In response to this, Thaddeus said that, on the contrary, when C_{60} is made in the lab, graphite fragments, having many unsatisfied valencies around their edges, bend in on themselves and form fullerenes rather than PAHs. To this Allamandola replied that when hydrogen is present, this "poisons" the formation of fullerenes, and leads to PAHs instead; thus in a hydrogen-rich region you should get PAHs instead of C_{60}. Herbst commented that if the molecules are ionized, it is very hard to add hydrogen to them; typically only one hydrogen atom can be added to carbon clusters having up to 9 carbon atoms. There appeared to be no clear consensus on this very important point.

Finally, on the question of formation and survival of carbon chains in the diffuse interstellar medium, Feldman noted that radio observations of "cloudlets" having A_V of about unity, seen in absorption against Cas A, show large abundances of small carbon chains such as C_4H, C_3N, HC_5N, and HC_7N. Since the physical conditions in these clouds are similar to those in which the diffuse bands appear, it seems entirely plausible that there are significant quantities of carbon chains present in diffuse band-forming regions. Furthermore, since the absorption cross sections of the small carbon chains can be very large, column densities as small as 10^{11} or 10^{12} cm^{-2} can be sufficient to produce even the strongest of the diffuse bands.

The general consensus was that carbon chains may very well be viable in terms of formation processes and ability to survive interstellar conditions, but that some important data, particular on photodissociation cross sections, are missing.

Is the model consistent with "cosmic" abundances? On this question Thaddeus asserted that there is no problem forming sufficient quantities of carbon chains with the amount of carbon that is observed to be depleted from the atomic gas; furthermore, he pointed out that the numbers needed for PAHs are roughly the same. There was no dissent from the audience.

Does the model explain the lack of polarization or emission wings? Again there was no dissent; clearly gas-phase molecular transitions are consistent with this constraint.

Is the model spectroscopically consistent with the observed diffuse bands? Thaddeus pointed out that spectroscopic identification of some diffuse band carriers might already have been achieved, based on Maier's statistical analysis showing that the number of wavelength coincidences between laboratory and astronomical measurements was greater than would be expected from random chance (see Maier, this volume). There was little direct comment on this assertion.

Later in the afternoon, Peter Jenniskens, in a sweeping criticism of *all* proposed carriers, commented that Maier's specific carbon chain model predicts diffuse bands that are not seen observationally. There was no particular reaction to this comment; it appears that few have examined the observational data relevant to this question as yet, but it clearly is an important question.

Bill Smith made two points related to spectroscopic consistency: (1) it is important to keep in mind that there are no strong diffuse bands below 4430Å; and (2) any kind of molecule must have very strong Rydberg transitions *somewhere* in the UV, and since no such transitions have been observed, this suggests that whatever forms the diffuse bands must have these transitions below 912Å. This may be regarded as an ultraviolet consequence that must be kept in mind for any molecular model.

On the question of consequences for wavelength regions other than the visible, John Mathis made the comment that if the 2175Å extinction bump is to be attributed to the same molecular species that forms the diffuse bands, then it is difficult to see how the UV transition could be broadened as much as observed without also shifting the central wavelength. Thaddeus concurred that this was an important consideration, since most of the candidate carbon chains should have strong UV absorption in this region. Léger stressed that there is also an important infrared consequence that should be tested: the carbon chain molecules, like the PAHs, should have strong bending and stretching modes resulting in IR transitions at specific wavelengths. Even for the same transition (e.g. C-H stretch), the precise wavelength will depend on the form of the molecule, as is seen for this transition in aromatics (where it occurs at 3.3 μm) and in aliphatics (where it occurs at 3.4 μm). Thus infrared spectroscopy, in principle, may allow distinctions to be made among various kinds of carbonaceous molecules that are candidates as the diffuse band carriers.

Is the model consistent with observed environmental dependencies of the diffuse bands? This question was not fully addressed, and must be considered an unknown at present.

Summary: No fatal flaws were found in the carbon chain hypothesis, but important questions remain. It was suggested that the most important area for future work to test this hypothesis is to carry out laboratory spectroscopy of candidate molecules *in the gas phase* (Maier particularly stressed this, saying that by now enough candidate species have been isolated on the basis of matrix experiments that it is time to move on to testing them definitively by making gas-phase measurements). Other important needs are to measure the photodissociation cross sections for candidate molecules, and to carry out infrared spectroscopy.

3. FULLERANES AS DIFFUSE BAND CARRIERS

This model has been proposed by Adrian Webster, who presented it at the conference (see Webster, this volume). The basic premise is that buckminsterfullerenes (carbon molecules of the form C_{50}, C_{60}, C_{70}, etc.) can acquire hydrogen atoms bonded to their outer structure (converting them to *fulleranes* as opposed to fullerenes), and that in their hydrogenated form these molecules should have rich optical spectra that could account for the observed diffuse interstellar bands. The spatial variations of the observed relative strengths that are observed are attributable to varying degrees of hydrogenation in different astrophysical environments. There is an enormous number of possible molecules, because there are very many potential sites on the surface of a fullerene where hydrogen atoms can attach. To date there are few laboratory data on these species.

Does the model produce invariant band wavelengths and profiles? Webster asserted that indeed it does; the fullerane model shares this property with other gas-phase molecular carriers.

Can the proposed carrier form and survive under interstellar conditions? Webster's response to this question was that it cannot form readily, but it certainly can survive. Quite high carbon densities, at high temperature, are required to make fullerenes in the laboratory, and these conditions would not exist in the diffuse interstellar medium. So certainly if fulleranes are the diffuse band carriers, they may form in circumstellar environments, such as carbon star envelopes or planetary nebulae. As evidence favoring this idea, Webster noted that the infrared emission features, which in his model are produced by fulleranes, are observed in these kinds of circumstellar environments.

Maier and Ted Snow raised the issue of overspecificity; i.e. there are very many possible hydrogenated fullerenes, so how do just a few get selected? Webster responded that in reality there are probably mechanisms that tend to form just a few species which are unusually stable. Shapiro asked whether this is borne out in the laboratory production of fulleranes: do you tend to get just one kind, or do you get a wide spectrum of species? Leach commented in response that in some experimental procedures you get just one fullerane at a time, but that other methods give you a variety.

Leach then addressed a related question to Thaddeus, asking whether the scheme Thaddeus had proposed (in which fullerenes build up from carbon chains that close in on themselves) works for neutral species? The laboratory experiments showing that this process occurs are all based on ions. Thaddeus replied that this is probably an artifact of the fact that the lab experiments are all done in drift tubes (which work only with ions), and that the process should work as well for neutrals. Leach expressed skepticism, or at least uncertainty, about this.

Thaddeus raised another question about the ability of fulleranes to survive in the interstellar medium, by asking why they would not react very rapidly with C^+ ions, which would rapidly remove the fulleranes. Webster replied that if this is the case, then all the C_{60} would be destroyed, and he would have to find something else to do. (Whereupon Thaddeus congratulated Webster for his "remarkable candor".)

Is the model consistent with "cosmic" abundances? Here, Webster confronts a problem, perhaps a very serious one. He reported that, given the low f-values for the broad transitions in fulleranes, quite high abundances are needed; about one-fourth of the total interstellar carbon budget must be in the form of fulleranes in order to explain the broad diffuse bands, and this amounts to virtually all of the carbon known to be depleted from the atomic gas. For the narrow diffuse bands in the yellow-red portion of the spectrum, however, the f-values are expected to be higher, and the abundance problem less severe. Léger asked about the f-values, and Webster clarified that while there are very few lab data, it appears that for the broad transitions, attributed to allyl radicals, the f-values are of order 10^{-3}, while for the narrow bands they may be as high as 10^{-1}. Léger commented that the f-values for ionized fulleranes should be higher than for the neutral molecules.

Does the model explain the lack of polarization or emission wings? Webster declined to discuss this specifically, commenting only that he considers this a minor question compared to the basic issue of whether fulleranes can produce diffuse bands at all. But it would seem that, as in the case of other gas-phase molecular carriers, neither polarization nor emission wings would be expected.

Is the model spectroscopically consistent with the observed diffuse bands? Webster responded by saying that in the absence of laboratory data, this is unknown. But, he did offer the consistency argument that in general the model is compatible with the general lack of diffuse bands below 4430Å, because the narrow bands known to exist in the spectra of non-hydrogenated fullerenes lie in the far red, and the addition of a few hydrogen atoms would only perturb the basic cnergy levels a little. So, this is consistent with the observation that the narrow diffuse bands tend to lie in the yellow-red portion of the spectrum. Similarly, the broad transitions due to allyl radicals lie above 4000Å, and the distortions introduced by fitting into the framework of C_{60} might shift the bands toward longer wavelengths, so this would also be consistent with observations.

Bernard Foing stated that in his view the fullerane model got a significant boost by the discovery (described elsewhere in these Proceedings) that two near-infrared diffuse bands have been found that appear to match spectral features expected from C_{60}^+. He also pointed out that in addition to the hydrogenated fullerenes proposed by Webster, there could be significant contributions from other fullerene compounds; i.e. elements other than hydrogen can bond to the expose valencies of a fullerene molecule. In his later comments addressed to all proposed models, Jenniskens raised the question of whether the two infrared bands of of C_{60}^+ that have been seen in the laboratory might have been due to the matrix in which the ions were embedded, rather than to the ions themselves. If so, then this would cast doubt on the identification of observed interstellar bands with C_{60}^+.

Snow asked about possible implications for ultraviolet and infrared spectroscopy. Webster replied that he views this as a strong point of his model: fulleranes could be the carriers of both the 2200Å bump (which he views as essentially a diffuse band) and the observed infrared emission bands, in addition to the optical diffuse bands. Detailed laboratory data to support these assertions do not exist, however.

On the question of consistency with infrared observations, Tom Wdowiak noted that laboratory data on fulleranes show four strong infrared bands, and asked why these have not been seen in the interstellar medium. He elaborated by saying that if *ISO* does not detect the four fullerane bands, then interstellar fulleranes can be relegated to the status of astrophysical silly putty: a remarkable substance with no practical value.

Is the model consistent with observed environmental dependencies of the diffuse bands? Webster had a ready answer here: the model expects the diffuse bands to weaken in regions, such as dense clouds where hydrogenation would increase. As hydrogen atoms are attached, the molecule becomes saturated, leaving fewer and fewer electronic bonds available. This can explain the observed weakness of the diffuse bands in dark clouds, and also the weakening of the 2175Å bump and the far-UV extinction rise.

Summary: As Webster himself put it, this is a model that is just in its infancy, and has a long way to go before it can be tested against the various criteria imposed on it in this discussion. It does appear that the model is in trouble on at least one account: it seems to require a greater share of the total carbon budget than can easily be allowed. It certainly appears that laboratory studies are justified, so that the speculation about whether hydrogenated fullerenes produce diffuse band-like features can be laid to rest. Leach commented that of the four hydrogenated fullerenes that have been observed in the laboratory (H_2C_{60}, H_4C_{60}, $H_{18}C_{60}$, and $H_{36}C_{60}$), the larger ones are relatively unstable in solution, and would likely be so in gas phase as well. This would make laboratory spectroscopy very difficult.

4. GRAIN IMPURITY ABSORBERS AS THE DIFFUSE BAND CARRIERS

This model was represented at the conference by Paul Shapiro, who had done considerable work on it some years ago. The model invokes impurity absorbers embedded in solid grains, and is generic in the sense that no specific absorber has been proposed. The model consists of a consideration of the effects of the host grain on the absorption by the embedded impurity, with an eye toward such questions as what the concentration of the absorbers is and how the size of the grain may affect the absorption profile.

Does the model produce invariant wavelengths and profiles? Shapiro addressed this, by first noting that there are many ways to cause the profiles and wavelengths to vary in this model. But calculations show that there is a "small concentration limit", a threshold in the concentration of absorbers, below which no further variation occurs. For other reasons, the model favors low concentrations of absorbers anyway, so the hypothesis may be viable so long as this is the case. Shapiro noted also that variations in the grain size would also produce variations in the diffuse bands, but that if the grain size distribution is truncated at 0.1 μm, then there are no significant variations. So, you need to constrain the model to small concentrations of absorbers within small grains, and you can satisfy the invariance condition.

But, as Shapiro noted, if you have a region where there are known variations in grain size, you ought to see variations in band profiles. In that connection he asked for comments from Greg Seab and Ted Snow, who were known to have been studying this question in the ρ Oph cloud, where there is evidence of grain size variations. Seab responded by showing a viewgraph of the results, which indicate that the diffuse band profiles within the ρ Oph cloud are perfectly matched over a range in λ_{max} for polarization from 5000 to 8000Å. Snow commented, with concurrence from Shapiro, that this means that the grains responsible for visible-wavelength extinction and polarization in the ρ Oph cloud are certainly not the carriers of the diffuse bands.

On the whole, the question of whether the grain impurity model is consistent with the observed invariance of the diffuse band wavelengths and profiles is close to being very troublesome for this model, because there are so many ways in which variations could be produced. But, at this point the model cannot absolutely be ruled out, because as Shapiro showed, it is still possible to construct conditions that satisfy the invariance criterion.

Can the proposed carrier form and survive under interstellar conditions? This was not discussed explicitly, but it would appear that the model can satisfy this condition. It is certainly expected that dust grains can form and survive, and there is no obvious reason to think that impurity absorbers within the grains could not do so as well.

Is the model consistent with cosmic abundances? Shapiro pointed out that there are two levels of concern here: the number of absorbing particles required cannot exceed what is available; and the amount of material required to be in the host grains also cannot exceed what is available. On both counts, according to Shapiro, the model appears to be OK. In the low-concentration limit required in order to avoid unacceptable wavelength or profile variations, it is not difficult to envision an impurity absorber that would have sufficient oscillator strength to produce the observed diffuse bands, while having a low-enough column density to fit within the constraints imposed by observations of abundances and depletions. Similarly, the quantity of material in the small grains that is allowed by the constraints on size variations can also be accounted for by abundances that are well within the constraints. Thus it appears that there is no problem for this model on the basis of cosmic abundances.

Does the model explain the lack of polarization or emission wings? Here, this model encounters a formidable challenge. Shapiro pointed out that polarization within the profiles of the diffuse bands should occur in this model, and that it would appear as a small percentage of the overall continuous polarization produced by the same grains. Current limits on polarization within the bands are sufficient to demonstrate that the diffuse bands cannot be formed by the same grains that create the observed continuous polarization. But the limits do not rule out the possibility that the diffuse bands could be produced within grains that have little or no contribution to the observed visual polarization. This is another way of saying that if the diffuse bands are produced by impurity absorbers in grains, the grains in question must be distinct from those that produce the observed visual extinction and polarization.

In this context, it is worth noting that the grains that produce far-ultraviolet extinction, which are presumed to be small grains and quite distinct from those responsible for visual extinction and polarization, do not produce ultraviolet polarization (as announced in recent papers on results from the Wisconsin polarimetry experiment on the *ASTRO* mission). This implies that the small grains that are responsible for far-ultraviolet extinction would not produce polarization structure in the diffuse bands if they were the carriers, and thus they are not ruled out as possible carriers of the diffuse interstellar bands.

On the question of emission wings, Shapiro explained that under certain conditions, namely the low-concentration and small grain-size assumptions already imposed by other constraints (see above), this model produces diffuse band profiles that do not have emission wings. This has been explored primarily for relatively narrow diffuse bands, however, and needs to be checked for the broader features. But at the moment it remains possible to reproduce the observed band profiles within the constraints of the model.

Is the model spectroscopically consistent with the observed diffuse bands? Here, Shapiro introduced a potential test of the model, based on the possible existence of "families" of diffuse bands. He pointed out that if there are multiple diffuse bands produced by the same carrier; i.e. by the same impurity absorber within the dust grains, there is a very straightforward prediction of how the band widths should vary with band strength. For a given absorber, the band widths must be directly proportional to the oscillator strengths of the transitions. Thus, if there is a group of diffuse bands which are shown from observations (presumably correlation studies) to arise from the same carrier, in this model they must have widths proportional to their strengths. This means that the question of diffuse band families, first raised by the work of Krelowski and Walker in 1987, may be very important in placing constraints on the impurity absorber model.

Another issue raised on the question of spectroscopic compatibility with the observed diffuse bands came from Léger, who asked whether the model can explain the pure emission profiles of the bands seen in the Red Rectangle. Shapiro replied that until this conference he had been unaware of the Red Rectangle results, but that it was possible in principle to produce emission instead of absorption in the grain impurity model. The shifts in emission wavelength with distance from the exciting star would require a gradient in the concentration of impurity carriers in the grains; Shapiro said that he has not computed any models to see how much variation in concentration would be needed.

There was no discussion of the possible infrared or ultraviolet consequences of this model. It seems clear that this is a secondary question, which depends entirely on what the impurity absorbers are. It is noteworthy, though, that the constraints on polarization structure and on the invariance of the band profiles in this model have forced it to rely on a population of small grains. These may be the same grains that are responsible for far-ultraviolet extinction, so it may be anticipated that there would be significant implications for ultraviolet observations from this model. For example, one might expect a very tight correlation between diffuse band strength and far-UV extinction, and no such correlations have been found, despite searches for them.

Is the model consistent with observed environmental dependencies of the diffuse bands? Again there was little explicit discussion. Such questions as the weakening of diffuse bands in dark clouds and in circumstellar environments did not come up. The only issue that was raised, the question of whether the diffuse band profiles vary with grain size in regions where such size variations are known to exist, was addressed above. Because so little is known about how the small grains may be affected by varying environmental conditions, this is a very open question at this point.

Summary: It appeared that the general mood of the conference participants was unfavorable toward the grain impurity model, based largely on the lack of polarization structure and the invariance of the observed diffuse band wavelengths and profiles. On the other hand, Shapiro was able to show that there remain some possible scenarios under which this model may be viable, and that it should not be ignored.

On the question of what tests ought to be made to pursue this model, there was general agreement that collecting accurate profiles of many more diffuse bands would be important, because we have seen that there are serious constraints to be derived from the behavior of the profiles; and that additional searches for profile variations, particularly in regions where environmental factors are known to vary, would be fruitful.

5. IONIZED PAHS AS THE DIFFUSE BAND CARRIERS

This model has been discussed by several authors, dating right back to the genesis of the PAH hypothesis in the mid-1980s. One of the strongest proponents has been Alain Léger, who presented this model at the conference (see Léger's article, elsewhere in these Proceedings). The essence of this suggestion is that, while neutral PAHs have rich ultraviolet and infrared spectra but no strong transitions in the visible, in contrast ionized PAHs should have complex optical spectra. Furthermore, PAH cations could be responsible for some or all of the infrared emission; ionization has very little effect on the bending and stretching modes that are presumed to explain the infrared emission features that led originally to the entire PAH hypothesis. Specific suggestions for possible PAH^+ diffuse band carriers have come from laboratory studies of the spectra of these features, although all measurements so far have been based on PAH ions embedded in solid-state matrices, and there remain uncertainties due to poorly-known matrix shifts.

Does the model produce invariant wavelengths and profiles? Léger pointed out that this is not a problem for the PAH^+ hypothesis, just as it is not for any other gas-phase molecular carrier that may be proposed.

Can the proposed carrier form and survive under interstellar conditions? Here Léger commented that one of the great strengths of the PAH hypothesis in general is the stability of these aromatic molecules against photodissociation; hence survival under interstellar conditions is not a problem.

On the question of formation, Léger noted that at present there is no complete model for producing PAHs in the interstellar medium. We can assume that there is a viable mechanism, since PAHs are widely believed to explain the infrared emission features, but we do not know what it is.

Is the model consistent with cosmic abundances? Here again, Léger argued that there is no problem: the abundance of PAHs needed to explain the diffuse bands is only about 1 percent of the quantity needed to explain the infrared emission features, if they are due to PAHs, because of the high f-values (around 0.1 typically) of the visible-wavelength transitions of ionized PAHs.

Does the model explain the lack of polarization or emission wings? Once again, this is not a problem, simply because gas-phase molecular transitions are not expected to produce either polarization structure or emission wings. This advantage is shared by all of the gas-phase molecular models for the diffuse bands.

Is the model spectroscopically consistent with the observed diffuse bands? Léger chose to address some of the detailed concerns. For example, does the model predict diffuse bands that are not seen? As he pointed out, this is unknown unless we have specific species as the proposed carriers of specific diffuse bands. Once we have particular PAH cations in mind, then we can see whether laboratory spectra would predict additional diffuse bands that are not observed. So the question of whether laboratory spectroscopy leads to predicted diffuse bands that are not seen is a secondary issue which must follow the primary step of identifying specific species as candidates.

In Jenniskens' general discourse near the end of the afternoon, he criticized one specific PAH^+ model; that of Léger *et al.*, which proposes methyl pyrene as the carrier of the 4430 and 7562Å diffuse bands. The laboratory measurements lead to an expectation that the strength ratio of these two features should be about 15 percent, but the observed ratio is closer to 4 percent.

On the question of ultraviolet or infrared consequences of the model, Léger made a strong statement that this is another strength of the PAH^+ model. He claimed that virtually all aromatics have strong absorption features near 2200Å, and similarly have steep far-ultraviolet rises in continuous extinction. Thus the PAH model is consistent with ultraviolet data. Regarding the infrared, of course this is the model that was originally inspired by its apparent ability to explain the observed infrared emission features, so indeed the PAH^+ model for the diffuse bands is consistent with current information on infrared observations.

Is the model consistent with observed environmental dependencies of the diffuse bands? Léger commented that the weakening of diffuse bands in dark clouds is easily explained by the PAH^+ model, since it would be expected that the ions would diminish with increasing density, as recombinations to form neutral PAHs would occur.

Léger then made a very significant comment about the apparent weakness of the diffuse bands in circumstellar regions: if the diffuse bands are weak in circumstellar environments where strong infrared features are known to arise and where the PAHs are expected to be ionized, this would represent a severe problem for the PAH^+ model. At this point Snow mentioned the case of HR 4049, a hot protoplanetary nebula central star having strong PAH infrared emission, but having no diffuse band absorption.

(Since the conference, Snow and colleagues have identified at least two additional regions where any PAHs that are present ought to be positively ionized and where strong infrared PAH emission is observed but the diffuse bands are weak; this would seem to add further weight to the problem that the PAH^+ model may not be consistent with the observed behavior of the diffuse bands in circumstellar environments.)

Summary: The PAH^+ model for the diffuse bands appears to be viable in many respects, but important questions remain. Of these, two seem outstanding: specific species need to be proposed, so that they can be tested in detail; and the question of how the behavior of the diffuse bands in circumstellar environments fits with the expectations of the PAH^+ hypothesis appears possibly to constitute a significant criticism of the model.

Léger commented that in his opinion the most important work that needs to be done to test this model is to conduct laboratory measurements, initially in solid-state matrices, of PAH ions, in order to select specific candidates for specific diffuse bands. He added that ultimately as many as 200 different species may need to be isolated, since there are at least 200 known diffuse bands. In his opinion, gas-phase measurements should follow; the first priority should be on finding specific candidates that appear viable, and this can be done in matrix.

6. PORPHYRINS AS CARRIERS OF THE DIFFUSE BANDS

The final model discussed during the "shoot-out" was the one proposed over 20 years ago by Fred Johnson, who presented it at this conference (see the article of Johnson in these Proceedings). Unlike the other models that have been proposed, this one invokes a very specific carrier that (in two separate forms) is claimed to account for *all* of the diffuse bands. The molecule in question is the tetrabenzoporphyrin molecule, which occurs in two forms (MgTBP and H_2TBP). The molecules are thought to be embedded within grains consisting of paraffin matrices; the nature of these matrices (called Shpol'skii matrices) allows both narrow and broad diffuse bands to be produced. Johnson claims a total of 187 coincidences in wavelength between laboratory measurements and observed diffuse bands.

Does the model produce invariant wavelengths and profiles? As in the case of other molecular proposals for the diffuse bands, this one is claimed by its author to produce invariant wavelengths and profiles. But in this case it is for a different reason: instead of having a free gas-phase molecule with no perturbations to alter its energy levels, here it is claimed that there are matrix effects, but that they are perfectly uniform. The MgTBP and H_2TBP molecules are embedded in a very specific, very consistent way in paraffin matrices, so that all you see in the interstellar medium are transitions that are matrix-shifted in exactly the same manner. Thus the diffuse band wavelengths and profiles do not vary from place to place.

Snow asked about the 2175Å extinction bump, pointing out that in his previous talk Johnson had stated that the bump wavelength should vary with the concentration of absorbers in the grains; the question now was whether the diffuse bands should also vary in wavelength. Johnson explained that the 2175Å bump is due to pyridine and therefore depends on the pyridine concentration within the grain, and that this could indirectly affect the diffuse band spectrum. He argued that in at least one star (HD

29647) with a very weak 2175Å bump, the diffuse bands spectrum is also unusual, supporting this view.

Can the proposed carrier form and survive under interstellar conditions? Johnson did not describe a proposed formation mechanism, but pointed out that the molecules are very stable, and can survive very easily. This is due to the protection offered by the host grain matrix; if the grain is destroyed, freeing the molecules, then they may not survive as readily (this was noted by Johnson in answer to a question from Wdowiak, who wondered why we do not see a completely different set of diffuse bands due to free MgTBP molecules).

Snow commented that formation is a key issue for this model, because the mechanism would have to be very specific in order to avoid creating a large population of other related species, each of which would have its own complex spectrum.

Is the model consistent with cosmic abundances? Johnson explained that this model requires column densities toward reddened stars such as HD 183143 that are comparable to those already known for other species (though much smaller ones). Thus, according to Johnson, there is no problem with abundances.

Does the model explain the lack of polarization or emission wings? This was not discussed; Johnson stated that there was no problem here, as indicated by laboratory measurements. But perhaps these issues require further thinking, since this is essentially a model invoking an impurity absorber in a grain.

Is the model spectroscopically consistent with the observed diffuse bands? Here Johnson made his strongest claims for the validity of his proposed carriers, while at the same time the audience made its strongest claims for the invalidity of the model. Johnson said that he now has a total of 187 wavelength coincidences, within 1Å of observed diffuse bands, between measured laboratory transitions and interstellar features. He also claimed that in certain lines of sight, where the absence of pyridine in the matrix alters the wavelengths, he has coincidences with a different class of diffuse bands (he cited the line of sight toward HD 29647, a well-studied star in the Taurus molecular cloud complex, for which observational data were supplied by Seab and Snow). He stated also that he had "discovered" several new diffuse bands, leaving the impression that he meant he had laboratory features that should correspond with interstellar bands, but not stating clearly whether these bands are seen. If not, he really was saying that he had several wavelengths where he would predict that diffuse bands should be detected. In his paper elsewhere in these Proceedings, Johnson lists many wavelengths from laboratory measurements; perhaps this list can form the basis of a thorough search for coincidences with interstellar features, as a definitive test of this model.

There is one specific prediction of which Johnson spoke: there should be a strong band, analogous to the 4430Å feature but weaker, at 4385Å in lines of sight having no pyridine in the grains (i.e. such as the direction toward HD 29647). He mentioned a spectrum published by Herbig, which shows a feature near this wavelength. Unfortunately, however, this feature is not a diffuse band, as noted by Jenniskens; it is a blend of stellar Mg II and Fe II lines. Apparently no high-quality data exist for the appropriate wavelength toward HD 29647, so at this point it is not possible to rule out the existence of a diffuse band at 4385Å toward that star.

Allamandola asked about infrared consequences of the model, noting that the paraffin matrices that are being proposed would certainly have mid-infrared bands that can be sought by *ISO*. He therefore pointed out that an elegant test of the model could be performed by making infrared measurements in the lab prior to the launch of *ISO*, so that *a priori* predictions could be made before observational data are available.

Leach pointed out a conundrum in the MgTBP model: it is claimed (based on laboratory measurements) that the rotational energy-level structure in excited vibrational states is the same as in the ground state, which, under the Franck-Condon constraint, means that you should not see vibrational transitions at all. Yet some of the claimed diffuse band transitions are indeed vibrational. This poses a spectroscopic puzzle.

Summary: This model was met with considerable skepticism from the audience. It appeared that the main bases for the skepticism were the precision required in duplicating the matrix environment for the carrier molecules under many different kinds of conditions (i.e. the invariance constraint), and the apparent prediction of diffuse bands, particularly the 4385Å feature, that are not seen in the interstellar medium (the spectroscopic consistency constraint). Johnson was challenged to make specific predictions that can be tested by observations; certainly further work towards seeking a feature at 4385Å should be done, since this was one specific prediction that was made. There was also a feeling expressed that it was very difficult to see how to make just the specific molecules in question, without also creating a wide variety of similar species having equally complex spectra.

To further test the model, it will be important to look for the 4385Å band in more sources, and to seek specific infrared and ultraviolet spectroscopic properties in the laboratory that might be sought observationally.

7. DIFFUSE BAND CARRIERS ON GRAIN SURFACES

This proposed type of carrier was not actually presented during the panel discussion, because its main advocate, Walt Duley, had already left. But it does represent a distinct class of hypotheses, and as such deserves some mention in this summary. Duley was present at the conference, and a more detailed description of his proposal is to be found elsewhere in the proceedings. Since there was no discussion of this model in the context of our adopted criteria, here we do not attempt a point-by-point test of the model against those criteria. But this does not stop us from making some comments about whether or not this hypothesis appears to be consistent with what is known from the observations.

The essence of this hypothesis is that large molecules (aromatics in particular) on the surfaces of small grains could produce absorption features comparable to the diffuse bands. The notion of the Shpol'ski effect, introduced also by Johnson (see above) is invoked in order to allow for transitions as sharp as the narrow diffuse features. Duley appears to refer to PAH cations as the actual carriers, presumably on the grounds that only the cations would have appropriate transitions in the visible. He envisions these molecular ions as being embedded within the surfaces of small carbonaceous grains, and cites scanning electron microscopy studies of hydrogenated amorphous carbon (Duley's infamous HAC) which show benzene-like structures to be present.

Because this is, in essence, a PAH^+ model, all of the comments made earlier about cosmic abundances and spectroscopic consistency should hold for this case as well. On the other hand, Duley makes no mention of the possibility of wavelength or profile shifts from region to region with changing grain size or density of absorbers within the grains, which leaves an open question as to whether this model is consistent with the general invariance of the diffuse bands.

This potential problem is solved by an alternative suggestion made by Duley: that small HAC grains are fragmented in weak interstellar shocks, and that the fragments, which would include PAHs, would then serve as carriers of the diffuse bands. In this version, Duley's hypothesis is essentially identical to the PAH^+ hypothesis discussed by Léger (above), with the addition of a proposed formation mechanism for the PAHs. It is clear that the entire question of how to make interstellar PAHs (or the other large molecules that have been proposed), will become increasingly important if these types of carriers continue to look attractive.

On the whole, the Duley proposal is too non-specific at this point to be thoroughly compared with the observational constraints. Much more must be known about the specific molecular species that might appear on the surfaces of HAC grains, and about the possible effects on the absorption-line wavelengths and profiles of changing grain sizes or concentration of the surface carriers. It would be helpful also if predictions could be made about how this mechanism might be affected by environmental considerations such as local radiation field intensity.

8. CONCLUSION

While few definitive findings were made as to which models are viable for the diffuse band carriers and which ones are not, significant benefit came from the discussions. It appears that some models (e.g. the dust impurity model, the hydrogenated fullerene proposal, and the porphryin hypothesis) are up against very serious constraints, and may soon be emphatically ruled out. It also appears that, despite the groundswell of opinion that PAHs are for real, there may be serious problems with the PAH^+ model for the diffuse bands. Only the carbon chain hypothesis seems to have escaped the afternoon largely unscathed, but even in this case there were significant questions (particularly about the photodissociation threshold). Even though the carbon chain model still encompasses a virtually unlimited number of specific candidate species, it is indeed useful to narrow the competition down to just the one class, if further work bears this out. But of course there remains the possibility (probability?) that some as-yet unrecognized new class of carriers will soon be proposed, and will raise an entirely new set of observational constraints and tests.

Perhaps the most significant aspect of the session was the apparent molecular flavor of most people's taste in diffuse band carriers. Looking back only a few years, there was a prolonged period when molecular hypotheses were few, and not thought viable in general. But in the late 1970s two things began to happen to change this: (1) the growing recognition of the importance of molecular processes in general in the interstellar medium, which was then brought to bear on the diffuse band problem; and (2) a growing family of observations, based on new technology such as CCDs and improved photopolarimeters, which began to erode the dust impurity models by showing that observations do not meet certain expectations of these models. More recently, the PAH hypothesis has stimulated many to believe that large organic molecules may be present in the diffuse interstellar medium, and this in turn has led to the widespread feeling that the solution to the diffuse band problem lies in the molecular arena, even if PAHs specifically may not be the answer. Now for many the problem seems to be more a question of *which* molecules, rather than whether the carriers are molecules or dust grains.

9. ACKNOWLEDGEMENTS

In preparing this summary, I have solicited comments on a draft version from all the main speakers during the panel session, and am grateful for the clarifications and corrections received. I particularly wish to thank Fred Johnson, Sydney Leach, Alain Léger, Pat Thaddeus, and Adrian Webster for their help in making this paper as accurate as possible.

SUMMARY

A.G.G.M. TIELENS
NASA Ames Research Center, MS 245-3, Moffett Field, CA 94035-1000

ABSTRACT.

Visible spectroscopy may be the best method for probing the molecular complexity of the diffuse interstellar medium. Electronic transitions are very characteristic for an individual species and hence can provide telling evidence for its presence. However, that is also its Achilles heel since it requires a good idea of the species involved before laboratory studies can be attempted. Hence, the actual identification of the large (> 10 atom) molecular species responsible for the diffuse interstellar bands will require Herculean labor in the areas of observational and theoretical astronomy to limit the countless possibilities before laboratory spectroscopy can provide the proof. As evidenced by this meeting, the first steps on this route have now been made.

1. INTRODUCTION

It was the best of times. It was the worst of times. I always wanted to start a paper this way and it is particularly apt here. Over the last decade, we have made vast strides in our understanding of the diffuse interstellar bands. Most workers in the field accept now that the carrier is molecular in origin and promising cases for various carriers have been assembled. Carbon chains, polycyclic aromatic hydrocarbons (PAHs), and fullerenes (C_{60}) are all exciting candidates and likely the age–old DIB mystery is just a rehash of the plot in *Murder in the Orient Express.* Yet, we now also understand that bringing the ultimate conclusive proof – the smoking gun acceptable to the community at large – will prove even more daunting and elusive than previously assumed – worthy of a real life Hercule Poirot. Observationally and experimentally, it now seems that every strong band corresponds to a unique molecular species and vice versa. While strong bands may be accompanied by weaker vibrational progressions, observationally there seems to be a weak DIB at almost any wavelength, making falsification of any proposed carrier very difficult. Coming to grip with this prospect may well be our first priority.

As evident from Eddington's famous quote made in connection with the DIB problem *atoms is physics but molecules is chemistry*, part of the problem may be that astronomers are physicists by training and tend to value simple and elegant solutions. Problems can be studied in isolation and "Grand Unification" schemes have some value. However, in chemistry, solutions tend to be far from pure and many factors (e.g., processes, species) contribute. In my view, the chemical universe is wondrous and well stocked and chemical diversity is the rule. Visible spectroscopy is one tool to probe this interstellar molecular complexity, but its view may well be different from that offered by the infrared or the millimeter windows. In the

A. G. G. M. Tielens and T. P. Snow (eds.), The Diffuse Interstellar Bands, 395–405.

remainder of this paper, I develop these themes a bit, while summarizing those parts of the meeting that can be molded into it.

2. OBSERVATIONS

2.1. THE DIB SPECTRUM

In recent years, the richness of the DIB spectrum has become increasingly clear. Herbig's 1975 list contained a few strong DIBs ($W_\lambda/E_{B-V} \simeq 1$; i.e. 4430, 5780, 6284Å) and about 50 medium ones ($W_\lambda/E_{B-V} \simeq 0.1$). The advent of CCD technology allowed deeper and more systematic surveys, permitting Jenniskens and Désert (1994; see also the cover and elsewhere in this volume) to identify about 150 DIBs. Finally, in deep surveys of very limited wavelength regions (around λ 4780 and 6284Å) a very high density of very weak DIBs ($W_\lambda/E_{B-V} \simeq 0.01$) was recognized (Jenniskens and Désert 1993; Krełowski and Sneden 1993). If that same density holds over the full DIB wavelength range, about 400 more DIBs at this strength would be present. Boldly extrapolating this trend to the confusion limit of 1 DIBÅ would imply that another 4000 or so very weak ($W_\lambda/E_{B-V} \simeq 10^{-3}$) DIBs remain to be discovered. While this shouldn't yet be taken too seriously, in surveying the literature I was left with the impression that, if studied deeply enough, there is indeed a band at every wavelength, albeit that most of these are very weak.

Integrating this trend, the total extinction due to DIBs is estimated to be about 0.5% of the "continuum" extinction of the dust. [Thank goodness, there is still some role for dust!]. Since the oscillator strength for electronic transitions in the DIBs and in the dust are probably comparable ($f \sim 0.1$), this implies that of the order of 1% of the elemental C is locked up in the DIB carriers. There is about equal absorption per decade of DIB strength and I did cut off the DIB contribution at the confusion limit [otherwise it could grow without end]. This, of course, means that a fair fraction of the DIB "extinction" is concentrated in the few strong DIBs. In terms of complex interstellar molecules, this implies an astounding abundance for these species, $\sim 10^{-7}$ (relative to hydrogen). Another striking aspect of the interstellar DIB spectrum is the wide disparity in strengths. This may either imply a large range in intrinsic strength of the transitions involved or a wide range in molecular abundances (or both).

2.2. FAMILIES

As reviewed by Somerville at this meeting, one of the striking "discoveries" over the past decade is that DIBs come in families (Chlewicki *et al.* 1986; Krełowski and Walker 1987; Josafatsson and Snow 1987; Krełowski 1989). The equivalent widths of the DIBs correlate reasonably well among themselves as well as with measures of total column density (i.e. E(B-V)). Yet, when examined in detail, variations in relative strength of DIBs exist from one sightline to another. This has led to the introduction of the concept of DIB families whose members correlate better with each other than with non–members (Krełowski and Walker 1987). There are three generally recognized families – the 4430, the 5780 and the 5797Å families – but membership in the individual families is not always agreed upon. Indeed, it seems that there is an exceptional sightline of poor correlation for every family member. Hence, these families are definitely of the nineties – dysfunctional at the core. At present, family studies have been largely restricted to the strongest DIBs. Extension to weaker DIBs will be important, but difficult to perform. While membership is confused, the message is clear: each DIB is on its own; i.e. carried by an individual species or compound.

Various origins for these families have been proposed. Each family may be due to a set of related chemical species; for example, compact PAHs versus non–compact PAHs. Alternatively, the different families may correspond to different charge states of similar species (i.e. cations/neutrals/anions) which will show a different environmental dependence (cf. Bakes and Tielens 1995, elsewhere in this volume). Alternatively, the different families may correspond to otherwise unrelated species (i.e. PAHs versus carbon chains or fullerenes), which react differently to varying interstellar conditions. It may well pay to consider each of these scenarios and develop limitations observationally and theoretically. Again, our progress is largely limited by the vast molecular playing ground.

2.3. THE VISIBLE EMISSION BANDS

One of the key observational pieces of evidence for a molecular, rather than grain–based, carrier of the DIBs is the realization that the narrow emission features observed superposed upon the extreme red emission in the spectrum of the Red Rectangle are close to the positions of diffuse interstellar bands (Fossey 1991; Sarre 1991; Scarrott *et al.* 1992; Schmidt and Witt 1991). Moreover, in going outwards in this nebula the peak positions and widths of these emission bands converge to those of the DIBs. This is the expected behavior for electronic transitions in molecules. The population of the higher rotational levels decreases as the molecule cools going outwards and the bands become narrower and shift to the blue (Scarrott *et al.* 1992). While the four most prominent emission bands can be readily traced back to DIBs, the weaker emission bands may correspond to very weak DIBs; or not, since at this level the DIB density is very high. As emphasized by Smith and Sarre at this meeting, the "missing" features might correspond to electronic decay to vibrationally excited states. This would be particularly interesting since it would provide another handle on the identification of the molecular species responsible (and we need all the handles we can get).

One key question remaining is the non–detection of these same emission features in the diffuse interstellar medium. The Red Rectangle observations show that fluorescence upon photon absorption is an important deexcitation channel and this should hold in the diffuse ISM as well. Clearly, the carrier of these features is much more abundant relative to the dust in the Red Rectangle than in the ISM. Assuming direct radiative excitation in this band, the absorption equivalent width per unit color excess, $W_\lambda/E(B-V)$, required to produce the 5797Å emission features relative to the dust scattering continuum in the Red Rectangle is $\simeq 4/\phi$ Å/mag where ϕ is the fluorescence yield and a (reasonable) dust albedo of 0.6 has been assumed. For comparison, in absorption, the measured $W_\lambda/E(B-V)$ of this DIB in the ISM is only 0.13 Å/mag. For some molecular ions, ϕ is very small ($\sim 10^{-5}$; Joblin *et al.* 1995, elsewhere in this volume). This might rule against them because of abundance considerations. However, this process is not well understood and further experimental studies are required to put this on a firm foundation. Furthermore, this is not an ironclad argument since these emission features are probably pumped by absorption into higher electronic states which then relax by rapid internal conversion to the first electronic state and fluoresce from there. The higher photon fluxes and larger molecular absorption cross sections at shorter wavelengths will then partly compensate these low ϕ's. Nevertheless, I would argue that ϕ is likely small and that therefore visible fluorescence is not an important deactivation channel in the ISM (i.e. reflection nebulae; diffuse galactic light). However, this immediately implies that the uniqueness of the Red Rectangle lies in a high abundance of the carrier.

We should take advantage of this inferred high abundance using other search techniques for large molecules. Besides mm and IR, one might consider UV fluorescence in reflection nebulae using HST, particularly if neutral PAHs are present. Neutral PAHs typically have $\phi \sim 0.2$ for their $S_1 - > S_0$ transitions. With a neutral fraction of 0.2 and about 2% of the stellar FUV flux going into all (small) PAHs, we calculate an equivalent width of all FUV fluorescence features of 10^{-3} relative to the dust, somewhat less than the absorption equivalent width of the DIBs. Another interesting approach would be to examine the unique characteristics of this source which allow such a large abundance of these specific carriers to be formed/survive - or perhaps prepared in the right charge or radical state. These emission features have now been observed in another post–AGB object (the R CrB star, V854 Cen; Rao and Lambert 1993) but they are not a common characteristic of such objects. Indeed, sources which inject freshly synthesized dust or carbonaceous molecules or which show strong PAH emission features do not, as a rule, show strong DIBs (cf. Seab 1995, elsewhere in this volume). I would say that the most puzzling aspect of these emission features and their relation to the DIBs is that they are present at all in the Red Rectangle – a unique object which in so many ways clearly does not resemble the ISM at all. Why are only the same molecules present in this object as in the ISM and not a set of completely different ones (as well)?

2.4. PROFILES

As mentioned above, in the study of the emission features in the Red Rectangle, DIB profiles can provide additional information and can constrain the carriers. DIBs range in width from about 0.5 to 50 Å; limited on both sides by measurement techniques (i.e. resolution/continuum placement). Their shapes range from shallow and broad (i.e. 4430Å), to narrow and symmetric (i.e. 5780Å), and asymmetric (red–shaded; 5797Å) features. In recent years, much progress has been made in this area. Much care has been expended to determine the detailed line shape along lines of sight with single clouds (cf. Krełowski and Sneden 1995, elsewhere in this volume). In general, no obvious rotational structure is present (but see Higgins *et al.* elsewhere in this volume, and Snow and Seab 1991).

The invariant linewidth and peak positions are two defining observational characteristics for the DIBs, and this has been used as one of the key arguments against a solid state carrier. To be fair, if you rap Peter, you should rap Paul as well and these characteristics also put strong constraints on the molecular viewpoint. As discussed by Leach at this conference, the linewidth may be dominated by (unresolved) rotational structure or by nonradiative intramolecular coupling. In the first case, the invariant linewidth and peak position must imply that the rotational population is largely independent of the physical conditions in the diffuse ISM since, in particular, the gas temperature is known to vary from 50 to 300 K. Well, the rotational population might result from a balance between radiative (net) "excitation" by UV photon absorption followed by IR fluorescence and radiative deexcitation through pure rotational transitions (Rouan and Léger 1995, elsewhere in this volume). Nonradiative molecular coupling would of course be independent of the gas density and temperature and this naturally leads to an invariant linewidth and position. However, note that the emission features in the Red Rectangle decrease in width and shift in peak position with distance from the central source and this has actually been cited as one of the strong arguments for a molecular carrier. In this view, these variations result from a decrease in rotational excitation; i.e. a cooling of the gas when flowing out (adiabatic expansion is the dominant gas cooling source in outflows). But the densities and temperatures in the regions where this emission originates are very similar to those in the ISM and we don't see these variations there. This suggests to me that intramolecular line broadening dominates the profiles of the DIBs (cf. Leach 1995, *ibid*). The contrast between the Red Rectangle variations and the ISM invariance must then reside in the difference between emission from excited molecules and the absorption by cold species. For example, FUV excitation through higher lying electronic states (i.e. S_2, S_3, ...) which interconvert to S_1 and whose contribution falls away with distance because of extinction may lead to this behavior (Smith 1995, elsewhere in this volume). Clearly, there are some interesting lab experiments to be performed to bear this out (cf. Joblin *et al.* 1995, *ibid*).

3. MOLECULAR CARRIERS

3.1. MOLECULAR IDENTIFICATION

Electronic transitions are very characteristic for individual molecules. Because of this uniqueness, visible (and ultraviolet) absorption features are very powerfull tools for the purpose of identifying specific molecules. Conversely, this implies that it will be difficult to pinpoint the one molecule responsible for an absorption feature from among the almost infinite possibilities. Essentially, you have to know the anwer before you can find it. It is precisely this point which has frustrated generations of spectroscopists.

There are some similarities and differences between IR, mm and visible spectroscopy which are worth pointing out. Like visible spectroscopy, rotational transitions at mm wavelengths are very characteristic for individual molecules. However, because many transitions are equally strong, patterns are easily recognized and identification is greatly aided. In some sense there is an embarrassment of riches: even a simple molecule such as CH_3OH can give rise to a multitude of spectral lines (cf. Herbst 1990). This is also its weakness as a molecular diagnostic, since the large partition function leads to very low column densities in any one rotational level and makes transitions weak. For example, the partition function for the 11 atom, linear molecule, cyanotetraacetylene (HC_9N) is $\simeq 12T^{1/2}$ while for the asymmetric top, 11 atom, ring molecule pyridine (C_5N_5N) it is $16T^{3/2}$; a difference of 10^2 at 100K. Thus, the detection of pyridine at the same level as cyanotetraacetylene requires a 100 times higher abundance. For even larger molecules this becomes comparatively worse. Moreover, the individual lines will disappear in the interstellar "grass" of unresolved rotational transitions, particularly if there is a very wide family of large interstellar molecules. It is for this reason that interstellar mm spectroscopy is limited to molecules much smaller than generally considered for the DIBs or for the IR emission features.

Vibrational spectroscopy is a powerful way of identifying classes of molecules. The fundamental vibrations of molecules with similar molecular structure (e.g., aromatics or aliphatics) line up at the same frequency. It is therefore undisputed that the interstellar IR emission features are due to aromatic hydrocarbons rather than aliphatic hydrocarbons or carbon chains (cf. Tielens 1993). However, this also implies that identification of specific molecules (say coronene rather than pyrene) within this class is exceedingly difficult. Indeed, only very limited, general statements about the molecular structure of interstellar PAHs can be made (condensed, ionized, hydrogenated; cf. Allamandola *et al.* 1989). An additional weakness, at least presently, is that a strong excitation source is required for the detection of vibrational emission in space. Such a strong FUV source may well influence the composition of the molecular mix present. Hence, we may be seeing an interstellar aromatic spectrum because we are looking at a bright FUV source and the more benign conditions in the diffuse ISM may well allow a richer chemical soup. *ISO* will tell what the real emission spectrum of the diffuse medium is.

Because of the molecular specificity of electronic transitions, spectroscopists trying to identify the carriers of the DIBs have generally used outside constraints to drastically limit the number of molecular species to be studied. Indeed, the cases for the three molecular carriers so dominantly mentioned throughout these proceedings are largely based upon outside arguments. PAHs are known to be abundant in the interstellar medium from the observed IR emission spectrum. Likewise, carbon chain molecules have been detected in the mm spectra of dark clouds. Our Greek heritage makes C_{60} and its derivatives irresistable candidates. But in making such arguments the general points made above on the strengths and pitfalls of different spectral regimes should be kept in mind and the possibility of a wider variety of large molecules should be considered.

3.2. ELECTRONIC SPECTRA OF MOLECULES

As reviewed by Leach at this conference, small molecules with closed electronic shells, typically, do not possess electronic transitions in the visible. Molecular bond energies are $\simeq$ 5 eV and hence require FUV photons to be excited. Open shell molecules, such as radicals or ions, will have lower energy transitions and consequently experimental studies of possible DIB carriers have largely concentrated on such species. Indeed, most molecules will be charged in the diffuse interstellar medium. While this has been most extensively modelled for interstellar PAHs, this conclusion has a very general applicability and applies equally well to C_{60} and carbon chains (Bakes and Tielens 1995, *ibid*). Large molecules (and radicals) have a high electronegativity. The high electron abundance in the diffuse medium can then favor a high anion abundance. Species with lower electronegativities (<1 eV) have low electron sticking coefficients and anionic states become irrelevant. All large molecules are readily positively charged by the strong interstellar FUV flux. The large FUV flux may also lead to loss of H (i.e. create radicals). Hence, inescapably, most of the large molecules present in the diffuse interstellar medium will have visible absorption bands.

Molecular visible absorption spectra show one or two strong transitions, generally corresponding to 0-0 transitions, plus weaker vibrational progressions (Salama *et al.* and Maier *et al.* elsewhere in this volume). With this in mind, the known, 50 or so, medium and strong DIBs will correspond to about 25–50 abundant interstellar molecules. There are a few hundred weak bands and probably a few thousand very weak bands (cf. § 2). These may be due to vibrationally assisted electronic transitions or much weaker forbidden transitions in these same molecules. However, their (extrapolated) sheer number suggests the presence of a large number ($\simeq$ 1000) of additional molecules with much lower abundance and/or much lower oscillator strength.

In my opinion, the family of interstellar molecules is very wide indeed.

As a corollary to this conclusion, it will not be easy to provide unambiguous proof for the identification of a DIB with a specific molecule. Even if a strong DIB coincides with a strong molecular transition, further supporting evidence will be needed. It will be difficult to separate the vibrational progression from the sea of unrelated weak lines. To phrase it differently, with a (weak) band at almost any wavelength, coincidence statistics is not sufficient identification proof. Lou Allamandola in his chapter presents an amusing story of how even the best can be led astray by coincidence statistics. The recent lab studies on carbon chains may be another case in point with good statistics. The species indicated by these experiments seem to me too small and too photochemically vulnerable to be very viable. Besides coincidence statistics, additional support will have to come by matching the line shape and relative strength of transitions. Emission spectroscopy may also develop into a powerful tool (cf. §2).

3.3. THE ISOMER CRISIS

While 1000 molecules may seem a lot, actually this is still a very small number and implies that nature is very selective. Let me illustrate that for small PAHs. The number of "hexamers" of small PAHs (i.e. with the same number of aromatic rings, h) grows exponentially with h ($n(h) \sim 0.07exp(1.2h)$). Hence, there are already 10,000 different PAHs with 10 rings (i.e. with typically 30 C–atoms), albeit most with non–conjugated structures. This doesn't even count the presence/absence of sidegroups such as H or CH_3. A similar isomer crisis exists for fulleranes where there are in excess of 9 zillion ways of decorating C_{60} with H (Webster 1995, elsewhere in this volume). This reemphasizes the question of how to select DIB candidates from among all the possible species. Otherwise we could be spending a lifetime examining molecular species without ever coming close to the right one. In other words, we have to understand how nature selected the 25 abundant DIB carriers just to have any hope of finding them. Some first attempts at this have been reviewed by Herbst at this conference.

This chemical specificity for the DIB carriers does not come as a total surprise. In general, nature is very selective. Again take the PAHs as an example. Sooting flames are well known to produce copious amounts of PAHs (up to several percent of the soot abundance by mass). Yet, among all the possible hexamers, the centrally condensed PAHs, such as pyrene, coronene, and ovalene are dominant (Stein 1978). This merely reflects their greater thermodynamic stability and consequently their role in the carbon polymerization route (i.e. soot formation). Now, while the chemical pathway for dust formation in carbon stars may resemble that in sooting flames (cf. Tielens 1989), the PAH population injected will be further modified by processes in the ISM. Among the important processes are peripheral H–loss after FUV photon excitation, dissociative recombination, and multiple charging followed by Coulomb explosion (cf. Leach, at this conference). All of these are likely to be more important for the smallest PAHs. For example, theoretical calculations show that H–loss only occurs for PAHs with less than about 25 C–atoms (Allamandola *et al.* 1989). While small PAHs are readily weeded out and only the most stable ones survive, those processes are likely less important for larger PAHs ($\simeq$ 100 C–atoms) and hence many more isomers are expected to be present. Thus, while there is a whole size distribution of PAHs, based upon these stability arguments, only a handfull of (small) PAHs will be abundant enough to show strong visual absorption bands (Crawford *et al.* 1985).

Similar stability arguments will apply for molecules other than PAHs. Carbon chains are of particular interest (cf. Maier 1995; Thaddeus 1995, elsewhere in this volume). Thermodynamics suggests that carbon clusters, C_n, with n <12 have linear structures while for 25> n >12 monocyclic rings are preferred. Graphitic structures may become important at n >25 and polyhedral structures become important at n >40 (e.g., C_{60}; Bernholc and Phillips 1986). Again according to thermodynamics, these chains would have cumulenic bonding (i.e. cumulative C=C bonds) rather than acetylenic bonding (alternating C≡C and C-C). However, unlike PAHs formed at high temperatures in stellar outflows, these chains are thought to form far from equilibrium in the ISM itself (cf. Thaddeus 1995, elsewhere in this volume). Hence, kinetics will play a far more important role. It has been suggested that, if these clusters are assembled one monomer at a time, the heat released upon formation is probably sufficient for the smaller chains to attain their thermodynamically most favored state (i.e. cumulenic chain). For larger chains and, particularly, rings, planes and polyhedral structures, this is unlikely to occur and kinetics should dominate. In the latter case, the number of possible isomers will also quickly grow very large with increasing cluster size. In contrast to the PAHs, the presence of sidegroups is of lesser concern, since these are likely limited to the caps and are probably dominated by H. The very limited number of isomers for small chains is certainly one attractive plus point for this proposed identification. Take the $C_{2n}H$ series as an example. Only the chains with n=2–6 will have absorption bands, typically one strong one, in the visible range (4000–9000Å ; Watson 1994; Leach 1995, elsewhere in this volume). Taking all the H–cap possibilities in account, the family of chain molecules would (if present) give rise to about 25–50 strong bands, in good agreement with the DIB spectrum. However, photochemical stability is a great concern for these chains, much more so than for PAHs (or for rings for that matter). This was addressed at this meeting by Thaddeus and by Herbst. While under favorable assumptions, (re)formation in the ISM may be possible, the models are not well validated experimentally at this point. In particular, the adopted photodestruction rate of $3 \times 10^{-11} N_c$ s^{-1} for small chains is much lower than measured carbon chain photodestruction rates ($\simeq 10^{-9} N_c$ s^{-1}). Hence, in reality the destruction rate will be much faster than the growth rate rather than the other way around and actual growth in the diffuse ISM (but not dark clouds) seems to me inhibited. Moreover, extrapolating from our studies on PAHs, with a typical bond energy of 5.6eV, we can expect that only species with more than 15 C–atoms are photochemically stable in the diffuse ISM, but those are not the ones thought to have visible absorption bands.

3.4. THE DIVERSITY OF ASTROCHEMISTRY

The diversity of astrochemistry reflects the wide range in physical conditions present in space and the variety of reactions made possible by that (cf. Herbst 1995, elsewhere in this volume). Thus, the chemistry of dark clouds is dominated by ion–molecule reactions driven by cosmic ray ionization and is characterized by high abundances of unsaturated molecules, radicals, and ions. In contrast, in the surface regions of such clouds, as well as in the diffuse ISM, and in regions around luminous O and B stars (PDRs), ion–molecule reactions are still important but the chemistry is now controlled by photoreactions (van Dishoeck and Black 1988; Tielens and Hollenbach 1985). The high temperatures of circumstellar outflows allow reactions with considerable activation barriers (i.e. involving closed shell species) to occur at appreciable rates. The possibility of non–solar abundances (i.e. C/O>1) opens up the range of reactions possible even further. In particular, carbon polymerization reactions leading to soot formation will be important in C–rich giants and the chemistry resembles that occuring in fuel–rich hydrocarbon flames (Frenklach and Feigelson 1989; Cherchneff *et al.* 1992). In that respect, soot formation around WC stars and RCrB stars (C/H>1) is an example of chemistry in an even more extreme environment (Cherchneff and Tielens 1995). At the surfaces of such outflows, the photochemical breakdown of molecular species now drives rapid neutral radical–molecule reactions (Cherchneff *et al.* 1993). Besides reactions associated with grain formation, reactions on grain surfaces can be important as well. In warm outflows, these may be catalytic in nature. In dark clouds, hydrogenation and oxidation reactions lead to ice formation (i.e. H_2O, CH_3OH, NH_3). Photochemistry of these ices by FUV photons produced by cosmic ray excitation of H_2 can initiate the formation of rather complex molecules (cf. Tielens and Allamandola 1987). The evaporation of these ices into the gas phase once a newly formed star turns on (i.e. hot cores in dark clouds) provides yet another twist on astrochemistry. In particular, the injection of alcohols leads rapidly to chemical complexity (Charnley *et al.* 1992; Hasagawa *et al.* 1992). Finally, on a somewhat more speculative note, the expansion of dense, warm vapor produced by vaporizing grain–grain collisions ($v > 20$ km/s) in interstellar shock waves may allow clustering, generally associated with circumstellar outflows, in the ISM itself.

The diversity of astrochemistry is testified to by the organic inventory of carbonaceous meteorites. A wide variety of molecular species have been detected in meteorites including more than 50 amino acids (more than on Earth), various carboxylic acids, alkyl-substituted cyclic alkanes, alcohols, amines, small PAHs (naphthalene, phenanthrene, pyrene, fluoranthene, perylene, coronene), and fullerenes. Now, many of these may have been affected by processing on the planetary body. Nevertheless, the initial chemical "soup", inherited from the molecular cloud in which the sun formed, must have been complex as well. Indeed, one defining characteristic of chemistry in general is the wide range of processes that play a role and the difficulty in isolating the dominant ones. Allowing for a multitude of processes in astrochemistry will be a big step for astronomers/astrophysicists who are blinded by Occam's razor on an almost daily basis.

4. FUTURE

A sense of exhiliration pervaded this conference. This was fed to no small extent by the announcement that two new near–IR DIBs had been identified with C_{60}^{+} (Foing and Ehrenfreund (1995, elsewhere in this volume). While some doubt still lingers on this identification – unfortunately, no gas phase data is available for comparison (cf. Maier 1994) – it solidified in everybody's mind that we are very close to solving the age–old DIB problem. Now, the actual carrier was not generally agreed upon by the participants; *au contraire*, various classes of molecules were defended and attacked with much fervor and this is reflected in these proceedings. Overall, the feeling is that the carrier is among those named and that if we only could measure the visible spectrum in the gas phase at low temperature, we could prove it. The actual proof will take more than that as summarized by Snow at this meeting. Nevertheless, as detailed in the preceding pages of these proceedings, a detailed plan of attack can be formulated.

Observationally, a deep survey of heavily extinguished lines of sight is a high priority, cataloging all DIBs and their strength down to the "grass" level. Extending the study of family ties to the weaker DIBs is important too. That will be much more difficult since it requires single–cloud sight lines and hence involves much lower optical depths. Visible fluorescence may be a new tool and should be pursued vigorously in reflection nebulae and the diffuse galactic light. Finally, environmental dependencies should be further studied. In particular, if the DIBs involve cations in the diffuse medium, then anions should dominate in dark clouds and they will have their own absorption spectrum. Theoretical efforts should concentrate on narrowing down the list of candidate classes of molecules, including charge state and dehydrogenation. Detailed models of formation and destruction, perhaps focusing on photochemical stability, will be very valuable. Of course, a key here will be experimental work on reaction rates involving large molecules, including photoionization and dissociation, electron recombination and attachment, and reactions with C^+ and H. The isomer crisis still worries me, particularly for the PAHs and this should be further explored. There is the question of the unique characteristics of the Red Rectangle. The problem could also be turned around. Under what conditions could some particular species be abundant? In a neo–Darwinian way, are there ecological niches where certain exotic species can fluorish? Can we identify these and search for them? The crux of the DIB problem will remain experimental; i.e. laboratory identification. As a preamble, matrix isolation studies have to narrow the field by selecting the promising species for the strong DIBs. This will have to be followed by gas phase studies in a molecular beam or in ion–traps. The latter would then have to be assisted by theoretical calculations to predict the line shape at low temperatures. Of course, this is a two-way street and laboratory studies of interesting species may drive observational studies. As evidenced by C_{60}^+, this may well prove the most promising avenue for direct progress.

REFERENCES

Allamandola, L.J., Tielens, A.G.G.M., & Barker, J.R. 1989, *Ap. J. Suppl.*, **71**, 733.
Bernholc, J. & Phillips J.C. 1986, *J. Chem. Phys.*, **86**, 3258.
Charnley, S.B., Tielens, A.G.G.M., & Millar, T.J. 1992, *Ap. J. Lett.*, **399**, L71.
Cherchneff, I., Barker, J.R., & Tielens, A.G.G.M. 1992, *Ap. J.*, **401**, 269.
Cherchneff, I., Glassgold, A., & Mamon, G.A. 1993, *Ap. J.*, **410**, 188.
Cherchneff, I. & Tielens, A.G.G.M. 1995, in *Wolf Rayet Stars: Binaries, Colliding Winds, Evolution*, eds. K.A. van der Hucht, & P.M. Williams, (Kluwer, Dordrecht), 346.
Chlewicki, G. *et al.* 1988, *A & A*, **173**, 131.
Crawford, M.K., Tielens, A.G.G.M., and Allamandola, L.J. 1985, *Ap. J. (Lett.)*, **293**, L48.
Fossey, S.J. 1991, *Nature*, **353**, 393.
Frenklach, M. & Feigelson, E.D. 1989, *Ap. J.*, **341**, 372.
Hasegawa, T.I., Herbst, E., & Leung, C.M. 1992, *Ap. J. Suppl.*, **82**, 167.
Herbst, E. 1990, in *Atoms, Ions, and Molecules: New Results in Spectral Line Astrophysics*, eds. A.D. Haschick and P.T.P. Ho, (ASP, SF), 313.
Herbig, G.H. 1975, *Ap. J.*, **196**, 129.
Jenniskens, P. & Désert, F.X. 1993, *A & A*, **274**, 465.
Jenniskens, P. & Désert, F.X. 1994, *A & A Supp. Ser.*, **106**, 39.
Josafatsson, K. & Snow, T.P. 1987, *Ap. J.*, **319**, 436.
Krełowski, J. 1989, in *Interstellar Dust*, eds. L.J. Allamandola & A.G.G.M. Tielens, (Kluwer: Dordrecht), 67.
Krełowski, J. & Sneden, C. 1993, *Pub A S P*, 105, 1141.
Krełowski, J. & Walker, G.A.H. 1987, *Ap. J.*, **312**, 860.
Maier, J.P. 1994, *Nature*, **370**, 423.
Rao, N.K. & Lambert, D.L. 1993, *M.N.R.A.S.*, **263**, 27P.
Sarre, P.J. 1991, *Nature*, **351**, 356.
Scarrott, S.M., Watkins, S., Miles, J.R., & P.J. 1992, *M.N.R.A.S.*, **255**, 11P.
Schmidt, G.D. & Witt, A.N. 1991, *Ap. J.*, **383**, 698.
Snow, T.P. & Seab. C.G. 1991, *Ap. J.*, **382**, 189.
Stein, S.E. 1978, *J. Phys. Chem.*, **82**, 566.

Tielens, A.G.G.M. 1989, in *Carbon in the Galaxy*, eds. J. Tarter, S. Chang & D. DeFrees, NASA-CP 3061, 59.
Tielens, A.G.G.M. 1993, in *Dust and Chemistry in Astronomy*, eds. T. Millar, D.A. Williams, (I.P.P: Bristol), 103.
Tielens A.G.G.M. & Allamandola, L.J. 1987, in *Interstellar Processes*, eds. D. Hollenbach & H. Thronson, (Kluwer, Dordrecht), 397.
Tielens A.G.G.M. & Hollenbach, D.J. 1985, *Ap. J.*, **291**, 722.
van Dishoeck, E.F. & Black, J. 1987, *Ap. J.*, **334**, 771.
Watson, J.K.G. 1994, *Ap. J.*, **437**, 678.

Addresses of Participants

Adamson, Andy
University of Central Lancashire
Corporation Street
Preston, UK PR12HE

Allamadola, Lou
NASA Ames Research Center
M/S 245-6
Moffett Field, CA 94035

Bakes, Emma
Princeton University
Peyton Hall
Princeton, NJ 08544

Beegle, Luther
U. of Alabama-Birmingham
310 Campbell Hall
1300 University Blvd.
Birmingham, AL 35205

Bierbaum, Veronica
University of Colorado-Boulder
Dept. of Chemistry, CB 215
Boulder, CO 80309

Boissel, Pierre
Labo Photophysique Molecule
Bat 213
Universite Paris
Sud Orsay, FRANCE 91405

Brechignac, Phillippe
University of Paris
Paris, France

Cardelli, Jason
Dept. of Astronomy
Univ. of Wisconsin-Madison
Madison, WI 53706

Clayton, Geoffrey
University of Colorado-Boulder
CASA, Campus Box 389
Boulder, CO 80309-0389

d'Hendecourt, Louis
IAS-CNRS Inst. of Astrophysics
Univ. Paris Lud, Bat 121
Orsay Cedex, France 91605

Danks, Anthony
Hughes Stx-GSFC
Code 683.0
Greenbelt, MD 20441

Depuy, Charles
Dept.of Chemistry/Biochemistry
University of Colorado-Boulder
Campus Box 215
Boulder, CO 80309

Dominik, Carsten
NASA-Ames Research Center
Mail Stop 245-3
Moffett Field, CA 96035

Donn, Bertram
Howard University and
University of Virginia
19 Woodland Way
Greenbelt, MD 20770

Duardo, Joe
UCLA, Chemistry Dept.
134444 Safari Dr.
Whittier, CA 90605

Duley, Walter
Physics Dept.,
U. of Waterloo
Waterloo, Ontario Canada N2L3Al

Ehrenfreund, Pascale
Leiden Observatory
Holland

Feldman, Paul
Herzberg Inst of Astrophysic
100 Sussex Dr., NRC of Canada
Ottawa, Ontario Canada K1AOR6

Foing, Bernard
European Space Agency
Solar Sys Div ISSD ESTEC(SO)
Noordwijh Netherlands 2200AG

Fossey, Stephen
University College London
553 Watford Way
Mill Hill Park
London, England NW72QS

Hadad, Christopher
Univ of Colo at Boulder
Dept. of ChemistryCampus Box 215
Boulder, CO 80309

Henning, Thomas
Max Planck Research Grp
Schillengabchen 2-3
Fena Germany D07745

Herbst, Eric
Ohio State University
Dept. of Physics
174 W. 18th Ave.
Columbus, OH 43210

Jenniskens, Peter
NASA-Ames Research Ctr.
Mail Stop 239-4
Moffett Field, CA 94035

Joblin, Christine
NASA-Ames Research Ctr.
Mail Stop 245-6
Moffett Field, CA 94035

Johnson, Fred
Dept. of Physics
California State University
Fullerton, CA 92634

Jones, Anthony
NASA Ames Research Center
Mail Stop 245-3
Moffett Field, CA 94035-1000

Kanik, Isik
Jet Propulsion Lab
4800 Oak Grove Dr.
Pasadena, CA 91109

Kerr, Tom
Dept. of Chemistry
University of Nottingham
Nottingham, UK NG7 2RD

Kolbuszewski, Marcin
Herzberg Inst. of Astrophysics
Natl Research, 100 Sussex Dr.
Ottawa, Ontario Canada K1AOR6

Koornneef, Jan
Sron-Groningen, PO Box 800
9700 AV Groningen
Netherlands

Krelowski, Jacek
Institute of Astronomy
N Copernicus University
Torun, Poland

Leach, Sydney
Meudon Observatory, DAMAP
O5 Place Jules Janssenbs de Paris
Meudon, France 92195

Léger, Alain
IAS Bat 121
University de Paris
Sud Orsay, France 91405

Little-Marenin, Irene
University of Colorado-Boulder
CASA, Campus Box 389
Boulder, CO 80309-0389

Maier, John
University of Basel
Inst. of Physical Chem.
Klingelbergstrasse 80
Basel CH Switzerland 4056

Martin, Peter
CITA
University of Toronto
Toronto ON M5S1A7

Mathis, John
Univ Wisconsin - Madison
Dept of Astronomy
475 N Charter St.
Madison, WI 53705

McIntosh, Alan
University of Edinburgh
Royal Observatory
Blackford Hill
Edinburgh UK EH93HJ

McNally, Derek
University of London Observ
Mill Hill Park
London UK NW72OS

Miles, Janet
University of Nottingham
Dept. of Chemistry, U. Park
Nottingham UK NG72RD

Moutou, Claire
Inst. d'Astrophysique
Spatial Bat 121, Universite Orsay
Cedex France 91405

Parisel, Olivier
Astrochimie Quantique Ecole
Normale Superieur
24 rue Lhomond
Paris, France 75231

Pfau, Werner
Universitats Sternwarte
SchillergaBchen 2
Jena Germany 07745

Porceddu, Ignazio
Cagliari Astronomical Obs
via Ospedale 72
Cagliari, Italy 09124

Rebrion-Rowe, Christiane
Dept. de Physique Atoque
et Universite de Rennes I
Campus de Beaulieu
Rennes Cedex France 35042

Robinson, Michael
Univ. of Alabama-Birmingham
310 Campbell Hall,1300 U.Blvd.
Birmingham, AL 35294

Rouan, Daniel
Observatoire de Paris-Meudon
DESPA
Meudon France 92195

Rowe, Bertrand
Departement de Physique
Universite de Rennes I
Campus de Beaulieu
Rennes Cedex France 35042

Salama, Farid
NASA-Ames Research Center
Mail Stop 245-6
Moffett Field, CA 94035

Sarre, P.J.
University of Nottingham
Dept. of Chemistry
University Park
Nottingham, UK NG7 2RD

Seab, C. Gregory
Dept. of Physics
University of New Orleans
New Orleans, LA 70148

Shapiro, Paul
Dept of Astronomy
University of Texas-Austin
Austin, TX

Shykula, Jeffrey
Rensselaer Polytechnic Inst.
Science Center,15th St.
Troy, NY 12180

Smith, William
Washington University
St Louis, MO

Snow, Theodore
University of Colorado-Boulder
CASA, Campus Box 389
Boulder, CO 80309-0389

Somerville, W.B.
Dept. of Physics&Astronomy
Univ. College London, Gower St.
London UK 6BT

Sorokin, Peter
IBM Research
TJ Watson Research Ctr.
P.O. Box 218
Yorktown Heights, NY 10598

Szczerba, Richard
University of Calgary
Dept. of Physics
2500 University Dr.
Calgary Alberta Canada T2N1N4

Thaddeus, Patrick
Harvard University
Ctr. for Astrophysics
60 Garden St.
Cambridge, MA 02138

Tielens, Alexander
NASA-Ames Research Center
Mail Stop 245-3
Moffett Field, CA 94035

Vetesnik, M.
Dept. of Astrophysics, Masaryk U.
CS-61137 Brno
Czech Republic

Volk, Kevin
U. of Calgary, Dept. of Physics
2500 University Dr. NW
Calgary, Alberta Canaca T2N lN4

Wdowiak, Thomas
U. of Alabama-Birmingham
310 Campbell Hall,1300 U.Blvd.
Birmingham, AL 35294

Webster, Adrian
Royal Observatory
Edinburgh, Scotland EH9 3AJ

Wegner, Walter
Pedagogical University
Chodkiewicza 30 85-028
Bydgoszcz, Poland

Williams, Richard
University of Colorado-Boulder
Campus Box 107
Boulder, CO 80309

Witt, Adolf
University of Toledo
Ritter Observatory
Toledo, OH 43606

INDEX

Be stars: 17-8
carbon
    chains: *see molecules, individual*
    cosmic abundance: 50, 150, 248, 250-1, 333-4, 360, 365, 381, 388, 396
    gunk: 181, 192-5, 328
    C_{60}: *see molecules, individual*
chemistry: 308-13, 370-4, 402-3
correlations of diffuse bands
    with atomic gas: 14, 16-7, 34-6, 107-9, 332
    with depletions: 14, 86, 89
    with each other: *see diffuse bands, families of*
    with extinction: 15, 32-3, 37, 73-4, 86, 88-93, 97-102, 106-11, 115-9, 138-40, 246-8, 331-2, 334
    with molecules: 15-6, 34-6, 86, 93, 109, 332
    with polarization: 15, 86
depletions: 14, 84, 86, 241, 247-51
diffuse bands
    circumstellar: 85, 129-40, 334-5
    in comets: 7-8
    correlations: *see correlations of diffuse bands*
    emission: 4, 9, 143-7, 162-3, 165-72, 335-6, 397
    emission wings of: 330, 380, 381, 383, 386, 388, 390
    energy diagram: 41, 43-4, 48
    families: 3, 14-5, 33-4, 37, 46-7, 49, 86-7, 332-3, 396-7
    infrared: 3, 4, 40, 65-72, 356-7, 403
    origin: *see diffuse band carriers*
    polarization of: 73-8, 269-70, 331, 380-1, 383, 386, 388, 390
    profiles: 4, 5, 14, 25-9, 42-3, 45-6, 48, 255-61, 301-2, 330, 331, 336, 380, 382, 385, 387, 389, 391-2, 398-9
    ratios of: *see diffuse bands, families of*
    ultraviolet: 11, 21, 48, 53-62, 190, 333
    weak features: 14, 20, 44
    widths: 42-3
diffuse band carriers
    abundance of: 4, 50, 56, 85, 333, 359-60, 364-5, 381, 383, 385, 388, 390
    destruction of: 6-7, 334

energy levels in: 6, 9-10, 29, 41, 48, 54-6, 110, 165-8, 282-7, 292, 294, 295, 400-1
formation of: 308-13, 334, 370-4
models, summary of: 326-9
observational constraints on: 329-36, 379-80
dust
graphite: 243-6, 271-6
as carrier of the diffuse bands: 74, 101, 137, 175-6, 253-61, 269-70, 326-7, 385-7, 391-2
polarization by: *see polarization*
emission features: *see diffuse bands, emission; and infrared, emission features*
ERE: *see extended red emission*
extended red emission (ERE): 149-54, 356
extinction
correlation with diffuse bands: *see correlations of diffuse bands, with extinction*
general law of: 241-3, 264
far-UV rise: 22, 85, 91, 97-100
2175Å bump: 4, 22, 73, 74, 78, 85, 87, 90, 93, 99-102, 136, 242-7,271-8, 353, 356
fullerenes: *see molecules, individual*
grains
impurity absorbers: 137, 253-61, 326-7, 360-1
size distribution: 245, 263, 267-9
HAC: *see hydrogenated amorphous carbon*
HD 183143: 33, 78, 107, 243
HD 197770: 74, 75-8, 276-7
hydrogenated amorphous carbon (HAC): 149-51, 153-4, 359-62
impurity absorbers: *see grains, impurity absorbers*
infrared
absorption features: 3, 4, 40, 65-72, 356, 403
emission features: 44, 364
interstellar clouds
dark: 83-6, 308, 402
physical conditions: 83-6
σ Sco type: 15, 18-9, 20-1, 93, 100, 115
ζ Oph type: 15, 18-9, 20-1, 93, 100, 115
interstellar medium, general: 83-6

matrix effects: 178, 203, 220, 295-7
molecules, individual
anthracene: 222, 224, 225, 231-8, 282-4
benzene: 318, 361
carbon chains: 181-6, 195, 199-204, 281, 300, 327-8, 369-78, 380-2, 349-57, 370, 376, 377, 382-4, 400, 403
coronene: 69, 187, 284, 400
fullerene, fullerane: 3, 6, 27, 68-9, 71, 154, 190-1, 195-6, 281, 291-8, 307, 316-9, 329, 349-57, 370, 376, 377, 382-4, 400, 403
methyl pyrene: 157, 187, 367
naphthalene: 157, 187, 209-10, 219, 225, 228, 231-8, 318
ovalene: 69, 187
pentacene: 316-21
perylene: 158-62, 209
phenanthrene: 158-62, 209-10, 283
polycyclic aromatic hydrocarbon (PAH): 26, 67, 69, 102, 110-1, 138, 157-63, 165-72, 186-90, 195, 207-17, 219-20, 221-8, 231,8, 254, 281, 282-91, 307, 312-3, 315-21, 328-9, 363-8, 387-9, 400, 401, 402
polyenes: 298-300
polyynes: 298-300, 313
porphyrins: 137, 191-2, 196-7, 307, 327, 341-6, 389-91
pyrene: 157, 187, 209-10, 224, 225, 228, 316-21, 367, 400
PAH: *see molecules, individual*
polarization: 15, 73-8, 86, 263-70, 271-6
polycyclic aromatic hydrocarbons (PAHs): *see molecules, individual*
porphyrins: *see molecules, individual*
Red Rectangle: 4, 9-10, 47, 143-7, 150, 157, 165-72, 335-6, 341, 397, 398, 399
UIRs: *see infrared, emission features*
ζ Ophiuchi: 15, 16, 54, 84

INDEX OF SPECIFIC DIFFUSE BANDS

1369Å: 53-62
4176: 41
4430Å (*aka 4428Å*): 11, 16, 39, 47, 87, 88-90, 92-3, 99, 100, 101, 115, 131, 134, 136, 157, 219, 253, 269, 330, 337, 352, 353, 356, 364-5, 367, 396
5762Å: 20
5766Å: 20
5769Å: 20
5778Å: 14, 20, 44, 47, 115
5780Å: 14, 15, 16, 18, 20, 22, 26, 31, 32, 34, 36, 37, 39, 45, 47, 74, 86, 89, 91, 92-3, 99, 100, 110, 115, 117, 131, 133, 138, 143, 145, 269, 351, 396
5795Å: 19
5797Å: 14, 15, 16, 17, 18, 19, 20, 22, 26, 31, 32, 34, 36, 37, 39, 45, 47, 53, 73, 74, 86, 89, 91, 93, 99, 115, 117, 131, 133, 143, 145, 336, 351, 352
5844Å: 15
5850Å (*aka 5849Å*): 15, 31, 34, 37, 115, 143, 336
6177Å: 91, 106-11, 330, 352
6196Å (*aka 6197Å*): 4, 5, 7, 15, 26, 42, 45, 47, 87, 91, 106-11
6203Å: 7, 15, 19, 87, 115, 117
6269Å: 15, 87, 91, 106-11
6284Å (*aka 6283Å*): 7, 15, 39, 45, 47, 86, 89, 91, 106-11, 131, 133, 269, 396
6379Å (*aka 6276-6279Å*): 143, 336
6614Å (*aka 6613Å*): 5, 7, 11, 17, 26, 27, 28, 29, 31, 34, 37, 45, 135, 143, 145, 336
7565Å: 367
9577Å: 3, 68-9
9632Å: 3, 68-9
11,797Å: 67
13,175Å: 67